The Emerald Handbook of Smart Cities in the Gulf Region

The Emerald Handbook of Smart Cities in the Gulf Region: Innovation, Development, Transformation, and Prosperity for Vision 2040

EDITED BY

MILTIADIS D. LYTRAS
Effat University, Saudi Arabia

AFNAN ALKHALDI
Arab Open University, Kuwait

AND

SAWSAN MALIK
Arab Open University, Kuwait

emerald PUBLISHING

United Kingdom – North America – Japan – India – Malaysia – China

Emerald Publishing Limited
Emerald Publishing, Floor 5, Northspring, 21-23 Wellington Street, Leeds LS1 4DL.

First edition 2025

British Library Cataloguing in Publication Data
A catalogue record for this book is available from the British Library

ISBN: 978-1-83608-293-4 (Print)
ISBN: 978-1-83608-292-7 (Online)
ISBN: 978-1-83608-294-1 (Epub)

Printed and bound by CPI Group (UK) Ltd, Croydon, CR0 4YY

We dedicate this book to the citizens of all ages, with the hope that the added value of smart cities will significantly enhance their quality of life and well-being. May these innovations foster a world where cultures are connected and leveraged for the happiness and prosperity of all.

Miltiadis, Afnan, and Sawsan

Contents

About the Editors

Dr Miltiadis D. Lytras is a world-class expert in advanced computer science and management, an Editor, Lecturer, and Research Consultant with extensive experience in academia and the business sector in Europe and Asia. Dr Lytras is a Visiting Scholar at Effat University, Kingdom of Saudi Arabia. His research focuses on smart cities strategies, cognitive computing, information systems, technology-enabled innovation, and knowledge management. Dr Lytras has co-authored more than 120 high-impact papers in Q1 and Q2 Web of Science and Scopus-indexed journals and has co-edited more than 80 books with international publishers. Dr Lytras's research promotes the utilization of the intersection of computer science and business research toward the promotion of sustainable development goals.

Dr. Afnan Alkhaldi is the Dean Assistant at the Arab Open University, Kuwait branch. She has a leading expert in smart cities, information systems, and e-Governance. Her research interests revolve around these areas, focusing on innovation, efficiency, and results-oriented approaches. She recently served as a consultant for the Al-Hareer Smart City project in Kuwait, where her expertise significantly enhanced economic development and urban planning. In her current role at the university, Dr. Afnan continues to contribute to shaping sustainable urban futures.

Dr Sawsan Malik is an Assistant Professor at the Arab-Open University, Kuwait branch, specializing in Smart City Management, gig and circular economies, eco-innovation, entrepreneurship, sustainable development, the informal economy, e-government, and digital transformation. She focuses on qualitative research methods, including grounded theory and ethnography. Malik serves as a peer reviewer for several notable journals in entrepreneurship, management, hospitality, and social sciences.

About the Authors

About the Contributors

Dr Ahmed Abdelkader has developed a solid and diverse track record of expertise and skillset. After obtaining his PhD in Marketing Management from the University of Huddersfield, GB, he has served as an Assistant Professor, Associate Professor, and Full Professor in universities in Egypt, Saudi Arabia, and Kuwait. He has over 20 years of academic and industry experience in regional and international enterprises in addition to research publications in indexed international journals.

Oualid Abidi is an Associate Professor of Business at The Australian University – Kuwait since February 2021. He cumulated over 15 years of teaching experience in Kuwait and Canada. Oualid Abidi received his PhD in administrative sciences from Université Laval (Canada) in 2011. He also holds a Master's degree in Project Management and a Bachelor's degree in Finance. Oualid Abidi has eclectic research interests in the fields of entrepreneurship, international business, human resource management, and higher education.

Ahmed Aboualam is an Assistant Professor of Business and Management with academic experience at the university level at different academic institutions in the Middle East coupled with Doctorate, Master and Bachelor in Commerce and Business Administration. Research interests in Business Models, Sustainability, Business Technology and Social Entrepreneurship. Memberships: The International Business and Economic Research Academy – Asian Council of Leaders, Administrators, Deans, and Educators in Business -Arab Business Administration Association.

Alice AlAkoum, PhD Student, Lecturer, an accomplished Lecturer at the Arab Open University's Deanship of Business Studies in Ardiya, Kuwait, and a diligent PhD candidate in Business Studies with a focus on Human Resources at Bucharest University of Economic Studies, Romania, stands out for her rich academic background. Holding a master's in business administration and a License in Law, Alice has cultivated a rich educational background. Her tenure at the Arab Open University has been marked by significant contributions to program development, execution, and evaluation, alongside assisting in strategic HR strategies and risk management studies. Alice engages in research that bridges theoretical knowledge with practical application in human resources, aiming to enhance organizational effectiveness and employee well-being. Her career is distinguished by a commitment to excellence, both in academia and in various professional roles that have leveraged her strong negotiation skills, ability to work under pressure, and capacity for building professional relationships.

Rashed Alhaimer holds a PhD in Electronic Marketing from Brunel University, a Master's in Business Administration and a Bachelor's In Management Information Systems. Currently, Rashed Alhaimer works as the Local Dean of Business faculty and Associate Professor at the Arab Open University – Kuwait. Previous work experience includes administrative roles at the SME Support Department. Rashed Alhaimer has a robust publication record, with recent focuses on corruption, virtual political campaigns, public health messages, genders' role in enterprise performance, social media's impact on innovation and performance, and online shopping behaviors during emergencies.

Dr Salah Alhammadi is an Assistant Professor in Economics and Finance at Arab Open University. He previously contributed to Islamic Economics and Finance education at ALM College Dundee and Durham University Business School. Dr Alhammadi is known for his research presentations at prestigious venues and publications in notable journals, earning him several best paper awards. An experienced PhD examiner and quality assurance evaluator, he holds fellowships with the Royal Society of Arts and the Higher Education Academy, along with academic certifications from Durham University. He also serves as a peer reviewer for top-tier ABS journals.

Munther Abbas Alkandari is the X-Director of the Higher Institute of Telecommunication and Navigation as well as the X-Director of the Nursing Institute in the Public Authority for Applied Education and Training. Alkandari holds Master's degree in Maritime Transport technology, Bachelor's degree in high seas navigation, High Diplomas in education as well as in Management, the author of the first book in the Arab world in the field of maritime business administration.

Dr Jarrah Fahad Al-Mansour is currently an Acting Dean for the Faculty of Business Studies, Arab Open University. He received his Bachelor's degree in Management Information System from Kuwait University, Kuwait. His Master's degree was in E-Commerce management from the University of Kent, United Kingdom, and his PhD from the Department of Marketing and Strategy at De Montfort University, United Kingdom in 2018. He has extensive research interests in public sector organizations, strategy process, knowledge sharing and strategy practices, social networking, digital strategy, and organizational practices. His current research interests include strategy transition process and organizational practices, knowledge sharing and social change, social practice theory, and dynamics of strategy formulation and implementation. In particular, my recent research concerns the practices of social actors in management and the consequent innovation.

Anmar AlObaidi, an Engineer at the Roads and Transport Authority (RTA) in Dubai, holds a Bachelor's degree in Civil and Environmental Engineering and is currently pursuing a Master's in Engineering Management. In his role as an Inspection Engineer, Anmar diligently ensures the quality and safety compliance of transportation infrastructure projects, showcasing expertise in team strategy management, utility inspections, traffic simulation, and consultancy over a three-year career. His versatile skill set spans infrastructure project management, final clearance certificate issuance, and significant contributions to digital transformation initiatives. Anmar has played a pivotal role in executing various Utilities Projects since July 2022, focusing on regulatory compliance oversight and navigating complex bureaucratic processes seamlessly. His responsibilities include collaborating with various departments for final clearance certificates, expediting NOC approval processes, and addressing project-related concerns. Anmar's meticulous project management, keen attention to detail, and commitment to safety and quality assurance contribute to the enhancement of transportation infrastructure and community safety. His oversight of electricity authority and water transmission lines projects ensures compliance with safety and quality standards in electrical distribution and water supply and distribution, respectively. Additionally, Anmar monitors and coordinates Dubai infrastructure projects to enhance public spaces while adhering to design and environmental guidelines.

Azzan Majid Al-Shukaili is a Research Assistant in construction management and smart cities in Middle East College, Oman. He participated in national research projects. The areas of his research include green technology, landscape of infrastructure, and the usability of urban spaces.

Dr Perfecto Gatbonton Aquino, Jr is a former Director of the Council for Research and Creativity/Institute for Research and Advanced Studies at the University of Cambodia, Cambodia. He was a former Research Associate cum Senior Lecturer at Duy Tan University, Vietnam. Dr Aquino graduated his PhD in Public Administration and Management at the Pontifical and Royal University of Santo Tomas de Manila Philippines in 2001. He has published several articles and books which were duly indexed in the Scopus database.

Dr Ahmed Badran is an Associate Professor of Public Policy at Department of International Affairs, College of Arts and Sciences, Qatar University. Dr Badran holds a PhD degree in Public Policy from the University of Exeter, the United Kingdom. Dr Badran's research interests extend to cover different areas of the regulatory governance and politics of regulation in liberalized public utilities. That includes telecoms, water, and energy sectors particularly in transition and developing economies.

Valentina Emilia Balas is Professor Habil at the Faculty of Engineering, "Aurel Vlaicu" University of Arad, Romania. She holds a PhD Cum Laude in Applied Electronics and Telecommunications from Polytechnic University of Timisoara. Dr Balas is the head of Intelligent Systems Research Centre and Director of the Department of International Relations, in Aurel Vlaicu University of Arad. Dr Balas is the author of more than 400 research papers, edited and co-authored more than 100 books in Elsevier, Springer, CRC Press where she is responsible for some book series. Her research interests are in Intelligent Systems, Fuzzy Control, Soft Computing, and Smart Sensors. She is the Editor-in Chief to IJAIP and IJCSysE journals in Inderscience, member in Editorial Board of several journals, evaluator for projects and PhD thesis. She is recipient of the "Tudor Tanasescu" Prize from the Romanian Academy (2019) and "Stefan Odobleja" Prize from Romanian Academy of Scientists (2023).

Gabriel Osório de Barros holds a Graduate degree in Economics, a Postgraduate degree in Public Sector Management from the University of Évora, a Master's degree in Financial Management from the University of Extremadura, a Master's degree in Management from Lusíada University, and a Master's degree in Economics and Public Policies from ISCTE. He has completed several advanced courses in Public Management, including the Advanced Course in Public Management (CAGEP), the Public Management Training Course (FORGEP), and the Diploma of Advanced Studies in Public Policy (academic part of the PhD in Public Policy). He is currently Director/Head of Economic Analysis in the Office for Strategy and Studies (GEE) of the Portuguese Ministry of Economy and Maritime Affairs.

Dr Neilson D. Bation is a distinguished figure with expertise in academia, research, statistics, and business. He holds esteemed fellowships with the Royal Institute of Management and the Royal Institute of Research in Singapore. Dr Bation is internationally acclaimed for his leadership in statistics and business research, receiving awards like the Most Outstanding Research Consultant at ETCOR Research Education Center. He currently serves as the Vice President for Research and Extension at Opol Community College and holds CEO positions at Digital Institute Training and Development Services and JHEk Upholstery Services. His career includes roles in cooperatives and healthcare service organizations, making substantial contributions to their growth. Dr Bation actively engages in research, presenting papers at conferences and publishing in both local and international journals, reflecting his dedication to academia, research, and entrepreneurship.

Prof Mohammed Ali Berawi is a Professor in the Department of Civil Engineering, Faculty of Engineering, Universitas Indonesia, and currently serves as the Executive Director of the Center for Sustainable Infrastructure Development (CSID) Universitas Indonesia and Director of ASEAN University Network for Sustainable City and Urban Development (AUN-SCUD). He also serves as the Working Group Leader for Smart Cities at the Association of Pacific Rim Universities (APRU) Sustainable Cities and Landscapes (SCL). He is the Chairman of the Indonesian Lecturers Association (ADI) for 2022–2027 and the Advisory Board Chairman of the Science and Technology Professional Organization Forum (FOPI) for 2023–2026. Stanford University listed Prof Ali Berawi as one of the World's top 2% scientists (2021–2023). Prof Berawi previously served as the Member Secretary of the Republic of Indonesia's Presidential Advisory

Council in 2020–2022. He is the Deputy for Green and Digital Transformation of Indonesia's Nusantara Capital City Authority (NCCA).

Assist. Prof. Dr. Ayat-Alah Bouramdane: Graduated from the International University of Rabat (IUR) as an Energy Engineer (with an exchange program at EEIGM, ENSEM, and FST of Lorraine University in Nancy), École Polytechnique (l'X) with a master's thesis conducted at Centrale Supélec-Université Paris Saclay-Sorbonne Université (GeePs lab), and from Institut Polytechnique de Paris (IP Paris) (LMD lab-l'X, IPSL), Ayat-Allah focuses on modeling and optimizing energy systems (renewables, storage, and spatio-temporal complementarities) in actual and future climate impacted by climate change scenarios. Her thesis was conducted in the framework of the creation of the Energy4Climate Interdisciplinary Center (E4C) of IP Paris and Ecole des Ponts ParisTech, supported by MED-OCEAN. Ayat-Allah is currently an assistant professor at the IUR (a researcher at the LERMA lab - a lecturer at the College E&A). Her research also focuses on resource management and hydrogen, based on a holistic approach that integrates technical, economic, environmental, and social dimensions. Her research has yielded peer-reviewed publications and conference presentations. For more information, visit: https://sites.google.com/view/ayat-allahbouramdane.

Davide Calandra, Assistant Professor in Business Administration and Accounting at the University of Torino – Department of Management. Co-lecturer in the PhD Program in Torino and didactic programs in several institutions such as Wroclaw University of Business and Economics, University of Palermo, and the University of Torino. His research interests cover new technology applications in accounting, auditing, and accountability, and the application and management of new technologies in smart cities, such as blockchain and artificial intelligence. Finally, he also deals with the Islamic economy and finance field.

Suwithda Charungkaittikul is an International Professor in the Non-Formal Education Division of the Department of Lifelong Education at Chulalongkorn University, Bangkok, Thailand. She also holds a Doctorate in Non-Formal Education from Chulalongkorn University. From 2010 to 2011, she was a Research Fellow in the Andragogy Doctoral Emphasis Specialty Instructional Leadership Program at Lindenwood University, MO, USA, where she completed her Post-Doctoral Training in Education Andragogy and Learning Society Development.

Xi Chen, PhD, is a distinguished machine learning professional with a robust background in software engineering, computational biology, and data science. With extensive experience at Meta, Dr Chen has led the development of simulation applications and has been instrumental in advancing ML/DL applications, synthetic data generation, and user data integrity. At Juvena Therapeutics, Dr Chen harnessed machine learning for biomarker suggestions and therapeutic target predictions, leading a dedicated team. His tenure at Verb Surgical was marked by significant contributions to surgical analytics software, where he dramatically improved model performance. As a Deep Learning Institute Certified Instructor at Nvidia, Dr Chen has educated developers and data scientists on artificial intelligence and GPU-accelerated computing, shaping the future of technology across various domains. Dr Chen's academic pursuits at the University of Kentucky resulted in groundbreaking research in Protein NMR, culminating in a PhD in Bioinformatics and a Master's certification in Statistics. Now, as a Machine Learning Faculty member at the University of California, Santa Cruz, Dr Chen brings his wealth of knowledge and experience to shape the next generation of innovators in the field of artificial intelligence and machine learning. His work continues to be at the intersection of academia and industry, driving forward the capabilities of machine learning technologies.

Anita Rani Chopra is a PhD in Education Management, focusing on core skills as an academician, including ensuring program and course curricula are current, relevant, and reflective of best practices within a professional/program discipline. She is an expert in designing, evaluating, and validating learning outcomes for courses and programs. Her strengths include creating learning environments that accommodate students' cultural diversity, education needs, experiential learning, and applying knowledge to professional life. Her core strength is designing appropriate strategies and methods for facilitating and assessing student learning that is invitational and reflective of best practices in teaching. Her expertise is producing multi-media materials, alternative delivery strategies, and incorporating various educational technologies. She has written many papers and book chapters reflecting her contribution to students' and adults' learning.

Ashok Chopra is a highly skilled, accomplished, and experienced Senior Academician with strong learning acumen in teaching, training, and development. He is a Registrar, Program Chair for the Ph.D. program, Member of the Research Committee at Amity University Dubai, and Associate Professor at Amity Business School. He has shown research success with multiple published articles/papers, books, and chapters. An expert in creating compelling environments for applied learning, accommodating students'/trainees' diverse cultural and educational backgrounds, experiences, and individual learning styles. Ashok has over 24 years of international industry experience in senior and leadership positions, over 10 years of entrepreneurial experience, and more than 13 years of experience as an academician and corporate trainer – a perfect blend of academia, industry, and entrepreneurship.

Ravee Chudasring is an Assistant Professor in the Non-Formal Education Division of the Department of Lifelong Education at Chulalongkorn University, Bangkok, Thailand. He also holds a doctorate in Non-Formal Education from Chulalongkorn University. His recent published works related to his research interests focused on aging population, social well-being, and community development.

Namho Chung is a Dean and Professor of the College of Hotel and Tourism Management, and the Director of Smart Tourism Research Center (STRC) at

Kyung Hee University, Republic of Korea. Also, He is a Co-Editors-in-Chief of *Journal of Smart Tourism*. He had been a Visiting Research Fellow at School of Hospitality and Tourism Management, University of Surrey in Guildford, UK. Currently, he leads smart tourism projects in the Ministry of Education of the Republic of Korea and the National Research Foundation of Korea over 10 years. His name is listed in the Hall of Fame at Kyung Hee University for his outstanding achievements.

Dr Edwin C. Du is currently Dean of Graduate School of Capitol University, Cagayan de Oro City, Philippines. He completed the degree of Doctor of Philosophy in Educational Management at Capitol University. For his master's degrees, he finished Master of Public Administration at Ateneo de Cagayan Xavier University and Master in Political Science at the University of San Carlos. He also graduated from Silliman University with a Bachelor's degree in Political Science. He has presented research papers at the International Conferences of the Korean Association for Public Administration (KAPA), Philippine Society for Public Administration (PSPA), Public Administration Association of Thailand (PAAT), Korean Association of Policy Studies (KAPS), the Asian Association for Public Administration (AAPA) and the Taiwan Association for Schools of Public Administration and Affairs (TASPAA). Dr Du is a Fellow of the Royal Institute of Educators, Singapore and an Associate Member of the National Research Council of the Philippines (NRCP).

Dr Taleb Eissa, Assistant Professor at Department of Civil Engineering, Florida Institute of Technology Melbourne, Florida, USA. His research interests and teaching are in the areas of transportation planning, transportation econometric modeling, freight mode choice, accident and safety analysis, urban transportation, and transport policy. Dr Taleb is also a Fellow of the Florida Board of Professional Engineers (PE), the American Society of Civil Engineers (ASCE), and the Transportation Research Board (TRB), and he got the Outstanding Student of the Year Award for 2019 (Florida Institute of Technology).

Zhengchao Fu graduated in Architecture and Civil Engineering at Wuyi University, currently a PhD student in the Heritage and Conservation of School of Housing, Building and Planning at Universiti Sains Malaysia, Penang, Malaysia. His areas of scholarly interests revolve around heritage conservation, heritage education, Overseas Chinese history, new museology, and community-based studies.

Dr Zeinab Gamal is Assistant Professor in Marketing at the Higher Institute of Computer and Business Administration & Onaizah Colleges, Saudi Arabia. Gamal holds PhD in Digital Marketing and MSc in Marketing (Innovation in SMEs). Participated in conducting marketing research and management training for many Egyptian and Arab companies by the Arab Bureau for Creativity & Development (ABCD). Gamal's research interests include sustainability, digital marketing, online customer experience, organizational ambidexterity, market orientation, firm growth, and SMEs.

Prisilla Jayanthi Gandam received her BE (CSE) from MVSR Engineering College, Osmania University, Hyderabad in 1993 and her M.Tech Degree in CN&IS from HITS College of Engineering, JNTUH in 2012. She presented her research papers at various national and international conferences (aboard). She has published over 80+ papers including 16 book chapters. She has received "Women Researcher Award" from International Scientist Awards (2022). She is an assistant professor at CMR Institute of Technology, Hyderabad. Prior to this organization, she worked as Faculty Associate at Mahindra University, Hyderabad from 2022 to 2023.

Dr Omer Hag Hamid is an Assistant Professor at the Arab Open University-Bahrain, a Senior Fellowship (SFHEA-UK), specializing in Economics, and Innovation & Technology Management. He holds a PhD in Economics from the OIU, and a candidate for a second Ph.D. in Innovation and Technology Management from the UNIKL, with a MSc (OIU) and BSc in Economics from the U of K. Dr Hamid is a renowned researcher in fintech and digital solutions in finance, with a focus on the impact of technology on financial services. His work has been published in prestigious journals, and he is a regular speaker at conferences and seminars. Dr Hamid is also certified by (INQAAHE), reflecting his commitment to quality education.

Dr Hend Sameh Hafez Hassan is an Assistant Professor at the Arab-Open University – Kuwait branch. She is also an Associate Professor at the Faculty of Commerce – Mansoura University, Egypt. She started her postgraduate studies at Cranfield University, United Kingdom in June 2007. She obtained her MPhil in the summer of 2008 before completing the programme of study in July 2011 and obtaining a PhD in Electronic Management following a programme of study that was applied to the Egyptian E-Government. Her main research interests are Experiential Marketing Dynamics, E- Customer Relationship Management, E-services, Digital Marketing, Brand Management, and Entrepreneurship.

Yingyilong Hu graduated in Adult Education and Community Development at the University of Toronto, currently a PhD Candidate in the Non-Formal Education Division of the Department of Lifelong Education at Chulalongkorn University, Bangkok, Thailand. His areas of scholarly interests revolve around lifelong education, heritage education, sustainable tourism, community-based studies, and placemaking.

Hamed M. Hussain – achieved a Bachelor's degree in Chemical Engineering and a Master's degree in Engineering Management at Abu Dhabi University, currently a PhD student in Chemical Engineering at United Arab Emirates University focusing on sustainability projects such as solar power, water desalination, and hydrogen production. Areas of previous publications revolve around PV panel utilizations and social acceptance in the GCC region.

Prof Dr Muralikrishna Iyyanki, received his PhD from the Premier Institute – Indian Institute of Science (IISc), Bangalore, was Dr Raja Ramanna Distinguished Fellow at Defense Research and Development Organization (DRDO), India, and the National Coordinator for Geospatial Public Health, which is National Networking Government of India Project. He was the Editor and Member of the editorial board of several S&T Journals. His present research focuses are on open innovation, artificial intelligence and geospatial data analytics and geospatial population and public health management. He was the Professor and Founder Head of the Centre for Spatial Information Technology (CSIT) at Jawaharlal Nehru Technological University (1990–2005), and Director of Research and Development Centre (2005–2008). He has served as a guest scientist at the German Space Research Institute (DLR) and GKSS Research Centre.

Dr Revenio C. Jalagat, Jr is presently employed in Al-Zahra College for Women as Associate Professor for almost 11 years. He taught at Gulf College as a Business Lecturer from 2011 to 2013. Prior to that, he had worked in various disciplines in both the academic and business industry with almost 9 years of accumulated experience in the Philippines acting as Accountant, Chief Account, Finance Manager, Assistant Professor, and Professor. In 2023, he received an award as among the Top 50 Outstanding Philippine Awardee in the field of Research and Publication and Distinguished Researcher in his institution. He has published articles, edited books, and chapters indexed in Scopus and Web of Science indexes and served as an Editorial Member and Reviewer in various international refereed journals in the field of business management. He was also awarded in 2021 the Certified Business Management Educator by the Chartered Association of Business Schools in the UK.

Kevin A. Jones teaches Software Engineering at Singapore Institute of Technology. He has 40 years of experience as a Canadian Signals Officer, IBM Consultant, and Academic. His research interests are instructional design, scholarship of teaching and learning, and learning motivation. He received his PhD in Educational Systems and Technology from the Nanyang Technological University in 2018.

Khider Hamid Khider from Sudan living in Kuwait and working at Arab Open University for 18 years working as Lecturer at Business Studies Faculty. Khider interest areas of study and research innovation are entrepreneurship and sustainable development. Khider has published articles one about knowledge-based economy in Gulf region, and the second one about "Explaining and Predicting of Business Students' Continuance Usage of E-Learning: Direct and Mediating Process Model-Extending Technology Acceptance Mode." Khider would like to continue in the field of sustainable development, and how could affect the life of human beings. The world today moving to digital platform and innovation also this area is interested area for research which Khider would like to approach. Khider had good experiences in the field of training and development as Khider participate in number of workshops and training session in local organization in state of Kuwait.

Dr Utku Kose received the BS degree in 2008 in Computer Education of Gazi University, Turkey as a faculty valedictorian. He received MS degree in 2010 from Afyon Kocatepe University, Turkey in the field of computer and DS/PhD degree in 2017 from Selcuk University, Turkey in the field of Computer Engineering. Between 2009 and 2011, he worked as a Research Assistant at Afyon Kocatepe University. Following, he has also worked as a Lecturer and Vocational School – Vice Director at Afyon Kocatepe University between 2011 and 2012, as a Lecturer and Research Center Director in Usak University between 2012 and 2017, and as an Assistant Professor in Suleyman Demirel University between 2017 and 2019. Currently, he is an Associate Professor in Suleyman Demirel University, Turkey. Kose also gave lectures at other higher education institutions such as Gazi University and Istanbul Arel University. He also works as a Visiting Researcher at the University of North Dakota, USA, and holds the Honorary Professor of Artificial Intelligence title at ITM (SLS) Baroda University, India. He has more than 300 publications including articles, authored and edited books, proceedings, and reports. He is also on the editorial boards of many scientific journals and serves as one of the editors of the *Biomedical and Robotics Healthcare* (CRC Press) and *Computational Modeling Applications for Existential Risks* (Elsevier) book series. His research interests include artificial intelligence, machine ethics, artificial intelligence safety, biomedical applications, optimization, the chaos theory, distance education, e-learning, computer education, and computer science.

Hyunae Lee is a Research Professor at the Smart Tourism Research Center (STRC) of Kyung Hee University, Republic of Korea. She received her PhD degree in Hotel Management from the same university. Her research interests focus on smart tourism and smart tourism city. Her research work has been published in *Journal of Travel Research, International Journal of Hospitality Management, International Journal of Contemporary Hospitality Management, Information & Management, Telematics and Informatics*, and others.

Andreia de Bem Machado, Federal University of Santa Catarina, Brazil. She holds PhD in Engineering and Knowledge Management. She is Professor at the Federal Institute of Santa Catarina and Postdoctoral Student at the Federal University of SantaCatarina. She is at the National Institute of Educational Studies and Research Anísio Teixeira (Ministry of Education of Brazil). She has been working in the educational field for over 25 years. Currently, her research interests are public policy and political science, innovation and business management issues: Innovation, education, digital transformation, management, hybrid education, digital technologies, active methodology, knowledge media, entrepreneurship, and knowledge management. She is the author of numerous articles and book chapters published in national and international journals.

Dr Ram Kishore Manchiryal is Associate Professor and Head of the Civil and Mechanical Engineering Department at Middle East College (MEC), Oman. His research interests and teaching are in the areas of material characterization, corrosion of steel in concrete, sustainable materials, renewable energy, innovative teaching and learning pedagogies, IoT and AI applications to civil engineering. He is actively engaged in research and publication areas of Civil Engineering and Higher Education. He is a member of the editorial board of well-known journals and extending his services as a reviewer for a few reputed Journals. He has received funding from various funding agencies in India, USA, and Oman. Dr Ram Kishore Manchiryal is also a Member of the Institution of Civil Engineers (ICE-UK), Fellow of Institution of Engineers (IEI), Sr. Fellow of the Higher Education Academy (HEA-UK) and Member of IFERP. He is professionally qualified as Charted Engineer from IEI (CEng (IEI, India)).

Dr Ali Ahmed Mohammed is a researcher in Traffic Engineering, Intelligent Transportation Systems, Traffic Safety, Sustainable urban design, Intelligent Urban Traffic, modelling, logit, Computational in Highway &Transportation, Remote Sensing GIS, and Urban Transportation in the Department of Civil Engineering, Lassonde School of Engineering, York University, Toronto, Canada.

Khairunnisa Musari is an Assistant Professor at Faculty of Islamic Economics and Business (FEBI) & Postgraduate Program, Kiai Haji Achmad Siddiq State Islamic University (UIN KHAS), Jember, Indonesia. She is a Deputy Coordinator of Central Indonesia for the Indonesian Association of Islamic Economist (IAEI) (2019–2023) and Secretary of the IAEI for Provincial Board of East Java since 2017. Most of her research concentrates heavily on issues related to sukuk, waqf, esham, fiscal & monetary policies, circular economy, halal logistics, sustainable finance, climate change, and Islamic microfinance & nanofinance. She is a reviewer for several national/international journals, international conferences, and also a reviewer of Research, Scientific Publications, and Community Service (Litapdimas) for Islamic higher education institutions under the Ministry of Religious Affairs (MoRA) of the Republic of Indonesia since 2018. She was listed as one of the Top 300 Most Influential Women in Islamic Business & Finance 2019, 2020, 2021, WOMAN*i* Significancica 2022, and WOMAN*i* Academica 2023 by Cambridge-IFA.

Elvira Nica, PhD, Full Professor at the Bucharest University of Economic Studies ranked 351–400th in the world, in the top 501–550 internationally, and second in Romania in the Business and Management subdomain, and the Dean of the Faculty of Administration and Public Management. A distinguished scholar in management and economics, she focuses on human resource management and smart economies, employing advanced research methodologies. PhD coordinator since 2016 and holds a Doctor Honoris Causa from the University of Zilina, Slovakia. Recognized as a Highly Cited Scholar in Web of Science and SCOPUS, with a WoS H-index of 12 and SCOPUS H-index of 29, she ranks among

Stanford University's Top 2% Scientists globally (2019-2022). She has contributed over 180 articles and 21 books in her fields, collaborating with top publishers and serving on several international editorial boards. She's also a Senior Research Fellow and a full member of Sigma Xi, The Scientific Research Honor Society.

Dr Khalil Rahi Assistant Professor of Engineering and Project Management at Abu Dhabi University, brings 16 years of global experience and a diverse background. With a PhD in industrial engineering, an MBA in IT management, and a Bachelor's degree in Electrical Engineering, his expertise lies in project resilience and management. Beyond academia, Dr Rahi has successfully managed multiple projects for corporations like ADP, Ceridian, and Eaton, leading software development and implementing technology-powered solutions. As a PMP certification holder, lecturer, researcher, and mentor, he modernizes the project management curriculum, emphasizing project management, innovation and entrepreneurship, and organizational resilience. With several published papers in top journals and the supervision of master's and PhD's students in cutting-edge research areas, Dr Rahi actively contributes to the academic field. He also volunteers in several regional and international competitions and reviews articles for project management journals, showcasing his commitment to nurturing the next generation of inventors, and advancing academic research.

Dr Wael A. Samad holds the position of Associate Professor of Mechanical Engineering and serves as the Chair of the Department of Mechanical & Industrial Engineering at the Rochester Institute of Technology – Dubai Campus. He obtained his PhD in Mechanical Engineering from the University of Wisconsin – Madison in 2013. Throughout his research endeavors, Dr Samad has delved into areas such as Experimental Solid Mechanics, Material Characterization of 3D Printed Artefacts and Smart Cities. His commitment is demonstrated in the extensive array of over 50 publications he has authored, encompassing peer-reviewed journals, book chapters, and conference proceedings. Among his contributions is an edited volume focused on smart cities in the Gulf. Since setting foot in the UAE almost 10 years ago, Dr Samad has been proactive in forging collaborations with prominent local industries. Notably, his work with Emirates Global Aluminium has played a pivotal role in projects that aim to extend the lifespan of ingot moulds, promoting a more sustainable aluminum production in the Emirates.

João Rodrigues dos Santos holds a PhD in Economics and Management, a Master's degree (with research in the area of Territorial Planning), and a Degree in Geography and Territorial Management (Universidade Nova de Lisboa). He is a Professor at the European University / IADE and he collaborates with the Portuguese Catholic University's Center for Studies and Surveys. His research interests currently are Economics, Political Economy, Political Economy of Communication, Public Policy, Territorial Planning, Environmental Sustainability, Innovation, Information Technology and Communication. He is the author of several articles and book chapters published in national and international journals. He is also the author of two books in the areas of economics, and territorial management, respectively. Finally, João Rodrigues dos Santos collaborates weekly with CNN Portugal, analyzing the national and international economic situation.

Dr Mustika Sari holds a PhD degree in Civil Engineering. Her primary research interests include infrastructure development and construction management, emphasizing innovations for economic and sustainable development. She is a Senior Researcher at the Center for Sustainable Infrastructure Development (CSID), Faculty of Engineering, Universitas Indonesia, and has been involved in many national and international research activities. She is the Managing Editor of the *CSID Journal of Infrastructure Development (CSID-JID)*. Currently, she is teaching at the Department of Civil Engineering, Faculty of Engineering, Universitas Indonesia.]

Silvana Secinaro, Professor in Business Administration and Accounting at the University of Torino, Department of Management. Chartered accountant and auditor, freelance journalist, and author of numerous publications on public and private accounting and new technologies impact on conservative businesses. Component of National Commission of Italian Ministry of Finance on public accounting. Finally, over the years, she has been the scientific coordinator of multiple European projects, such as H2020 Avangard, to study proper business models for electric mobility in smart cities and an Erasmus + on social entrepreneurship.

Dr Andreea Claudia Şerban is Professor at the Department of Economics and Economic Policies, Faculty of Theoretical and Applied Economics and Director of Doctoral School of Economics I. She holds a PhD in Economics from Bucharest University of Economic Studies and a post-PhD in Economics from Romanian Academy of Science. Her research activity consists in publication of seven books and numerous economic studies in highly ranked journals and participation in many international conferences. She is Associate Editor for *International Journal on Semantic Web and Information Systems* (IJSWIS – WoS indexed journal) and Editor in Chief for *Global Journal of Business, Economics and Management: Current Issues* (journal indexed in many international databases). She is reviewer for prestigious international journals and member in editorial board or organizing committee of many international conferences. She is expert for Lithuania Research Council. Her areas of research interest include: sustainable development, smart development, knowledge economy, labor market, education, demographic issues.

Ravi S. Sharma is Professor of Information Systems & Technology Management at Zayed University, a federal university in the United Arab Emirates. Prior to this he held faculty appointments in New Zealand, Singapore and Canada. He had also spent a decade in industry as an ICT consultant with Deutsche Telekom and IBM Global Services. Ravi received his PhD in Management Sciences from the University of Waterloo, Canada and is a Fellow of the IET and Senior Scholar / Distinguished Member of the AIS. His research interests are in the application of design science methods to use-cases of digital platforms and services. His research publications may be found at https://orcid.org/0000-0002-8235-5344.

Maria José Sousa, Iscte, Portugal. University Professor, with a PhD in Management and a Research Fellow at ISCTE/IUL and of CIEO (Algarve University). Her research interests currently are public policies and political science, innovation, and business management issues. She is a best-seller author in ICT and People Management and has co-authored over 70 articles and book chapters and published in several scientific journals (e.g. *Journal of Business Research, UAIS, Future Generations Computer Systems, Journal of Medical Systems, I. J. of Knowledge, Culture and Change Management, Knowledge Management, Wseas Transactions on Business and Economics, Information Systems Frontiers*, and others), she has organized and peer-reviewed international conferences. She is also External Expert of COST Association – European Cooperation in Science and Technology.

Dr Bambang Susantono is the Chairman of Indonesia's Nusantara Capital City Authority (NCCA). He was the Vice-President for Knowledge Management and Sustainable Development of the Asian Development Bank (ADB). He holds a PhD in Infrastructure Planning and has two master's degrees in Transportation Engineering and City and Regional Planning from the University of California, Berkeley. Susantono was the Acting Minister, Vice-Minister of Transportation of Indonesia (2010–2014), and Deputy Minister for Infrastructure and Regional Development at the Office of Coordinating Ministry for Economic Affairs (2007–2010). He chaired several research institutes and taught in universities. He is actively doing research, publishing several books, and writing articles on infrastructure, transportation, urban and regional planning, sustainable development, and climate change.

Dr Hussin A. M. Yahia, Assistant Professor in Transportation Engineering and Traffic Management in the Department of Civil and Mechanical Engineering, Middle East College affiliated with Coventry University. He was the principal investigator of several national and international studies, among which Governing the smart cities, Growth of Green Buildings in the Oman Construction Industry, supported by the (Ministry of Higher Education, Research and Innovation, Oman) and financed with the TRC (The Research Council – Oman). Total amount founded; 18,000 OMR (46,800 USD). His main research interests are focused on issues in smart cities; IoT smart transportation, intelligent transportation systems, road safety, and green technology in buildings and roads.

Preface

The Emerald Handbook of Smart Cities in the Gulf Region delves into the multifaceted development of smart cities across the Gulf, showcasing how technology, innovation, and sustainability converge to transform urban landscapes. The book is structured around contributions from experts who provide comprehensive insights into various aspects of smart city development, ranging from governance frameworks and technological advancements to sustainability initiatives and societal impacts. Through detailed case studies and empirical research, the handbook explores key themes such as smart infrastructure, intelligent transportation systems, renewable energy integration, and digital governance. It also examines the socio-economic benefits of smart city initiatives, including improved quality of life, economic growth, and enhanced public services. By highlighting best practices and emerging trends, the handbook is an essential resource for policymakers, urban planners, and researchers aiming to understand and implement smart city strategies in the Gulf region.

Chapter Summaries:

Chapter 1: A Holistic Framework for Smart Cities Governance in the Gulf Region: From Hype to Sustainable Impact
Authors: Miltiadis D. Lytras, Afnan Alkhaldi, Sawsan Malik, and Andreea Claudia Serban

This chapter introduces a framework for governing smart cities in the Gulf region, focusing on governance, strategy, technology, and infrastructure. The framework aligns with the Sustainable Development Goals (SDGs) to guide smart city initiatives. The authors emphasize the importance of data integrity, regulatory resilience, and civic engagement in ensuring ambitious and sustainable strategic initiatives. This approach provides a robust model for smart city projects, which can be adapted and applied to various urban settings globally.

Chapter 2: Smart City Innovation and Sustainability: A Comparative Study in GCC Countries
Authors: Revenio C. Jalagat, Jr, Edwin C. Du, Neilson D. Bation, and Perfecto Gatbonton Aquino, Jr

This chapter evaluates the benefits and challenges of smart city technologies in the UAE, Saudi Arabia, and Qatar and offers policy recommendations for sustainable development. The authors conducted a comparative analysis using documentary evidence to identify best practices and areas for improvement. Their work emphasizes the need for collaborative efforts among stakeholders to sustain smart city operations. This chapter provides valuable perspectives on how innovation drives sustainable urban development.

Chapter 3: The Science of Mobile Pedestrianism Through Smart Cities
Authors: Ashok Chopra and Anita Rani Chopra

This chapter explores how walkability and IoT enhance urban living, promoting health, safety, and environmental sustainability through strategic planning. The authors discuss the role of walkable communities and technology integration to support pedestrianism. Their findings suggest that well-planned, walkable cities can significantly improve residents' quality of life. The chapter highlights the importance of designing urban environments prioritizing human mobility and health.

Chapter 4: Innovating Urban Environments: The Impact of Smart City Technologies on Employee Performance, Quality of Life, and Service
Authors: Alice AlAkoum and Elvira Nica

The chapter discusses how smart city technologies improve employee performance, quality of life, and service delivery, emphasizing the need for ongoing innovation and policy support. The authors analyze the interplay between advanced technologies and urban living conditions. Their work illustrates the potential of smart cities to create efficient and sustainable urban environments. This chapter emphasizes the critical role of technology in shaping future workplaces and urban spaces.

Chapter 5: Democratization and Transformation of Higher Education in the Era of Smart Cities
Authors: Kevin A. Jones and Ravi S. Sharma

This chapter explores the shift toward personalized and diversified education modes in smart cities, advocating for adaptable, learner-driven approaches. The authors discuss the impact of digital platforms on higher education and the need for institutions to reimagine their roles. Their analysis reveals the potential of smart city environments to enhance educational accessibility and relevance. The chapter emphasizes the importance of innovation in education for sustainable urban development.

Chapter 6: The Rebirth of Museum Education Echoing of Smart Education
Authors: Yingyilong Hu, Zhengchao Fu, Suwithda Charungkaittikul and Ravee Chudasring

This chapter explores digital platforms in museum education and presents the Danxiashan Geo-Museum case study, illustrating innovative educational approaches through smart technologies. The authors highlight the potential for museums to become dynamic educational environments. Their work suggests that integrating digital tools can enhance cultural and educational outreach. This chapter demonstrates the transformative power of technology in museum education, aligning with smart city goals.

Chapter 7: The Synergy Between Advanced Innovative Water Management and Smart Cities Development: Enhancing Sustainability and Quality of Life
Authors: Dr Hend Sameh Hafez Hassan, Dr Ahmed Abdelkader, and Dr Oualid Abidi

The chapter outlines a framework for adopting ICTs to support sustainable water management in smart cities, emphasizing citizen engagement and interdisciplinary collaboration. The authors present case studies from the GCC region to illustrate best practices. Their findings highlight the critical role of technology in managing urban water systems sustainably. This chapter contributes to the discourse on how smart cities can address environmental challenges effectively.

Chapter 8: AI's Role in the Development and Transformation of Smart Cities
Authors: Andreia De Bem Machado, Gabriel Osório De Barros, João Rodrigues dos Santos, Silvana Secinaro, Davide Calandra, Maria José Sousa

This chapter maps AI applications in public safety, transportation, energy, and more, emphasizing AI's role in enhancing urban living conditions and supporting Vision 2040 initiatives. The authors use qualitative narrative analysis to explore AI's impact. Their research highlights AI's potential to revolutionize smart city services. This chapter underscores integrating AI into urban planning for innovative and sustainable growth.

Chapter 9: The Impact of Innovation & Sustainable Development in Smart Cities
Author: Khider Hamid Khider

This chapter focuses on the integration of IoT, AI, and smart grids and discusses how innovation and sustainable practices address urban challenges and improve quality of life. The author analyzes the role of advanced technologies in fostering sustainable urban environments. The findings suggest that continuous innovation is essential for maintaining urban resilience. This chapter highlights the need for strategic planning and investment in sustainable technologies.

Chapter 10: Digital Intersections: Synergies Between the Gig Economy and Evolution of the Smart City
Authors: Sawsan Malik, Afnan Alkhaldi, Jarrah Fahad Al-Mansour, and Miltiadis D. Lytras

The chapter explores how the gig economy integrates with smart city frameworks, enhancing urban living through technology-mediated work arrangements. The authors provide case studies from global and regional examples. Their analysis reveals the potential for digital platforms to create efficient urban ecosystems. This chapter emphasizes the importance of digital literacy and inclusivity in leveraging the benefits of the gig economy for smart city development.

Chapter 11: Enhancing Prosperity through Smart Tourism: A Competitive Analysis of Gulf Cities Using STCCI
Authors: Hyunae Lee and Namho Chung

Introducing the STCCI, this chapter evaluates Gulf cities' smart tourism capabilities, providing strategic insights for fostering innovation and development. The authors benchmark cities like Dubai, Abu Dhabi, and Doha using the index. Their findings highlight the strengths and areas for improvement in each city. This chapter emphasizes the role of strategic enhancements in driving smart tourism and urban development.

Chapter 12: Innovative Digital Transformation of Issuing Final Clearance Certificates for Infrastructure Projects in Dubai
Authors: Anmar AlObaidi and Wael A. Samad

This chapter presents a case study on digitizing clearance processes for infrastructure projects, highlighting the eNOC portal's role in enhancing efficiency and transparency. The authors analyze the process using the DMAIC methodology, demonstrating how digital transformation can streamline application submissions and improve project management. Their findings emphasize the importance of technological innovation for regulatory compliance and customer satisfaction, offering valuable insights for smart city administrators aiming to optimize urban infrastructure projects.

Chapter 13: Innovating Urban Policy Development: Lusail Smart City's Role in Qatar's Transition to a Knowledge-Based Economy
Author: Dr Ahmed Badran

The chapter discusses Lusail Smart City's contributions to Qatar's knowledge-based economy, emphasizing strategic initiatives for innovation and sustainable urban planning. By presenting a detailed case study of Lusail City, the author illustrates how smart infrastructure, and advanced technologies can foster urban living and economic growth. The analysis highlights the need for policy support and stakeholder engagement, providing a model for other cities aiming to achieve similar transformational goals.

Chapter 14: The Role of Government in Ensuring Sustainability in Smart Urban Development: A Case Study of Kuwait
Authors: Afnan Alkhaldi, Sawsan Malik, Salah Alhammadi and Miltiadis D. Lytras

Using Kuwait as a case study, this chapter explores the government's role in promoting sustainability through strategic initiatives and policy support. The authors highlight successful models from the GCC and beyond, demonstrating effective combinations of technology, infrastructure, and sustainable practices. Their findings illustrate the importance of public-private partnerships and robust legislative frameworks in achieving sustainability goals, significantly contributing to smart city research and practice.

Chapter 15: Innovating Sustainable Urban Development: Evaluating the Social Acceptance of Residential Solar Integration for the Prosperity of Abu Dhabi
Authors: Hamed M. Hussain and Dr Khalil Rahi

This chapter analyzes the potential for residential solar integration, highlighting the importance of public awareness and supportive policies for sustainable energy adoption. The authors present survey results from Abu Dhabi residents, revealing a high interest in solar energy and a preference for government financial support. Their findings emphasize the need for comprehensive public awareness campaigns and strategic incentives, contributing valuable insights into promoting renewable energy in smart cities.

Chapter 16: Strategic Planning for the Development of Smart Cities in Oman
Authors: Hussin A. M Yahia, Azzan Majid Al-Shukaili , Ram Kishore Manchiryal, Taleb Eissa, and Ali Ahmed Mohammed

Focusing on Muscat, this chapter discusses strategies for transitioning traditional cities into smart urban areas, emphasizing legislative support and community engagement. The authors outline principles of sustainable development and urban policy formulation, presenting practical solutions for integrating advanced technologies. Their findings suggest that strategic planning is essential for successful smart city implementation, highlighting the importance of aligning urban development with national visions like Oman Vision 2040.

Chapter 17: Toward Kuwait Seaport's Sustainability 2040: The Implication of Digital Twins Technology on Terminal Containers' Transformation
Authors: Zeinab Gamal, Ahmed Aboualam, and Munther Abbas Alkandari

The chapter explores digital twins' technology in enhancing Kuwait's seaport sustainability, providing a framework for managing logistics operations and future demands. The authors discuss the role of digital twins in improving efficiency and reducing environmental impact. Their analysis highlights the correlation between population growth and logistics needs, emphasizing the importance of advanced technologies in managing urban infrastructure. This chapter underscores the potential of digital technologies in transforming seaport operations for sustainable development.

Chapter 18: Shifting Toward Digitalization and Smart Cities: Evidence from Kuwait
Author: Dr Rashed Alhaimer

This chapter examines Kuwait's strategic investment in digital infrastructure, exploring its impact across various sectors, including political campaigns, public health, and corporate innovation. The author highlights Kuwait's $10 billion commitment to enhancing its digital capabilities, demonstrating how digitalization fosters innovation and economic growth. The findings emphasize the transformative role of social media and digital platforms in shaping urban development, contributing to Kuwait's Vision 2035 and the broader smart city agenda.

Chapter 19: United Arab Emirates: Integration of Smart Technology – An Innovative Solution for Climate Change Mitigation
Authors: Prisilla Jayanthi Gandam, Xi Chen, Muralikrishna Iyyanki, Utku Kose, and Valentina Emilia Balas

The chapter highlights UAE's initiatives in using smart technologies for climate change mitigation, showcasing AI, IoT, and digital twins in sustainable urban development. The authors present case studies from Abu Dhabi, illustrating how smart systems enhance energy efficiency, public health, and environmental monitoring. Their analysis illustrates the UAE's digital transformation leadership and commitment to achieving net zero emissions by 2050. This chapter provides valuable insights into the role of smart technologies in addressing global environmental challenges.

Chapter 20: The Social, Economic, and Sustainability Agenda of Smart Cities in the Kingdom of Bahrain
Author: Omer Hag Hamid

This chapter presents a comprehensive framework for advancing smart cities in Bahrain. It focuses on integrating technological advancements with sustainable practices to enhance societal well-being, economic growth, and overall development. The author aligns the framework with Bahrain's 2030 Vision, emphasizing social impact, healthcare access, and renewable energy projects. The analysis highlights initiatives such as smart parks and community centers that foster social cohesion. This chapter underscores the importance of a holistic approach to creating inclusive, prosperous, and sustainable smart cities.

Chapter 21: Transforming Ecological Footprint Through Circular Economy Innovation toward Sustainable Well-Being in the Development of (Islamic) Smart Cities in GCC Countries
Author: Khairunnisa Musari

The chapter explores how Islamic Smart Cities can leverage circular economy innovations to mitigate environmental impacts and foster sustainable development in the GCC region. The author discusses the high ecological footprint of high HDI countries, and the potential of circular economy principles supported by technology and community engagement. The analysis highlights the strategic role of Islamic Smart Cities in global green transitions, emphasizing their capacity to set an example for sustainable urban development. This chapter contributes to understanding how cultural and technological integration can drive environmental sustainability.

Chapter 22: Understanding and Optimizing Hydrogen Transport Technologies for 21st Century Smart Cities: Innovation for Morocco's Green Interconnections in Hydrogen, Ammonia, and Kerosene Production
Author: Ayat-Allah Bouramdane

This chapter evaluates hydrogen transport methods, using the Analytic Hierarchy Process (AHP) tool to optimize technologies for sustainable energy solutions in

Morocco's smart city initiatives. The author explores various hydrogen transport methods, assessing their advantages and process equations. The analysis identifies barriers and opportunities for hydrogen transport, providing strategic insights into optimizing these technologies. This chapter aligns with Morocco's vision for sustainable development, highlighting the potential of hydrogen transport in reducing carbon emissions and enhancing energy security.

Chapter 23: Smart City Framework for Nusantara Capital City Development
Authors: Bambang Susantono, Mohammed Ali Berawi, and Mustika Sari

The chapter outlines a comprehensive framework for developing Nusantara, Indonesia's new capital city, as a smart city. The authors integrate advanced technologies across six key domains: governance, transportation, living, resources, industry, and infrastructure. Their detailed insights into strategic planning and implementation processes highlight the importance of technological innovation and sustainability. This chapter provides a model for future urban development, emphasizing the need for community engagement and environmental preservation in building world-class smart cities.

Chapter 24: Artificial Intelligence for Smart City Vision 2040 in Gulf Region
Authors: Miltiadis D. Lytras, Afnan Alkhaldi, Sawsan Malik, and Andreea Claudia Serban

This concluding chapter explores the transformative role of AI in smart city development, emphasizing its integration with governance, strategy, and infrastructure. The authors discuss strategic areas such as LLMs, AI-driven social network analysis, and predictive analytics, showcasing how AI can enhance smart city services and foster new markets. Their holistic approach ensures that AI improves the quality of life and economic prosperity and positions the Gulf Region as a global leader in smart city innovation. This chapter calls for leveraging AI's potential to transform urban development sustainably.

Concluding Remark:

This book serves as an essential reference point and a comprehensive guide to the groundbreaking Smart Cities initiatives and projects that are shaping the future of the Gulf region. The transformation occurring within these cities and countries is a testament to the region's vision and ambition, offering a unique perspective on economic and social development. Through a deep exploration of the smart city concept, this volume showcases the incredible potential for positive change and the profound social footprint that can be achieved.

The academic community, policymakers, government officials, and citizens will find valuable insights and knowledge within these pages. It highlights the dedicated work and contributions of academics, practitioners, and students who strive for a world of peace, prosperity, and mutual respect. Their collective efforts are leading us toward sustainable and inclusive urban environments, and their impact will be felt for generations.

This book not only celebrates their dedication but also invites further collaboration and innovation. It is a call to action for continued progress, inspiring all readers to embrace the potential of smart cities and the positive impact they can have on our world. We believe this volume will spark new ideas, foster cooperation, and drive the vision of smart cities forward, creating a better future for all.

Miltiadis D. Lytras, Afnan Alkhaldi, and Sawsan Malik

Organization in Sections

Section 1: Governance and Strategic Frameworks
This section focuses on the foundational frameworks, governance strategies, and policy implications for developing and sustaining smart cities. It covers holistic frameworks, the role of government, and strategic planning necessary for smart city initiatives.

Chapter Number	Chapter Title
1	A Holistic Framework for Smart Cities Governance in the Gulf Region: From Hype to Sustainable Impact
2	Smart City Innovation and Sustainability: A Comparative Study in GCC Countries
3	The Science of Mobile Pedestrianism through Smart Cities
4	Innovating Urban Environments: The Impact of Smart City Technologies on Employee Performance, Quality of Life, and Service
13	Innovating Urban Policy Development: Lusail Smart City's Role in Qatar
14	The Role of Government in Ensuring Sustainability in Smart Urban Development: A Case Study of Kuwait
16	Strategic Planning for the Development of Smart Cities in Oman
18	Shifting Toward Digitalization and Smart Cities: Evidence from Kuwait

Section 2: Technological Innovations and Digital Transformation
This section delves into the technological advancements driving smart city development, including AI, IoT, digital twins, and other innovations. It highlights the transformative impact of these technologies on various urban sectors.

Chapter Number	Chapter Title
5	Democratization and Transformation of Higher Education in the Era of Smart Cities
6	The Rebirth of Museum Education Echoing of Smart Education
7	The Synergy between Advanced Innovative Water Management and Smart Cities Development: Enhancing Sustainability and Quality of Life
8	AI's Role in the Development and Transformation of Smart Cities
9	The Impact of Innovation & Sustainable development in Smart Cities
10	Digital Intersections: Synergies between the Gig Economy and evolution of the Smart City
12	Innovative Digital Transformation of Issuing Final Clearance Certificates for Infrastructure Projects in Dubai
22	Understanding and Optimizing Hydrogen Transport Technologies for 21st Century Smart Cities: Innovation for Morocco's Green Interconnections in Hydrogen, Ammonia, and Kerosene Production

Section 3: Sustainability and Environmental Impact
Description: This section addresses the environmental aspects of smart city development, focusing on sustainability practices, renewable energy integration, and climate change mitigation. It includes case studies and strategic approaches for sustainable urban living.

Chapter Number	Chapter Title
11	Enhancing Prosperity through Smart Tourism: A Competitive Analysis of Gulf Cities Using STCCI
15	Innovating Sustainable Urban Development: Evaluating the Social Acceptance of Residential Solar Integration for the Prosperity of Abu Dhabi
17	Toward Kuwait Seaport's Sustainability 2040: The Implication of Digital Twins Technology on Terminal Containers' Transformation
19	United Arab Emirates: Integration of Smart Technology – An Innovative Solution for Climate Change Mitigation
20	The Social, Economic, and Sustainability Agenda of Smart Cities in the Kingdom of Bahrain
21	Transforming Ecological Footprint through Circular Economy Innovation toward Sustainable Well-Being in the Development of (Islamic) Smart Cities in GCC Countries

Section 4: Case Studies and Regional Insights
Description: This section compiles specific case studies and regional analyses, offering insights into smart city implementations across different Gulf countries. It includes discussions on strategic urban projects, regional frameworks, and localized smart city solutions.

Chapter Number	Chapter Title
13	Innovating Urban Policy Development: Lusail Smart City's Role in Qatar's Transition to a Knowledge-based Economy
14	The Role of Government in Ensuring Sustainability in Smart Urban Development: A Case Study of Kuwait
16	Strategic Planning for the Development of Smart Cities in Oman
18	Shifting Toward Digitalization and Smart Cities: Evidence from Kuwait
23	Smart City Framework for Nusantara Capital City Development
24	Artificial intelligence for smart city vision 2040 in gulf region

Recommended Reading

Alhalabi, W., Lytras, M., & Aljohani, N. (2021). Crowdsourcing research for social insights into smart cities applications and services. *Sustainability (Switzerland)*, *13*(14). https://doi.org/10.3390/su13147531

Alkhaldi, A., Malik, S., Alsadeeqi, A., & Lytras, M. D. (2024). Transformative leadership in Kuwait Direct Investment Promotion Authority: Investing in talent, innovation, and the next generation. In S. Baroudi & M. D. Lytras (Eds.), *Transformative leadership and sustainable innovation in education: Interdisciplinary perspectives* (pp. xx-xx). Emerald Publishing Limited. ISBN 9781837535378

Alkhammash, E. H., Jussila, J., Lytras, & Visvizi, A. (2019). Annotation of smart cities twitter micro-contents for enhanced citizen's engagement. *IEEE Access*, *7*, 116267–116276. https://doi.org/10.1109/ACCESS.2019.2935186

Balfaqih, M., Balfagih, Z., Lytras, Alfawaz, K. M., Alshdadi, A. A., & Alsolami, E. (2023). A blockchain-enabled IoT logistics system for efficient tracking and management of high-price shipments: A resilient, scalable and sustainable approach to smart cities. *Sustainability (Switzerland)*, *15*(18). doi:10.3390/su151813971

Chui, K. T., Lytras, & Visvizi, A. (2018). Energy sustainability in smart cities: Artificial intelligence, smart monitoring, and optimization of energy consumption. *Energies*, *11*(11). https://doi.org/10.3390/en11112869

Chui, K. T., Ordóñez de Pablos, P., Shen, C.-W., Lytras, & Vasant, P. (2022). Towards sustainable smart city via resilient Internet of Things. In *Contributions to Management Science* (pp. 117–135). https://doi.org/10.1007/978-3-030-85954-1_8

Jones, K. A., & Sharma, R. S. (2019). An experiment in blended learning: Higher education without lectures?. *International Journal Digital Enterprise Technology*, *1*(3), 241–275.

Lytras, & Şerban, A. C. (2020). E-Government Insights to Smart Cities Research: European Union (EU) Study and the Role of Regulations. *IEEE Access*, *8*, 65313–65326. https://doi.org/10.1109/ACCESS.2020.2982737

Lytras, & Visvizi, A. (2018). Who uses smart city services and what to make of it: Toward interdisciplinary smart cities research. *Sustainability (Switzerland)*, *10*(6). https://doi.org/10.3390/su10061998

Lytras, & Visvizi, A. (2021b). Information management as a dual-purpose process in the smart city: Collecting, managing and utilizing information. *International Journal of Information Management*, *56*. https://doi.org/10.1016/j.ijinfomgt.2020.102224

Lytras, Alsaywid, B., & Housawi, A. (2023). Digital transformation and smart cities: Insights from the healthcare domain. In *Smart cities and digital transformation: Empowering communities, limitless innovation, sustainable development and the next generation* (pp. 319–325). https://doi.org/10.1108/978-1-80455-994-920231015

Lytras, Housawi, A. A., & Alsaywid, B. S. (2023d). *Smart cities and digital transformation: Empowering communities, limitless innovation, sustainable development and the next generation*. https://www.scopus.com/inward/record.uri?eid=2-s2.0-85163420912&partnerID=40&md5 = 3df36699895f7ba6f8f30eebd76342a9

Lytras, M. D., Malik, S., & AlKhaldi, A. (2024). Transformative leadership and sustainable innovation in higher education: Setting the context. In S. Baroudi & M. D. Lytras (Eds.), *Transformative leadership and sustainable innovation in education: Interdisciplinary perspectives* (pp. xx-xx). Emerald Publishing Limited. ISBN 9781837535378.

Lytras, Visvizi, A., & Jussila, J. (2020). Social media mining for smart cities and smart villages research. *Soft Computing*, *24*(15), 10983–10987. https://doi.org/10.1007/s00500-020-05084-3

Lytras, Visvizi, A., Chopdar, P. K., Sarirete, A., & Alhalabi, W. (2021). Information Management in Smart Cities: Turning end users' views into multi-item scale development, validation, and policy-making recommendations. *International Journal of Information Management*, *56*. https://doi.org/10.1016/j.ijinfomgt.2020.102146

Lytras, Visvizi, A., Sarirete, A., Torres-Ruiz, M., & Daim, T. U. (2021). IEEE access special section editorial: Future generation smart cities research - Part II: Services, applications, case studies, and policymaking considerations for well-being. *IEEE Access*, *9*, 27298–27303. https://doi.org/10.1109/ACCESS.2021.3057809

Lytras, Visvizi, A., Torres-Ruiz, M., Damiani, E., & Jin, P. (2020). IEEE access special section editorial: Urban computing and well-being in smart cities: services, applications, policymaking considerations. *IEEE Access*, *8*, 72340–72346. https://doi.org/10.1109/ACCESS.2020.2988125

Lytras. (2023e). Future smart cities research: Identifying the next generation challenges. In *Smart cities and digital transformation: Empowering communities, limitless innovation, sustainable development and the next generation* (pp. 1–11). https://doi.org/10.1108/978-1-80455-994-920231001

Marouli, C., & Lytras, M. (2018). Smart cities and internet technology research for sustainable and inclusive development: An integrated approach of best practices for policy makers and educators. In *E-Planning and Collaboration: Concepts, Methodologies, Tools, and Applications* (Vol. 1–3, pp. 434–458). https://doi.org/10.4018/978-1-5225-5646-6.ch021

Nikiforova, A., Flores, M. A. A., & Lytras. (2023). The role of open data in transforming the society to society 5.0: A resource or a tool for SDG-compliant smart living? In *Smart cities and digital transformation: empowering communities, limitless innovation, sustainable development and the next generation* (pp. 219–252). https://doi.org/10.1108/978-1-80455-994-920231011

Serban, A. C., & Lytras. (2020). Artificial intelligence for smart renewable energy sector in Europe – Smart energy infrastructures for next generation smart cities. *IEEE Access*, *8*, 77364–77377. https://doi.org/10.1109/ACCESS.2020.2990123

Sharma, R., Jones, K., Anderson, W., Inthiran, A., & Tabatabaee, M. (2022). The digital transformation of higher education - "uni for nothin', MOOCs for free"?. *Journal of Information Technology Case and Application Research*, *24*(1), 34–60. https://linkprotect.cudasvc.com/url?a=https%3a%2f%2fdoi.org%2f10.1080%2f15228053.2021.1889741&c=E,1,5fKEqW_Q_rbXyPjdeN900clftr3IsVsUFTqreSSRdDFUp44dxV1qdH_oYzupodYiwGfSEkgodbt-O5Vv_yo-mdMvmwd0Lok3lwoubcMOQ,,&typo=1 https://doi.org/10.1080/15228053.2021.1889741

Sharma, R. S., Kshetri, N., & Nesbit, T. (2018). Revisiting turing's imitation game: A commentary on replacing teachers with internet devices. *Communications of the Association for Information Systems*. *43*(24). DOI: 10.17705/1CAIS.04324. https://linkprotect.cudasvc.com/url?a=https%3a%2f%2faisel.aisnet.org%2fcais%2fvol43%2fiss1%2f24&c=E,1,zV_aYS2OGknMfrIUQWV2KG69trXGbVpW1eRBGja9JObpmzYhx5NAV6H7eIJ3Xk5j5xTnYytf-hUBwoA26RbA1LjvD6JCRXkwsE18M0j7Fw,&typo=1 https://aisel.aisnet.org/cais/vol43/iss1/24

Visvizi, A., & Lytras. (2018a). It's not a fad: Smart cities and smart villages research in European and global contexts. *Sustainability (Switzerland)*, *10*(8). https://doi.org/10.3390/su10082727

Visvizi, A., & Lytras. (2018b). Rescaling and refocusing smart cities research: From mega cities to smart villages. *Journal of Science and Technology Policy Management*, *9*(2), 134–145. https://doi.org/10.1108/JSTPM-02-2018-0020

Visvizi, A., & Lytras. (2019a). Preface. In *Smart cities: Issues and challenges mapping political, social and economic risks and threats* (pp. xv–xvi). https://doi.org/10.1016/B978-0-12-816639-0.05001-8

Visvizi, A., & Lytras. (2019b). Smart cities research and debate: What is in there? In *Smart cities: Issues and challenges mapping political, social and economic risks and threats* (pp. 1–14). https://doi.org/10.1016/B978-0-12-816639-0.00001-6

Visvizi, A., & Lytras. (2020). Sustainable smart cities and smart villages research: Rethinking security, safety, well-being, and happiness. *Sustainability (Switzerland)*, *12*(1). https://doi.org/10.3390/su12010215

Visvizi, A., Lytras, Damiani, E., & Mathkour, H. (2018). Policy making for smart cities: innovation and social inclusive economic growth for sustainability. *Journal of Science and Technology Policy Management*, *9*(2), 126–133. https://doi.org/10.1108/JSTPM-07-2018-079

Wu, S. M., Chen, T.-C., Wu, Y. J., & Lytras, M. (2018). Smart cities in Taiwan: A perspective on big data applications. *Sustainability (Switzerland)*, *10*(1). https://doi.org/10.3390/su10010106

Chapter 1

A Holistic Framework for Smart Cities Governance in the Gulf Region: From Hype to Sustainable Impact

Miltiadis D. Lytras[a], *Afnan Alkhaldi*[b], *Sawsan Malik*[b] *and Andreea Claudia Serban*[c]

[a]*College of Engineering, Efrat University, Saudi Arabia*
[b]*Arab Open University - Kuwait Branch, Kuwait*
[c]*Bucharest University of Economic Studies, Romania*

Abstract

This introductory chapter lays the foundation for a Holistic Smart City framework organized around five pillars namely: Smart City Governance – Strategy – Utilization – Technology – Infrastructure (GSUTI) model. Smart cities governance establishes the overarching structure for effective management, emphasizing data integrity, regulatory resilience, and civic engagement. The strategy pillar integrates vision, mission, and strategic initiatives aligned with sustainable development goals, ensuring resource allocation and innovation resilience. The Utilization Framework encompasses diverse domains like transportation, energy, and healthcare, thus prioritizing initiatives and monitoring impact, enabling technologies that drive smart city innovation, leveraging artificial intelligence (AI) (Chui et al., 2018), Internet of Things, blockchain, and cybersecurity to enhance efficiency and service delivery. Infrastructures and capabilities encompass physical and data ecosystems, from buildings and transportation hubs to big data streams and social media content. The chapter sets the context for the scientific discussion, offering practical insights into strategic consultation, performance monitoring, technology management, and data governance. It outlines directions for smart city development in the Gulf Region, including bold visions, megaprojects, national-level initiatives, data-driven

The Emerald Handbook of Smart Cities in the Gulf Region: Innovation, Development, Transformation, and Prosperity for Vision 2040, 1–12

doi:10.1108/978-1-83608-292-720241005

services, and AI-driven excellence. This holistic framework provides a roadmap for smart city development, fostering sustainability, innovation, and societal well-being across the Gulf Region and beyond.

Keywords: Smart cities; artificial intelligence; digital transformation; smart cities governance; Gulf Region; Gulf Cooperation Council Countries; Vision 2040

1. Introduction: A Holistic Framework for Smart Cities Governance and Strategy

The debate on smart cities research and its utilization in the Gulf Region, requires the introduction of a "holistic" context for the discussion and analyses of the diverse value layers of smart city concept.

In our *Emerald Handbook on Smart Cities Research in the Gulf Region*, we decided in our introductory chapter to provide a quite abstract and concurrently integrative context. For this purpose, in Fig. 1, we summarize our Smart City Governance – Strategy – Utilization – Technology – Infrastructure (GSUTI) framework. The basic logic of this abstraction is to provide at a glance the critical pillars of a holistic smart city strategy. In our approach, we decided to include in our framework five pillars or dimensions of unique smart city-value proposition namely: smart city governance, smart city strategy, smart city utilization framework, enabling smart city technologies, and enabling infrastructures & capabilities.

Smart Cities Governance: This is the overarching layer of the entire smart city value proposition. In this context, a high-level smart city governance framework is needed which will provide a thorough description and identification of roles and responsibilities, and integral data and processes management need to be secured

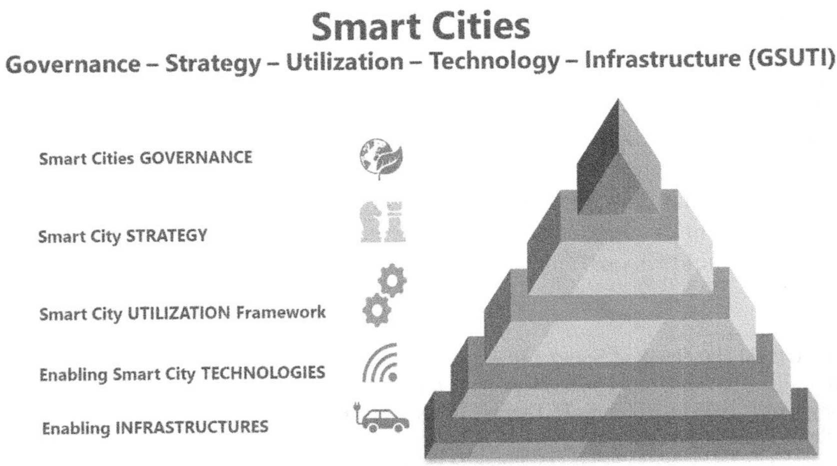

Fig. 1. Smart City GSUTI Framework. (*Source*: The Authors).

toward the fulfillment of all smart city objectives and vision (Nikiforova et al., 2023). It is also necessary at this level to build bold, resilient, and sustainable regulatory frameworks and also civic engagement platforms that will multiply the impact and the social and economic values of the smart city governance (Lytras, 2023e; Lytras, Visvizi, Chopdar, et al., 2021; Visvizi & Lytras, 2019a). Critical actions are also required at this level in relevance to risk management strategies and also overarching privacy, security, and ethics governance aiming to offer a fully trusted, and a secured smart city ecosystem for the benefit of the people and their well-being.

Smart Cities Strategy: The detailed design of smart city strategy is a bold effort toward the specification of vision and mission to be accomplished by the utilization of the diverse value components of smart city. It incorporates also the prioritization, selection, and implementation of strategic initiatives and projects, within a robust roadmap and execution plans that need also to be clarified and supported fully by resources (Lytras & Visvizi 2018; Lytras, Visvizi, Chopdar, et al., 2021; Lytras, Visvizi, & Jussila, 2020; Lytras, Visvizi, Sarirete, et al. 2021; Lytras, Visvizi, Torres-Ruiz, et al. 2020; Visvizi et al., 2018). smart city strategy has also to be anchored and aligned to sustainable development goals (SDGs) of United Nations, with reference to well-defined KPIs and analytics. For this purpose, a complementary resources allocation framework and a resilient innovation, research and development strategy have to be documented and disseminated. In the same context, capacity-building initiatives are also critical and crucial for the success of the smart city strategy.

Smart City Utilization Framework: The utilization framework is a comprehensive abstraction that covers the diverse and complementary aspects of smart city agenda (Alkhaldi et al., 2024; Lytras, Visvizi, Sarirete, et al., 2021). In our GSUTI model, we list these areas of smart city utilization but it is true that many other domains can be identified and integrated: smart transportation, smart energy (Serban & Lytras, 2020), smart education, smart environment, smart waste management, smart economy, smart governance/E-Governance (Lytras & Şerban, 2020), smart resilience, smart healthcare (Lytras & Alsaywid, 2023), smart social inclusion, smart civic engagement, smart public safety (Visvizi & Lytras, 2020), smart waste management, and smart infrastructure. For each of these areas, a systematic effort to prioritize initiatives, projects, impact and cost is required. The overarching monitoring of the execution of the diverse initiatives is a critical task for this layer and has to be supported by innovative managerial and monitoring applications, that will have also the capacity to measure the social and the economic impact of the utilization.

Enabling Smart City Technologies: In the current evolution in the technology ecosystem, diverse technologies have proved their capacity to enable strategic initiatives and to promote multi-dimensional digital transformation. In the context of smart cities, the following list represents just a sample of the diverse pool of technologies that can be utilized and integrated: artificial intelligence (AI) and machine learning (ML), cloud computing, the Internet of Things (IoT), sensors (Balfaqih et al., 2023), analytics and dashboards, open and free source technologies, social networks, the metaverse and augmented reality, blockchain, digital twins, cybersecurity, edge computing and real-time data processing, biometric identification systems, smart grids for energy distribution, and 5G and 6G networks (Lytras, Alsaywid, et al., 2023; Lytras & Housawi, 2023; Lytras, Housawi, et al., 2023; Sarirete et al., 2022).

Enabling Infrastructures & Capabilities: The enabling infrastructures refer to both the physical spaces and data-processes ecosystems that are constituting the context of smart city context. In our models, we summarized items such are buildings, public infrastructures, schools, urban areas, hospitals and healthcare facilities, transportation hubs (such as airports and train stations), parks and recreational areas, community centers and cultural institutions, government offices and administrative buildings, utilities infrastructure (water treatment plants, sewage systems, etc.), commercial and industrial zones, housing areas. For sure are part of the infrastructure that can be expanded to items such as: roads and highways, bridges and tunnels, railways and metro systems, ports and harbors, airports and airstrips, waterways and canals, traffic signals and streetlights, parking facilities and garages, telecommunication networks including fiber optics and cellular towers, waste management facilities such as landfills and recycling centers, water supply and distribution systems, sewage and wastewater treatment plants, electrical grids and power distribution networks, gas pipelines and distribution systems, green spaces and urban parks. One of the most significant aspects of the infrastructure is also related to data and big data streams (Wu et al., 2018), social media clickstreams, and huge data lakes that complement structured data warehouses within the context of the smart city. In this really complicated environment, the following data examples are few of the cases that have to be utilized in the entire smart city governance framework: personal data including identification information and preferences, location data from GPS and geolocation services, health data from wearable devices and medical records, transportation data such as traffic conditions and public transit schedules, environmental data including air quality and weather forecasts, energy consumption data from smart meters and appliances, financial data for transactions and payments, social media data related to citizens opinions and ideas (Lytras, Visvizi, & Jussila, 2020), educational data for learning and skill development, public safety data, government data for accessing public services and information, urban planning data for infrastructure development, community engagement data for citizen feedback and participation.

In the next section, we elaborate further on the key determinants of our framework and provide ideas on how this can be exploited on real-world smart city projects at national, international, municipality, and local contexts.

2. Translation of the GSUTI Smart City Framework: Enabling Smart City Efficiency and Performance for Social and Economic Value

In a fast-advancing area, like the Gulf Region, where the innovations are embraced and enhanced, it is also critical to understand that such Smart City GSUTI Framework, has to take into consideration micro, meso, and macro levels and also has to support all the diverse contexts of interoperability, operation, and exploitation including smart city strategies at cross-national, international, national, municipality, and local levels. This is though something that goes beyond the scope of this chapter, and we will discuss it in another volume.

In Fig. 2, we provide an abstraction of the expanded Smart City GSUTI framework. In the next paragraphs, we elaborate further on the aspects of the model that can guide the development of a holistic smart city governance strategy in mega-projects of smart cities in the Gulf Region and globally (Visvizi & Lytras, 2018b). These aspects can also support any strategic consultation on national, municipal, and local projects.

Our model can be exploited in diverse directions:

- Strategic Consultation: The five pillars and the diverse dimensions integrated into each pillar can serve as a reference master document for drafting detailed strategies for smart city management in national and local contexts.
- Priority Matrixes: Our model can be also utilized for the identification of smart cities initiatives and their prioritization within the strategic and philosophical directions of our model.
- Performance Monitoring: Our model has also the capacity to serve as a holistic performance and efficiency monitoring tool, with the deployment of additional guides and resources.
- Sophisticated Technology Management: In another strategic direction, our model can be utilized for the detailed mapping of technologies, infrastructures, and smart city applications and services that are implemented and used.
- Integral Data Governance and Management: Our model can also be exploited for the integrated data governance and management of all smart city applications that fall under the capacity of the smart city governance and strategy pillars. It is one of the really important parts of the usability of the framework to enable smart city owners and stakeholders to deploy a full understanding and exploitation of the smart city data ecosystem.

Given the high-level items that are included in our framework, it is rather difficult within a single chapter to discuss all its aspects and components. For this reason, in the next paragraphs, we elaborate on some indicative areas of interest,

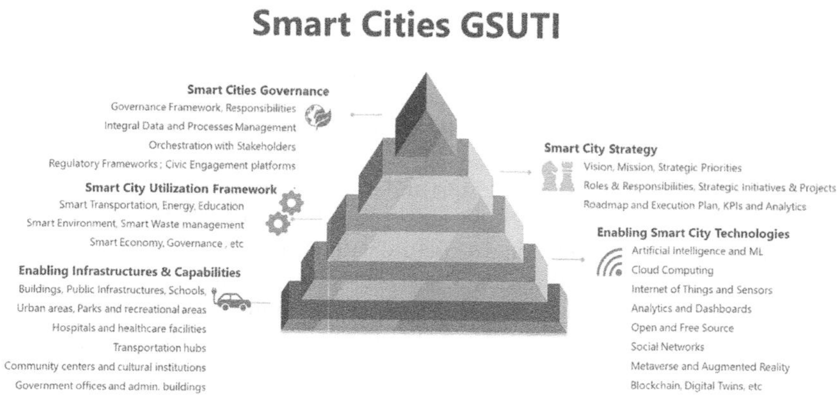

Fig. 2. *The Expanded Smart City GSUTI Framework.* (*Source*: The Authors).

that significant effort has to be allocated. We will present selectively a few items from each of the five pillars below:

Pillar 1: Smart Cities Governance

- Governance Framework: The governance framework is a masterplan for the holistic management of all the resources utilized within the smart city context.
- Integral Data and Processes Management: One of the most important aspects of the governance pillar is the integrated management of data and processes. It requires enormous effort and systematic business processes and data governance modeling tools and methodologies.
- Orchestration with Stakeholders and Partners: The smart city governance offers guidelines, procedures, and services for the systematic dissemination of the diverse policies, strategies, and regulations. From this perspective, the governance layer provides all the overarching mechanisms that secure an integral, unified, and global approach to smart city that has also significant social and economic impacts.
- Regulatory Frameworks: These should cover a wide range of significant aspects of the smart city model including data privacy and protection AI and ML utilization, taxation, sustainability, environmental, social, and governance (ESG) aspects, SDGs compliance, etc.
- Civic Engagement Platforms: The smart city governance has also the pivotal role on the provision of platforms for civic engagement. These promote the co-design and the co-responsibility of the smart city services. They have also to exploit a unified data ecosystem that meets all the standards for high-value quality information.

Pillar 2: Smart Cities Strategy

- Vision and Mission: The smart city strategy has to be reflected in a compact, high-level, abstract strategy house with reference to vision and mission reflecting the intended social, economic, and sustainable impact.
- Strategic Priorities: The same strategy house should also integrate meaningful strategic objectives and priorities that are also directly linked to the smart city utilization areas (e.g., energy, mobility, education, economy, governance, environment, healthcare, waste management, civic engagement, public security, etc.).
- Roles & Responsibilities: The design and execution of smart city governance and Strategy also requires a modern organo-diagram, to reflect an efficient management and organizational structure with clear roles and responsibilities.
- Strategic Initiatives and Projects: It is crucial for smart city governance at the national level to maintain a robust framework to manage effectively the identification and execution of strategic initiatives and projects. This will provide a sufficient mechanism for the continuous monitoring of efficiency and performance and also will serve as a key managerial instrument.

- Roadmap and Execution Plan: The road mapping process is another high-level component of the smart city strategy. The attachment of milestones, deadlines, priorities, and deliverables in the execution of the smart city governance and Strategy is crucial.

Pillar 3: Smart City Utilization Framework

The smart city concept serves as an umbrella mechanism for diverse, complementary, and integrated domains of human and urban and rural space co-existence. It is important though to understand that the utilization of technologies in order to design and implement socio-technical systems has the ultimate objective to add value to the society and to promote the well-being and the quality of life for people. The following list of sub-domains of smart city applications is indicative. For each of these areas, there are numerous processes and workflows that can be enhanced, supported, or re-invented with the deployment of modern technologies. In the concluding chapter of this volume, we discuss how artificial technology as a disruptive technology can provide new unforeseen opportunities for systems, services, and applications within these domains:

- Smart Transportation
- Smart Energy
- Smart Education
- Smart Environment
- Smart Waste management
- Smart Economy
- Smart Governance/E-Governance
- Smart Resilience
- Smart Healthcare
- Smart Social Inclusion
- Smart Civic Engagement
- Smart Public Safety
- Smart Waste Management
- Smart Infrastructure

Pillar 4: Enabling Smart City Technologies

- AI and ML
- Cloud Computing
- IoT and Sensors
- Analytics and Dashboards
- Open and Free Source Technologies
- Social Networks
- Metaverse and Augmented Reality
- Blockchain
- Digital Twins
- Cybersecurity

- Edge Computing & Real-time Data Processing
- Biometric Identification Systems
- Smart Grids for Energy Distribution
- 5G and 6G Networks

3. Bold Smart City Strategic Directions in the Gulf Region

The current investments in the Gulf Region in the current period, are huge and have the potential to transform the business, economic, and cultural environments, with diverse services and benefits for individuals, businesses, and the countries. They lead also the region in the frontend of innovation, and leadership in the areas of Smart cities. In our volume, different chapters discuss big projects and initiatives. In this section (Lytras et al., 2022), we just comment on the type of smart cities strategic directions in the Gulf Region. These could be for sure titles for future volumes.

In this short section, we decided to comment in only five characteristic directions:

1. **Bold Visions for each country integrating smart city aspects**: In this context, the smart city vision is materialized in bold visions, where smart city pillars are integrated in diverse strategic distinct directions of the visions including areas such as vibrant society, enhanced quality of healthcare and quality of life for all, smart transportation and various others.
2. **Smart cities megaprojects**: All the countries in the Region, have designed and implemented huge megaprojects aiming to transform the entire region. These megaprojects serve also as flagships of the economic development and the leadership of the nations in the region. The required know-how, and the deployment of local and international resources, serve as huge knowledge dissemination and knowledge transfer best practices. The utilization of these mega projects has also to take into consideration the impact effect of these on the societies and the economies. In our volume, various contributors discuss these projects, in the next chapter. Saudi Arabia for example is investing multibillion budgets to move forward ambitious smart city megaprojects such as the Neom (https://www.neom.com/en-us) and the Line (https://www.neom.com/en-us/regions/theline). In the same context, Masdar City, Abu Dhabi, UAE (https://masdarcity.ae/), the city of tomorrow, the King Abdullah Economic City (KAEC), in Saudi Arabia (https://www.kaec.net/), the Lusail City in Qatar (https://www.lusail.com/), the Smart Muscat in Oman are few examples of visionary, big transformational smart city megaprojects.
3. **Smart cities projects at national level promoting digital transformation**: The quest for digital transformation in diverse aspects of the society, economy, and life promotes in the region significant projects related to smart cities agenda (Lytras & Alsawid, 2023; Lytras & Housawi, 2023). Projects such as the South Saad Al-Abdullah City in Kuwait (https://southsaadkw.com/), the XZero Smart city, that represents Kuwait's model for a sustainable smart city, the Dubai Blockchain Strategy (https://www.digitaldubai.ae/initiatives/blockchain). Other initiatives in the same context include projects like Tasmu

in Qatar (https://tasmu.gov.qa/) and AlNama in Saudi Arabia (https://urb.ae/projects/alnamasmartcity/).

4. **Smart cities projects aiming to utilize social media content for value-adding services**: In this context, the target is the utilization of data lakes in social media to exploit collective intelligence for the promotion of the civic engagement and the quality of life. Crowdsourcing applications and location aware services are used to filter content and to offer useful insights for the city life (Alhalabi et al., 2021; Alkhammash et al., 2019; Lytras et al., 2022).
5. **Robust data governance smart cities projects**: The unique data ecosystem that has to be exploited for the design and support of smart city applications and services, requires a robust data governance strategy and overarching mechanisms to secure the quality and the security of data. Organizations like SDAIA, Saudi Authority for Data and AI (https://sdaia.gov.sa/en/default.aspx) play a pivotal role on the digital transformation. It is also interesting in this are to investigate opportunities for cross-national interoperability and collaboration. Organizations like GCCC have to promote this vision, where the entire region will be valued as a unique hub of smart city excellence and increased efficiency.
6. **Huge projects of digital transformation in all areas of human activity**: In all the countries of the region, the strategic planning for the digital transformation, has drafted agendas with bold items in most of the critical domains of economic and smart city framework we presented in a previous section. Domains like healthcare, energy (Chui et al., 2018; Serban & Lytras, 2020), innovation, mobility, environment, waste management and civic engagement are in the top priorities. Various projects in all these areas are evident in the region. Several of the contributors on our handbook focus on such initiatives and projects in their discussion in next chapters.
7. **Investments on AI-next generation** smart city **excellence**: This area is an emerging one with a great potential impact. Investment funds in the region and local governments invest a lot on new startups, related to AI. The potential of the AI to develop a brand-new value chain for smart city unique proposition is currently shaped in new AI-driven projects. In the last chapter of this book, we elaborate further on this area, by providing our initial thoughts.

4. Conclusions

The promotion of the Smart City Framework we presented in this introductory chapter can serve as a contextual instrument for the evolution of the smart city strategy and governance. It is required to be understood from the beginning that smart city at national level is a very complicated and demanding priority that required resources, strategic planning and staffing to utilize human talent and smart services. The Gulf Region serves as a huge test bed for the capacity of diverse visions to guide the transformation of the region to a global lighthouse of smart cities for the promotion of SDGs. In the execution of the visions, it is also evident that human leadership and human talent make the difference. Emerging technologies like AI and Metaverse serve as catalysts for the form of innovation and transformation but still remains to leaders and professionals to cultivate a culture of added value for the benefit of the societies and the people in the region. In our humble opinion, the current initiatives in the region, in terms of smart cities and digital transformation projects, move the entire region to a leading position. It must be a priority in the next decade the utilization of resources, financial and human, toward a new sustainable era of economic development and social well-being. Our volume is a small piece of contribution to the relevant debate.

5. Suggested Teaching Assignments

Case Study: The Use of Smart City GSUTI Framework for Strategic Consultation

Review the Smart City GSUTI framework. Select one of the countries in the Gulf Region. Choose one of the pillars of the framework. Develop a comprehensive presentation and report on how the selected country is promoting the dimensions of the pillar. Refer to trusted information that is available in strategic documents and web portals.

Use generative AI tools such as ChatGPT, Gemini, Claude, in order to propose some bold projects to promote the pillar objectives.

Prepare a 30-minute presentation and discuss your strategic proposition.

Titles for Research Essays

1. IoT application for next-generation smart city application in the Gulf Region.
2. The impact of the Generative AI and Artificial General Intelligence (AGI) on Smart City Megaprojects in the Gulf Region.
3. Crowdsourcing applications for civic engagement toward the increased quality of life in the Gulf Region.
4. Digital Transformation in Healthcare in the Gulf Region: Initiatives, projects and impact.
5. Transforming Tourism Industry in GCCC with Smart Tourism Projects.

References

Alkhaldi, A., Malik, S., Alsadeeqi, A., & Lytras, M. D. (2024). Transformative leadership in Kuwait Direct Investment Promotion Authority: Investing in talent, innovation, and the next generation. In S. Baroudi & M. D. Lytras (Eds.), *Transformative leadership and sustainable innovation in education: Interdisciplinary perspectives* (pp. 27–38). Emerald Publishing Limited.

Alhalabi, W., Lytras, M., & Aljohani, N. (2021). Crowdsourcing research for social insights into smart cities applications and services. *Sustainability (Switzerland)*, *13*(14). https://doi.org/10.3390/su13147531

Alkhammash, E. H., Jussila, J., Lytras, & Visvizi, A. (2019). Annotation of smart cities twitter micro-contents for enhanced citizen's engagement. *IEEE Access*, *7*, 116267–116276. https://doi.org/10.1109/ACCESS.2019.2935186

Balfaqih, M., Balfagih, Z., Lytras, Alfawaz, K. M., Alshdadi, A. A., & Alsolami, E. (2023). A Blockchain-enabled IoT logistics system for efficient tracking and management of high-price shipments: A resilient, scalable and sustainable approach to smart cities. *Sustainability (Switzerland)*, *15*(18). https://doi.org/10.3390/su151813971

Chui, K. T., Lytras, & Visvizi, A. (2018). Energy sustainability in smart cities: Artificial intelligence, smart monitoring, and optimization of energy consumption. *Energies*, *11*(11). https://doi.org/10.3390/en11112869

Lytras. (2023e). Future smart cities research: Identifying the next generation challenges. In *Smart cities and digital transformation: Empowering Communities, limitless innovation, sustainable development and the next generation* (pp. 1–11). https://doi.org/10.1108/978-1-80455-994-920231001

Lytras, & Alsaywid, B. S. (2023). Next generation healthcare education and research: Utilizing the talent, skills, and competencies for value-based healthcare. In *Active learning for digital transformation in healthcare education, training and research* (pp. 211–220). https://doi.org/10.1016/B978-0-443-15248-1.00005-9

Lytras, Alsaywid, B., & Housawi, A. (2023). Digital transformation and smart cities: Insights from the healthcare domain. In *Smart cities and digital transformation: Empowering communities, limitless innovation, sustainable development and the next generation* (pp. 319–325). https://doi.org/10.1108/978-1-80455-994-920231015

Lytras, & Housawi, A. (2023). Active learning in healthcare education, training, and research: A digital transformation primer. In *Active learning for digital transformation in healthcare education, training and research* (pp. 1–11). https://doi.org/10.1016/B978-0-443-15248-1.00006-0

Lytras, Housawi, A. A., & Alsaywid, B. S. (2023). Preface. In *Smart cities and digital transformation: Empowering communities, limitless innovation, sustainable development and the next generation*, xxvii–xxxvii. https://www.scopus.com/inward/record.uri?eid=2-s2.0-85163484376&partnerID=40&md5 = 19c3211d29913a9673b7f03d-3d12eb20

Lytras, & Şerban, A. C. (2020). E-Government insights to smart cities research: European Union (EU) study and the role of regulations. *IEEE Access*, *8*, 65313–65326. https://doi.org/10.1109/ACCESS.2020.2982737

Lytras, Serban, A. C., Ruiz, M. J. T., Ntanos, S., & Sarirete, A. (2022). Translating knowledge into innovation capability: An exploratory study investigating the perceptions on distance learning in higher education during the COVID-19 pandemic – The case of Mexico. *Journal of Innovation and Knowledge*, *7*(4). https://doi.org/10.1016/j.jik.2022.100258

Lytras, & Visvizi, A. (2018). Who uses smart city services and what to make of it: Toward interdisciplinary smart cities research. *Sustainability (Switzerland)*, *10*(6). https://doi.org/10.3390/su10061998

Lytras, Visvizi, A., Chopdar, P. K., Sarirete, A., & Alhalabi, W. (2021). Information Management in Smart Cities: Turning end users' views into multi-item scale development, validation, and policy-making recommendations. *International Journal of Information Management*, *56*. https://doi.org/10.1016/j.ijinfomgt.2020.102146

Lytras, Visvizi, A., & Jussila, J. (2020). Social media mining for smart cities and smart villages research. *Soft Computing*, *24*(15), 10983–10987. https://doi.org/10.1007/s00500-020-05084-3

Lytras, Visvizi, A., Sarirete, A., Torres-Ruiz, M., & Daim, T. U. (2021). IEEE access special section editorial: Future generation smart cities research - Part II: Services, applications, case studies, and policymaking considerations for well-being. *IEEE Access*, *9*, 27298–27303. https://doi.org/10.1109/ACCESS.2021.3057809

Lytras, Visvizi, A., Torres-Ruiz, M., Damiani, E., & Jin, P. (2020). IEEE access special section editorial: Urban computing and well-being in smart cities: Services, applications, policymaking considerations. *IEEE Access*, *8*, 72340–72346. https://doi.org/10.1109/ACCESS.2020.2988125

Nikiforova, A., Flores, M. A. A., & Lytras. (2023). The role of open data in transforming the society to society 5.0: A resource or a tool for SDG-compliant smart living? In *Smart cities and digital transformation: Empowering communities, limitless innovation, sustainable development and the next generation* (pp. 219–252). https://doi.org/10.1108/978-1-80455-994-920231011

Sarirete, A., Balfagih, Z., Brahimi, T., Lytras, & Visvizi, A. (2022). Artificial intelligence and machine learning research: Towards digital transformation at a global scale. *Journal of Ambient Intelligence and Humanized Computing*, *13*(7), 3319–3321. https://doi.org/10.1007/s12652-021-03168-y

Serban, A. C., & Lytras. (2020). Artificial intelligence for smart renewable energy sector in Europe – Smart energy infrastructures for next generation smart cities. *IEEE Access*, *8*, 77364–77377. https://doi.org/10.1109/ACCESS.2020.2990123

Visvizi, A., & Lytras. (2018b). Rescaling and refocusing smart cities research: From mega cities to smart villages. *Journal of Science and Technology Policy Management*, *9*(2), 134–145. https://doi.org/10.1108/JSTPM-02-2018-0020

Visvizi, A., & Lytras. (2019a). Preface. In *Smart cities: Issues and challenges mapping political, social and economic risks and threats*, xv–xvi. https://doi.org/10.1016/B978-0-12-816639-0.05001-8

Visvizi, A., & Lytras. (2019b). Reflecting on oikos and agora in smart cities context: Concluding remarks. In *Smart cities: Issues and challenges mapping political, social and economic risks and threats* (pp. 333–339). https://doi.org/10.1016/B978-0-12-816639-0.00018-1

Visvizi, A., & Lytras. (2021). Guest editorial. *Transforming Government: People, Process and Policy*, *15*(2), 193–196. https://doi.org/10.1108/TG-05-2021-315

Visvizi, A., Lytras, Damiani, E., & Mathkour, H. (2018). Policy making for smart cities: Innovation and social inclusive economic growth for sustainability. *Journal of Science and Technology Policy Management*, *9*(2), 126–133. https://doi.org/10.1108/JSTPM-07-2018-079

Wu, S. M., Chen, T.-C., Wu, Y. J., & Lytras, M. (2018). Smart cities in Taiwan: A perspective on big data applications. *Sustainability (Switzerland)*, *10*(1). https://doi.org/10.3390/su10010106

Chapter 2

Smart City Innovation and Sustainability: A Comparative Study in GCC Countries

Revenio C. Jalagat, Jr[a], *Edwin C. Du*[b], *Neilson D. Bation*[c] *and Perfecto Gatbonton Aquino, Jr*[d]

[a]*Al Zahra College for Women, Oman*
[b]*Capitol University, Philippines*
[c]*Opol Community College, Philippines*
[d]*Independent Researcher, Manila, Philippines*

Abstract

Moving into digital transformation, the development and implementation of smart cities have gained global attention, including in Gulf Cooperation Council (GCC) countries. The aim of this chapter primarily rests on the innovation and sustainability of the implemented smart cities, considering the advantages and benefits gained from the utilization of the latest technologies. It also tackles the challenges faced by these countries in managing smart cities and the processes involved in addressing these challenges toward their resolution. The main contribution of this chapter is to highlight the innovative platforms initiated by Gulf nations on smart cities by modeling transformation and development toward Vision 2040. In the form of a narrative account, the contribution extends to assessing the present status of innovativeness and development through smart cities as a baseline for future development directions in the Gulf region. Documentary evidence was solicited from government reports, private agency documents and files, journals, and newspapers to address the chapter's objectives. Findings revealed that sustainability issues are prevalent and may impact future management and control of its operations. Key recommendations include the intensification of policies and regulations toward smart cities through information and communication technology maximization in the areas of transportation, education, healthcare, government

The Emerald Handbook of Smart Cities in the Gulf Region:
Innovation, Development, Transformation, and Prosperity for Vision 2040, 13–34

doi:10.1108/978-1-83608-292-720241001

services, infrastructure, and urban planning. Heightened laws and policies to address privacy and security, water, energy, technological infrastructure, overcrowding and traffic congestion, and funding. More collaborative efforts should be made by the different stakeholders, such as the government, private companies, non-government organizations, and individual experts, to ensure the smart city's sustainable operations.

Keywords: Smart cities; ICT development; innovation and sustainability; GCC countries; digital transformation; urbanization

Introduction

Urbanization has increased progressively worldwide amidst the insurgence of population growth. The United Nations (2018) Population Division reported that the population living in urban areas accounted for more than 50% of the worldwide population and is expected to increase to three-quarters in 2050. Furthermore, under the World Urbanization Prospects released in 2018, the urban population has been estimated to reach 300,000 inhabitants for the period 1950–2035. The rate of urbanization is becoming unprecedented and creates pressure in terms of the constantly growing population, quality of living, consumption, environmental pollution, traffic issues, infrastructures, etc. At the height of these complexities, many countries have introduced the concept of a smart city in line with the advancement of technology. As evidence, Barth et al. (2017) stressed that the emergence of information urbanism has become visible in urban zones and found that smart cities have a higher potential for resolving increasing population pressures. For instance, it is estimated that half of the population in the Middle East and North Africa (MENA) and the Middle East lives in urban areas, while over 80% of the whole populace in Europe also lives in urban cities (Constro, 2021; Ringel, 2021). Economically, studies by Asmyatullin et al. (2020) and Bufetova (2013) depict that nations that are classified as high-income will reach 88.4% in 2050, along with the expected rate of 90% in the GCC countries.

Anchored on the above-mentioned need for urbanization, the role of smart practitioners and planners is vital to the implementation of innovative smart cities that would eventually improve the quality of life (QoL) of their citizens through having a smart city structure that provides a sustainable environment, resiliency, and efficient provision of the needed services to its citizens. Asmyatullin et al. (2020) mentioned the different challenges faced by businesses in the GCC region as urbanization is rapidly growing. These challenges comprise the lack of diversification and high dependence on oil products, fast-rising regional urbanization growth, concerns about infrastructure related to transportation, education, and healthcare, climate issues, quality provision and upgrades of government and private services, sustainable economic development, and others. Initiatives were implemented to address these issues that led to the initiation of smart cities and digitalization within the GCC countries based on vision-oriented plans and urban management. However, the extent of the implementation is relatively in its infancy stage which requires further development, innovation, transformation, and inclusion in the future nationwide policies and schemes throughout the GCC nations.

Dating back to the 1930s–1960s, GCC countries' development trend was centered on oil exportation attributed largely to urbanization, population growth, technological advancement, sociocultural change, and global cultural exchange (Almulhim, et al., 2022). In other words, the Arabian Gulf has developed into a composition of highly urbanized nations while faced with challenges in managing its cities and sustainable development (Zaidan & Abulibdeh, 2020). Undeniably, urbanization improves the inhabitant's QoL but at the same time poses a challenge to natural resources.

The term "smart city" refers to an idea that can address the social, economic, and environmental issues that contemporary cities are currently and will likely confront (Venkatachalam, 2020). Even if the term "smart city" isn't yet widely understood, cities are adopting the idea of becoming smarter to promote economic growth and sustainable development while also raising the standard of living for their populace. A "smart city" is, to put it simply, a city that makes use of cutting-edge technologies to accomplish its objectives. These objectives can vary based on the city's location, socioeconomic circumstances, available resources, and the opinions of the relevant local authorities. They could include financial development, education, social equality, eradication of poverty, enhanced citizen security, tourism, cultural education, intercultural physique, environment, and quick citizen service in public services.

Worldwide, the concept of the Sustainable Smart City (SSC) is boosted to be the prime resolution for urbanization and its associated issues (Wu & Chen, 2021). Hanania (2020) supported the concept that there is a global need to upgrade digitally through smart technologies to meet the pressures of digital revolution and address the growing challenge of pollution and congestion. Attaining sustainability is viewed worldwide as many nations emphasize the economic, environmental, and social sustainability that can be achieved through the implementation of the smart city (Patrao et al., 2020). According to Pantic et al. (2021), being smart and sustainable is significantly unavoidable as smart mechanisms empower people to livable, wise use of energy, and stay in an environment using the resources efficiently and effectively. In the Gulf region, sustainability and smartness have been endeavored as evidenced for example by the development of smart cities such as Zayed Cities, Masdar Cities, and Dubai smart cities (Digital UAE, 2022), and the NEOM city in Saudi Arabia (Tomorrow City, 2022).

This chapter contends that while traditional industrial and service societies are ideal cities of the past, due to the technological revolution the significance and importance of utilizing the smart or digital city have become more relevant and responsive to global demands including but not limited to the GCC countries which are in line with their strategic vision and goals. The countries in GCC are optimistic about the need for smart cities and they are in one view that implementing it provides future sustainable benefits as embodied in their national strategies. While the studies of smart cities in the Gulf region are relatively scarce compared to other studies outside the region, the study tries to foster the dissemination of the status of smart implementation beyond the GCC countries. Elian and Kisswani's (2024) recent study exclaimed that smart city development is still less attended in the MENA and GCC countries although urbanization is increasing, and technological advances are being experienced. This study tries to investigate the smart city application in GCC countries taking into consideration the comparison of the experiences of these countries, the advantages of adopting the smart city platform, the challenges faced by each country in smart city implementation, and its proposed resolutions.

Moreover, this chapter also examines the extent of the application of innovative smart cities in the Gulf region and the development agenda for sustainability. It practically deals with the advantages, challenges, and initiatives related to smart city innovation by focusing on the GCC countries. One of the major contributions of the study is that it adds value to the recent studies conducted about smart cities in the GCC and inputs on each country's challenges, experiences, and proposed resolutions for further investigations considering that studies of this kind are still limited in the Gulf region. Sustainability measures and related information are discussed in the chapter, and it is expected that they can add to the existing academic literature and studies.

The study starts with introductory information about the concept of smart cities and urbanization at the global level and cascades it down to the GCC region with a discussion on the global concept of sustainable smart cities. It is followed by a discussion of the GCC's experience of smart city utilization. The next section then explains the advantages and benefits of smart cities and is followed by a section about the challenges that each country in the GCC experienced. Then proposed resolutions were offered for each country followed by concluding statements to summarize the whole discussion in the chapter.

Utilization of Innovative Smart City in the GCC

Smart city is relatively a new concept in the GCC countries but this initiative has been already crafted in the strategic vision and plans (Asmyatullin et al., 2020; Saxena & Al-Tamimi, 2018). Several innovative and development initiatives from these countries have been implemented for the past years noting for example in Bahrain, the country implemented an e-government program that is aligned with the government's Vision 2030. It also launched the Economic Vision 2030 which emphasized the intensification of innovation activities and more advanced infrastructures, upgrading the quality of services while implementing cost reduction strategies, enabling an environment of security and safety, and continuous improvement of technologies. Evidence of the implementation of the e-government is the application of the "Cloud First" policy wherein all the activities and affairs of the government are transmitted to the cloud. Moreover, the government of Bahrain has initiated the development of 10 smart cities based on their national strategic plan and intensified the application of information and communication technology (ICT) through online conversion of government services by the public utilization of e-government networks including but not limited to open data portal, smart cards, website portal, call centers, mobile applications, and booths.

The authors continued that in Saudi Arabia, the country's National Transformation Program (NTP) and Vision 2030 planned to convert most of the cities into a smart city platform wherein the high quality of service is demonstrated, adapt and sophistication of digital infrastructure, and building a "digital economy." Projects anchored on Vision 2020 had been initiated, for instance, the employment of modern infrastructures and new technologies and the application of the digital traffic control system. Specifically, Saudi Arabia endeavored to implement the smart city in the 17 cities throughout the country wherein it has been inhabited by approximately 72% of citizens and residents. One of the highlights for example is the development of NEOM City situated in Tabuk, a new construction that has an estimated cost of $500 billion. The emphasis of the NEOM Smart City rests on the proliferation of digital technology, biotechnology, tourism, healthcare, education, water and energy supply, robots, entertainment, etc. The government aims to collaborate with the private sector in implementing and sustaining the smart city by providing training and enhancements to the contributors to designing and developing the initiatives on smart city. Another smart city project that was launched is the King Abdullah Economic City which was primarily designed to address the issues of energy consumption, security, home automation program, and smart lighting.

In the case of Qatar, several initiatives have been implemented in the country starting with its upgrading of the services through the e-Government 2020 which aimed at improving the government services using various platforms and initiatives such as cloud infrastructures and digital applications, mobile applications, and other related activities. An example of these initiatives is the Lusail Smart City which offers technological advancement in terms of managing disasters and crises for the benefit of the citizens, residents, and visitors using centralized cooling system installation, waste management using smart pipes for transmitting waste to recycling plants, fully automated metro, wastewater processing, etc. The country has also envisioned in 2030 the residents' QoL needs, digital economy, and larger technological infrastructure. Other initiatives implemented in Qatar are the Qatar Rail Development Programs, Qatar Oasis, and the Msheireb Downtown Doha. To ensure the acceptability of the smart initiative to the public, the government has delivered workforce training platforms that are linked with hardware and software applications. Qatar's vision of a smart city as planned is believed to augment the QoL of the people through the utilization of the integrated ICT infrastructure that enables companies to produce quality products and service outcomes.

Relatively, Oman has crafted the Digital Oman Vision 2030, and in 2017, it launched new technologies and obtained digital literacy and skills. The country's forefront of development is the e-Oman strategy that focuses on the three main platforms namely the e-government services, society, and IT industry development. Oman's Vision 2030 is driven by the objectives of government online applications, eLaws, digital literacy, and ICT enhancements, accessing public services through mobile applications, and open-source and free software utilization.

Oman gains its pride with the establishment of the Duqm Special Economic Zone which was established in 2011 to address the concerns on smart building management, tourism, smart port solutions, security, road lighting, utilities, waste management, and traffic lights. It also envisioned meeting the demands of digital transformation through ICT progressive implementation with the initiatives from the Information Technology Authority (ITA) mainly on strategies that include but are not limited to the Digital Oman Strategy, smart government, e-Health portal, e-transformation plans, education portal, e-payment gateway, and others.

Notably, the leading initiator of the smart city is the United Arab Emirates throughout the Gulf Cooperation Council (GCC) region. The country has a smart state strategy with the "Smart Dubai" initiative started in 2014 that caters to the efficient promotion and delivery of services coupled with safety and people's memorable and impressive experiences. Smart Dubai strives toward becoming the happiest city on earth and is anchored on four pillars: efficient, seamless, safe, and impactful (Smart Dubai, 2016). It has a smart navigator that focuses on Smart Dubai, Dubai Data, CX Lab, Dubai Now, Dubai ID, and Happiness Meter. Moreover, the government's smart Dubai 2021 plan is centered on these platforms including digital transportation, social services digital access, sustainability, technological economy advances, private sector partnerships, and a clean environment. Another example is the Smart Dubai Index which measures smart performance and achieves the economic diversification plan of the UAE. Furthermore, the "Wi-Fi UAE" system serves the entire UAE with public Wi-Fi services through the provider, Dubai Telecommunications while the Dubai Silicon Oasis (DSO) provides charging stations for electronic vehicles, and this is evidence of artificial intelligence (AI) infrastructure applications. Abu Dhabi on the other side has taken pride in Masdar City as a smart city initiative that centers on digital applications, solar energy systems, and green building infrastructures. In 2018, the Zayed smart city project was established and financed by the Abu Dhabi Municipality, which is intended to convert digitally the financial, environmental, social, and QoL of its citizens and residents (EMergy Consultancy, 2018).

Like other GCC countries, Kuwait has its share of smart city formation considered as the country's strategic concern in line with the New Kuwait Vision 2035. Saxena and Al-Tamimi (2018) stressed that the vision has seven strategic dimensions which are "*global positioning, developed infrastructure, creative human capital, effective public administration, high-quality health care, sustainable diversified economy, sustainable living environment.*" The government is committed to developing the so-called "Information Society" wherein collected and processed data are modernized in consonance with the overall vision to transform Kuwait into regional cultural, financial, and commercial centers. Most initiatives are introduced through e-government services that include conversion into online publicly offered services through infrastructure, education, health, and others. Schemes on intellectual mobility were also established that emphasized monitoring of infrastructures and managing traffic along with the installation of smart water and electricity meters to facilitate the outright transformation of data and information.

In addition, one of the prime initiatives related to smart city based on the new Kuwait 2035 vision is the establishment of Saad Al-Abdullah city in 2019 with the main goal of fostering economic growth in the coming years, reducing reliance on oil, and transforming the country as regional hub for commercial, tourism, financial, and trade to encourage domestic and regional investments (Smart City Hub, 2017; Urban Gateway, 2018). According to Jazzar (2019), the $4 billion Saad Al-Abdullah city is the pioneer smart city in the country and will operate as a green and smart city that is environmentally friendly and is estimated to house 400,000 citizens and residents through utilization of ICT as linking systems for roads and services. The country has combined the classic and smart city as compliments to each other. The structure of the city is devised to prevent pollution and install solar energy for electricity consumption. It also emphasized the utilization of particular building materials according to the people's affordability (METenders, 2022). Continuing plans are also in place to develop eight additional smart cities to be launched in the future to add value to the already established Saad Al-Abdullah smart city project.

Analyzing closely the extent of smart city applications in GCC countries, comparable outcomes show that the concerns are similarly focused on ICT smart city initiatives, particularly in e-government and e-service initiatives. However, these countries differ in the degree of development in terms of digitalization and smart city implementations. Table 1 shows the smart city initiatives of the GCC countries.

Table 1. Smart City Initiatives in GCC Countries.

Country	National Strategy	National Strategy	Initiatives and Programs	Country Vision Focus
Bahrain	Bahrain Economic Vision 2030	National Development Strategy	National eGovernment Strategy	Infrastructure innovation, quality of public services, business environment, and e-government networks
Kuwait	New Kuwait 2035	Kuwait National Development Plan	eGovernment Program	Advanced IT infrastructure, human capital, effective public services, sustainable diversified economy
Oman	Vision 2020, Vision 2040	National Program for Enhancing Economic Diversification	Oman	IT industry development society, and e-government services

(*Continued*)

Table 1. (*Continued*)

Country	National Strategy	National Strategy	Initiatives and Programs	Country Vision Focus
Qatar	National Vision 2030	National Development Strategy	Qatar e-government 2020, Lusail Smart City vision	Efficiency of public services, QoL requirements, empowering businesses, and digital economy
Saudi Arabia	Vision 2030	National Transformation Program	Smart Cities Program	Quality of public services, advanced digital infrastructure, and adopting a digital economy platform
United Arab Emirates	UAE Vision 2021	UAE National Agenda 2021	National Innovation Strategy; Dubai Smart Government; Smart Dubai 2021; Abu Dhabi Economic Vision 2030	People's QoL standards, advanced e-government formation, energy high-tech sustainable economy, digital transportation, and environment

Source: Adapted from Asmyatullin et al. (2020) and Saxena and Al-Tamimi (2018) as cited in the study of Elian and Kisswani (2024).

Advantages and Benefits of Innovative Smart City

Smart city has been implemented in many countries like Singapore as one of its pioneering implementors of this initiative. With the advent of fast technological advances, many studies have been conducted to assess its benefits, and one of the recent affirmations is in the study of Warner and Mansker-Young (2023) appraising the major benefits of implementing a smart city. The utilization of technology can transform cities into efficient and effective communities where the urban populace experiences a QoL. The benefits it offers include but are not limited to the following:

- *Safer communities* – Adopting new technology through public/private collaborations can lessen crimes such as the use of gunshot detectors, connected crime centers, body cameras, next-generation 911, license plate recognition, and others. These devices can help enforcers perform their jobs better.

- *Improved transportation* – According to SmartCity Press, there is an increasing investment of over 25% in the next 5 years wherein tracking train or bus locations can be performed through smart technologies. While the rapid population continues to increase, better services can still be provided to the citizens. An example is the intelligent traffic signals that can enhance traffic flow, especially when operating during peak traffic situations. Alternative systems like smart parking management, vehicle routing, and others.
- *Improved government and citizen encounters* – Data provided by the government needed to be as accessible as possible to the citizens as they expect friendly digital services to enhance their quality of living, with all attractions and connected citizen experience. Government ministries and agencies have developed self-service portals, modern websites, collaboration digital mechanisms, mobile applications, etc. Smart cities can augment community engagement and city official's trust.
- *Increased data-driven decision-making* – The use of big data plays a central role in creating a well-designed data analytics approach that enables government and individual users to retrieve and evaluate information that contains actionable and substantial insights. Monitoring, for example, required real-time metrics is expected to increase service quality levels. Another example, effectively applying big data to help a city with many crime-related incidents identify police staff profiles that will extend instant and quick response whenever a crime happens. The use of smart cities in turn aids the citizens and residents in upgrading the current services and cutting costs, stronger decisions with big data and along with the Internet of Things (IoT). For the government, smart city-driven data increases evidence-based decisions, encourages reliable public service administration, and increases accountability and transparency in governance (Nijhuis et al., 2016).
- *Improved digital parity* – A smart city can grant digital equity that enables the provision of affordable yet high-speed internet facilities like the installation of Wi-Fi hotspots throughout a city that can cater to the needs of all citizens and residents through a reliable internet service. The COVID-19's aftereffects resulted in a growing demand for high-speed internet for people that may indicate future implications.
- *Economic development* – The development of smart cities is an economic booster for economic development because it can generate job opportunities, and build new businesses industries, and innovation centers (United Nations Human Settlements Programme, 2016). It also invites more investors and entrepreneurs toward diversified economic activities.
- *Optimum resource utilization* – Capitalizing on the latest technologies like IoT and data analytics with the smart city helps optimize the effective and efficient use of resources particularly water, energy, and transportation results into cost savings, increased waste management, and encouraged sustainable use of resources (Alawadhi & Al-Atawi, 2019).
- *Environmental sustainability* – Environmental conservation is one of the high priorities in establishing smart cities. This can be evidenced by applying efficient waste management, eco-friendly city planning strategies, enhancing

clean energy sources, and green buildings that lead to better air quality and minimized carbon emissions (United Nations Economic Commission for Europe. 2019).

- *Social inclusion* – Platforms of smart cities are focused on providing technological facilities that benefit all sectors and society in all spheres to fill the gap of the digital divide (Cohen, 2012). Communities that are classified as marginalized will receive equal benefits from others in terms of digital literacy programs, easy accessibility, and inexpensive use of digital technologies.
- *Empower employees for engagement at work* – The utilization of smart city encourages employees to effectively function in their tasks as management of work would be done digitally daily. The employment for instance in mobile devices, sensors, and bots enables the employees to increase their efforts to be strategic in their workplaces by minimizing manual tasks and replaced by digitalization.

Challenges of Innovative Smart City Implementation in the GCC

UAE

The country UAE has been proven to be the leader of smart city platforms in the Gulf with its sophisticated technologies. However, challenges are inevitable, and the nation also encounters these realities that need to be attended to as the progress in smart city applications. Firstly, maintaining data privacy and security. UAE's connection digitally tremendously increased which signals a crucial concern for data sensitivities and personal data safeguards. The real challenge is how to strike a balance between data usage for enhanced services and safeguarding individual privacy. The second challenge rests on accessibility and inclusivity wherein there is a need to ensure that the smart city initiatives should provide advantage to all segments of the population. Overcoming the challenge of the digital divide is a necessity to ensure that all groups will benefit from the smart city implementation. The third challenge is technological integration, and this is centered on integrating various technologies and systems into an interoperable and cohesive structure to meet the ideals of having a seamless experience as proposed by the smart city Dubai. The fourth challenge is sustainability. The UAE with its Dubai smart city initiatives is committed to achieving sustainability. However, this may not be realized without the government and the people's continued efforts and cooperation. Long-term environmental preservation and resource efficiency becomes a real challenge.

Kuwait

The prevalent challenge that Kuwait is facing is the transformation of the e-government system into a smart government. The IoT based smart-government advancement has the same goal as other technological advances however it faces a higher level of difficulties for governments especially in the sustainability of this initiative amid economic uncertainty and societal changes (Howard, 2013). The challenges that the country faces rest on landscaping, construction, and maintenance. A survey was conducted of 60 participants to determine their perception of landscaping in smart cities, and results show that it is a main challenge for the Kuwaiti government. The country also lacks the coordination and collaboration of stakeholders such as providers and experts of smart cities, private, and the government. Sustainability is the third concern which emphasizes the security and digital sustainability of the city in the long run. In the study of Alghais and Pullar (2018), they pointed out the major concerns of rapid urban expansion in the country and found out whether establishing a smart city would help address these issues. Qualitative research was conducted through interviews with government executives and representatives from the private sector and consensus outcomes revealed that the two eminent issues are housing shortages and traffic congestion. The result implies the negative consequence of urban population growth based on the above-identified issues. AlEnezi et al. (2018) emphasized that the key element of applying to a smart city is the utilization of the IoT to accelerate urbanization, industrialization, and communication concerns.

Qatar

One of the eminent challenges encountered by Qatar is found in the Telecommunication sector. Technically, implementing the smart city adds a burden for the telecom sector's regulations requiring high-speed connectivity and applications like the machine-to-machine communicative environment. An interview conducted with a policy expert in the field of telecommunications shared that transitioning from traditional cities into smart cities brings risk factors and issues from regulatory viewpoints, technical, and resistance from telecom stakeholders (Badran, 2023). Resistance level if remains unmanaged would hinder the smooth implementation of the smart city format that requires change management strategies. According to Gartner (2022), changing the mindset from what was traditionally done into a long-term focused smart city development initiative is timely and important in addition to the people's social behavior alteration and resource consumption. Another major issue confronted by the Qatar government is cybercrime in different features that include hacking information networks and websites, information systems, and information programs; content crimes; electronic card-related crimes; and intellectual property rights violations.

Bahrain

Like other countries in GCC, Bahrain experienced challenges while implementing the smart city format and one of these is traffic congestion. Another problem encountered even after the implementation of the AI ecosystem is securing smart cities from cyberattacks, privacy interferences, and the lack of laws and regulations to protect citizens from these attacks. Although the AI ecosystem potentially uplifts people's lives, cybersecurity remains to be an enormous concern. The third challenge of the smart city implementation in Bahrain is the different development risks. The study of Mamlook et al. (2019) stressed in their study findings that the country of Bahrain has faced challenges in energy, water, climate change, and food security.

Saudi Arabia

The country has identified several challenges encountered in the implementation of smart city and these include data privacy and security, regulatory and legal concerns, technological challenges, skills gap, social and cultural issues, etc. Alshehri and Elyas (2018) opined that in the aspect of privacy, voluminous data or information gathered from different sources raised issues in safeguarding and maintaining privacy for example, ensuring that sensitive information is protected and secured in transmitting information from stored data and creating a vigorous cybersecurity procedure. Maintaining public trust and cyber threats have become a major concern for smart city implementation. Another challenge rests on legal and regulatory concerns as Alshehri and Seifi (2017) pointed out that recent developments in urbanization and emerging technologies related to the smart city are not substantiated by the existing laws and regulations in the country. A need for developing suitable legal structures to address privacy, protection, and legal issues. Technical and interoperability challenges may hinder the implementation of the smart city like the compatibility of various technologies and mechanisms. Alawadhi and Al-Atawi, (2019) emphasized the importance of determining the appropriateness of different technologies as they need standardized protocols, data formats, and communication interfaces to support uniform connectivity and information exchange.

Moreover, the skills gap is another major concern for smart city implementation considering that it requires skilled manpower and experts in the field of digital governance, data analytics, urban planning, and the IoT. The present workforce lacks the capacity and skills to sustain the implementation of the smart city resulting in the recruitment of competent workforce that is usually hard to find. And smart cities pose a challenge to the country's social and cultural environment. According to the European Commission (2021), the introduction of the latest technologies poses a challenge to the traditional practices and norms of the people that led to resistance. Citizens are heavily engrossed with the traditional system and become reluctant to embrace new technologies for fear of losing privacy and incompatibility with adopting the new system. Investment in technology is another challenge for smart city implementation as it requires a huge outlay.

Oman

Oman has experienced growth in terms of population growth from 31.3% in 1971 to 86.3% in 2020 and population-wise, the increase has seen from 723,851 in 1971 to 5,106,626 in 2020 (Bakici et al., 2012). Dutta and Lanvin (2019) assessed that the changes in social and demographic environments have been associated with challenges in security, safety, energy use, climate change, and others. As cities worldwide are inclined to build sustainable smart cities, Oman responded to the challenge by envisioning a future sustainable plan that is founded on the 12 national priorities that will empower the Omanis to be actively involved in the field of city management (Omanuna, 2017). The country's recent vision is moving forward toward "Oman Vision 2040" which emphasizes the country's future and sustainability efforts which are also in line with the global sustainable development goals (SDGs). According to this vision, the ninth priority nationwide dwelt on "Development of Governorates and Sustainable Cities" which takes due consideration the global and national challenges related to inequity and poverty, converting the traditional cities into sustainable smart cities providing all citizens equal access to technology and an IT infrastructure that enhances social justice and societal development with prosperous livable communities. In the study of Oraba and Jalagat (2023), the results of their interview with experts on the smart city resulted in the prevailing challenges in the smart city including economic costs, especially for SMEs, the complexity of the requirements to establish the smart city, data privacy and protection, coordination and governance, lack of competent data providers, and sustainability of the smart city. Changing rules and regulations and administration of smart cities remains to be a real challenge.

Initiatives to Address Challenges Faced in Implementing Innovative Smart City

Qatar

The country has instituted appropriate measures to address the challenges encountered with smart city implementation. Firstly, the Ministry of Information and Communications Technology (2014) maximizes the use of smart technologies through ICT for continuous development in transportation and mobility, education, infrastructures, healthcare, urban planning, and government services. Laws and regulations are in place for instance, committing cybercrime is punishable by the law and will face three years imprisonment and pay a fine of 500,000 Qatari Rials. Unlawful access to websites owned and managed by government entities and any companies that are affiliated with the government or its counterparts. Since 2004, the government has created the Supreme Council of Information and Communication Technology known as "ictQATAR" to lead in the regulation of ICT in the country which was later changed through Emiri Decree no. 42 in 2014 and the creation of the Ministry of Transportation and Communication. The Ministry has been tasked to formulate a regulatory framework that is transparent and consistent with the establishment of a digital society and a postal sector that caters to the economic and social benefits of the country's progress. Moreover, it is also responsible for promoting sustainable competition and fostering equality for all telecom stakeholders. For telecom customers on the other side, assurance is given by the Qatari regulatory office for smart availability, high-quality services, and innovative production (The Communications Regulatory Authority, 2019).

Kuwait

Anchored on the Kuwait Vision 2035, the country has laid down the plans and program for transforming the city into a smart and sustainable place. Converting and sustaining the smart city in the future into a sustainable digitalized city, some requirements should be met. One of those is substantial collaboration with all the stakeholders including the private sector, government agencies, smart city contributors, and the communities. Moreover, the stakeholders should be informed about clear advantages derived from implementing a smart city and made aware of the challenges, privacy and security issues, capacity, and coverage that a smart city can provide. Another proposition is to establish applicable regulations, policies, models, and funding to improve the existing transportation systems, water, energy, and technological infrastructure. Proposed eight future smart city platforms were initiated to overcome the challenges associated with overcrowding and traffic, improve public safety, augment economic development, and attain transparency, environmental and economic sustainability, and uplift the QoL of the people. Achieving a smart infrastructure is one of the keys to effectively implementing a smart city with ICT. Kuwait is deliberating its urban development plan to emphasize the initiation of smart cities to lessen the urban population tensions and as the strategic vision 2035 suggests, the government aims to transform the country as a hub for commercial, financial, and cultural activities.

Bahrain

While urbanization is progressing in the Kingdom of Bahrain, it carries with its negative consequences and impacts. The country implemented the SDGs pioneered by the UN which was patterned from the year 2000 Millenium Development Goals (Information & eGovernment Authority, 2022). In addition, the latest application of ICT enables laws and policies related to urbanization to be more inclined to modern technologies that utilize communication, urban computing, and data sensing (Bibri & Krogstie, 2017). The country has embraced a technological development that caters to the need for an urban environment which led to the creation of the 2016 Bahrain Smart Cities Summit. For smart cities to be sustainable in the Kingdom of Bahrain, a society called the "Bahrain Smart City Society" (BSCS) has been formed to showcase the significance of smart cities nationwide. Various measures were implemented to ensure the sustainability of the smart city, and these comprised of online complaint system, e-financial transactions, healthcare appointments, billing, etc. Efforts were rewarded because of the above-mentioned measures that made the country the second among the GCC countries and top 15 throughout Asia in terms of the E-Government Development Index (United Nations, 2020). Using ICT to establish a smart infrastructure is very important in achieving smart city innovations. As an example, the utilization of ICT and green economy are significant in offering energy security and in climate change risk reduction.

Sustainability initiatives were presented by the country and one of these is the program on urban planning, housing, and other related innovations. Evidence of this is the government's provision of 40,000 housing units to citizens with the establishment of five new cities funded by the GCC Development Program and jointly managed by the private entities including the East Sitra, Salman City, Al Ramli, East Hidd, and Khalifa City. The program called "Mazaya" is a housing financing that provides citizens an opportunity to construct their houses based on subsidized loans, especially for those who lack the resources to build their homes. Sustaining transportation is another priority of the country in line with the opening of the new Bahrain International Airport in 2021 and this has been consistent with the action plan of boosting renewal energy efficiency and other sustainability initiatives like waste management, and carbon emission reduction. Through these efforts, the airport has received various sustainability awards including the ACI Green Airports Recognition Silver Award 2023. The country has also strengthened its eServices and systems implemented in the following services: Eskan Bank Services, Ministry of Housing, Benayat, Askar Landfill Management, and RERE Licensing (Ministry of Sustainable Development Bahrain, 2023). Finally, preserving the culture and heritage has been also considered for sustainability measures with the continuous promotion of the three UNESCO World Heritage sites under the management of the Bahrain Authority for Culture and Antiquities.

Saudi Arabia

Al Jaafreh and Allouzi (2023) stressed from their study findings that there is a need for multi-facet strategies to overcome the challenges in smart city implementation and these include effective governance systems, vigorous policy structures, coordination and cooperation among stakeholders, and massive public awareness campaigns. The government has crafted policies and initiatives to ensure the adoption of smart cities anchored on Vision 2030 as the basis for social development and economic diversification initiated in 2016 (World Bank, 2017). Major initiatives are focused on NTP, Public–Private Partnerships (PPP), Regulatory Reforms, and Smart Cities and Urban Management Initiatives. The *NTP* focuses on defining definite goals and strategies for different industries including city development where smart city is a priority, improving city infrastructure, efficient public services, and digital connectivity. On the other hand, through the *PPP* the government urged private companies to collaborate to promote joint management and handling of smart city projects. The partnership's scope covers bringing together resources, funding, and expertise to carry out the smart project. *Regulatory Reforms* are also essential in smart city implementation by formulating policies and regulations that encourage private sector investments. These policies are composed of offering rewards for innovation, easy processing of licenses, and establishing supportive legal frames. The initiative called *Smart Cities and Urban Management* launched by the Ministry of Municipal and Rural Affairs emphasized the application of smart cities nationwide. It encompasses sustainable development promotion, enhanced transportation systems, community-based decision-making, and improved urban planning. It is also vital for the government to increase digital literacy, awareness, and prompt response to addressing concerns on smart city utilization. Through Vision 2030 the emphasis is on achieving sustainability of resources in the future while providing the quality of services and the welfare of the country's citizens (Vision 2030, 2016). Moreover, smart reforms are directed to the five cities namely Riyadh, Al-Madinah, Makkah, Jeddah, and Al-Ahsa followed by the cities of Yanbu and NEOM. The government has embraced smart governance under Kingdom of Saudi Arabia's designed Smart Government Strategy (2020–2024) which also contains Robotics and Automation, AI, IoT, Big Data, and Blockchain.

Evidence of the initiatives taken by the government is the platform initiated by the Natural Resources and Environmental Research Institute at King Abdulaziz City for Science and Technology (KACST) with the creation of five stations for managing air quality in Riyadh. To improve water efficiency, water strategies, and databases were established by using the Smart Water Meter system, and in Medina, the Pollution Measurement Laboratory was also initiated as a smart center to control air quality and other pollutants built on global indicators of the smart environment. Addressing environmental issues, the country has utilized intelligent technologies that depend on ICT, IoT, and Wireless Sensor Network (WSN). Applicable solutions include wastewater management and water distribution management.

UAE

The country has made a remarkable effort to sustain the smart city and the cities of Abu Dhabi and Dubai have become recipients of the award as the smartest cities in the MENA region in the "Smart City Index 2021" (Telecommunications and Digital Government Regulatory Authority, 2023). The Economic Vision 2030 of Abu Dhabi is transitioning from natural resources to exporting to cutting-edge technologies, innovations and knowledge, and economic diversification. The implementation of Abu Dhabi's smart city is one of the responses to the vision through the establishment of Masdar City. Masdar City is designed to operate as a smart sustainable city not only nationwide but worldwide. The establishment of a well-designed urban environment throughout the city with world-class transport facilities and traffic congestion solutions. The city is expected to minimize the consumption of water, energy, and waste generation leading to environmental conservation by utilizing a combination of solar power, technology, and architectural design. Another project is the Zayed Smart City in Abu Dhabi which started as the pilot phase five-year plan for Smart Cities and AI (2018–2022). Its focus is on the implementation of IoT and information technology to supervise the infrastructure. In Dubai, the implementation of the Dubai Plan 2021 is a commitment to enabling the city to be a smart and sustainable place through the Dubai Smart City project. Its main emphasis is to transform the 1,000 government services that highlight the major industries such as economic services, transportation, communications, urban planning, infrastructure, and electricity. Under these six industries, various initiatives were developed and categorized as follows: smart transport, smart parks and beaches, new master control room, energy maximization, easy accessibility of data, and police smartphone applications. In line with these activities are the recent projects that include smart tourism, smart life, and smart economy. Other sustainable cities include the Sustainable City (Dubai Land Department), DSO, Dubai South District, Desert Rose City, and Sharjah Sustainable City.

Oman

Oman like other countries in the GCC aspires to implement smart city and be sustainable in its application. The Telecommunications Regulatory Authority (2023) conducted a public consultation to enable the different stakeholders and the public to share their views and opinions on the Smart Cities Regulatory Policy Framework. Public engagement and participation are vital to the successful implementation of the smart city. The study of Sameer et al. (2022) stressed in his study the two objectives that can be achieved the first objective is to understand and highlight the alignment of local government regulations to the SSC with the utilization of Digital Participatory Planning (DPP) in assessing the country's readiness of the government and the public in attaining a SSC. It also evaluates recent local policies and procedures in Oman that support the creation of a generic framework to assess the readiness for DPP. According to Bouzguenda et al. (2020), DPP plays a major role in determining the success and failure of SSC initiatives or platforms. Oman has initiated the SSC anchored on the Oman Vision 2040 such as the portal Omanuna and the Oman smart city platform wherein the participation of the Oman citizens is vital for its implementation. Omanuna portal enhances citizens' participation and interaction with the government through e-services while the Oman smart city platform promotes innovative ideas from citizens toward smart city resolutions (Omanuna, 2017; Smart City Platform, 2021). Essential to the smart city implementation is the digital citizen's participation in urban planning, the Oman National Spatial Strategy (ONSS) initiated by the Ministry of Housing and Urban Planning envisioned to encourage the citizens' participation in urban planning where they have already conducted a series of orientations and workshops with various stakeholders Omanis and Non-Omanis although the participation is still in the strategic level (Sameer et al., 2022). Driven by the Oman Vision 2040, the involvement of the private sector is given importance and fair participation of the different stakeholders and partners. The integration of DPP into the government's rules, policies, and regulations is essential and requires participation from all stakeholders such as citizens, expatriates, workers, business owners, etc.

Conclusion

The increasing trend in the utilization of smart cities is unstoppable in congruence with population growth, urbanization, congestion, and other related activities. The world, including GCC countries, is moving toward the adoption of smart cities and their sustainability. This study evaluates the state of application of smart cities in the GCC and provides information for comparability among the six GCC nations in terms of the recent experiences they experienced initiating the smart city, the challenges faced, and the proposed resolutions to attain sustainable smart cities. Experiences in the use of smart cities and the concepts of establishing it revealed that smart city innovation initiatives and platforms have been incorporated into their national strategic vision and goals (GCC countries) and have initiated the smart city implementation, especially in UAE, Saudi Arabia, and Qatar while other countries developed the ICT ready for smart city transition and transformation. The status of implementation differs between these countries however, all acknowledged the need for smart city platforms for economic and environmental sustainability and more efficient and effective provision of public services. Furthermore, the advantages and benefits that a smart city can offer include improved transportation, safer communities, improved government and citizen encounters, increased data-driven decision-making, improved digital parity, economic development, optimum resource utilization, environmental sustainability, and social inclusion. In the aspect of challenges, the consensus from the prevailing challenges faced by the GCC countries includes sustainability issues, difficulty transitioning from traditional cities to smart cities, lack of coordination of the government and all other stakeholders, insufficient laws and regulations that support and sustain smart city implementation, cybercrime, data privacy and security, technological integration, digital divide, landscaping in smart city, housing shortages and traffic congestion, skills gap, social, and cultural issues.

Using the comparative approach to dealing with the extent of smart implementation in the GCC nations, different stakeholders, such as planners, sponsors, government authorities, and others, will be provided with relative information on how smart cities are utilized at the regional level, particularly in the Gulf region, and this will serve as a baseline input for further development of smart city innovation. The outcome of this study further justifies the reasonableness of countries in the Gulf region in their efforts to constantly improve smart city applications and benchmark their developments along with neighboring GCC nations.

Summary of initiatives taken to address the above-mentioned challenges include intensification of policies and regulations toward smart cities through ICT maximization in the areas of transportation, education, healthcare, government services, infrastructure, and urban planning. Heightened laws and policies to address privacy and security, water, energy, technological infrastructure, overcrowding and traffic congestion, and funding. Collaboration and collaboration among different stakeholders such as the government sector, private sector, experts, and communities are vital to smart city implementation and the establishment of smart infrastructure, smart government system, well-designed urban environment by providing world-class transport facilities, traffic congestion solutions, and activities that focus on smart tourism, smart life, and smart economy. Finally, the integration of related policies that promote smart cities such as DPP to assess the readiness of the smart city implementation and other initiatives.

The contributions provided by this chapter are enormous and worth taking. First, the findings of the comparative study practically provide an avenue for each country in the Gulf nations to re-evaluate and examine their strategic vision and plan to ensure the attainment of innovative and prosperous implementation of smart cities in line with Vision 2040. It also adds value to the recent theories and literature on smart cities in the GCC. Second, comparing the status of the application of innovative smart cities among the countries in the GCC will aid the planners and implementors of each nation in designing, implementing, and governing smart cities toward continuous development and transformation. And for future research, the ethical aspect of innovation can be integrated into an expansion study for different stakeholders to assess the ethical impact of smart city innovation initiatives.

However, the outcome of the study is limited to the gathered published secondary data such as journals, articles, news, and government information and the authors acknowledge that the study can be expanded to consider quantitative studies to further validate these findings. Although the scope of the study covers the GCC countries, the information provided through this research may require more detailed data for researchers for an expanded study.

References

Al Jaafreh, M. B., & Allouzi, M. A. (2023). Smart cities adoption in Saudi Arabia: A comprehensive review and future drivers. *British Journal of Multidisciplinary and Advanced Studies*, *4*(5), 20–39. https://doi.org/10.37745/bjmas.2022.0300

Alawadhi, S., & Al-Atawi, A. (2019). Smart city initiatives in Saudi Arabia: A review. In *2019 International conference on computer and information sciences (ICCIS)* (pp. 1–6). IEEE.

AlEnezi, A., AlMeraj, Z., & Manuel, P. (2018). Challenges of IoT based new generation smart government. *Journal of Informatics and Mathematical Sciences*, *10*(3), 53–544.

Alghais, N., & Pullar, D. (2018). Projection for new city future scenarios: A case study for Kuwait. *Heliyon*, *4*, e00590. https://doi.org/10.1016/j.heliyon.2018.e00590

Almulhim, A. I., Bibri, S. E., Sharifi, A., Ahmad, S., & Almatar, K. M. (2022). Emerging trends and knowledge structures of urbanization and environmental sustainability: A regional perspective. *Sustainability*, *14*(20). https://doi.org/10.3390/su1 42013195

Alshehri, M., & Elyas, T. (2018). A systematic review of smart city initiatives in Saudi Arabia. In *2018 1st MEC international conference on big data and smart city (ICBDSC)* (pp. 1–5). IEEE.

Alshehri, M., & Seifi, A. (2017). Smart city initiatives in Saudi Arabia: Vision 2030. In *2017 International conference on sustainable energy, engineering, materials and environment (SEEME)* (pp. 1–5). IEEE.

Asmyatullin, R., Tyrkba, K., & Ruzina, E. (2020). Smart cities in GCC: Comparative study of economic dimension. *IOP Conference Series: Earth and Environmental Science*, *459*, 062045. https://doi.org/10.1088/1755-1315/459/6/062045

Bakici, T., Almirall, E., & Wareham, J. (2012). A smart city initiative: The case of Barcelona. *Journal of the Knowledge Economy*, *4*(1), 135–148.

Badran, A. (2023). Developing smart cities: Regulatory and policy implications for the state of Qatar. *International Journal of Public Administration*, *46*(7), 519–532. https://doi.org/10.1080/01900692.2021.2003811

Barth, J., Fietkiewicz, K., Gremm, J., & Hartmann, S. (2017). Informational urbanism. A conceptual framework of smart cities. In *Proceedings of the 50th Hawaii international conference on system sciences*. http://hdl.handle.net/10125/41496

Bibri, S. E., & Krogstie, J. (2017). Smart sustainable cities of the future: An extensive interdisciplinary literature review. *Sustainable Cities and Society*, *31*, 183–212. https://doi.org/10.1016/j.scs.2017.02.016

Bouzguenda, I., Alalouch, C., & Fava, N. (2020). Examining digital participatory planning: Maturity assessment in a small Dutch city. *Journal of Cleaner Production, 264*, 121706. https://doi.org/10.1016/j.jclepro.2020.121706

Bufetova, A. (2013). Structure of urbanization and trends in intraregional differentiation of standard of living. *World of Economics and Management, 13*, 57–66.

Cohen, B. (2012). *What exactly is a smart city?* https://www.fastcoexist.com/1680538/whatexactly-is-a-smart-city

Constro. (2021). *Smart city-elements, features, technology and government approach*. https://www.constrofacilitator.com

Digital UAE. (2022). *Smart sustainable cities*. https://u.ae/en/information-and-services#

Dutta, S., & Lanvin, B. (2019). *The Network Readiness Index 2019*. https://networkreadinessindex.org/wpcontent/uploads/2020/03/The-Network-Readiness-Index-2019-Newversion-March-2020.pdf

Elian, M. I., & Kisswani, K. M. (2024). Smart cities: GCC and Kuwait experience. In F. Belaïd & A. Arora (Eds.), *Smart cities. Studies in energy, resource and environmental economics* (pp. 339–358). Springer. https://doi.org/10.1007/978-3-031-35664-3_18

EMergy Consultancy. (2018). *Abu Dhabi smart city pilot project*. https://www.emergy.global/project/abu-dhabi-smart-city-pilot-project/

European Commission. (2021). *Smart cities*. https://ec.europa.eu/info/eu-regional-and-urban-development/topics/cities-and-urban-development/city-initiatives/smart-cities_en

Gartner. (2022). *Hype Cycle for Smart City technologies and solutions*. https://www.gartner.com/en/documents/4016805

Hanania, P. A. (2020). Briefing: Building smart cities to enhance lives. *Proceedings of the Institution of Civil Engineers-Urban Design and Planning, 173*(6), 211–213. https://doi.org/10.1680/jurdp.20.00073

Howard, R. (2013). *Smart government key initiative overview*. https://www.gartner.com/doc/2520516/smart-governmentkey-initiative-overview

Information & eGovernment Authority (IGA). (2022). *Bahrain's efforts to achieve sustainable development goals 2030*. http://www.sdgs.gov.bh/

Jazzar, M. (2019). Smart clean city – Kuwait. *International Research Journal of Innovations in Engineering and Technology (IRJIET), 3*(9), 1–7.

Mamlook, R., Al-Jayyousi, O., Alabyooki, S., & Attallah, S. A. A. (2019). Smart City and technology transfer: Towards sustainable infrastructure in Bahrain. In A. Visvizi & M. Lytras (Eds.), *Research & Innovation Forum 2019. RIIFORUM 2019. Springer proceedings in complexity*. Springer. https://doi.org/10.1007/978-3-030-30809-4_24

METenders. (2022). *Middle East project intelligence & tenders*. METenders.com

Ministry of Information and Communications Technology. (2014). *Emerging ICT trends in Qatar*. http://www.motc.gov.qa/sites/default/files/emerging_ict_trends_in_Qatar.pdf

Ministry of Sustainable Development Bahrain. (2023). *Goal Eleven: Make cities and human settlements inclusive, safe, resilient and sustainable*. https://www.sdgs.gov.bh/Goal/?ID=vuSn/qaCWuyJjx/mF/WvaA==

Nijhuis, S., Beukers, E., van de Ven, M., & Peeters, P. (2016). The governance of smart cities: A systematic literature review. In *Transforming governance: New values, new systems, new institutions* (pp. 177–202). IOS Press.

Omanuna. (2017). https://omanuna.oman.om/

Oraba, R. M. B., & Jalagat, R. (2023). The challenges of smart city application in Oman. The Al Irfan Project Experience. *European Academic Research, 10*(12), 4291–4308.

Pantic, M., Cilliers, J., & Cimadomo, G. (2021). Challenges and opportunities for public participation in urban and regional planning during the COVID-19 pandemic – Lessons learned for the future. *Land, 10*(12), 1379. https://doi.org/10.3390/land10121379

Patrao, C., Moura, P., & de Almeida, A. T. (2020). Review of smart city assessment tools. *Smart Cities, 3*(4), 1117–1132. https://doi.org/10.3390/smartcities3040055

Regulatory Authority. (2023). *Public consultation on Smart Cities Regulatory Policy Framework*. https://www.tra.gov.om/En/ViewPublicConsultations.jsp?code=50

Ringel, M. (2021). Smart city design differences: Insights from decision-makers in Germany and the Middle East/North-Africa region. *Sustainability, 13*, 2143. https://doi.org/10.3390/su13042143

Sameer, N., Alalouch, C., Saleh, M. S., Al-Saadi, S., & Saleem, A. (2022). Towards sustainable smart cities: A study on digital participatory planning in Oman. *Proceedings of the Institution of Civil Engineers: Urban Design and Planning, 175*(2), 72–90. https://doi.org/10.1680/jurdp.21.00040

Saxena, S., & Al-Tamimi, T. (2018). *Visioning "smart city" across the Gulf cooperation council (GCC) countries*. Emerald Publishing Limited.

Smart City Hub. (2017). *South Korea exports smart city concept into Kuwait*. https://Smartcityhub.com

Smart City Platform. (2021). http://en.smartoman.om

Smart Dubai. (2016). *Towards becoming the happiest city on earth*. https://www.itu.int/net4/wsis/forum/2016/Content/AgendaFiles/document/13267dca-e1af-4833-a722-dccef7630f27/Towards_Becoming_The_Happiest_City_on_Earth.pdf

Telecommunications and Digital Government Regulatory Authority. (2023). *Smart sustainable cities*. https://u.ae/en/about-the-uae/digital-uae/digital-cities/smart-sustainable-cities

Tomorrow City. (2022). *Neom, the Smart City build from scratch in the Arabian desert*. https://www.tomorrow.city/neom-saudi-arabia/

United Nations. (2018). *Urbanization*. https://www.un.org/development/desa/pd/content/urbanization-0#:~:text=The%20world%20is%20becoming%20increasingly,around%20two%2Dthirds%20in%202050.

United Nations. (2020). *UN E-Government Survey 2020*. https://publicadministration.un.org/egovkb/en-us/Reports/UN-E-Government-Survey-2020

United Nations Economic Commission for Europe. (2019). *Smart sustainable cities profile: Saudi Arabia*. https://www.unece.org/fileadmin/DAM/hlm/documents/Publications/smart-city-profile-Saudi-Arabia.pdf

United Nations Human Settlements Programme. (2016). *Urbanization and development: Emerging futures*. https://unhabitat.org/books/urbanization-and-development-emerging-futures

Urban Gateway. (2018). *Transforming Kuwait's cities into secure & digitalized sustainable cities*. https://urbangateway.org/event/transforming-kuwait%E2%80%99s-cities-secure-digitalized-sustainable-cities

Venkatachalam, S. (2020). *Re-thinking the role of citizens in evaluating quality of life in the Smart City*. Deft University of Technology.

Vision 2030. (2016). *Vision 2030: Kingdom of Saudi Arabia*. https://www.vision2030.gov.sa/media/rc0b5oy1/saudi_vision203.pdf

Warner, M., & Mansker-Young, S. (2023). *Thinking about becoming a smart city? 10 benefits of smart cities*. https://www.plantemoran.com/explore-our-thinking/insight/2018/04/thinking-about-becoming-a-smart-city-10-benefits-of-smart-cities

World Bank. (2017). *Saudi Arabia's vision for 2030*. https://www.worldbank.org/en/country/saudiarabia/publication/saudi-arabia-vision-2030

Wu, Y. J., & Chen, J. C. (2021). A structured method for smart city project selection. *International Journal of Information Management, 56*(31), 101981. https://doi.org/10.1016/j.ijinfomgt

Zaidan, E., & Abulibdeh, A. (2020). Master planning and the evolving urban model in the Gulf cities: Principles, policies, and practices for the transition to sustainable urbanism. *Planning Practice & Research, 36*(2), 193–215. https://doi.org/10.1080/02697459.2020.1829278

Recommended Readings

Arif, M., & Aldosary, A. S. (2023). Urban spatial strategies of the Gulf Cooperation Council: A comparative analysis and lessons learned. *Sustainability*, *15*, 13344. https://doi.org/10.3390/su151813344

Belaïd, F., Amine, R., & Massie, C. (2024). Smart cities initiatives and perspectives in the MENA Region and Saudi Arabia. In F. Belaïd & A. Arora (Eds.), *Smart cities. Studies in energy, resource and environmental economics*. Springer. https://doi.org/10.1007/978-3-031-35664-3_16

Mutambik, I., Lee, J., Almuqrin, A., & Zhang, J. Z. (2023). Transitioning to smart cities in Gulf Cooperation Council Countries: The role of leadership and organisational culture. *Sustainability*, *15*(13), 10490. https://doi.org/10.3390/su151310490

Trindade, E. P., Hinnig, M. P. F., Costa, E. M., Marques, J. S., Bastos, R. C., & Yigitcanlar, T. (2017). Sustainable development of smart cities: A systematic review of literature. *Journal of Open Innovation: Technology, Market, and Complexity*, *3*(3), 1–14.

Zaidan, E., Al-Saidi, M., & Hammad, S. H. (2019). Sustainable development in the Arab world – Is the Gulf Cooperation Council (GCC) region fit for the challenge? *Development in Practice*, *29*(5), 670–681. https://doi.org/10.1080/09614524.2019.1628922

Chapter 3

The Science of Mobile Pedestrianism Through Smart Cities

Ashok Chopra[a] *and Anita Rani Chopra*[b]

[a]*Amity Business School Dubai, Amity University Dubai, Dubai, United Arab Emirates*
[b]*Faculty of Management Studies - Wisdom, Banasthali University, Rajasthan, India*

Abstract

COVID-19 not only created havoc in bringing retardation of the global economy and massive loss of humankind but also caused rapid urbanization, which unprecedentedly pressured sustainability. Smart cities, using several technologies, interfaces, software, and especially Internet of Things (IoT), deliver integrated solutions in helping achieve SDGs, especially the goal of good health, clean energy, innovation, infrastructure, sustainable cities, helping environmental protection. Intelligent cities deploy various software, interfaces, and state-of-the-art communication tools, to communicate and exchange data with the public to achieve better health. Today, smart cities can map emergencies and hazards, impeding crime in real time. Smart cities can provide efficient city systems and development by deploying a network of intelligent infrastructure. Walkability is the talk of the town because of its ability to improve the urban health economies. Walking in the urban community provides quality of life in technology-oriented modern life. Today's communities provide facilities, including parks, amenities, stores, shops, and restaurants, civic and cultural facilities within walking distance. Baby Boomers and Millennials, representing adult communities in metro cities, prefer seeking walkable places to spend quality time with their families. Bike riding is today's hot topic among youth, especially after community developers have developed high-quality bike infrastructure, leading to excellent mobility within the community. The solution providers and planners of smart cities using transit screen provide information to walkers who prefer to avoid involving with mobile technology. At the 93rd Transportation

The Emerald Handbook of Smart Cities in the Gulf Region:
Innovation, Development, Transformation, and Prosperity for Vision 2040, 35–51

doi:10.1108/978-1-83608-292-720241002

Board Meeting in Washington D.C., speakers stressed that walking is vital to getting to the car, bike sharing, or even getting to transit; thus, walking has a vital role in engaging all planned transport services.

Keywords: SDGs; sustainability; technology; quality of life; smart cities; connectivity; Internet of Things; citizens

Introduction: What Is a Smart City?

> A smart, sustainable city is an innovative city that uses information and communication technologies (ICTs) and other means to improve the quality of life, the efficiency of urban operation and services, and competitiveness while ensuring that it meets the needs of present and future generations concerning economic, social and environmental aspects. (Toh, 2022)

World Urban Forum (WUF) says more than 10,000 cities are worldwide today. Looking at the two most populous countries of the world, i.e., India and China, one would find that India has over 5,000 innovative city projects; in contrast, China has more than 800 intelligent city pilot projects. There is no doubt regarding explosive growth and the undivided and undeniable attention smart cities are getting worldwide. G20 nations and other countries governing the world economy have some ongoing intelligent city projects (Chen, 2023).

As per International Institute for Management Development (IMD) which surveyed more than 120 residents from 118 countries worldwide, Abu Dhabi (Capital of United Arab Emirates) ranked 13th globally out of 141 that were part of the current (Year 2023) ranking. The critical factors included in the survey were the current infrastructure and the host of digital services available to the residents. This survey was based on five essential pillars: mobility, health and safety, opportunities, activities, and governance.

"The Science of Mobile Pedestrianism Through Smart Cities"

The basic proposition a smart city should offer is improving the quality of life. Using technology, smart cities offer within-reach access that improves, accommodates, and necessitates fulfilling individual needs and elevates the welfare and comfort of individuals using the Internet of Things (IoT). As of 11th July 2022, United Nations released its updated projections for global, regional, and national population change up to 2100. Its key conclusion is that there will likely be another 2.4 billion people on the planet by the 2080s, but for the first time, it predicts that the world population will peak during this century and then slowly start to decline (https://populationmatters.org/). This means that, on the one hand, industrial production would go up using technology and robots, and on the other, the net birth rate would come down. This poses the severe economic problems of deflation, i.e., a demand reduction. Smart cities can improve per capita income by spending on the disposal income of Generation Z, improving the quality of life, and offering them state-of-the-art smart cities that offer sustainable living and contribute to SDGs.

After COVID-19 , the demand for intelligent cities has continuously risen as more and more people migrate to cities. Smart cities offer the quality of life through design for improved health care of habitats. Today, cities and developers worldwide have not even touched the tip of the iceberg, as smart city technology offers endless benefits. Today's smart cities can fetch and share information on residents' and visitors' welfare by having information about sleeping habits, stress management, nutrition, physical well-being, stress management, exercise, and mental health. The precautionary methods and collected data (through IoT) can help deploy limited and precious assets and human resources effectively and provide vital information on probable threats, crime, city hazards, and emergencies. Through a sustainable environment, intelligent cities also help decrease carbon dioxide (CO_2), help prevent air pollution, and improve vehicular traffic. Besides reducing the side effects of polluting processes, smart cities bring a behavioral shift in people's minds toward better, greener, and healthier living.

Intergovernmental Panel on Climate Change (IPCC) (Field, 2012) also emphasized what people and administrations need to do to mitigate the negative impacts of global warming. IPCC always encouraged civilian participation in doubling down on carbon dioxide reduction processes.

As per (Appio et al., 2019), technologies and inventiveness bring about behavioral changes and motivate communities to engage in sustainable achievement citizen behavior, which plays a crucial role in governmental sustainable drives.

An effort by Pavegen (Diamond, 2020) – a developed array of piezoelectric bright flooring and GO ENGAGE app. – pairs in encouraging noticeable behavioral change in inspiring sustainability through rewards and offers is a classic example of driving people toward sustainability through a series of initiatives and reward offers by tracking generated data toward better and environmentally friendly living.

Data Focused Healthcare Through Smart City Tech

The Unique Selling Proposition (USP) of a smart city is Data Focused Healthcare, the distinct feature by which a smart city can pull vital information from comprehensive sources, such as IoT devices, infrastructure laced with a network of intelligent hubs, mobile device sensors, vehicle mounted networks which offers an array of bright networked state of the art infrastructure able monitor and assess efficacy of city systems and practices. Such an array of state-of-the-art infrastructure through machine learning, data support, and communicative insights offers the ability to tackle health and welfare inequities efficiently and develop progressive community healthcare and well-being. By carefully evaluating IoT devices-based data, health professionals can identify individuals and groups who may be at risk and should be offered preventive measures. Smart city developers and infrastructure planners should invest in systems capable of intelligently collecting, communicating, and dispersing vital health data of occupants of intelligent cities to respond to medical and health emergencies that lead to a better quality of life for occupants. Savvy city planners are also incorporating a host of other IoT-based infrastructures, which include but are not limited to connected public transportation, analytics for video surveillance, water-related smart metering, weather forecasting, water quality monitoring, integrated street lighting, traffic control, and observation, monitoring of pollution and air quality, analytics for video surveillance, and detection of smoke and fire.

Here is a graphic sketch of the Niagara Digital System installed at City Walk, a premium smart city in Dubai that can offer lighting control, chiller water, fresh-water handling, generators, uninterrupted power supplies, and fire alarm systems by Niagara System architecture. Such premium features are offered in 34 low-rise residential buildings, all part of one of the city's most prestigious and tourist-centric locations in the Safa district of Dubai. Conceived and built by Meraas, a premier, high-end developer, the 14 buildings comprising City Walk offer tenant views of the world's tallest tower (Burj Khalifa) and easy access to many premium dining, shopping, and entertainment experiences. Meraas (The Developer) envisioned City Walk as a place where inhabitants could experience a more futuristic lifestyle in other ways, too; it explored how City Walk residents could leverage current technology to personalize their spaces and incorporate the use of their mobile devices (see Fig. 1).

Another classic example of offering smart infrastructure: Dubai, a prestigious city in UAE was in the Year 2018 when DEWA (Dubai Electricity and Water Authority), a provider of clean electric power and water, adopted G.E.'s IoT platform "Predix" to regulate and analyze data in real-time for DEWA entire operation which powering more 3.6 million residents.

In 2018, the DEWA adopted General Electric's IoT platform Predix to enable the regulator to collect and analyze data in real-time to enhance the speed and

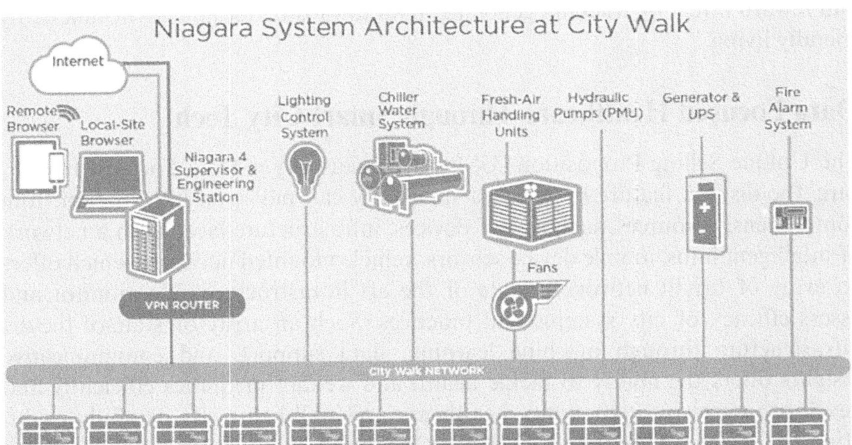

Fig. 1. Graphic Adapted from Niagara System.

efficiency of its operations. Dubai is a city that saw over 8.55 million visitors in the first half of the year 2023. Continuing to take the lead for intelligent cities, the United Arab Emirates Roads and Transport Authority (RTA) has planned to host the World Congress of Intelligent Transport Systems (ITS) in 2024. ITS, incubated in 1994 for the first time, chose an Arab world country to host its world congress. Gulf nations have also paved the way for nine intelligent city projects to be achieved by 2025 – two brownfields and seven built from scratch. Furthermore, in the race, UAE and Saudi Arabia are expected to invest more than $49.00 billion in Smart City Projects by 2025 (Salem, 2016).

ABU DHABI, 1st May 2023 (WAM) – For the third consecutive year, Abu Dhabi ranked as the most innovative city in the Middle East and North Africa (MENA) region for 2023, according to the IMD Smart City Index released by the International Institute for Management Development (IMD) in Switzerland in collaboration with the Singapore University of Technology and Design (SUTD).

Santclassiccity in Los Angeles is a classic example of measuring city happiness. Here, city planners achieved happiness and better health of citizens by integrating a proven structure and method for measuring the success of occupants by offering access to mobility, education, housing, and community health by being sustainability-focused. A similar example is also seen in City Walk in Dubai. City Walk has also ensured happiness by providing infrastructure for hosting public and occupants' entertainment and well-being. For several years, town planners were busy thinking and providing connectivity infrastructure from point A to point B through automobiles, their parking ensuring connectivity to offices and commercial centers. In contrast, intelligent city planners have a paradigm shift in thinking, where they think of providing quality of life as intelligent engineers and behavioral psychologists rather than regulators. The new equation of smart cities for providing connectivity from point A to point B is through an innovative and sustainable way.

Walkability is the talk of the town in striving to improve the urban economies. Making Millennials, Baby Boomers, and Generation Z walk through the urban fabric is the dominant theme of smart cities. Such an objective is achieved by having facilities like, Parks, Civic niceties, amenities, and cultural and social infrastructure within walking distance in smart cities. Distance of the world seeks walkable places to live, and this feature is a hot topic of the town in smart cities. Thus, riding a bike is a hot topic for youth and is becoming infectious daily. Thus, buzzwords like sidewalks, shared streets, and protected lanes are essential for smart cities. In this race, metros have also become a necessity for connectivity (see Fig. 2).

(a) (b) (c)

Fig. 2a, 2b, 2c. Smart City Infographics Showing Facilities Planners Keep in Mind.

Paradigm Shift in Mobile Pedestrianism

Introduction of the paradigm shift in the way the Millennials, Baby Boomers, and Generation Z walk through using technology, which includes Walkscore ("A Score that helps in finding walkable living space from work). Score of 90–100 is termed Walker's Paradise, and it does not require a car for achieving daily necessities. A score of 70–89 is termed very walkable (most shopping can be accomplished on foot). Score of 50–69 is termed somewhat walkable meaning some daily requirement can be achieved through walking. Score of 25–49 means most daily requirement need a car and score of 0–24 means almost all daily requirement need cars. Ride Scout is "a mobile app that benefits one to get from point A to point B faster and smarter." Walkonomics is "a free app for iPhone, Android and the web that measures the pedestrian-friendliness (walkability) of streets and communities." For forward-thinking city leadership, such initiatives should be looked at as forward-thinking. Such initiatives offer platforms that offer a boon for health, economic, and environmental welfare to pedestrianism, i.e., the practice of walking. Such forums also provide a strong base for a demographic shift from automobiles to intelligent and sustainable forms of transportation, encouraging "Foot Power," a common denominator, bringing paradigm among people on investigating their mobility (see Fig. 3 and Table 1).

(a) (b) (c)

Fig. 3a 3b, 3c. View of City Walk at Dubai and Freedom House in Saudi Arabia, Bahrain Bay Smart City Encouraging Walkability.

Table 1. How Planner of Smart Cities Strive to Offer Urban Design Dimension Offering Features of Walkability (*Source*: Information by Authors).

S. No.	Smart City Design Dimension	Feature of Walkability
1	Visual	Outdoor eating joints, open views, timely maintenance
2	Connectivity	Wise accesses may be potential hurdles
3	Density	Building concentrations and altitudes
4	Form/Arrangement	Street scape discontinuity
5	Pedestrian Amenities	Curb outs, street fixtures and pavements
6	Personal Safety	Scrawls, meshed windows, and brood

Table 1. (*Continued*)

S. No.	Smart City Design Dimension	Feature of Walkability
7	Physical Activities	Frivolous tools like exercise uses
8	Proximity of Users	Non-residential acreage in uses in neighborhood
9	Public Spaces Parks	Marketplaces, playing parks, and playing grounds
10	Traffic Procedures	Comforting traffic
11	Technology	Integration of technology through infrastructure
12	Social Learning	Strengthening human arrangement
13	Institutional Improvement	Through civilian engagement
14	Urban Economics	Through walking in and among community
15	Mode of Travel	Presence of bike facilities through paths, lanes, and routes

Changing Mindshift

The new age, on-the-go youth, aged between 18 and 34, purchasing an automobile has come down by 30% as per (Ross D 2014) reported from 2007 to 2011. Furthermore, the number of automobile miles run by American youth in their twenties has decreased from 20.8% to 13.7% on average. The crux is that Millennials, Baby Boomers, and Generation Z, choosing to settle in urban areas, prefer walking and bicycling with other non-traditional modes of physical commuting over automobiles. So, the new equation is that autos are out, and two feet are in. It is also seen that Millennials prefer walking, and Empty Nesters are also jumping on the bandwagon in large numbers. They flaunt physical movement as a cure and medicine for common disorders in an aging population.

Flooding of Mobile Apps

Infiltration of mobile phones has led to the flooding of mobile apps offering walkability info, with the lead taken by MapQuest and Google Maps, recognized as the talk of the town among people of all ages and walks of people. These and other apps are known for delivering the best in class for just-in-time driving directions and providing state-of-the-art walkability functions, offering point-to-point map guides for reaching a destination by foot.

Both Walkscore and Walkonomics are new-generation apps that are a boon for pedestrian friendliness in an area, converting the information to a score and gamifying the walking among all the classes of society. It uses the personality trait of people wanting to convert any achievement into numbers for comparison and helps individuals quench their urge to be competitive. A state-of-the-art app called Foursquare, a social media app that provides locals and visitors information on finding the right spot for entertainment, food, and other conveniences truly supports Walkscore and Walkonomics. Foursquare is popular among millennials seeking social connections and sharing their social experiences.

Walkability Preferences

Riding on the success of getting people out of their automobiles through leveraging walkability apps, transport planners and municipality leadership, in collaboration with tech solution companies, are offering an array of new products and solutions to encourage the public to get walkability (Lorinc, 2018).

RideScout CEO Steve Carroll is encashing on walking on foot or through other similar alternatives by providing real-time information to users and enthusiasts. He thinks municipalities and private companies can benefit from the current wave where youth are inculcating themselves to healthy living through independent car lifestyles. Walking enthusiasts are further finding that RideScout offers excellent walking directions, making them reach target destinations much faster than any other alternative transportation system (Gebresselassie & Sanchez, 2018).

Fundamentals of Human Manners

People who collaborate and form alliances with Tech Savvy Mobile App developers should keep in mind the human mannerisms of beginners (Who just got into walking). The technology intimidates many; thus, instead of great aid, walking apps are a barrier for beginners who need more mobile savviness. In fact, as per (Krivý, 2018), walking in smart cities is more of togetherness (walking with others) rather than getting lost into so as per (Greenfield, 2010). "This is the unlooked-for downside of app-supported pedestrian activity."

Paradigm Shift in Mobile Walking Boosting

Governments in city development must keep and communicate awareness about the latest mobile solutions that are already developed and are under development (Agbali et al., 2019). Government leaders also must promote pedestrianism among visitors and locals alike. In any intelligent city, it is essential for main stakeholders, from developers, planners, and municipalities to public park officials, to orchestrate pedestrianism through coalescing efforts.

In a classic example, the president of WalkDenver (a non-profit activism group), Shaffer, witnessed the advocacy groups make a noticeable difference while developing smart cities in Denver. Dedication and strong advocacy by non-profit groups facilitate and support walkability in communities. Shaffer has further cited work and partnership among WalkDenver, WALKscope, and PlaceMatters to beta test the walkability apps to collect and analyze data promoting walkability. Such partnership also helps create a repository of side walking conditions, street information, infrastructure solicited, pedestrian counts, and intersections required to improve the walkability infrastructure.

Non-Technology Solutions Efforts

In developing Denver, intelligent city developers and architects stressed creating practical communication methods with-non-technology savvy people, too. It is stressed that directional sign boards, gateways, and prominent buildings are the conventional means that help non-tech-savvy pedestrians maneuver through intelligent cities. In Washington D.C., TransitScreen Company, through its state-of-the-art screens, provides information on the interconnection between walkers and other conventional transportation to make life easy for walkers going through marketplaces, government, and residential buildings. Such efforts help non-technology walkers reach destinations efficiently by promoting walkability. At the 93rd Annual Meeting of the Transportation Research Board in Washington D.C., the award recipient, Ryan Croft (The founder of Transit Screen), emphasized the crucial role of engaging all stakeholders in connecting to alternative modes of transport through promoting walkability.

Innovatory Mobile Pedestrianism

Walking is not just an act of the most justifiable and equitable means of getting around but is also connected to many other factors that are not in the control of individuals, including safety, comfort in walking, physical and social abilities, and the presence of a host of infrastructural abilities. Thus, walking in cities is not only merely reaching point A to point B but also a manifestation of being healthy, proactive, and liveable in the urban environment by learning where humans live. Such convergence of techno-pedestrianism using technology offers a host of possible collaborations and syncing of newer mobile technologies in urban metro liveliness. Embracing such pedestrianism can help curb the poisonous impact of automobile pollution and the ill effects of traffic jams. Taming the convergence of technology and pedestrianism for building a healthy environment is the target of ultra-build smart cities. This would help achieve meaningful returns for people moving to cities from the current trend of moving away from congested cities. Factors defining urban walkability include convenience, safety, efficiency, pleasure, and achieving goals. Thus, walkability has a noticeable impact on urban walkability.

Thus, the planners of smart cities must consider factors like commerce, recreational amenities, while designing such cities. These factors must have the equation of having parks, recreational activities, restaurants, walking connectivity, amusement, resting, and fun areas in smart cities. Savvy city planners should consider making people think about preferring walking over automobiles. Planners should envisage integrating the crafting of sidewalks and combining best practices for improving and encouraging walking to improve the quality of life. Urban planners are building rethinking to boost Walkability as a fundamental characteristic of upcoming cities like Zürich, Dubai, Bahrain, Helsinki, Riyadh, Muscat, Hamburg, Copenhagen, Amsterdam, and many more.

Smart Cities in the Gulf and Their Essential Features

1. Vision and Approach:

   Smart city projects in the Gulf Cooperation Council (GCC) are often government-motivated projects; thus, governments target ambitious objectives for transforming urban landscapes into technologically oriented, green, and sustainable cities. The classic examples are Saudi, UAE, and Egypt's Vision 2030, emphasizing sustainability through modernization, transformation, and economic expansion.

2. Connectivity Through Infrastructure:

   They are creating and investing heavily in digital infrastructure to pave the way for smart cities. Thus, GCC has invested heavily in the IoT, high-speed Internet, and 5G networks. Such investments are a backbone for smart cities' support applications for health care, tourism, energy management, and intelligent transportation systems.

3. Mobility and Transport:

   Transportation is a vital ingredient and essential feature of smart cities; thus, one can see Dubai's Smart transportation system, including driverless Metro, shared streets to support city transportation supporting e-scooters, delivering medicine through drones, soon launching driverless public transportation, hyperloop, water transportation, and flying cars. Abu Dhabi, too, is focusing on eco – an eco-friendly public transportation system to decongest city traffic.

4. Energy Efficacy:

   Keeping environmental challenges, pollution, and clean air crucial, governments focus on innovative city initiatives through renewable energy, energy management systems (EMS), and curbing efficient energy use to reduce carbon footprint.

5. E-Digital Services:

   Smart cities offer residents various online and digital services, including digital and online platforms for assessing a range of government services and transactions, including intelligent healthcare services, digital payment platforms, and e-governance.

6. Sustainability and Environment Protection:

   The intelligent and innovative solutions include solar-powered clean energy, data-driven decision-making, bright waste treatment and management, water conservation for a sustainable future, and urban development.

7. Data Analytics:

   Governance and maintenance of essential services through the collection of data and its analysis using data analytical tools. Data-driven decision capability helps waste resources and provides predictive data decision capability. This helps occupants with the capability to make informed decisions. Data-driven methods enable predictive maintenance through optimized resource allocation and healthier urban development.

8. Citizen Engagement:

   Smart cities offer a participatory forum for receiving, participating, and making decisions through a citizen feedback mechanism. Such forums help run the best-suited awareness campaign toward sustainability, an essential feature of smart city.

9. Citizen Security Through Privacy:

   Since intelligent cities are data-driven, ensuring residents' data protection, offering cybersecurity, and maintaining privacy and confidentiality through regulated platforms are essential to intelligent city strategy.

   As intelligent cities rely heavily on data collection and sharing, ensuring cybersecurity and protecting citizen privacy are paramount. Robust cybersecurity measures and data protection regulations are crucial to any smart city strategy.

10. Cooperation and Innovation:

    The partnership between government entities, private sector companies, research organizations, and start-ups is an essential feature of the success of smart cities. Promoting innovation through hackathons, incubators, and funding opportunities offers groundbreaking solutions and participation by many stakeholders.

Dubai–United Arab Emirates The LOOP

Dubai Announced Huge 93 km Environment-Friendly Project Offering Healthy Living

As per El Khatib et al. (2023), a massive 93-km-long project named "The LOOP" was unveiled by Dubai, which would offer a healthy transportation mode connecting over 3 million people. Residents and visitors would be offered access to critical services and localities through cycling or walking in minutes through state-of-the-art sustainable architecture. The promoter URB emphasized the all-round year temperature-controlled environment, which is in line with Dubai's ambition of becoming a 20-minute city, where residents can access daily needs and destinations within 20 minutes by foot or bicycle, a unique initiative promoting living healthy by a primary mode of transport being foot or bicycle.

The unique features offered by LOOP are:

1. Food Security through vertical farming
2. 93-km long
3. It would use only 100% renewable energy through kinetic power
4. Zero emissions transportation system
5. 100% recycled water for irrigation purposes
6. Unique and state-of-the-art additional amenities for residents
7. Promoting health and well-being.

LOOP project would be a testimony of urban living and mobility targeting to make Dubai the best-connected city on the planet by bike or foot. LOOP is the epitome of urban mobility infrastructures, offering a unique, sustainable transport system irrespective of weather conditions (see Fig. 4).

LOOP would include face-lifting urban centers, preserving heritage, urban vertical farming, and a "Pedestrian Network Master Plan." LOOP would further

Fig. 4. Map of The LOOP (a New Environment-Friendly Project Offering Healthy Living in Dubai, UAE).

Fig. 5a, 5b, 5c, 5d. LOOP the Sustainable Highway.

uplift urban spaces such as bridges, alleyways, squares, and underneath bridges by supplementing critical elements of well-being. "LOOP would include key projects like Urban Heritage Plan, Urban Farming Plan, Preserving Urban Heritage Plan and Pedestrian Network Master Plan." Dubai is on the path to achieving urban infrastructure through eco-friendly and pro-pedestrian cities, focusing on high-yield urban agriculture that supports sustainability and self-sufficiency. Dubai 2040 urban plan includes six levels, including intelligent transportation, clean energy, innovative government services, recreational services, and flexible and sustainable mobility options (see Fig. 5).

KSA 2030 Vision Project NEOM

Kingdom of Saudi Arabia (KSA) announced a flagship project being developed on the Red Sea Coast to diversify the economy. It is a $500.00 billion project spread across 26,500 sq. km consisting of intelligent towns, cities, research centers, sports and entertainment venues, and tourist centers. NEOM is designed to offer an alternative to environmental challenges faced by urban living. NEOM would also have the world's largest international airport (see Fig. 6).

Qatar Lusail and Msheireb City

Qatar is spending USD 50 billion in developing Qatar's Future Cities called Lusail and Msheireb. Projects include residential and commercial schools and service centers incorporating digital technologies and services. The project would include intelligent transportation, autonomous vehicles, safety, and security systems (see Fig. 7).

(a) (b) (c)

Fig. 6a, 6b, 6c. Project NEOM KSA.

(a) (b)

Fig. 7a, 7b. Qatar Future City Lusail and Msheireb.

Sultan Haitham City, Oman's First-Ever Smart City

The first-ever Smart City project in Oman would be in the Al Seeb district of Muscat (the capital of Oman). Sultan Haitham City would offer 2.9 million sq. m. of public space with 20,000 new homes with smart urban mobility across 14.8 million sq. m. Design is conceived with a growing population and demand by locals, expatriates, and visitors. Sultan Haitham City would focus on transforming the existing dry – river called a Wadi, which runs through the proposed site. The city's promoters would convert the dry river by harnessing the water of flooding (common in that part of Oman). This would help reinvent Wadi as a speculator of green land dedicated to being a public park. Once ready, the focus will be on setting up plazas, outdoor activities, and open spaces for people to spend quality time on social exchange (see Fig. 8).

Bahrain Smart Cities

Based on the New Urban Agenda by the United Nations HABITAT, Bahrain has also initiated an intelligent city agenda to be efficient, sustainable, and liveable for the short and long term. Bahrain is working on involving citizens, the public, and the administration in an accelerated manner to achieve SDGs, especially SDG 11, dealing with sustainable communities and cities.

Smart Cities in Kuwait

Kuwait works simultaneously on two projects in Smart Cities: XZero City and South Saad Al-Abdullah Smart City. The USP of the XZero project is its sparkling landscape, designed to focus on health, well-being, and biodiversity.

Fig. 8. Skidmore, Owings & Merrill's (SOM) Design of Sultan Haitham City.

The smart city is based on three essential pillars, namely sustainability: social, economic, and environmental conditions. The project focuses on low energy requirements through orientation, form, and density (see Fig. 9).

The holistic approach addresses the three key pillars of sustainability: social, economic, and environmental conditions. Design decisions such as orientation, density, and form were considered in the design's early stages to reduce the energy demand with a relatively low financial investment. The project was optimized with various digital tools, such as energy modeling and microclimate analysis. The city focuses on multiple farming methods, including bio-domes, community gardens, vertical farming, and bio-saline agriculture. The city would also focus on scalable food energy and waste farms. The city proposes installing temperature and light condensers depending on climatic conditions by reducing energy consumption and maintaining a living standard.

South Saad Al-Abdullah Smart City Project

(see Fig. 10)

South Saad Al-Abdullah (SSAA) Smart City Project focuses on developing smart cities based on international standards through eco-friendly solutions. The project is spread over 64.4 sq. km, housing 30,000 units accommodating over 400,000 people. This city depends on ultra-modern technology, connecting services through efficient energy sources. It uses eco-friendly construction materials, keeping target occupants in mind. An MOU with Korea Land and Housing Corporation has already been achieved to ensure a timely and energy-efficient city.

(a) (b)

Fig. 9a, 9b. XZero City Is Kuwait.

(a) (b) (c)

Fig. 10a, 10b, 10c. South Saad Al-Abdullah Smart City Project.

Conclusion

Intelligent cities have the potential to offer country citizens energy-efficient, well-connected communities living with better amenities and infrastructure, maintaining the health and well-being of citizens and the public. Planners can offer occupants a healthy life and ideal citizen engagement by incorporating efficient uses of resources and data through modern, well-defined, and well-connected infrastructure. By designing a modern, energy-efficient, healthy lifestyle, we can improve our planet's well-being and achieve UN SDGs well before the end of year 2030 and worldwide well before the year 2050.

Key Findings

Smart cities in the Gulf can provide valuable insights for city planners, policymakers, and stakeholders involved in the development and implementation of smart city initiatives in the region through integrated infrastructure, smart mobility, sustainable energy management, community engagement innovation hubs, and smart economy; digital healthcare and education; cybersecurity measures; urban resilience and safety; E-Government services and IoT and data analytics.

Using a data and reference approach, this chapter steers through initiatives taken by GCC nations by referencing advanced economies that have already benefited their citizens and economies by incorporating smart city initiatives in more robust and more resilient economies. Various stakeholders would find the chapter a valuable contribution to urban planners and infrastructure developers, government officials and policymakers, citizens and communities, industry and technology partners, environmental and sustainability advocates, academic and research communities for potential for ongoing collaboration, adaptive strategies, and the evolving role of stakeholders in the future development of smart cities in the Gulf Region.

Furthermore, all stakeholders, as mentioned above, by incorporating cultural sensitivity in the local context, government leadership and collaboration, necessary regulatory framework into the approach, infrastructure readiness, data governance, cybersecurity, sustainability, and environmental considerations, stakeholders can work collaboratively to create a sustainable and technologically advanced urban environment in the Gulf Region. This approach ensures that smart city initiatives align with local needs, promote inclusivity, and contribute to the overall well-being of residents.

Teaching Perspective

- Search the Internet for articles on the Future of Smart Cities in the Gulf Region and how smart cities would help achieve SDGs.
- Case study: How does each country in the Gulf Region look at smart cities from a cultural perspective?
- Write an essay titled smart citied and Gen. Z point of view.
- List and briefly describe projects that developed out of bold initiatives of smart cities in any of the Gulf countries.

References

Agbali, M., Trillo, C., Ibrahim, I. A., Arayici, Y., & Fernando, T. (2019). Are innovative innovation ecosystems seeking to meet citizens' needs? Insights from the stakeholders' vision on smart city strategy implementation. *Smart Cities*, *2*(2), 307–327.

Appio, F. P., Lima, M., & Paroutis, S. (2019). Understanding smart cities: Innovation ecosystems, technological advancements, and societal challenges. *Technological Forecasting and Social Change*, *142*, 1–14.

Chen, C. W. (2023). Can smart cities bring happiness to promote sustainable development? Contexts and clues of subjective well-being and urban livability. *Developments in the Built Environment*, *13*, 100108.

Diamond, K. E. (2020). The Yoga Analogy: Scaling up the U.S.'s renewable energy sector mindfully with new technologies, evolving standards, public buy-in, data sharing, and innovation clusters. *Fordham Environmental Law Review*, pp. *32*, 381.

El Khatib, M., Beshwari, F., Beshwari, M., Beshwari, A., Alzoubi, H. M., & Alshurideh, M. (2023). Can better capabilities lead to better project and program governance? Cases from Dubai. In *The effect of information technology on business and marketing intelligence systems* (pp. 1295–1313). Springer International Publishing.

Field, C. B. (Ed.). (2012). *Managing the risks of extreme events and disasters to advance climate change adaptation: Special report of the intergovernmental panel on climate change*. Cambridge University Press.

Gebresselassie, M., & Sanchez, T. W. (2018). "Smart" tools for socially sustainable transport: A review of mobility apps. *Urban Science*, *2*(2), 45.

Greenfield, A. (2010). Everywhere: The dawning age of ubiquitous computing. *New Riders*. https://populationmatters.org/news/2022/07/un-2-4bn-more-people-but-is-end-of-growth-in-sight/

Krivý, M. (2018). Towards a critique of cybernetic urbanism: The smart city and the society of control. *Planning Theory*, *17*(1), 8–30.

Lorinc, J. (2018). Promise and peril in the smart city: Local government in the age of digital urbanism. *Institute on Municipal Finance and Governance*.

Ross, D. (2014). *Millennials do not care about owning cars, and car makers cannot determine why* (p. 19). Fast Company, Co. Exist.

Salem, F. (2016). A *smart city for public value: Digital transformation through agile governance – The case of 'Smart Dubai.'* World Government Summit Publications.

Toh, C. K. (2022). Smart city indexes, criteria, indicators, and rankings: An in-depth investigation and analysis. *IET Smart Cities*, *4*(3), 211–228.

Chapter 4

Innovating Urban Environments: The Impact of Smart City Technologies on Employee Performance, Quality of Life, and Service

Alice Alakoum[a b] *and Elvira Nica*[a]

[a]*Bucharest University of Economic Studies, Bucharest, Romania*
[b]*Arab Open University, Ardiya, Kuwait*

Abstract

This chapter delves into how smart city innovations positively affect workforce efficiency, residents' quality of life (QoL), and the delivery of services, particularly within the dynamic context of smart cities: innovation, development, transformation, and prosperity. It discusses the role of technologies like cyber-physical systems, the Internet of Things, and intelligent transport systems in creating efficient, sustainable urban spaces that benefit the workforce and the broader community. The chapter highlights strategies for improving urban environments, ensuring workforce well-being, and fostering sustainable growth by examining the interplay between these technologies and urban living. The narrative emphasizes the necessity of ongoing innovation, policy support, and workforce adaptation, underscoring the importance of tailoring smart city initiatives to regional needs for maximal impact on employee performance, QoL, and service delivery. Additionally, it introduces a comprehensive framework designed to guide the development of next-generation smart cities. This framework integrates advanced technologies for optimized urban management and service provision, directly linking to enhanced employee performance through improved urban infrastructure and services. The strategic

The Emerald Handbook of Smart Cities in the Gulf Region:
Innovation, Development, Transformation, and Prosperity for Vision 2040, 53–68

doi:10.1108/978-1-83608-292-720241003

application of this framework aims to elevate economic prosperity and societal well-being, ensuring workforce efficiency is central to the urban development agenda. The enhanced employee performance, catalyzed by smart city innovations, is pivotal in driving economic vibrancy, social inclusivity, and environmental sustainability, shaping the future of urban development. This analysis will offer valuable insights for smart cities research and development in the Gulf Region, suggesting pathways for implementing these concepts to address the region's urbanization and development challenges.

Keywords: Urban innovation; workforce efficiency; sustainable growth; technology integration; digital infrastructure; urban well-being

Introduction

The advent of smart cities marks a significant shift in the paradigm of urban development, connecting the power of advanced technologies and data analytics to significantly improve the living standards of its populations, enhance the operational efficiency of urban services, and stimulate sustainable economic advancement (Gracias et al., 2023). The concept of smart cities encompasses an extensive spectrum of applications, from the integration of renewable energy systems and transportation management to the advancement of smart healthcare and public safety, all directed toward forging more liveable, sustainable, and efficient urban landscapes (Al-Yousfi et al., 2019).

The incorporation of technologies associated with smart cities has been identified as a catalyst for enhancing workforce performance. Through the optimization of service delivery and equipping employees with immediate access to data and state-of-the-art tools, smart cities are pivotal in enhancing decision-making capabilities, boosting productivity, and reducing the time and resources allocated for task completion (Hendy et al., 2017). This enhancement directly contributes to superior service delivery and heightened customer satisfaction, as the workforce becomes more adept at fulfilling the requirements of the community it serves (Sarkar et al., 2021). Furthermore, the integration of smart technologies cultivates a workforce that is both versatile and capable of promptly adapting to the evolving demands and challenges within urban environments. The embedded connectivity and data exchange capabilities inherent in smart cities underscore the promotion of collaborative workspaces, further augmenting employee efficiency and fostering innovation (Hendy et al., 2017).

Moreover, the well-being of residents in smart cities is intimately connected to their quality of life (QoL). Through services such as intelligent transport systems (ITS), smart energy grids, and e-governance, smart cities contribute to an urban living experience that is markedly more comfortable, convenient, and efficient (Sarkar et al., 2021). The continuous monitoring and assessment of these services play a critical role in ensuring their meaningful impact on the QoL and overall satisfaction among citizens (Aleksic et al., 2019). This elevated standard of living invariably influences employee well-being, which is a fundamental factor in enhancing productivity and job satisfaction. Employees benefit from a healthier work-life equilibrium facilitated by the efficient urban systems and technologies that mitigate daily stressors and streamline routine tasks.

In summary, smart cities are at the forefront of urban development, leveraging technological innovations to improve both employee performance and the QoL. The interconnection between smart city technologies and workforce efficiency is evident, offering the essential infrastructure and tools for efficient and effective service provision. As urban areas continue to progress, embracing smart city solutions will be crucial in developing sustainable, liveable communities and supporting a high-performing, satisfied workforce (Alenezi, 2023).

Defining Smart Cities: Foundations and Impacts on Workforce Efficiency, Urban Living, and Municipal Services

Smart cities emerge as a visionary model for urban development, employing cutting-edge digital, communication technologies, and data analytics to elevate resident well-being, boost workforce efficiency, and streamline municipal operations. According to the European Commission, smart cities are characterized by the strategic use of technology and infrastructure to improve sustainability, resource management, and inhabitants' QoL. A smart city utilizes technological infrastructure to gather, analyze, and leverage data from diverse sources, such as sensors, mobile devices, and existing urban systems, to optimize assets, resources, and services. This approach facilitates informed decision-making and urban planning. Furthermore, a comprehensive definition of smart cities includes six dimensions: smart people, smart economy, smart living, smart mobility, smart environment, and smart governance (Sifah et al., 2020).

At the heart of smart cities lies the integration of digital and communication technologies along with data analytics, involving key stakeholders like citizens, city officials, and businesses (Midor et al., 2021). These cities are distinguished by their ability to employ advanced ICT for real-time monitoring and management of urban infrastructures and services. They leverage big data and analytics to process vast amounts of information, leading to more efficient urban operations and planning. Sustainability and enhanced QoL are central goals, achieved through tech-driven solutions and practices that reduce environmental footprints and improve living conditions. Moreover, smart cities promote participatory governance by encouraging citizen involvement in urban decision-making processes.

The role of digital technologies and data analytics is pivotal in smart cities, facilitating an interconnected, intelligent urban framework that improves services and enhances the QoL for its citizens (Tong et al., 2021). These technologies ensure continuous information flow, which is essential for the effective functioning of urban services. The comprehensive application of information and communication technology (ICT) and real-time data analysis supports sustainable economic growth and the efficient delivery of urban services (Toli & Murtagh, 2020). For example, IoT devices collect data on different city life aspects from traffic patterns, energy utilization, etc., which is analyzed to refine services and address inefficiencies, thereby enhancing urban sustainability and liveability.

The evolution of smart cities markedly influences employee performance by providing an environment conducive to improved QoL and efficient service delivery. This leads to heightened job satisfaction, well-being, and work-life balance, vital for sustaining workforce productivity and engagement. The alleviation of daily stressors, such as traffic and inefficient public services, enables employees to devote more attention to their professional roles. Technologies automating routine tasks and optimizing resource use not only elevate productivity but also foster creativity. For instance, smart traffic systems reduce commuting time, enhancing employee efficiency. The integration of smart technologies also supports flexible working arrangements, increasing job satisfaction and retention rates. For instance, the implementation of blockchain-based decentralized employee performance assessment systems can enhance smart city governance and contribute to improved employee productivity and accountability (Sifah et al., 2020). These systems utilize the fundamental features of blockchain technology, including decentralization, unchangeability, and openness, to overcome the shortcomings of conventional centralized performance evaluation approaches.

Furthermore, smart cities significantly enhance the living standards of their residents by revolutionizing sectors like healthcare, transportation, and agriculture. Smart healthcare systems improve access to services and enable personalized care, impacting employee well-being positively. Smart mobility solutions enhance public transport and reduce congestion, benefiting daily commutes and overall QoL. Smart agriculture technologies contribute to more efficient farming, better resource management, and food security, further elevating urban employees' living standards (Talari et al., 2017).

Smart city initiatives also transform work processes by introducing efficiencies and innovations that reshape service delivery and task execution. The digitization and automation of services streamline operations and enhance service quality (Abiad & Ionescu, 2020), benefiting both employees and residents. Insights from data analytics drive operational innovation and the development of new services, promoting economic growth and creating a workforce skilled in exploiting technological advancements. The growth of smart cities necessitates investment in human and social capital, communication infrastructure, and participatory governance, essential for sustainable development and improved urban life quality (Arduin et al., 2016).

In conclusion, smart cities signify a transformative shift in urban development, harnessing technology, and data analytics to foster efficiency, sustainability, and enhanced urban living. Through strategic technology and data utilization, smart cities not only improve employee performance but also revolutionize work processes and citizen engagement, setting new benchmarks for urban living. As technological advancements continue, the potential of smart cities to redefine urban landscapes and enrich residents' lives becomes increasingly significant, heralding a future where cities are smart, inclusive, resilient, and vibrant.

Technological Paradigms in Smart Cities: Enhancing Efficiency and QoL in Urban Areas

Smart cities represent a comprehensive framework that integrates a wide array of technologies to elevate urban living standards, focusing on augmenting efficiency and productivity, and enhancing both the QoL and quality of service (QoS). These innovations are instrumental in transforming urban areas into environments that are more habitable, manageable, and sustainable.

Central to the technological advancements within smart cities are ITS, which stands as a testament to the convergence of technology with urban mobility. ITS are designed to streamline urban traffic flow, alleviate congestion, and bolster the safety and reliability of public transportation through a balanced approach to urban traffic supply and demand, underpinned by thorough analysis and research into traffic patterns (Guo & Guo, 2023). Such improvements not only optimize transportation but also positively affect employee performance by minimizing commute times and stress, thereby facilitating a healthier work–life balance. Moreover, the application of technologies such as blockchain and artificial intelligence (AI) within ITS facilitates cross-departmental collaboration in traffic management, enhancing the synchronization of urban traffic supply with demand through adaptive learning mechanisms (Guo & Guo, 2023). Additionally, the employment of technologies for sensing, analysis, control, and communication within ground transportation systems furthers safety, mobility, and efficiency (Hou, 2023). Following the COVID-19 pandemic, the issue of the digital divide has become a prominent challenge for urban populations, underscoring the urgency for smart cities to ensure digital inclusivity. Alkhaldi (2022) outlines the responses of developed countries in supporting their citizens, illuminating strategies that can promote digital inclusion and tackle disparities within smart city infrastructures.

The link between ITS and employee performance is nuanced, mediated by the numerous benefits ITS provides, including enhanced transportation efficiency, reduced commute times, and decreased employee stress (Babu, 2023). ITS's contributions to road safety through innovative technologies decrease traffic-related absences, fostering a more consistent employee presence and facilitating flexible work models like telecommuting, thereby maintaining productivity levels (Mathew, 2019). Despite the scarcity of direct evidence linking ITS to employee performance, the indirect benefits suggest its substantial potential to improve workplace productivity by refining commuting processes and safety measures. This relationship invites further scholarly inquiry to unpack the full impact of ITS on-employee performance, considering employment nature, organizational roles, and transportation infrastructure.

Smart digitalization, encompassing the digitization of services and automation, marks a broader trend toward enhancing public services via digital platforms (Klarić, 2023). This digital transformation streamlines city operations, making processes more efficient and freeing employees to focus on complex and rewarding tasks. Such shifts not only boost productivity but also elevate job satisfaction, as employees engage in more meaningful work (Başar & Arkan, 2023).

Additionally, the digitization of public services simplifies interactions, enhancing transparency, satisfaction, and the functional interaction between authorities and citizens (Bolgova et al., 2021).

Cyber-physical systems (CPS) and the Internet of Things (IoT) form the technological foundation of smart cities, merging digital technology with physical systems. CPS combines computing, networking, and physical operations (Nair et al., 2021), and when linked with IoT, which interconnects devices through the internet, a robust network for data sharing and enhancing system efficiency emerges (Armoogum et al., 2021). Utilizing data analytics, CPS offer a dynamic feedback loop, facilitating the instantaneous assessment and modification of physical systems to confirm they function according to their intended design parameters. This instant feedback is crucial for enhancing the efficiency of operations within the workplace. It supports the refinement of processes, potentially boosting productivity and operational efficiency. Additionally, the capability to oversee and modify systems in real-time plays a significant role in promoting employee well-being, by maintaining a working environment that is both safe and favorable for productivity.

This integration, through the use of digital twins and universal sensors, facilitates the optimization of city resources, thereby improving both QoL and QoS for residents. Smart cities utilize these technologies across various domains such as energy management, urban mobility, and smart buildings, contributing to sustainable economic growth and well-being (Kodjiku et al., 2023). The interconnectedness of CPS and IoT fosters situation awareness and supports smart grids and transportation, enhancing urban safety and efficiency (Bhushan et al., 2022). The incorporation of CPS and IoT has been pivotal in sectors including healthcare, transportation, and agriculture, significantly enhancing employee QoL. Smart healthcare systems, for example, improve access to services and enable personalized care (Manwal, 2019). Smart agriculture technologies increase farming efficiency, sustainable resource management, and food security, further enhancing urban employee living standards.

The deployment of efficient technological solutions within smart cities markedly improves QoS by addressing urban challenges effectively. IoT and digital infrastructure advancements in areas like waste management, energy efficiency, and environmental sustainability contribute significantly to QoS improvements (Nguyen et al., 2023). Furthermore, smart governance and digital twin technologies refine city planning and resource allocation, leading to enhanced service delivery and QoS (Faheem & Butt, 2022).

The integration of big data, intelligent sensing, advanced communication networks, cloud computing, and AI within smart cities further enriches QoS by bolstering connectivity, energy efficiency, and overall service quality (Huang et al., 2022). Innovations such as drones, robotics, and the IoT amplify the sophistication of smart cities, consequently enhancing QoS (Alsamhi et al., 2019). Demonstrations of successful smart city projects in locales such as Helsinki and China highlight the transformative impact these technologies have on city ecosystems and resource management (Hämäläinen, 2019; Huang et al., 2022).

Blockchain technology's decentralization, immutability, and transparency offer substantial improvements over traditional centralized methods for employee performance within smart governance contexts (Fu, 2022), addressing centralization's inherent issues and enhancing evaluation processes' reliability and fairness. Traditional models are frequently critiqued for their vulnerability to failures concentrated at a single point and the possibility of data being manipulated by authoritative figures, undermining trust among public sector employees. By dispersing data throughout a network, blockchain's decentralized architecture significantly diminishes the chance of data interference and ensures no singular authority monopolizes system control (Antal et al., 2021). As smart cities evolve, the integration of blockchain technology presents a transformative approach to urban governance and public value creation. Alkhaldi et al. (2023) demonstrate the utilization of blockchain technology to build smarter cities in Kuwait, offering insights into the potential for similar innovations to enhance transparency, security, and citizen engagement across urban settings. Implementing a blockchain-driven approach to employee performance appraisal in the realm of smart governance could significantly bolster security, trustworthiness, accountability, and operational efficiency. This approach addresses the limitations inherent in centralized systems while capitalizing on blockchain's unique features to foster a more dependable and equitable assessment environment.

In summary, the integration of advanced technologies in smart cities catalyzes significant improvements in urban efficiency, employee performance, and the quality of urban living. Through reducing inefficiencies and fostering a balanced work-life dynamic, these technologies contribute to heightened productivity and employee satisfaction. Smart city technologies not only revolutionize healthcare, transportation, and agriculture, directly benefiting employee QoL but also enhance QoS through innovative responses to urban challenges. This comprehensive approach to integrating technology within urban environments underscores the potential of smart cities to create more efficient, sustainable, and liveable communities, setting a new standard for urban development.

Advancements in Smart Urban Development: Elevating Workforce Efficacy and Streamlining Operations

The launch of smart cities reveals a revolution in the way urban areas evolve, with a focus on using cutting-edge technology to improve work environments and enhance employee productivity. This shift toward incorporating a digital framework at the heart of urban planning has led to notable advancements in operational productivity, decision-making efficiency, workplace satisfaction, lifelong learning, and safety measures. Smart cities bring forth a slew of benefits that impact far more than just the living standards within urban boundaries, affecting the broader economic and social fabric of the communities they serve. Key among these benefits is the capacity to draw in highly skilled professionals, uplift the productivity of the workforce, and make both transactional and administrative workflows more efficient. These elements play a crucial role in steering urban areas toward a more dynamic, productive, and welcoming future. Central to achieving these outcomes is the strategic use of digital technologies, the adoption of evidence-based decision-making, and the streamlining of city-wide services.

The appeal of smart cities as hubs for skilled talent is evident. By blending the latest in technology, such as IoT, AI, and seamless connectivity, with a superior living experience, these cities become magnets for individuals seeking to engage in transformative projects within ecosystems that prioritize both innovation and sustainability (Alkhaldi et al., 2024). This influx of skilled individuals brings in a dose of creativity and propels economic vitality, boosting the city's image on the global stage.

Smart city technologies facilitate the recruitment, nurturing, and retention of skilled personnel by offering the requisite infrastructure and amenities sought after by such workers. This includes the establishment of educational and training facilities within communities to cultivate a workforce proficient in necessary skills (Sleezer & Denny, 2004). Additionally, the provision of a high QoL through efficient services emerges as a critical element in retaining adept employees and enhancing organizational efficacy (Joof & Olanipekun, 2023; Memon, 2021). Smart cities engender an innovation and learning-centric culture among the workforce, as the continual integration of advanced technologies presents abundant opportunities for skill enhancement and adaptation to technological evolutions (Parubochaya et al., 2020). Furthermore, the implementation of heightened safety measures in smart cities contributes to a more secure working ambiance, culminating in a more engaged and productive workforce (Parubochaya et al., 2020).

The infusion of smart technologies, including IoT-enabled workplace solutions, digital communication platforms, and data analytics, significantly bolsters workforce performance. Employees in smart cities enjoy enhanced connectivity and automation, which translates to superior productivity and efficiency. Automation facilitated by AI-driven applications alleviates workload, enabling staff to focus on intricate job facets (Parubochaya et al., 2020). The exploitation of data-driven insights and algorithmically mediated technologies not only unveils novel efficiencies in city services but also fosters a conducive environment for workforce development (Botero Arcila, 2022). The rich data milieu of a smart city facilitates informed decision-making, efficient resource allocation, and swift adaptability to fluctuating circumstances, engendering a responsive and versatile workforce.

Moreover, smart cities' emphasis on sustainability and efficient public utilities fosters a healthier work setting, yielding increased job satisfaction and performance (Parubochaya et al., 2020). ITS mitigate commute times and stress, fostering work-life balance and overall employee contentment. Smart city initiatives frequently encompass the development of digital platforms that enable remote and flexible working arrangements, thus accommodating employees' productivity preferences. Furthermore, the application of data analytics and automation in the workplace streamlines operations, diminishes manual errors, and liberates employees to engage in higher-value activities. This not only enhances individual performance but also boosts organizational productivity. Additionally, improved safety protocols in smart cities, including enhanced surveillance and rapid emergency responses, ensuring a safer working environment, contributing to higher employee engagement and productivity.

Smart cities are also at the forefront of making bureaucratic processes more streamlined through digitalization. By adopting blockchain and other secure digital technologies, these urban settings efficiently conduct transactions, minimize paperwork, and reduce bureaucratic delays, significantly speeding up service delivery and enhancing record precision. This efficiency not only simplifies access to services for residents and businesses but also fosters a more open and reliable interaction between the government and its citizens.

The transformation of urban centers into smarter, technologically enhanced entities yields substantial benefits. The technological foundation of smart cities crucially transforms the work environment, with a direct positive impact on employee performance (Parubochaya et al., 2020). By improving efficiency, enhancing decision-making, enriching work environments, offering opportunities for ongoing education, and bolstering safety, smart cities elevate urban living standards and empower the workforce toward more significant achievements. This evolution enhances the attractiveness of cities as places to live and work for talented professionals boosts workforce performance, and simplifies administrative tasks, creating a dynamic, efficient, and inclusive urban ecosystem. Thus, the advantages of smart city technologies extend beyond infrastructural upgrades to embrace comprehensive aspects of talent management and workforce performance enhancement. By leveraging the transformative potential inherent in smart cities, this research elucidates strategic frameworks that utilize state-of-the-art technologies – such as the IoT, AI, and blockchain – to significantly redefine urban environments and workforce dynamics. Central to this study is the innovative application of these technologies to enhance work environments and boost employee productivity. By presenting a comprehensive digital framework for urban planning, this work offers actionable insights and practical strategies that are instrumental in fostering more dynamic, productive, and inclusive urban futures.

The findings from this research provide invaluable insights for various stakeholders involved in smart cities, such as urban planners, architects, policymakers, faculty, students, education administrators, and government officers. By highlighting the critical role of smart technologies in enhancing work environments and employee productivity, these insights offer a robust framework for decision-making and strategic development across various sectors. Urban planners and architects can leverage this knowledge to design efficient and sustainable infrastructures.

At the same time, policymakers can formulate supportive policies that propel smart city initiatives forward, particularly optimizing employee performance. These can advocate for establishing digital literacy and workforce training programs, implementing collaborative governance models that include a broad spectrum of stakeholders, and prioritizing sustainable urban infrastructure development. These strategic policies aim to equip employees with the necessary skills to utilize smart technologies effectively, foster an inclusive environment for engaging stakeholders, and create workspaces that promote health, safety, and productivity. Together, these measures aim to improve workforce productivity and foster innovation, securing the full actualization of smart cities' transformative potential.

Moreover, integrating these findings into educational curriculums can prepare students for future challenges in urban development, fostering an environment of innovation and interdisciplinary collaboration. This holistic approach informs and inspires stakeholders, driving the development of smart cities that are sustainable and effective, thereby improving the QoL for all residents.

Addressing Challenges and Ethics in Smart City Development: Workforce and Governance

The transition toward smart cities heralds a significant transformation in the confluence of urban development and technology, aimed at elevating the standard of living for inhabitants and optimizing municipal functionalities. Yet, this progression toward seamlessly integrated smart urban environments is laden with complex challenges. The integration of technological advancements into the urban landscape and its workforce necessitates a concerted effort to ensure adaptability and comprehensive training. Additionally, achieving a symbiotic relationship between innovative solutions and pre-existing infrastructure and systems presents a substantial challenge, necessitating a strategic approach to workforce skill enhancement and development to prevent potential skill disparities that could impede the successful realization of smart city projects (Clever et al., 2018).

The endeavor to blend new technological interventions with traditional systems often disrupts established workflows, necessitating significant adjustments within the workforce. This complex interplay between innovation and legacy systems, while vital for maintaining continuity in urban operations, can be resource-intensive and demands a high level of adaptability from employees (Dembski et al., 2020). The imperative to balance cutting-edge developments with existing frameworks underscores the necessity for a workforce proficient in navigating both realms to sustain productivity during periods of transition.

Beyond the logistical and operational challenges, the ethical dimensions of smart cities and digital transformation introduce critical considerations. Concerns regarding privacy, data security, and the equitable application of technology and data within urban settings emerge as pivotal issues with direct implications for employee efficacy and societal equity. The task of mitigating social and economic disparities further compounds these challenges, emphasizing the need for a delicate balance between technological advancement and social responsibility (Clever et al., 2018). The intensive data-centric nature of smart cities escalates concerns around privacy and security, placing additional responsibilities on employees to manage sensitive information while adhering to strict regulatory standards, thereby influencing their performance and well-being.

Furthermore, the ethical deployment of technology and data, including surveillance and bias in algorithmic decision-making, necessitates continuous employee education to navigate and address these ethical dilemmas effectively.

The digital divide, highlighted by smart city initiatives, accentuates the urgency for inclusive strategies to ensure equitable access to the benefits of urban technological advancements, tasking employees with integrating technical expertise with a commitment to social equity.

Reflecting these challenges within organizational contexts, the adoption of blockchain and other emergent technologies introduces complexities associated with cybersecurity, workforce retraining, and change management, necessitating a multifaceted response to navigate the digital transformation effectively. The multifaceted nature of the challenges facing smart cities, ranging from technological integration to addressing ethical and social considerations, calls for a comprehensive and strategic approach to ensure the efficient functioning of smart urban systems and the well-being of their workforce (Dembski et al., 2020; Omayer et al., 2022).

According to Batty et al. (2012), implementations of smart city concepts, as demonstrated in NEOM, Saudi Arabia, and the Old Town of Semarang, Indonesia, have illustrated the efficacy of smart city technologies in improving the conditions of urban living and enhancing the QoL and services for employees. Nevertheless, such advancements bring forth complexities including concerns over data security, the demands for digital infrastructure, and the need for robust governance and collaborative efforts among stakeholders, as outlined by Evertzen et al. (2019). The fruitful recognition of smart city frameworks necessitates tackling pivotal issues such as engaging citizens, fostering business partnerships, and establishing decisive leadership, which are identified as critical factors for success in the evolution of smart cities (Evertzen et al., 2019).

The role of cooperative governance is underscored as a determinant for the successful deployment of smart cities, supported by research on systems rooted in local wisdom and the significance of stakeholder collaboration for smart city achievement (Novita & Supranoto, 2019; Triyanto et al., 2022; Wahyuni et al., 2022). Moreover, the architecture of governance in smart city ventures and the institutional efforts toward collaborative practices are vital for enduring partnerships and governance in the smart city context (Broccardo et al., 2019; Mills et al., 2021; Ooms et al., 2020).

The involvement and analysis of stakeholders stand as crucial elements for the triumph of smart city initiatives, highlighting the necessity for effective engagement and cross-sectoral alliances (Jayasena et al., 2019; Pereira et al., 2017). Additionally, smart governance, representing a collaborative framework, emerges as a foundational pillar of smart cities, necessitating a seamless integration of technology, human capital, and governance mechanisms (Meijer & Bolívar, 2015; Pereira et al., 2018). Factors such as participatory approaches, the design of smart cities, and the journey of digital transformation further enrich the discourse on successful smart city projects (Hämäläinen, 2019; Myeong et al., 2018; Spil et al., 2017).

Smart cities, while promising a more efficient and habitable future, present a complex array of challenges that can influence employee performance. Addressing the intricate dynamics of integrating technology into the urban fabric and workforce, fostering workforce adaptability, aligning innovation with existing infrastructures, and navigating the ethical and social intricacies, is imperative for the realization of smart cities' full potential. This comprehensive approach, prioritizing workforce development, ethical technology use, and the harmonization of innovation with existing urban frameworks, is crucial for cultivating smart cities as inclusive, equitable, and supportive ecosystems for all community members and contributors.

Conclusion

This study sheds light on the intersection between smart city innovations and their effects on workforce dynamics, with a particular focus on the evolving urban landscapes. It articulates a novel theoretical perspective by integrating the functionalities of CPS, the IoT, and ITS to elucidate their collective impact on enhancing workforce productivity, QoL, and service efficiency. This contribution deepens the academic discussion on smart cities. It paves the way for practical application, offering a strategic blueprint for smart urban environments' future development, management, and governance.

From a practical standpoint, the research outlines a comprehensive framework that promises to revolutionize smart city initiatives, driving technology-led improvements in employee performance and urban welfare. It highlights the crucial role of sustained innovation, tailored policy formulation, and workforce adaptability in maximizing the benefits of smart city projects for economic development, social inclusivity, and environmental conservation.

Smart city projects significantly enhance employee performance, work-life balance, and overall QoL, marking a pivotal shift in urban development and management. By leveraging digital and communication technologies, data analytics, and the IoT, smart cities create ecosystems that support efficient service delivery, sustainable economic growth, and an enhanced urban living experience. These technological advancements equip employees with access to real-time data and advanced tools within a collaborative work environment, thereby increasing productivity, fostering innovation, and improving job satisfaction. Additionally, smart cities' enhancements in QoL – from ITS to smart healthcare – directly contribute to a more satisfying work-life balance for employees.

For industry application, the study's insights into smart city design, implementation, and governance highlight the transformative potential of these urban models. It advocates strategically aligning technological advancements with regional specificities to propel smart cities toward achieving their intended socio-economic and environmental outcomes.

Smart cities are actively gearing toward sustainable development and ongoing innovation. Emerging technologies promise to improve employee performance further and ensure urban development is in harmony with environmental sustainability and resilience to future challenges. However, these initiatives' success relies on the seamless integration of technology into the urban fabric and the workforce's ability to adapt to these changes. Ongoing education, training, and skill enhancement are crucial for preparing employees to excel in this evolving landscape.

Collaboration between technology developers, city planners, and the workforce becomes essential as cities evolve into smarter ecosystems. This involves implementing new technologies and seamless integration with existing systems and ensuring that employees can effectively use these tools. Addressing privacy, data security, and ethical technology use challenges will also be critical in maintaining trust and ensuring fair access to smart city benefits.

In summary, smart cities offer a promising avenue for enhancing employee performance and urban life. Through continuous innovation, effective technology integration, and a focus on sustainable development, smart cities can act as catalysts for creating more liveable, efficient, and inclusive urban environments. The road ahead is filled with opportunities to transform the urban experience, placing employees at the core of this change, where they benefit from an environment that supports their growth, well-being, and productivity.

Future research should integrate this study's findings with broader academic efforts, particularly in tackling emerging challenges around data privacy, security, and ethical technology use within smart cities. Investigating blockchain technology's role in enhancing smart governance and workforce efficiency opens a promising path for extending the current insights. Such forward-thinking research directions aim to enrich the theoretical base and refine practical approaches for leveraging smart cities as engines for sustainable urban development and improved resident well-being.

References

Abiad, M., & Ionescu, S. (2020). Building an algorithm for predictive maintenance. *UPB Scientific Bulletin, Series D: Mechanical Engineering,*82(4), 337–348.

Aleksic, A., Ljepava, N., & Ristic, M. (2019). Smart city transportation services, customer satisfaction and impact on the quality of life. In *Smart technologies and innovation for a sustainable future: Proceedings of the 1st American University in the Emirates International Research Conference – Dubai, UAE 2017* (pp. 333–338). Springer International Publishing.

Alenezi, M. (2023). Investigating the software engineering roadmap for smart city infrastructure development: Goals and challenges. *arXiv preprint arXiv:2305.05574.*

Alkhaldi, A. N. (2022). Digital exclusion during the COVID-19 pandemic: A review of how developed countries responded to support their citizens. *International Journal of Electronic Government Research (IJEGR), 18*(1), 1–19. http://doi.org/10.4018/IJEGR.306231

Alkhaldi, A., Alrashidi, H., Alhasan, K., Alsadeeqi, A., & Alshami, A. (2023). The use of blockchain technology to build smart cities: Creating public value in Kuwait. *Global Knowledge, Memory and Communication*, ahead-of-print. https://doi.org/10.1108/GKMC-11-2022-0263

Alkhaldi, A., Malik, S., & Lytras, M. D. (2024). Transformative leadership in Kuwait direct investment promotion authority: Investing in talent, innovation and the next generation. In S. Baroudi, & M. D. Lytras, (Ed.), *Transformative leadership and sustainable innovation in education: interdisciplinary perspectives (Emerald studies in higher education, innovation and technology*, pp. 27–38). Emerald Publishing Limited.

Alsamhi, S., Ma, O., Ansari, M., & Almalki, F. (2019). Survey on collaborative smart drones and internet of things for improving smartness of smart cities. *IEEE Access, 7*, 128125–128152. https://doi.org/10.1109/access.2019.2934998

Al-Yousfi, W. M., Yahia, H. A., & Kishore, R. (2019). Instigating smart city infrastructure in Oman: Considering social benefits and economic values. *Journal of Student Research.*

Antal, C., Cioara, T., Anghel, I., Antal, M., & Salomie, I. (2021). Distributed ledger technology review and decentralized applications development guidelines. *Future Internet*, *13*(3), 62.

Arduin, P., Negre, E., & Rosenthal-Sabroux, C. (2016). Knowledge and decision for smart cities initiatives: Cases of Paris and nice. In *2016 International conference on industrial informatics and computer systems (CIICS)*. https://doi.org/10.1109/iccsii.2016.7462421

Armoogum, S., Khonje, P., & Li, X. (2021). Digital forensics of cyber physical systems and the Internet of Things. In *Crime science and digital forensics: A holistic view* (pp. 117–148). CRC Press.

Babu, A. M. (2023). Smart cities and intelligent transport systems. *International Journal of Applied and Structural Mechanics (IJASM) ISSN: 2799-127X*, *3*(01), 22–25.

Başar, P., & Arkan, G. (2023). The relations between digitalization and employee satisfaction in aviation. *Journal of Aviation*, *7*(3), 388–392.

Batty, M., Axhausen, K., Giannotti, F., Pozdnoukhov, A., Bazzani, A., Wachowicz, M., Ouzounis, G., Portugali, Y., (2012). Smart cities of the future. *The European Physical Journal Special Topics*, *214*(1), 481–518. https://doi.org/10.1140/epjst/e2012-01703-3

Bhushan, B., Anushka, Kumar, A., & Katiyar, L. (2022). Security magnification in supply chain management using blockchain technology. *Blockchain Technologies for Sustainability*, 47–70.

Bolgova, E. V., Bolgov, S. A., & Haitbaev, V. A. (2021). Digital transformation of government services: A longitudinal case study of tax service. In *European Proceedings of Social and Behavioural Sciences*.

Botero Arcila, B. (2022). Smart city technologies: A political economy introduction to their governance challenges. In Justin B. Bullock, and others (Eds.), *The Oxford Handbook of AI Governance*. Oxford. https://doi.org/10.1093/oxfordhb/9780197579329.013.48

Broccardo, L., Culasso, F., & Mauro, S. (2019). Smart city governance: Exploring the institutional work of multiple actors towards collaboration. *International Journal of Public Sector Management*, *32*(4), 367–387. https://doi.org/10.1108/ijpsm-05-2018-0126

Clever, S., Crago, T., Polka, A., Al-Jaroodi, J., & Mohamed, N. (2018). Ethical analyses of smart city applications. *Urban Science*, *2*(4), 96. https://doi.org/10.3390/urbansci2040096

Dembski, F., Wössner, U., Letzgus, M., Ruddat, M., & Yamu, C. (2020). Urban digital twins for smart cities and citizens: The case study of Herrenberg, Germany. *Sustainability*, *12*(6), 2307. https://doi.org/10.3390/su12062307

Evertzen, W., Effing, R., & Constantinides, E. (2019). The internet of things as smart city enabler: The cases of Palo Alto, Nice and Stockholm. In *Advances in smart cities: Smarter people, governance, and solutions* (pp. 293–304). https://doi.org/10.1007/978-3-030-29374-1_24

Faheem, M., & Butt, R. A. (2022). Big datasets of optical-wireless cyber-physical systems for optimizing manufacturing services in the internet of things-enabled industry 4.0. *Data in Brief*, *42*, 108026.

Fu, X. (2022). A unified view of IoT and CPS security and privacy. *arXiv preprint arXiv:2209.13793*.

Gracias, J. S., Parnell, G. S., Specking, E., Pohl, E. A., & Buchanan, R. (2023). Smart cities – A structured literature review. *Smart Cities*, *6*(4), 1719–1743.

Guo, X., & Guo, X. (2023). A research on blockchain technology: Urban intelligent transportation systems in developing countries. *IEEE Access*, *11*, 40724–40740. https://doi.org/10.1109/access.2023.3270100

Hämäläinen, M. (2019). A *framework for a smart city design: digital transformation in the Helsinki smart city* (pp. 63–86). https://doi.org/10.1007/978-3-030-23604-5_5

Hendy, M. E., Miniaoui, S., Atalla, S., & Hashim, K. F. B. (2017, March). A survey on smart city technologies, initiatives and global technology providers. In *Proceedings of the second international conference on Internet of Things, data and cloud computing* (pp. 1–7).

Hou, W. (2023). Applications of big data technology in intelligent transportation system. *Highlights in Science, Engineering and Technology*, *37*, 64–71. https://doi.org/10.54097/hset.v37i.6040

Huang, Y., Peng, H., Sofi, M., Zhou, Z., Xing, T., Ma, G., … Zhong, A. (2022). The city management based on smart information system using digital technologies in china. *IET Smart Cities*, *4*(3), 160–174. https://doi.org/10.1049/smc2.12035

Jayasena, N., Mallawaarachchi, H., & Waidyasekara, K. (2019). Stakeholder analysis for smart city development project: an extensive literature review. *Matec Web of Conferences*, *266*, 06012. https://doi.org/10.1051/matecconf/201926606012

Joof, S. E., & Olanipekun, W. D. P. (2023). *Dynamics of talent management in enhancing employees' performance in the Gambian banking sector*.

Kodjiku, S. L., Han, T., Fang, Y., Aggrey, E. S. E., Sey, C., Asamoah, K. O., ... & Wang, X. (2023). WQCrowd: Secure blockchain-based crowdsourcing framework with multi-tier worker quality evaluation. Journal of King Saud University-Computer and Information Sciences, *35*(10), 101843.

Klarić, M. (2023). Smart digitalization and public services in the EU. *EU and Comparative Law Issues and Challenges Series (ECLIC)*, *7*, 151–174.

Mathew, E. (2019, October). Intelligent transport systems and its challenges. In *International conference on advanced intelligent systems and informatics* (pp. 663–672). Springer International Publishing.

Meijer, A., & Bolívar, M. (2015). Governing the smart city: A review of the literature on smart urban governance. *International Review of Administrative Sciences*, *82*(2), 392–408. https://doi.org/10.1177/0020852314564308

Memon, N. A. (2021). *Need for talent management and investigating its impact on organizational performance of higher education institutes*.

Midor, K., Kuzior, A., Płaza, G., Molenda, M., & Krawczyk, D. (2021). Reception of the smart city concept in the opinion of local administration officials – A case study. *Management Systems in Production Engineering*, *29*(4), 320–326. https://doi.org/10.2478/mspe-2021-0040

Mills, D., Izadgoshasb, I., & Pudney, S. (2021). Smart city collaboration: A review and an agenda for establishing sustainable collaboration. *Sustainability*, *13*(16), 9189. https://doi.org/10.3390/su13169189

Myeong, S., Jung, Y., & Lee, E. (2018). A study on determinant factors in smart city development: An analytic hierarchy process analysis. *Sustainability*, *10*(8), 2606. https://doi.org/10.3390/su10082606

Nair, M. M., Kumari, S., & Tyagi, A. K. (2021, January). Internet of Things, cyber physical system, and data analytics: Open questions, future perspectives, and research areas. In *Proceedings of the second international conference on information management and machine intelligence: ICIMMI 2020* (pp. 325–339). Springer Singapore.

Nguyen, P. H., Sen, S., Bretones-Cassoli, B., Jourdan, N., & Magnanini, M. C. (2023). Software engineering and AI for data quality in cyber-physical systems/Internet of Things-SEA4DQ'22 Report. *ACM SIGSOFT Software Engineering Notes*, *48*(1), 108–111.

Novita, D., & Supranoto, S. (2019). Smart city's success: The importance of stakeholder collaboration. *The International Seminar Series on Regional Dynamics Proceeding*. https://doi.org/10.19184/issrd.v1i1.13732

Omayer, H., Mobarak, M., & Selim, O. (2022). Digital transformation: Towards sustainable and smart new cities in upper Egypt (new Qena case study). *IOP Conference Series*

Earth and Environmental Science, 1113(1), 012027. https://doi.org/10.1088/1755-1315/1113/1/012027

Ooms, W., Caniëls, M., Roijakkers, N., & Cobben, D. (2020). Ecosystems for smart cities: Tracing the evolution of governance structures in a Dutch smart city initiative. *International Entrepreneurship and Management Journal, 16*(4), 1225–1258. https://doi.org/10.1007/s11365-020-00640-7

Parubochaya, E., Piskunov, N., & Drinova, E. (2020). *The "smart cities" concept in the European Union and the Russian Federation: From project to practical implementation.* https://doi.org/10.1007/978-3-030-59126-7_108

Pereira, G., Cunha, M., Lampoltshammer, T., Parycek, P., & Testa, M. (2017). Increasing collaboration and participation in smart city governance: A cross-case analysis of smart city initiatives. *Information Technology for Development, 23*(3), 526–553. https://doi.org/10.1080/02681102.2017.1353946

Pereira, G., Parycek, P., Falco, E., & Kleinhans, R. (2018). Smart governance in the context of smart cities: A literature review. *Information Polity, 23*(2), 143–162. https://doi.org/10.3233/ip-170067

Sarkar, N. I., Iasepi, L. S., & Gul, S. (2021). Designing a smart city: Technologies, challenges and solutions. In *Green computing in smart cities: Simulation and techniques* (pp. 51–70).

Sifah, E. B., Xia, H., Cobblah, C. N. A., Xia, Q., Gao, J., & Du, X. (2020). Bempas: A decentralized employee performance assessment system based on blockchain for smart city governance. *IEEE Access, 8*, 99528–99539. https://doi.org/10.1109/access.2020.2997650

Sleezer, C. M., & Denny, D. (2004). Strategies for developing a high-skilled workforce. *Performance Improvement Quarterly, 17*(1), 41–55.

Spil, T., Effing, R., & Kwast, J. (2017). *Smart city participation: Dream or reality? A comparison of participatory strategies from Hamburg*. Berlin & Enschede. https://doi.org/10.1007/978-3-319-68557-1_12

Talari, S., Shafie-khah, M., Siano, P., Loia, V., Tommasetti, A., & Catalão, J. P. (2017). A review of smart cities based on the Internet of Things concept. *Energies, 10*(4), 421. https://doi.org/10.3390/en10040421

Toli, A. M., & Murtagh, N. (2020). The concept of sustainability in smart city definitions. *Frontiers in Built Environment, 6*. https://doi.org/10.3389/fbuil.2020.00077

Tong, Z., Feng, Y., Yan, M., Liu, H., & Basodi, S. (2021). A survey on algorithms for intelligent computing and smart city applications. *Big Data Mining and Analytics, 4*(3), 155–172. https://doi.org/10.26599/bdma.2020.9020029

Triyanto, D., Warsono, H., & Herawati, A. (2022). A study collaborative governance in Bengkulu city, Indonesia. *Journal of Madani Society, 1*(1), 59–63. https://doi.org/10.56225/jmsc.v1i1.128

Wahyuni, S., Alwi, A., & Indar, N. (2022). Implementation strategy smart governance in Makassar city (case study on the missing capil program for the population and civil record). https://doi.org/10.2991/assehr.k.220101.010

Chapter 5

Democratization and Transformation of Higher Education in the Era of Smart Cities

Kevin A. Jones[a] *and Ravi S. Sharma*[b]

[a]*Singapore Institute of Technology, Republic of Singapore*
[b]*Zayed University, United Arab Emirates*

Abstract

This chapter is a retrospective commentary on the efficacy of teaching and learning in a higher education space that embraces the incredible diversity of delivery modes available in the post-Covid-19 era of "Smart Cities." The current reality of widespread and leading-edge experimentation with online learning necessitates that existing brick-and-mortar institutions reimagine their places as providers of higher education in this new age of digital disruptions that will resonate with all stakeholders a future of endless possibilities. The authors, with four decades between them of practice and field research at leading universities and colleges in Australia, Canada, New Zealand, Singapore, and UAE, advocate an approach to higher education that is personalized for learning effectiveness, industrial operations, and institutional evolution; that is, a higher education that is democratized. They warn that the wasted opportunities of meaningful digital transformation pre-Covid-19 have led to an urgency of transformation at the present time. While randomized control trials continue to be the "elephant in the room"; scholars, leaders, technocrats, and regulators must drive the quest for the growth and relevance of a diversified and learner-driven higher education in the years ahead. The platform of a "smart city" may just be the catalyst for such a radical innovation.

Keywords: Higher Education 4.0; online learning; democratization; personalization; MOOCs; smart universities

The Emerald Handbook of Smart Cities in the Gulf Region:
Innovation, Development, Transformation, and Prosperity for Vision 2040, 69–97

doi:10.1108/978-1-83608-292-720241004

Introductory Reflection and Chapter's Trajectory

Life is the sum of its experiences, but the same cannot be said of a person's mind. It grows as a consequence of understanding life's events, which more often than not is uncertain at best. Consider this all-too-common situation: when things happen around us, there seems to be no making sense of some of them, ... no forming a connection between the recognition of the event and the perception of what it means, ... or even, no realization that it has a significance that if known would meaningfully add to one's current understanding of some as yet incomplete idea. Nevertheless, with the slow and persistent piecing together of glimpses of clarity interspersed with moments of serendipity, one can eventually create a blessed nugget of pure original thought that suddenly changes everything for that person. For one author, in relation to the vastness of education, this has happened twice. One of these, we will be talking about in this chapter. It is the democratization of higher education.

We will take on this task of elucidating all the major aspects of this proposition to you, dear reader, as follows.

First, we will set the stage by laying out all the terminology applied in the proposition. This will comprise the concepts, factors, and practices that are relevant to a deep and holistic understanding of the proposition. To that end, we need to dispel what we believe are misunderstandings of these concepts, factors, and practices. Additionally, we'll provide suitable references for follow-up by the reader at their discretion.

With the terminology suitably elucidated, we will introduce our proposition. We will clarify that democratization is about not politics. It is about an individual choosing which learning journey he/she will take, and industry plus society recognizing the legitimacy of said journey by virtue of its accomplishment and not its fits into an entrenched and overtly narrow bureaucratic system.

Now that you understand what the proposition is, it seems only fitting that we step back in time to the take-in the history of the precedents that lend credence to the proposition. And, if by now in the chapter the reader has not yet been able to guess what said precedents are and what their positive impact on past societies was, then, this part of the chapter will be especially, dare we say, ... fun! Please, do look up some of the references, and do further research in this area. It is really fascinating what you can and will find with a modicum of tenacity.

Returning to the present, we want to highlight a prevalent factor about online education's acceptance from students that we knew was probably true, but nevertheless disillusioning when it was verified. After all, this is the very condition that our proposition depends on, it being the fact that a proportion of students have a deep-seated need, borne from their enthusiastic inculcation of that siren call, to follow a new and yet tacitly unsanctioned path of doing their higher learning online without lack of favor of the physical institutions in place.

From there, we will analyze how the higher education space was turned on its head by Covid-19.[1] It cannot be emphasized enough how societies' remedies to the virus revealed deep fissures hidden below the surface in the physical-institution-based system. Ergo, for the authors, it is these fissures that are an impetus for enacting a "sustaining disruption" upon the norms and expectations of the workings of the higher education space. Here we will talk briefly about the elephant in the room, that is randomized control trials (RCTs).

Next, we will review the present, that is post-Covid-19, state of usage in the world of online learning. Although the online offered by some higher education institutions has settled back down to its pre-Covid-19 levels, the trend is that innovation in online usage is on the rise; perhaps not dramatic, but certainly continual.

Finally, we will wrap our chapter up with a brief outline of some areas of research that could enhance our understanding and confidence in resolving the known contentious items concerning the implementation of online education as well as uncovering some unknown consequences before they are encountered as unexpected obstacles after implementations have been started.

So, without further ado, let's dive into the content of this chapter.

Establishing the Context for the Proposition

In this section, we will clarify and build a common understanding of the terminology that is relevant to our proposition. We will begin our elucidation of the terminology with two terms that we consider the bedrock of the chapter, namely education and smart cities. The first term dictates how the proposition will be applied, and the second sets the context for the proposition. Following the bedrock terms, we will explain the remaining terms in this order: higher education, online, online learning, online delivery of education, and online education. So, let's have at it.

Education

Education is one of those terms that nearly everyone thinks they intrinsically know its meaning, and hence, have established their personal definitions for it. After reviewing 20 unique definitions, the authors realized that trying to retrofit their take on the term was futile. Instead, we opted to reveal our own definition, one that we reckon is simple, professional, centered enough that it is agreeable for most readers, and adaptive enough to be consistent with smart cities. Our

---

[1]Is it COVID-19 or Covid-19? Per the rights of discovery, the World Health Organization writes the disease and the virus that causes it in all caps. On the other hand, media ruled it a proper noun that is spoken as a word and so writes it in lowercase except for the first letter being capitalized (English Language & Usage, 2020). We employ the latter as it seems to have widely acceptance now with a reduced prominence within the text.

definition of education is, *a system transcending age, gender, race, and creed, that manages access to and delivery of a specific curriculum*[2] *of interest to persons suitably motivated to acquire it all via the process of learning*. This means it is man-made and intended for use by humans only. Its size and scope are completely open to any and all possibilities. The system must not pick and choose who can seek access, nor dictate what of the available knowledge and skills each person can learn once access is granted. Finally, the system cannot force the transfer of knowledge and skill; the system can only make the offer, and it is up to the learner to enact the transfer to whatever degree: success, failure, complete, partial, long-lasting or short-lived.

There are three universally-recognized levels of education, namely primary, secondary, and tertiary. These are delineated by their curriculums which are modeled against a fairly subjective interpretation of the tenets of Piaget's theory of cognitive development (Lefa, 2014). All levels of education are comprised of physical or brick-and-mortar institutions that practice mostly traditional teaching and learning: the physical presence of the students and teacher in a physical classroom.[3] The first two education levels are mandatory in most countries, while the third is optional. We will revisit the tertiary level later; let's move on.

Education organizes and operates according to two properties, namely in-charge and qualification. "In-charge" specifies the party that sets the goals, plans, executes, reviews, and remedies the whole of the educational program. "Qualification" specifies the award or not of an official accreditation for the successful (adjudicated) completion of the educational program. Each property has two mutually exclusive states. For in-charge, they are "institution/teacher" and "learner." Institution/teacher means the holder of the structure of rules and norms that shape and constrain a person's learning journey; the contrast to this is that the learner shapes and constrains his/her own learning journey. For qualification, "formal" means an official qualification is awarded (a) the education has been successfully completed, and (b) the educating party is accredited; "informal" means there is no official qualification. The four states can be plotted on a simple four-quadrant chart, with the in-charge property on one axis and the qualification on the other (see Fig. 1).

Quadrant 1 represents the education provided by brick-and-mortar institutions that are accredited. The learners become registered participants designated one and all as students, and are grouped into cohorts. These students select their program of study from a suite of prescribed ones, and are awarded an official qualification when that program is completed. Also represented in this quadrant are programs of study that include correspondence courses, eLearning, part-time

[2]The totality of the student experience in education, based on some expansive suite of knowledge and skills.

[3]It is euphemistically known as the "sage on the stage" because the delivery is one way, teacher to student. Over the past century, there has been a gradual shift to incorporate other approaches such as student-centric, self-directed, and interactive. For a detailed and expansive definition, see Jones and Sharma (2021).

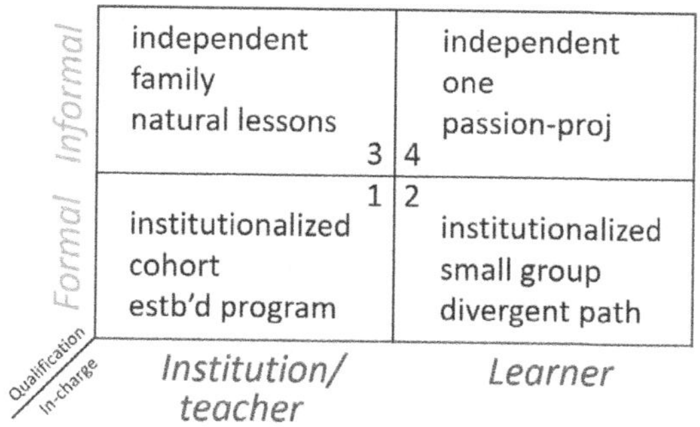

Fig. 1. Four-Quadrant Chart for In-Charge and Qualification Properties.

courses, and home schooling, so long as these are sanctioned by an accredited institution. On the whole, this quadrant represents the education that has been in operation for millennia, and applies to the vast majority of persons who live in countries with sufficient funding for an inclusive education.

Quadrant 2 represents the education provided by the accredited brick-and-mortar institutions, but that authorize special requests from a small group of students in a cohort to follow a selectively personalized education that is clearly understood to be an exception and not belonging to the standard. For example, taking a double degree, post-graduate degree, extended education courses, part-time courses for additional credit. So, while these students still belong to an institution because they desire the official qualification, nevertheless, they embark on a learning journey that they shape and constrain. This is what makes this quadrant learner in-charge. While such divergent programs of study have existed for long time, they are generally not popular as they entail an onerous amount of effort – in essence it is additional to the current workload of the learner – that few are willing to undertake.

Quadrant 3 represents the education whose providers are not accredited and so there is no award of any official qualification. This includes many commercial educational service providers that provide "certificates of completion" which are not equivalent to degrees or diplomas. It also includes seniors, mentors, or role-models of the learner who engage in unsolicited sharing in a family or casual setting, of their experiences or thoughts on anything that might be useful in growing up or handling some of life's more difficult aspects. In all cases, the learner is not taking a learning journey that is personalized. Sure, the life-lessons are meant for you, but they are generally ad hoc and often or not either too late or irrelevant for the journey that the learner is planning. It is the same with the unaccredited commercial providers. They fit an immediate or emergency need for upgrading, but they too are more ad hoc than part of a meaningful learning journey. So, while the education in this quadrant has been experienced by many, it definitely wouldn't be most learners' first choice.

Quadrant 4 represents the totally-personalized education, one in which the curriculum is completely conceived by the learner, and the subsequent learning journey is completed to the will of the learner. Unfortunately, there are so many barriers to this education takes it is rarely undertaken. The lack of an official qualification is a huge disincentive, as is the courage needed to embark on such a risky journey. The nature of personalization dictates that the curriculum would always be a one-off original that could never work for anyone else except the creator. In fact, it will likely be unworkable if the creator were to repeat it at a later period in his/her life. Nevertheless, there are learners that have the boundless confidence and drive to succeed, such as entrepreneurs, technopreneurs, captains of industry, and eminent leaders, and this is the education that they would undertake.

Smart Cities

Now that a common – and hopefully to the majority of the readership, suitable – understanding of education has been established, let's apply it to the concept of smart city. In other words, what are the implications for education when it is put into the context of smart cities, or, how would the tenets of smart cities shape the features and functions of education. "Smart cities" are *technologically modern urban areas that are governed at a distance via innovational types of computational urban governance, including monitoring and manipulating their citizens as data objects*. Essentially, the smart city has a massive quantity of sensors placed in it, plus the skills and knowledge needed to proficiently and thoroughly manage it. By necessity, such cities will be populated by "smart citizens" who have the technical algorithmic and data-mining skills to participate in "computational urbanism," that is the coding of programmable urban processes, also known as "civic coding." Hence, smart cities will need an education system that (1) exposes its students to innovative analytics that enable using big data to anticipate and shape the city's future, and (2) inculcates the citizenry with computational thinking fundamental to learning how to govern the smart city through civic coding (Williamson, 2015; Wolff et al., 2015).

The writing is on the wall for all to see: delivery of education figures prominently in the transformation of current urban areas into smart cities. It is because the "smart learning" settings needed for smart cities to flourish will be real-time, ubiquitous spaces that detect the real-world learning status of students and their environmental contexts. In the minds of the smart city planners, smart learning will require education that uses **both** physical and online delivery methods and activities. Already, there are learning labs operated remotely entirely using IoT processes supported by mobile wireless and sensing technologies, that are poised to enable smart learning (Jones & Sharma, 2020, May 5; Nikolov et al., 2016). Plus, one of several well-documented strengths of online delivery of education, is the rapidness whereby learning material can be enhanced and/or updated to stay current and relevant (Cooke, 2022; Emma, n.d.; Olmstead, 2023; Pacific College, 2022; Wilson, 2023); a vital trait for the furtherance of computational urbanism. So, by its very nature, online delivery of education figures into the concept of smart cities at least as prominently as the traditional if not more. A point of clarification: the term "education" is specifically used in the smart city references, so we believe that aspects of computational urbanism and civic coding will be spread across all three levels of education, from very small and simple in the first level, and increasing in amounts and complexities when progressing through the subsequent levels.

Higher Education

So far, so good. Now, let's explain some other terms that relate to our proposition. Doing so is a bit of a reveal of the proposition, but it's always best to build the base first before the superstructure. Our first reveal is that the proposition is exclusive to higher education, so we need to be clear about what that term is referring to. To do so, we return to the tertiary level of education. That level is comprised of two groups of institutions distinguished by a single characteristic, namely the awarding of a degree; one group awards degrees, while the other awards diplomas.[4] Those institutions *with the vested authority to confer degrees* upon their graduates are collectively designated as "higher education," and include mostly universities, colleges, and some polytechnics. The remaining tertiary-level institutions, which include some polytechnics and institutes of technology, academies, seminaries, conservatories, vocational schools, trade schools, and further or continuing schools, cannot confer degrees (Higher Education, 2002). Of note in the discussion about the proposition, is that in the past decade or so a few higher education institutions have enacted eLearning plus online delivery of some of their courses, and in very tiny numbers for very specific disciplines, have they deigned to award degrees for programs they own and whose courses are delivered entirely online.

Online, Online Learning, Online Delivery, and Online Education

Our second reveal is that the proposition involves being online and how it is paired with education.

Universally, "online" means that *an electronic device is powered up and connected by the* **Internet** *to other electronic devices that are also powered up*. The Internet is key to the definition. Connectivity in original definition of online was more inclusive, that is to any network. But, with the usage of the Internet transcending all other networks – just under two-thirds of the Earth's human population are now using the Internet – the semantics of being online shifted to its present form (Moore et al., 2011; Rouse, 2012). The definition stops short of dictating the kind of work being conducted across the connection. The opposite

---

[4]Degree and diploma are official qualifications that only accredited education providers can award. The degree curriculum contains more courses, a lot of which are multidisciplinary, and takes more years to complete than a diploma. There is also a tier system for degrees, that is bachelor's, master's, associate's, and doctoral; the diploma has less, graduate and postgraduate. Of the two, a degree is widely considered by the more valuable of the two.

of online is "offline," meaning that the electronic device is disconnected from the Internet, though it might still be powered up. By extension, the phrases online learning, online delivery of education, and online education are all associated with a connection to the Internet.

"Online learning" is any activity that the electronic device operator conducts within his/her Internet connection such that the condition of learning is achieved.[5] For your edification, the term "eLearning" may come to mind as a synonym for online learning, but this is a misconstruction. Recall that the original semantics of online included all networks, but had shifted to include only one, namely the Internet. Connectivity to all networks is encompassed in the definition of eLearning. eLearning is accessing educational curriculum outside of a traditional classroom, through either any network or portable storage devices like CD-ROM or video tape, to achieve learning (Bonk & Graham, 2005); typically, the learner works alone in an eLearning session (Uddin, 2023). In contrast, online learning doesn't specify the kind of work done or how many learners are involved in the connection. Ergo, the two terms are similar but not identical.

"Online delivery of education" refers to the use of the Internet to conduct any and all activities according to the curriculums of any of the three levels of education; naturally, these activities involve students who are properly registered with those institutions.

Unlike the other phrases, we have appropriated "online education" especially for use in this chapter to refer to the use of the Internet to conduct any and all activities according to the curriculums of **higher** education only. Doing so will increase the readability of the chapter by substantially reducing the requirement to qualify for every usage of online education that pertains to higher education.

Statement of the Proposition

Essentially a derivation of the chapter's title, our proposition is as follows

> Higher education must be democratized to meet the unique challenges posed by smart cities.

One might be tempted to think that the proposition is about politics and voting. No, it has nothing to do with voting, but it is about **choice**, … specifically the choice between implementations of education that are **equal in stature and availability**. The education implementations we envision are modeled after the smart city's education requirement of both physical and online. Naturally, the physical corresponds nicely to the "traditional" learning journey in which the whole of the

[5]There are multiple definitions of learning, but we prefer the following from Ormrod (2011). Learning is the processing and inculcation by a person of hitherto unrealized and possibly unknown knowledge, such that the person has gained the new-found ability to function in his/her environment either thinking- or behaviour-wise. For a detailed and expansive definition, see Jones and Sharma (2021).

learning is directed by the institution and teacher in sync. As for the online, some institutions already engage in eLearning; some even deign to give a few credits for online courses. We believe this is not enough. If online is to be truly equal in stature and availability, then we must accept that the institution cannot be dictating a standard curriculum to cater to the explosion of non-standard knowledge and skills required now and in the future. In short, a choice of the online option must fully entail the necessity of the student establishing his/her own curriculum. Therefore, the choice that our vision of democratization intentions is between the education implementation designated as "traditional," a.k.a., physical, where the curriculum is established by the institution in conjunction with the teacher, and an "innovative and highly adaptive" one, a.k.a., online, where the curriculum is established by the student. If it helps in understanding, we can re-phrase the two implementations as follows: institution/teacher-directed and learner-directed. Crucially, both education implementations would **co-exist** as co-mainstream paths in higher education, plus, upon completion of their respective pre-established academic activities to some measurable pre-agreed standard, each would result in the student being conferred with a **nationally-recognized degree**.

Pretty radical, right? Actually, not at all! It is probably the most common-sense idea given all the relevant factors. If smart cities become a reality, then online education would certainly be consistent with the projected education needs for "constant improvements in and re-evaluation of the curriculum taught to the learners have to be done regularly to keep the learners up-to-date in fulfilling the requirements of these industries and corporations" (Nikolov et al., 2016, p. 1). There are already partnerships between industry, universities, and online course developers, for example About Udemy Business (n.d.). This could rapidly lead to whole industries being built up for conducting the research and development of new and better online learning micro-module stacks. It could also lead to new revenue streams for universities that can no longer bring in enough students to pay the costs of staying open, for example, as online content creators or as online learning assessors. Universities whose main revenue is from the assessment of another institution's students are not a new idea.[6] If there is one thing that governing bodies are really good at, it is developing and promulgating standards. Without the slightest doubt, such bodies are capable of designing airtight standards for adjudicating the completion of online academic activities against nationally recognized degrees. Is assessing the large number of students participating in online education prohibitive? Well, it's being done right now with large-scale online learning platforms such as Massive Open Online Courses (MOOCs) and Khan Academy. Plus, there are promising innovative technologies on the horizon in bio-medical and computing fields that will transform assessing a person's learning, presently done by experienced teachers making subjective rulings using rubrics, into an automatic and objective process that can be undertaken by lab

[6]University of London began in 1836 as a purely examining and degree awarding body, to which approved institutions could submit students for examination (University of London, n.d.).

assistants (Swiecki et al., 2022). With all the tools at our proverbial fingertips, it's puzzling that our proposition hasn't been muted earlier.

Try this brief thought experiment. Imagine that online education has been running for many years alongside the physical classrooms as a co-mainstream mode of life-long learning. Its academic, business, and technical machinery would run like clockwork. It would be arranged into levels of competency and reputation, from "ivy-league" to professional to community, their programs could be updated every six months to keep pace with the pace of industrial and societal innovation, and the portfolios of qualifications earned by their graduates would be as acceptable as if not more than those from any physical institution. Learners would choose what, where, and when they would take the online programs; their degree of motivation would be palpable. Being able to personalize and inject real purpose into their learning, learners would ultimately be as close to achieving Sir Ken Robinson's vision of "escaping from education's Death Valley" (Sir Ken Robinson, n.d.) as any system we could possibly devise.

The last point to make about our proposition is, to confirm in which quadrant of the In-charge and Qualification chart (Fig. 1) that the innovative and highly adaptive implementation would fit. First and foremost, we reiterate that the innovative and highly adaptive implementation must result in a recognized degree. Clearly, this means that the appropriate property state is, formal. It is also indicated that the student would be establishing their own curriculum, so the property state is, the learner. Thus, the quadrant in the property chart that the innovative and highly adaptive implementation fits is at the intersection of formal and learner, namely quadrant 2.

Review of a Legacy of Success: Distance Education

At its core, online education is teaching and learning conducted 100% outside of the physical classroom, and instead, inside virtual "rooms" on the Internet or other such computer networks. Typically, the virtual room is either a webpage or an application programming interface. Getting online is achieved via a computer, worldwide web software, and a broadband connection. Clearly, online education is inexorably bound to recent technologies in computing and communications. Does this mean that education outside of the physical classroom has never occurred before the inception of this technology?

Educational Correspondence

In fact, it does not. The earliest state-sanctioned occurrence of education outside of the physical classroom was "distance education." With the technology of long-distance postal systems becoming reliable and affordable in the early 1700s, there came "correspondence colleges" that offered education programs such as shorthand to less affluent students who did not have the means to attend the brick-and-mortar educational institutions. Enterprising teachers advertised in newspapers their offers of education, and upon receiving positive responses, thereafter engaged their students in half-duplex discourse via postcards (Kentnor, 2015;

Wheeler, 2016). In 1858, the University of London became the first established university to offer degree programs by parcel post. Learning materials used in existing courses were re-packaged for delivery to registrants, followed by invigilated examinations conducted at various national centers in England. Soon after, examination centers were established internationally, starting with Commonwealth nations, and then expanding to some of the remaining nations (Distance Education, 2001).

Educational Radio

The invention of the radio in 1894 heralded the next version of distance education: "educational broadcasting." It started as a process of transmitting learning material to schools and colleges for continuing education. But it naturally grew to serve individuals learning at home. Of course, two-way interaction between the learner and teacher was mostly not possible, even with transceivers at both ends, because the broadcaster was often not the subject matter expert. Additionally, broadcasting revealed that the existing teaching practice was not effective for that mode of education. It was observed that "the first years of university broadcasting was generally ineffective because many a professor repeated his classroom lecture before the microphone without realizing that a good lecturer was not necessarily an effective broadcaster" (Hokanson & Hooper, 2000, p. 542). Also, it was realized that

> mere "listening-in" may be a profitable expenditure of time in connection with an occasional program; but the more significant and more enduring benefits can come, if contemporary educational theory teaches us the truth, only when the learner is inspired to some effort on his own. (Bagley, 1930, p. 256)

These lessons learned inspired a spate of innovative learning theories pertaining to educational instruction that exists to this day (Haworth & Hopkins, 2009).

Educational Television

Television, a.k.a., TV, existed in experimental form by the late 1920s and was marketed to consumers a few years later (Television, 2001). Accordingly, educational broadcasting began to shift away from radio to TV. The earliest known televised courses – oral hygiene and star constellations – were created by the University of Iowa in 1933–1934 (Sleator, 2010). However, educational broadcasting on TV would not gain traction until the development in the late 1940s of an affordable, moderately sized, home TV. The first televised college classes were offered by the University of Houston in 1953. These classes comprised a series of 13 five-hour segments that were aired weekly in the evening to make them accessible to learners who worked during the day (Visual Academy. 2014). While TV was exciting, convenient, and immediate, there was no hiding the non-interaction of the medium due to its innate complexity and the dearth of duplex communication channels in the homes of most learners. Initially, a parcel post between the content creator and the student provided a degree of interaction. It was later replaced by the telephone making the interactivity near real-time. In an identical format, one author commenced his teaching career at the University of Regina in the mid-1980s delivering distance education over the cable TV network.

Textbooks

Textbooks are a powerful medium for storing information usable in education curriculums. Some of the earliest textbooks date back to the ancient Greeks in the 12th to 9th centuries BC. There were at least two major obstacles preventing the widespread use of textbooks. The first is that they were extremely tedious to produce due to being handwritten, so typically, there were very few copies of a textbook produced. If textbooks were going to be used in education, for the longest time it was a case of the owner lending his/her own copy. Second, most textbooks before the 1700s were written in Latin, which was not the language spoken by most of the population. So, even if the textbook was sufficiently available, it couldn't be read by the populous. The first shortcoming was eliminated with Johannes Gutenberg's invention of the changeable type printing press in 1455; multiple copies of a book could now be produced. However, textbooks were probably near the bottom of the list of books to be printed first. It was only when education began to be compulsory in countries, starting in the mid-1700s, did textbooks began to appear in significant numbers. Coincidently, compulsory education provided the impetus to eliminate the second problem of language of the text. The adoption of compulsory has been a process drawn-out to today; even now, there are still four countries that continue to holdout.

From the 19th century to today, textbooks have been a mainstay in education curriculums globally. They organize knowledge, facilitate teaching and learning of that knowledge, and safeguard it in a static but permanent form (Bierman, 2006; Textbook, 2003). One would expect that textbooks would figure prominently in distance education, but they didn't due to their high costs and copyright issues. This has been resolved recently, around 2010, with the introduction of the "open license," a regulation that can make textbooks available online for free use by students, teachers, and members of the public. Examples of such licenses are Creative Commons and GNU Free Document.

Portable Storage

Another variation of distance learning is the use of portable storage media in analog or digital formats to distribute educational material that was contained therein. Obviously, this approach is not that spectacularly different from the educational correspondence of three centuries ago; in fact, the media being physical necessitated their distribution using parcel posts. Nevertheless, it quickly became popular for several reasons. First, it greatly increase the speed of producing and upgrading the distance education curriculum, second, it would reduce the archival footprint while increasing the information holding capacity, and third, it would last longer. These expectations would not be met initially, but eventually, they would be exceeded immensely.

The earliest media was analog format on magnetic tape in 7-inch reels. Although developed in 1928, it would come into popular use until the 1950s for recording educational radio. However, it was prone to disintegration after around 10 years. An improvement on the reel magnetic tape was the Compact Cassette, that went into production in 1963. Its hard protective shell and greater storage capacity made it the better and lower-cost choice for distance education. The last storage media using magnetic tape was the Videocassette which went into production in 1982. For the first time, images and moving pictures could be included in the curriculum. The learner could now see as well as hear the teacher. For all these media, the interaction with the teacher was still minimal, basically one-way. Requests for more tapes, or submission of work for feedback by post.

As the sun was setting on magnetic tape, it was rising on the new technology of floppy disks. By the early 1970s, digital format floppy disks were available on the market for distance education. Lighter, with more storage capacity, and not prone to disintegration, this media became an immediate success for distance education, so long as you had the computer to read it. This was followed up by the digital compact disk produced in 1982, versions of which could hold the insane 700–800 MB of data. Now, the media could be fully interactive, with built-in assessments, learning aids, teacher demonstrations, and tons of references. As fantastic as was, even this media has been eclipsed by the USB flash drive, or thumb drive as it's commonly known, produced in 2000 with its mindboggling holding capacity of terabytes of data while being the size of an eraser. Finally, there is no longer a need for distance education to even use these portage storage media, as the Internet and world-wide web have elevated distance education to an instantaneous, totally interactive system that is as real-time as the operation of physical classrooms.

Older Forms of Distance Education Are Still Going Strong

We have seen how distance education has evolved over three hundred years, how one form was superseded by the next. But there is one last aspect of the story that we have not yet revealed, and that is what happened to the forms that were superseded. Probably, you think that the new forms needed to be utilized in distance education because the older forms were failures that garnered little participation over their periods of service? Well, if you are thinking that, then we must correct your thinking. Not only were they successful for their day, but they paved the way for the next technology that followed. Even more astounding is that forms of them are still in operation today.

Several universities worldwide still offer distance education by post, including the University of London and the Open University. They do so because there are people in the Commonwealth and other organizations that that don't have Internet connectivity, or can't afford a computer and broadband connection into their home; for them, correspondence courses are a lifeline to reputable education (Hall, 1996). Educational radio broadcasts are still being made in the hinterlands of Canada and Australia. Radio in education, a pioneer of educational technology, created a legacy for itself by setting the stage for the development of other technologies still used in education to date and perhaps will continue with other technologies in the future. "Radio still captures the imagination, too. As a child once said, he preferred radio over television because 'the pictures are better'" (Haworth & Hopkins, 2009, p. 8). Japan, Britain, France, Italy, Sweden, Germany, Netherlands, Brazil, Latin America, and most of the countries in Africa, all operate educational radio to combat illiteracy and provide advice on public health and agriculture innovations (Camacho, n.d.).

Considerable educational television broadcasts continue to be made around the globe. The biggest in the United States, namely the Public Broadcasting Service has even increased the amount of educational and thought-provoking material it airs. A pioneer in educational television in Europe, the BBC offers primary and secondary students upwards of 40 television series. One such series is analyzed in Koumi's book *Designing Video and Multimedia for Open and Flexible Learning* (2006). It also offers a variety of biweekly programs supplemented by publications and correspondence for degree courses with the Open University. NHK from Japan airs programs for primary, secondary, and higher education, plus "social education," that reach more than 90% of Japan's primary and secondary schools (Camacho, n.d.). Finally, educational broadcasts are assisting relief efforts to increase the quality of life of residents in Tanzania (Dearden et al., 2023).

Online Continues the Success of Its Predecessors

A similar success is evident in their descendant, online education, only it is to a greater magnitude. Here are some key statistics.

> In 2019, student enrolment into online learning was 7.4 million, the global online learning market was $188 billion (400% increase over the amount just six years ago), and the global educational technology investment was US$19 billion. In 2020, 5.4 million college students (74%) took at least one class online, and 90% of college students rated their online education about the same and better than the education in their physical classroom. From 2011 to 2021, the number of learners taking massive open online courses (MOOCs) increased from 300,000 to 220 million.

How does online education measure up in academic performance with that in the physical classroom? It is an overachiever that shows no sign of declining any time soon if at all. The majority of two decades of research into online education has found that online is nominally if not substantially better than the physical classroom. Even one author's research into online education – a five-year longitudinal study of engineering students in a highly rated national university's degree program – yielded the same finding: online can render better academic performance in the higher education setting. More telling is research has found that students in online education require far less time to learn and retain far more of the learned content than in physical classrooms.

After considering all of its successes, historical, business-wise, and academic, we believe we can safely conclude that the implementations by established and experienced providers of online education, are well planned, high quality, and highly effective. If they weren't, distance education and educational broadcasting would never have survived to this day. Also, the numbers for online education would paint an entirely different picture, one of decreasing: enrollments, satisfactions, investments, and markets share. Keep in mind that all three are not state mandated; if they weren't reliable and effective, then very few dedicated learners would volunteer their time to participate.

Personal Experience with Distance Education

The following is a personal recollection by one of the authors that outlines the experiences in his life that involved distance education. It is quite feasible that you, dear reader, will also recall similar experiences. The authors' intentions are to lead the reader to the realization that distance education a constant and constructive companion over the past 60 years, and there is no reason we can think of why its descendent, online education, would be any different.

> I have vivid memories of many experiences in my past that were variations of distance education, but that I did not at the time perceive them as such. The following is a short list of these experiences. A public service message on TV about "stoopy-doops" (in 1961; Keith Milford, n.d.), mail-order kits for building electronic devices at home (in 1970; Popular Mechanics, 2007), televised first steps on the moon (20 July 1969), video teleconferencing during war games in Germany (1989), four-player "team battles" over serial cable links setup in our building (in 1992; Multiplayer, 2005), televised collapse of the twin towers (11 September 2001) and my consequent search for the design flaws that were to blame, VHS tapes in our multi-media library that I found and used in my classes (circa 2005), remote connection for outstation info exchange, completing eLearning training for my university's qualification policies (circa 2010), and YouTube. It was only with the emergence of MOOCs that I finally able to recognize all the instances I mentioned, and more, as experiences I had not then perceived as variants of distance education. Having internalized the new concept, I was sufficiently curious about the possibilities of online learning, that I created two eLearning courses that were broadcast to my students over some ten semesters via the university's Blackboard infrastructure. One was a series of home-made videos on the five major diagrams belong to the Unified Modeling Language specification. The second was a 13-week blended learning course in Software Engineering that was 30% online and 60%

> physical, but no lectures (Jones & Sharma, 2021). I know both of these were successful because my student feedback was still far above most of school's faculty. After I retired from full time employment and took up part-time teaching with local universities, I suddenly was able to have a much more open exchange with the students concerning their expectations and aspirations in their higher education. It was during this late period in my life that the revelation of online education serendipitously came to me.

So Compelling Yet So Unready

In the previous section, we made the case for the feasibility of online education given its historical precedents. So much, in fact, that there are remnants of centuries-old distance education and educational broadcasting still in operation today, and the business and participation aspects of online education are growing at a phenomenal pace. Suitably enthused, we wanted to know if undergraduate students understood the potential enhancements to their education that an online one would offer. So, we tried a thought experiment, to ask the cohort of final-year undergraduate students that one of the authors was currently teaching, to critique the journal article "A constructivist approach to online college learning" (Rovai, 2004). Largely and unmistakably, they indicated incredulity that a learning session belonging to a program offered by a formal education institution can be effectively conducted using an online instructional design. We expected that this might be the outcome. Why? We have seen it among members of the formal education community, that is, students and faculty, as well as in segments of society that have some stake in formal education, for example., politicians and parents; it seems widespread and deeply ingrained.

As we reflect upon this condition, the question is not just "why?" The answer to that question is, because for the past hundreds of years, learning has always been in a physical classroom, and it's worked just fine from the perspective of the powers that be; in other words, if it ain't broke then don't fix it. Well, if one ignores the inexorable march of technological innovation in almost every other human endeavor, then yes, there is a certain logic to that supposition. Yet, even if that were so, it doesn't take much digging to discover a glaring conundrum that once seen, cannot be erased from the zeitgeist: why do the students exhibit that?

Proceeding with the thought experiment, not only are the vast majority of students extremely unmotivated to learn while in the formal education system, but they also spend the greatest proportion of their lives online and a lot of that time learning. Keep in mind as we proceed with this argument, that formal education systems of today are operated as mass-production factories that produce graduates. It is an unarguable tenet of mass-production systems that their outputs be majorly correct and according to specification; quality control and strict adherence to quality standards are mandatory aspects of any mass-production. Thus, if a mass-production system's output is majorly not correct, then that system is considered broken and in need of a complete overhaul; if it can't be overhauled, then it is shut down and abandoned.

To the first point, the best that can be said about the students' attitudes toward formal learning opportunities is that they tolerate them. Their attendance at learning sessions is minimal. Even when attendance is declared to be mandatory, students can and do always find the barest of excuses to absent themselves. For those who do attend, their attention is upon matters that are more than not antithetical to the material being taught, such as catching up on gossip, exchanging notes about another subject, or arranging an outing for the evening. Further to this, students do not what is necessary to learn the material, meaning they do not review and/or revise the material from their daily sessions. Instead, they cram just before the final exam, and once the exam is over, they make no effort to retain any of that knowledge for use in whatever follows. If there is anything that can be pointed to as being the source of motivation for the students, it is exclusively their grades. From our observations, this is the case across the globe and disciplines. The achievement that the majority of students aim for is the highest possible grade; the highest degree of learning does not yet enter into their educational objectives.

To the second point, most academics will admit to the veracity of the notion that the majority of people alive today spend the vast part of their lives online. Consider that the world population is expected to reach eight billion by the year 2023. According to a broad global survey by datareportal (Kemp, 2023), all households worldwide will have at least one standard desktop computer at home, the number of mobile phone users will be 7.26 billion, and the number of Internet daily users will be 5.3 billion. Although owning a computer does not necessitate being "online" per se, owning a mobile phone certainly does; the key enabler of being online is a broadband connection. When household and personal computing devices started to be mainstream and affordable, learning researchers identified that the people online were primarily young. But now, online users are not distinguished by age, only by financial means, a condition called the "digital divide." Does this pervasive access mean that all of the time spent online is in the pursuit of learning? No, but humans are "learning machines" by nature. The brains of infants are "designed" to be thirsty for new knowledge and skills. Thus, setting brain abnormalities aside, all people can learn anywhere, anytime, anyhow; it can be deliberate, but it can also be ad hoc and automatic. Whether the new knowledge is the latest technological breakthrough, science discovery, specification of every warplane in World War One, or even the latest scandal of a favorite celebrity, it's all learning. So, yes, it is fair to assert that the majority of time spent online can be utilized for learning.

With the above suppositions now broadly and clearly articulated, the conundrum is vividly evident. Students know at some level, even if it is just intuitively, that they can and do learn online; it is impossible to believe otherwise. So, why then do they almost universally harbor such incredulity at the thought of practicing online learning in higher education? After all, online learning is entirely consistent with what the late Steve Jobs envisaged as the digital lifestyle and bet his company on such a transformation (Fig. 2). How could they be so unready?

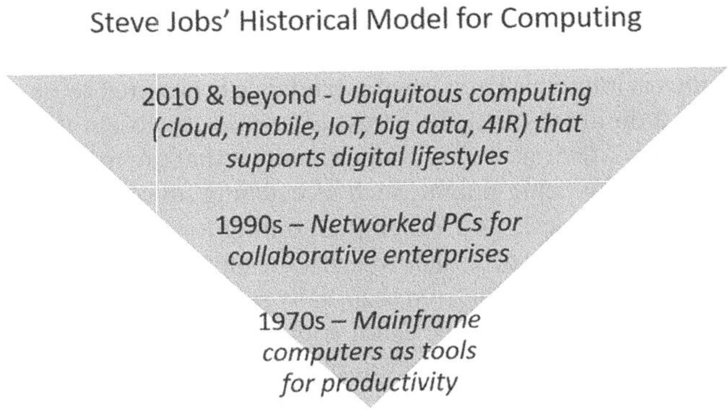

Fig. 2. Steve Jobs' Historical Model for Transformation to Digital Lifestyles.

We believe it is the occurrences during the Covid-19 lockdown period that are the major cause of the students' almost unanimous negativity concerning online education.

Online Education Upset the Status Quo of Physical Institutions

In this section, we are not going to give a chronological breakdown of the events. Instead, we will give our personal retrospective and arm-chair analysis of our experiences and how were affected.

We begin by stating as forthrightly as possible that the application of online education was an emergency activity necessitated by Covid-19; there was nothing normal about it. Starting about the second month of 2020 and lasting for about 12 months, we witnessed the unprecedented lockdown of many brick-and-mortar institutions, and the replacement of physical face-to-face learning encounters with online learning in what has become the largest live experiment of a technology platform in human history.

It was an unmitigated disaster. Not only was it poorly carried out, little or nothing has been put in place to eliminate the same thing from occurring again. In our minds, this was an opportunity lost to conducting large, replicable RCTs across the globe (Sharma et al., 2022) which could have benchmarked the efficacy of learning without lectures against traditional modes of delivering higher education. We believe the ruinous outcome – students' scepticisms of online education – can be boiled down to the following in Fig. 3.

First, the unplanned and rapid move to online education because of Covid-19 was very poorly executed. Essentially, the institutions attempted to replicate their physical educational artefacts and processes in the virtual world. They rushed into service untried web conferencing technologies for connecting teachers and students only to discover that their communication backbones would not handle the sharp and enormous usage increases; lesson outages were regular occurrences. While a lot of teachers received adequate training in the new technologies before

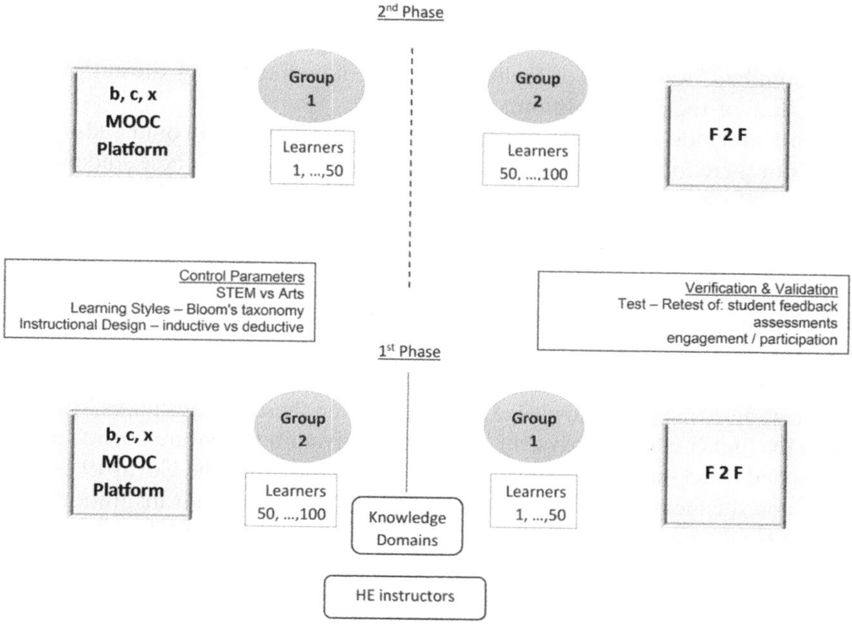

Fig. 3. RCTs Research Methodology.

going live, most students did not; confusion and apprehension about operating the technologies were rampant. In the years preceding Covid-19, a lot of institutions had "eLearning week" on the schedule, but mostly it was conducted as a spare or self-study week that the teachers absented themselves from; hence, when they needed it most, teachers had no practical experience in conducting effective online lessons. Likewise, the institutions' educational excellence departments that possibly had the necessary corporate knowledge of online education practice, had inadequate capacity to pass on that knowledge en masse to all the faculty. Undeniably, the immediate legacy of the spasmodic implementation of online education is one of chaos and ineptitude, mixed with a deep longing to return to the physical classroom.

Second, there was constant and immense pressure from the brick-and-mortar institutions and political leaders that online education was only an expediency brought on by the Covid-19 crisis, and that it was not and never would be mainstream. Their mantra of, "once the Covid-19 crisis passed, education would return to 'normal' (physical classroom)" was a not-so-subtle condemnation of online education as being the root cause of all the problems being encountered; thus, conveniently diverting attention away from the true culprits. Are we hyperbolizing the intent? Not at all; we can't recall any counter-messaging that extolled the research-supported virtues of online education, or that the lessons learned from the Covid-19 crisis would become the foundation for a new co-mainstream mode of education and ensure future generations would never experience the same chaos and ineptitude with respect to expedient online education? As far as can be expected, the students' attitudes of "physical classroom good, online classroom bad" were more of them naturally being influenced by the authorities, rather than of their own volition.

Third, and most important, the motivation for participating in online education is just not there for the majority of higher education students. These students are in higher education for only one thing: the qualification that they have been told, by business leaders and society in general, will give them employment and security. Furthermore, the formal qualification must come from only authorized brick-and-mortar higher education institutions. Sadly, the learning for the degree has the reputation of being conceptual rather than practical, and once the ex-student has been hired, he/she will have to "learn the ropes." Often, this requires the ex-student to forget the learning associated with the formal qualification and replace it with new, more relevant knowledge and skills. Given these circumstances, students inside the higher education system have little motivation to venture outside of the sanctioned brick-and-mortar institutions. Couple this with the Covid-19 experiences, and the fact that a lot of teachers in the brick-and-mortar institutions still have no training or proficiency in online education, and the incredulity that higher education students harbor about online education is entirely understandable.

When the Covid-19 crisis abated, and the physical classrooms re-opened, did the students' motivations for learning in the higher education system become stronger? Yes and no. As expected, there was an initial euphoria in faculty and students upon re-engaging socially with friends and peers, but then it soon shrank back to the pre-Covid-19 levels. Yet, the numbers show that online education in the private and for-profit sectors continued to grow; higher education students are a proportion of that growth. So, how do we make sense of this?

Well, after a few years in their first job, these same ex-students who didn't take their higher education institutions' attempts at online education seriously, will suddenly view online education in a whole new light. This is because an employee's time for upgrading knowledge and skills is at a premium, especially for the recently hired or low men on the totem pole. Undertaking higher education like that done in the brick-and-mortar institutions is a rarity at best. Yet, professional progression and climbing the corporate ladder both require continuous learning of new knowledge and skills. Fortunately, online education provides a way forward for employees that is much more conducive: it is focused and relevant, short duration, with convenient access, and affordable. Knowing this, it is easy to understand that one of the most popular courses in online education is the Masters of Business Administration.

Herein we have the key to unraveling the conundrum, of why students who can and do learn online nevertheless harbor incredulity at online learning when it is associated with higher education. It has got nothing to do with online education per se, it being as effective as any mode of learning can be when implemented correctly. The real problem is the low motivation that the higher education system instills into many of its students. Compound that with the manner in which many bricks-and-mortar institutions implement their online education programs, and it becomes a foregone conclusion that the students would rather stick with the physical classrooms they know, than venture into the virtual rooms that they, the faculty, and the institutions are all unfamiliar with and likely ineffective at using.

Readiness for Online Education Development

If there is any silver lining to come from any major crisis in our lives, it is that we come out the other end with some meaningful and positive change. At the very least, there would be a removal of the conditions that, if left in place, would cause a re-occurrence of the crisis. Sadly, this appears to not be the case for the Covid-19 crisis (Jones & Sharma, 2020, April 17).

Why? Because, in the minds of policy-makers for national education, the physical institution, particularly of higher education, symbolizes a history of teaching and research that cannot possibly be challenged. There is no room for any other mainstream education implementation, particularly an online one.

In our view, online has already demonstrated that it can take on the mantle of national education. Despite the missteps in the emergency deployment, online education has earned its "street cred" in responding to the crisis, plus what is being achieved with MOOCs, Khan Academy, YouTube, and the myriad of other platforms that offer online courses.

Earlier, we published two online opinion pieces on medium.com, and that were later re-released under SSRN. It is this paragraph from one of them that expresses our thoughts on this matter most precisely and with the eloquence to make it memorable.

Academics and practitioners who understand online learning know it has significant challenges not only for instructors but also for the non-digital-native students. Teachers who are very good at capturing the attention and engagement of students in the physical classroom will not suddenly transform into similarly great online instructors. This innate truism would be a factor – the most prominent factor is undoubtedly expediency – in throwing a myriad of online resources at instructors to prepare them for their instant digital transformation; the irony of this is not lost. Especially, when the cherished time and cost savings do not materialize. Online learning is neither cheaper nor easier. (Jones & Sharma, 2020, April 17)

One of the most promising entrants into the online education space are MOOCs. As their name implies, these are online courses that are open to anyone, and as such, are designed and built to accommodate a massive number of participants. The first of its kind is traced back to 2008, and so, its rise, fall, and rise again have been well-recorded. This abundance of data points exactly aligns with the sequence of peaks, troughs, and rises that are modeled in Gartner's hype cycle line graph (Fig. 4; Fenn et al., 2017). This two-decades-old graph is a powerful tool in the arsenal of trendsetters, innovators, and influencers because it shows the authentic maturation of new technologies over time. Our point is that the exactness of the correspondence of online courses, as typified by MOOCs, to the hype cycle strongly suggests the relevance of Gartner's hype cycle for planning the development and deployment of online education.

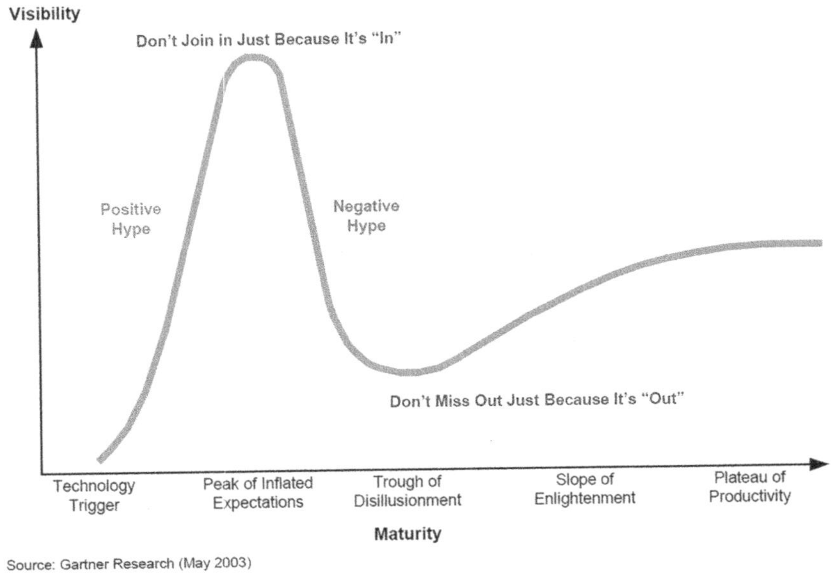

Fig. 4. Gartner's Hype Cycle.

MOOCs are a storehouse full of innovations, from videographic to interactivity to assessment. But there is one more innovation that may be the most impactful for online education. That is the knowledge center called "Class Central" incepted by Dhawal Shah to provide free access to all information related to MOOCs. For example, if you visit the website https://www.classcentral.com/ right now, you can discover how many courses are available, what are their rankings, which universities are building MOOCs, what platforms are airing or running MOOC, and a report with a wide assortment of statistics on users, qualifications such as degrees, and course usage (Shah, 2021, December 1). Having something similar in place at the time work in researching and developing online education begins, would be a powerful management tool at the city and national levels.

Expanding upon this idea, one of the authors earlier muted the need for a national organization tasked to track the latest innovations in online learning and apply that to the development of an online education program. With the launching of new educational technology being constant and even accelerating, this national organization would probably become a permanent fixture in the government, similar to cyber security and other tech-intensive services. More to the point of this chapter, such an organization would of great advantage in the education envisioned for smart cities.

We recognized that MOOCs could give the world a useful and accurate portrayal of what might be a map that the stakeholders in online education might need to follow. Accordingly, we developed a business case analysis of nine possible direct monetization approaches for the institutions' revenues from students and other orgs (Jones & Sharma, 2019, April).

1. MOOC certification – Part-time learners pay fees for the award of a recognized achievement certificate for recently completed MOOCs.
2. Online-learner course – Part-time learners pay fees for learning effectively online with MOOC. Full-time students have this bundled into the freshman orientation of their online degree program.
3. Traditional degree with accredited MOOCs – Full-time students pay discounted tuition fees for degree due to reduced physical delivery.
4. MOOC-based online degree – Full-time students pay competitive tuition fees based on new cost structures. Rates are lower in online orgs than in brick-and-mortar institutions.
5. Professional development via MOOCs – Upwardly-mobile professionals pay fees for management of their MOOC-based training paths so that they remain competitive in their profession.
6. MOOC production course – Novice video production learners pay tuition fees for a course on producing effective high-quality MOOCs.
7. Premium audience service – Institutions pay fees to distribute their MOOCs on a proprietary channel that has outstanding participation rates due to its content and quality.
8. Pedagogy in online courses – Institutions new to online learning pay fees for enhancing their online pedagogy and distinguishing itself in its sector.
9. Student data archival service – MOOC-providers pay outsourcing fees for the lifetime storage and protection of their students' MOOCs performance data.

We believe that these would be a starting point for higher education institutions to make their own plans as to how they fit into the operation of online education for smart cities according to our proposition.

The Road Ahead: A New Normal Shall Emerge

> A senior colleague with over 10 years of online delivery captured it succinctly: *it will require much more work on the instructor's part to deliver the same learning outcomes as a traditionally-delivered face-to-face course. In both cases, online delivery will cost more than face-to-face delivery; either there will be more costs to deliver the same level of quality, or costs of lost or unachievable learning outcomes. There are other issues too, for instance, research has shown that students will often report that they're learning more in an online course than their face-to-face traditional counterparts, though test scores and other in-kind evaluations contradict that belief. However, online delivery may be the only option in some cases, such as working professionals who cannot afford to leave work to attend classes, stay-at-home parents, students living in rural communities or with other life commitments, or… during a pandemic; and for these students, online learning is better than nothing.* (Dr Steve Wingreen, 7-April-20, cited in Jones & Sharma, 2020)

We believe that the linchpin of any transformation is getting the people involved to change their mindsets so that they can accept the change. After that, it is all work and all uphill, but when there is a clear objective there is also a clarity of mind; in other words, if you know well what you have to do, then you are already halfway to success. During the lockdown period of the COVID-19 crisis, we wrote a TEDx narrative hoping to do exactly that, called "Does online learning provide a 'real' college education?" Its purpose was to calibrate the thinking of influential people toward a positive mindset for the digital transformation of college education in the post-COVID-19 era (Jones & Sharma, 2020, April 17).

We identified four areas that would need calibrating in order for online to be truly seen as collegiate. They are (a) synchronous lecture or equivalent sessions, (b) authentic lab experiments, (c) summative assessments, and (d) academic calendars. In the short space of our narrative, we believe that we were able to present reasonable and conducive approaches to handling all five. Again, what is important is identifying a breakdown of the key factors that all interested parties can agree to, and then resolve them, one after another.

Of course, other researchers and academics have been coming up with their own list of factors that need either addressing or consideration as part of the larger effort to seriously collaborate in the design and implementation of some version of online education. This is the set of factors that Adamas University has proposed (Sarkar, 2020, May 16):

- Artificial intelligence in the learning space as both colleague and competitor.
- Cloud-based solutions.
- Massive investments by major market players.
- Use of virtual reality technology to extend the virtual space into the real one.
- Application of the Internet of Things to the knowledge repositories.

Giving a distinct perspective on achieving the new normal, Nolasco (2022) proposes that readiness for online education has to be uniquely tailored for each element of the classic didactic triangle (Fig. 5; Kansanen & Meri, 1999); ergo, three unique readiness models. For current eLearning and online courses, the basic requirement for readiness is orientation for students, technology training and refreshers for teachers, and a needs assessment for the institution. However, for a full-blown online education to be successful, serious investment will need to be put toward specialized processes for each element, for example 24-hour support systems and workforce reassignment for students and teachers respectively. However, it is the institutions that will experience the brunt of the changes. Presently, institutions enjoy a wide latitude in academic bundling: teaching, coercion, qualification, alumni, and counseling (Wang, 1975). In the new normal, some of the surviving institutions will see a sharp reduction of their bundling.

Finally, we have a reflection from Forbes, giving us some business-centric insight into how online education could be supported by business plans in the future

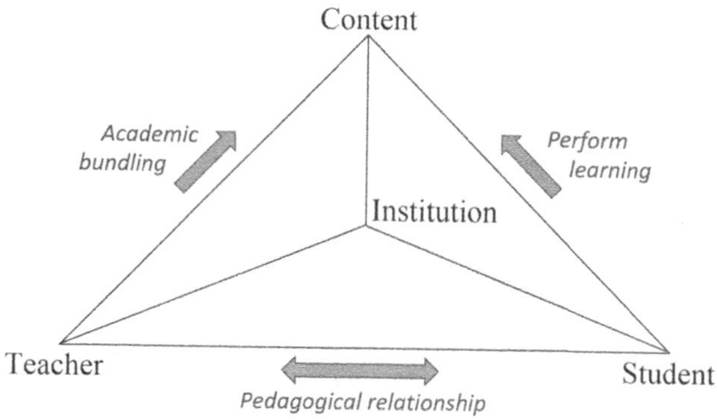

Fig. 5. Classic Didactic Triangle.

(Craig, 2015, June 23). There are three ways that a higher education institution can differentiate its online programs: (a) branding, (b) specialized programs, and (c) connecting students to jobs. Where the institution does not have a prestigious brand, it is recommended they partner with those that do. Although it seems counter-intuitive, the best specialized online programs might be blended. Finally, although becoming a "full-stack" higher education company via online programs is complicated for both the company and the potential employee doing the program, there are several studies ongoing on the basics, like contributing projects to the courses, monitor student work, and directly recruit students after the course has concluded.

Closing Thoughts

We have no illusion that our main thesis of the democratization of higher education will not receive universal support. There are too many players that have something at stake, for a grandiose declaration of change to sit well with every one of them. There needs to be a crisis or other unignorable precedence for change, either happening now or looming in the near future, for any kind of *kumbaya* moment, or a circling of the wagons to occur. That is why we have proffered several small iterative options for change that can be acted upon without upsetting the whole apple cart. Naturally, the first implementations of any of these options would face an uphill battle of putting in place all the resolutions for known contentious items while catering for unexpected consequences that will probably only become evident after such change has been rendered. To be sure, this is not a proposition we take lightly. We are keenly aware that we are talking about real people with real lives who are affected by such democratization of this key piece of society. So, it is with this in mind that we dare to share our belief that this proposition will eventually bring benefits and advantages so as to propel civilized society to heights befitting those engendered in the era of smart cities.

References

About Udemy Business. (n.d.). *Udemy*. https://teach.udemy.com/teaching-on-udemy/about-udemy-for-business/

Bagley, W. C. (1930). Radio in the Schools. *The Elementary School Journal, 31*(4), 256–258. https://doi.org/10.2307/996158

Bierman, P., Massey, C. & Manduca, C. (2006). *Reconsidering the textbook*. Wiley Online Library. Retrieved from https://agupubs.onlinelibrary.wiley.com/doi/pdfdirect/10.1029/2006EO310004

Bonk, C. J., & Graham, C. R. (2005). *The handbook of blended learning: Global perspectives, local designs*. John Wiley & Sons.

Camacho, J. A. (n.d.). Educational broadcasting. *In Encyclopaedia Britannica*. Retrieved November 6, 2023, from https://www.britannica.com/technology/broadcasting/Educational-broadcasting

Cooke, G. (2022, July 5). Online learning vs face to face training. *Elucidat*. https://www.elucidat.com/blog/online-learning-vs-face-to-face-learning/.

Craig, R. (2015, June 23). A brief history (and future) of online degrees. *Forbes*. https://www.forbes.com/sites/ryancraig/2015/06/23/a-brief-history-and-future-of-online-degrees/?sh=75a76aca48d9

Dearden, K., Mulokozi, G., Linehan, M., Cherian, D., Torres, S., West, J., Crookston, B., & Hall, C. (2023, January 10). The impact of a large-scale social and behavior change communication intervention in the Lake Zone Region of Tanzania on knowledge, attitudes, and practices related to stunting prevention. *International Journal of Environmental Research and Public Health, 20*(2), 1214–1227. https://doi.org/10.3390/ijerph20021214

Distance Education. (2001, December 17). *Wikipedia: The free encyclopedia*. Retrieved November 6, 2023, from https://en.wikipedia.org/wiki/Distance_education

Emma. (n.d.). Online learning vs. in-person classes – what's better? *EF Education First Go blog*. https://www.ef.com/wwen/blog/language/online-learning-vs-in-person/

English Language & Usage. (2020, March 25). Is it CoViD? Or COVID? Covid? How should the word be spelled? *Stack Exchange*. https://english.stackexchange.com/questions/528487/is-it-covid-or-covid-covid-how-should-the-word-be-spelled

Fenn, J., Raskino, M., & Burton, B. (Analysts) (2017, January 4). Understanding Gartner's Hype Cycles. *Gartner*, first published 2013, July 2, now refreshed. https://wiki.harvard.edu/confluence/download/attachments/213389796/Shaffer_Understanding%20Gartners%20Hype%20Cycles.pdf?version=2&modificationDate=1485889905000&api=v2.

Hall, P. (1996, October). Distance education and electronic networking. *Information Technology for Development*, *7*(2), 75–89. https://idl-bnc-idrc.dspacedirect.org/server/api/core/bitstreams/54177a1f-7277-4940-808252517cc38547/content.

Haworth, M., & Hopkins, S. (2009, October 26). On the air: Educational radio, its history and effect on literacy and educational technology implementation. *Faculty of Education: Language & Literacy Education,* University of British Columbia. https://blogs.ubc.ca/etec540sept09/2009/10/28/on-the-air-educational-radio-its-history-and-effect-on-literacy-and-educational-technology-by-michael-haworth-stephanie-hopkins/

Higher Education. (2002, February 25). *Wikipedia: The free encyclopedia*. Retrieved November 6, 2023, from https://en.wikipedia.org/wiki/Higher_education

Hokanson, B., & Hooper, S. (2000). Computers as cognitive media: Examining the potential of computers in education. *Computers in Human Behavior, 16*, 537–552.

Jones, K. A., & Sharma, R. (2019, April). Sustaining or disruptive: Future of MOOC innovation in tertiary education still up for grabs. In *2019 IEEE global engineering education conference (EDUCON)*, pp. 900–909. IEEE.

Jones, K., & Sharma, R. S. (2020, May 5). Does online learning provide a 'real' college education? *First posted on medium.com, now SSRN*. https://ssrn.com/abstract=3593216

Jones, K., & Sharma, R. (2020). Reimagining a future for online learning in the post-COVID era. Available at https://medium.com/@rs.sharma.sg/in-a-narrative-style-q-a-dr-kevin-jones-and-i-think-aloud-about-blended-learning-25d415ea1058

Jones, K. A., & Sharma, R. S. (2021). *Higher Education 4.0: The digital transformation of classroom lectures to blended learning*. Springer Nature. https://doi.org/10.1007/978-981-33-6683-1

Kansanen, P. & Meri, M. (1999). The didactic relation in the teaching-studying-learning process. In B. Hudson et al. (Eds.), *Didaktik/Fachdidaktik as Science(s) of the Teaching Profession* (pp. 107–116). The Thematic Network of Teacher Education in Europe

Kemp, S. (2023, January 26). Digital 2023: Global overview report. *Datareportal*. https://datareportal.com/reports/digital-2023-global-overview-report

Kentnor, H. E. (2015). Distance education and the evolution of online learning in the United States. *Curriculum and Teaching Dialogue*, *17*(1), 21–34.

Koumi, J. (2006). Screenwriting for video in educational multimedia. In *Designing video and multimedia for open and flexible learning* (Chapter 5, pp. 95–132). Routledge.

Lefa, B. (2014). The Piaget theory of cognitive development: Educational implications. *Educational Psychology*, *1*(1), 1–8.

Milford, K. (n.d.). "Don't be a stoopy-doop". *Keith Milford's vintage Detroit TV and movie board*. https://www.tapatalk.com/groups/vintagedetroittvandmovieboard/don-t-be-a-stoopy-doop-t1223.html

Moore, J. L., Dickson-Deane, C., & Galyen, K. (2011). e-Learning, online learning, and distance learning environments: Are they the same? *The Internet and Higher Education*, *14*(2), 129–135.

Multiplayer (2005, January 16). *DoomWiki*. https://doomwiki.org/wiki/Multiplayer

Nikolov, R., Shoikova, E., Krumova, M., Kovatcheva, E., Dimitrov, V., & Shikalanov, A. (2016). Learning in a smart city environment. *Journal of Communication and Computer*, *13*(2016), 338–350. https://doi.org/10.17265/1548-7709/2016.07.003

Nolasco, C. E. (2022, February 1). Online distance learning: The new normal in education. *eLearning Industry*. https://elearningindustry.com/online-distance-learning-the-new-normal-in-education

Olmstead, L. (2023, September 28). Online Learning vs. Face-to-Face Learning: Which Is Best? *Whatfix blog*. https://whatfix.com/blog/online-learning-vs-face-to-face-learning/.

Ormrod, J. E. (2011). *Human learning* (6th ed.). Prentice Hall Inc.

Pacific College. (2022, July 31). *Online classes vs. traditional classes: What's the difference?* https://www.pacific-college.edu/blog/traditional-vs-online-classes

Popular Mechanics. (2007, March 12). *Wikipedia: The free encyclopedia*. Retrieved November 6, 2023, from https://en.wikipedia.org/wiki/Popular_Electronics

Rouse, M. (2012). Online. *Techopedia: The techDictionary*. https://www.techopedia.com/definition/658/online

Rovai, A. P. (2004). A constructivist approach to online college learning. *The Internet and Higher Education*, *7*(2), 79–93.

Sarkar, S. (2020, May 16). *A brief history of online education*. Adamas University. https://adamasuniversity.ac.in/a-brief-history-of-online-education/

Sharma, R., Jones, K., Anderson, W., Inthiran, A., & Tabatabaee, A. (2022). The digital transformation of higher education – "uni for nothin', MOOCs for free"? *Journal of Information Technology Case and Application Research*, *24*(1), 34–60. https://doi.org/10.1080/15228053.2021.1889741

Shah, D. (2021, December 1). By the numbers report: MOOCs in 2021. *Class central*. Retrieved November 6, 2023, from https://www.classcentral.com/report/mooc-stats-2021/

Sir Ken Robinson. (n.d.). Author, educator, speaker for *TED: Ideas worth spreading*. Retrieved from https://www.ted.com/speakers/sir_ken_robinson. Also, see video released on 13 August 2014 "Ken Robinson: How to escape education's death valley". TED conference, April 2013.

Sleator, R. D. (2010). The evolution of eLearning: Background, blends and blackboard… Science Progress (1933-), *93*(3), 319–334.

Stella, A., & Gnanam, A. (2004). Quality assurance in distance education: The challenges to be addressed. *Higher Education*, *47*(2), 143–160.

Swiecki, Z., Khosravi, H., Chen, G., Martinez-Maldonado, R., Lodge, J. M., Milligan, S., Selwyn, N., & Gašević, D. (2022). Assessment in the age of artificial intelligence. *Computers and Education: Artificial Intelligence*, *3*, 1–10. 100075.

Television. (2001, November 4). *Wikipedia: The free encyclopaedia*. Retrieved February 2, 2022, from https://en.wikipedia.org/wiki/Television

Textbook. (2003, June 27). *Wikipedia: The free encyclopedia*. Retrieved November 6, 2023, from https://en.wikipedia.org/wiki/Textbook

Uddin, N. (2023, November 7). eLearning vs online learning: What's the difference? *Fuse | Classroom*. https://insights.fuseclassroom.com/elearning-vs-online-learning-whats-the-difference/

University of London. (n.d.). *History of distance education at University of London*. https://www.london.ac.uk/centre-online-distance-education/who-we-are/history-distance-education-university-london

Visual Academy. (2014). *The history of online schooling*. OnlineSchools.org (A Red Ventures company). https://www.onlineschools.org/visual-academy/the-history-of-online-schooling/

Wang, W. K. S. (1975). The unbundling of higher education. *Duke Law Journal*, *24*(1), 53–90.

Wheeler, S. (Interviewer) (2016, September 17). Who are the founding fathers of distance education? *Online Learning and Distance Education Resources*. https://www.tonybates.ca/2016/09/17/who-are-the-founding-fathers-of-distance-education/

Williamson, B. (2015). Educating the smart city: Schooling smart citizens through computational urbanism. *Big Data & Society*, *2*(2), 1–13. 2053951715617783.

Wilson, D. (2023, April 20). Why Online Learning is the Future of Education. *Wondershare*. https://democreator.wondershare.com/elearning-tips/why-online-learning-is-the-future-of-education.html?

Wolff, A., Kortuem, G., & Cavero, J. (2015, April). Towards smart city education. *Sustainable Internet and ICT for Sustainability (SustainIT), IEEE 2015*, pp. 1–3.

Further Readings

Afanasiev, M., & Lysenkova, M. (2019). How university acts in the development of "Smart Cities". *SHS Web of Conferences*. https://doi.org/10.1051/shsconf/20197105011

Ahad, M., Paiva, S., Tripathi, G., & Feroz, N. (2020). Enabling technologies and sustainable smart cities. *Sustainable Cities and Society*. https://doi.org/10.1016/j.scs.2020.102301

Anttila, J., & Jussila, K. (2018). Universities and smart cities: The challenges to high quality. *Total Quality Management & Business Excellence*, *29*, 1058–1073. https://doi.org/10.1080/14783363.2018.1486552

Bhatia, S. (2018). Sustainable smart universities for smart cities. *Journal of Economics, Management and Trade*. https://doi.org/10.9734/JEMT/2018/44521

Caragliu, A., Bo, C., & Nijkamp, P. (2011). Smart cities in Europe. *Journal of Urban Technology, 18*, 65–82. https://doi.org/10.1080/10630732.2011.601117

Chichernea, V., & Smedescu, D. (2016). Campus information systems for enhancing quality and performance in a smart city high education environment. *eLearning and Software for Education*. https://doi.org/10.12753/2066-026x-16-007

Kutty, A., Abdella, G., Kucukvar, M., Onat, N., & Bulu, M. (2020). A system thinking approach for harmonizing smart and sustainable city initiatives with United Nations sustainable development goals. *Sustainable Development*. https://doi.org/10.1002/sd.2088.

Lyapina, I., Sotnikova, E., Lebedeva, O., Makarova, T., & Skvortsova, N. (2019). Smart technologies: Perspectives of usage in higher education. *International Journal of Educational Management*. https://doi.org/10.1108/IJEM-08-2018-0257

Molnar, A. (2019). Education in smart cities: Practices and challenges. *Proceedings of the 16th International Conference on Applied Computing 2019*. https://doi.org/10.33965/ac2019_201912c032

Oh, D. (2019). Building inter-personal competence in architecture and urban design students through smart cities at a higher education institution. *Sustainability*. https://doi.org/10.3390/su11247179

Sadeh, A., Feniser, C., & Dusa, S. (2020). *Technology education and learning in smart cities* (pp. 78–95). https://doi.org/10.4018/978-1-7998-1591-4.ch005

Semenova, N., Svyatkina, E., Pismak, T., & Polezhaeva, Z. (2017). *The realities of smart education in the contemporary Russian universities* (pp. 48–52). https://doi.org/10.1145/3129757.3129767

Chapter 6

The Rebirth of Museum Education Echoing of Smart Education

Yingyilong Hu[a], Zhengchao Fu[b], Suwithda Charungkaittikul[a] and Ravee Chudasring[a]

[a]*Department of Lifelong Education, Chulalongkorn University, 10330, Bangkok, Thailand*
[b]*School of Housing, Building and Planning, Universiti Sains Malaysia, 11800, Penang, Malaysia*

Abstract

The contemporary view of museums differs markedly from the traditional perception of scattered institutions, with urbanization and globalization reshaping their definitions and functions. This chapter explores the evolving roles of museums, particularly within the context of smart city development, with a focus on China. The chapter employs a structured methodology, first examining the positive aspects of museums, including their role in safeguarding natural heritage, promoting social responsibility, and facilitating education. Conversely, it scrutinizes the negative influences on museums, such as political and colonial interventions that impact pedagogical practices. A significant aspect of the discussion is the intersection of museum education with smart and adult education, heralding a renaissance in museum learning. A case study on the Danxiashan Geo-Museum illustrates the integration of cultural and heritage institutions into smart city initiatives, serving as a prototype for future endeavors. The chapter underscores the importance of incorporating smart and adult education into museum initiatives, along with smart innovation projects involving local museums and heritage sites. The implications extend to next-generation smart cities, offering insights into their societal, economic, and developmental impacts in the Gulf Region. Museum education emerges

The Emerald Handbook of Smart Cities in the Gulf Region:
Innovation, Development, Transformation, and Prosperity for Vision 2040, 99–116

doi:10.1108/978-1-83608-292-720241006

as a crucial yet understudied component within the framework of smart cities and education, providing a lens for reimagining future educational paradigms. By embracing the resurgence of museum education and integrating socio-cultural changes, including the rise of heritage education, smart cities can enhance their educational and societal contributions nationally and globally.

Keywords: Museum education; smart education; smart cities; heritage education; natural education; China

1. Introduction

This chapter examines museums' evolving roles in the context of urbanization and globalization, particularly analyzing the smart education city development in China. It discusses both the positive aspects, such as preserving natural heritage and promoting education, and negative influences like political interventions. The chapter also explores the potential of integrating smart and adult education with museum initiatives, exemplified by the Danxiashan Geo-Museum case study, showcasing how museums can enhance smart city fabric. It highlights the importance of integrating socio-cultural changes and promoting heritage education to advance museums' roles nationally and globally, offering valuable insights for next-generation smart cities in the Gulf Region. Ultimately, the revitalization of museum education serves as a lens to envision future education paradigms and informs smart city development efforts.

2. Positive Roles of Museums

Museums have played many roles that positively promoted the development of culture and the protection of natural heritages. First, museums were the traditional "protectors" to store and protect tangible natural heritages. Knowing the history of the development of museums and the definitions of museums, it will be easier to understand its protective role since the old times. Referencing a Chinese blog that introduced European history, the writer believed museums had their collecting and displaying purpose. Under the influential development of Greek and European culture, in the third century BC, the city of Alexandria in Egypt established a huge museum focused on studying learning, set up a library, and collected valuable artifacts. More than a hundred years later, the Roman Empire sacked Greece. All works of art were moved to Roman palaces, temples, and the homes of princes and nobles. In effect, fine art became the private property of princes and nobles. In medieval Europe, Christian monasteries became places where fine art was preserved. The rise of navigation in Europe in the 16th century spurred the development of private collections, and the Renaissance stimulated the study of antiquities. In the 17th century, private collectors began opening their treasures for public viewing. Private museums sprang up, as did public ones.

From the 18th and 19th centuries, museums sprang up all over the United States, as did other Asian countries like China, India, Japan, and Korea. After knowing the development of museums, generations were taught the history and the story of the rise and fall of museums. Besides the wars and killings, the fate of arts had changed while museums' collective and pedagogical purpose was also seen in its protective role. There was another aspect taken from the "history," which was the developed and found museums had changed from private institutions to public institutions. This transition allowed citizens to access museums more easily than in the "past" when the royal family (Empires) or private collectors only greeted a few number people or visitors to watch their collections and arts. Wherever, both private and public insinuations showed museums' responsibility to protect and guard the collections with technologies and patrol parties. At this moment, museums are the public space serving the needs of valued objects and humans.

Secondly, museums assumed a social role in everyday life. Janes (2016) reviewed museums' social responsibility without borders. The world's issues and challenges have dramatically involved the development of museums and their social role. The decline of public funding was one of the serious concerns for Canadian museums in the 1990s and "led to a preoccupation with visitor numbers, high-profile exhibitions, and burgeoning marketing departments to increase earned revenues" (p. 187). So did other museums worldwide. The pressures now facing all museums went far beyond money and popularity, to recognize social needs and issues from the emergence of climate change to social justice. Museums act its social promotion and advertising to citizens interdisciplinary knowledge and good manners for how to be a "good citizen" or even a global citizen. To achieve the collective goal and museums' social responsibility, museums enlarged the context of work and increased the number of considerations in various aspects. If citizens deeply trusted museums, as social institutions in civil societies, "museums are essential in fostering public support for decisive action to address these topical issues. Among these pressing issues are climate change, water scarcity, overpopulation, habitat destruction, loss of topsoil, rising toxin levels, and increasing financial inequality" (p. 187). Similarly, Gray and McCall (2020) studied museums and societies and looked at the key question of "what are museums for?." While they approached to examine the justifications for both social and individual, they found museums undertaking their functional roles within communities and different societies. Because of museums' building themselves, the social role is incorporated with both the physical tangibility of the space and how the internal space of buildings is shaped and used. Museums were often set in a specifically chosen geographical location and environment. Many of them had a series of collections to display or a specific objective to define and introduce local traditions, and culture, and reinforce a process of citizenship development. Therefore, museums were involved in an interaction of place and space that was seen as a shared sphere that was linked to their role within local communities and groups. On the other hand, the social role of museums has also been consistently linked to debates over value formation and strengthening ideology, where the activities and gathering of museums often promoted social cohesion, cross-cultural understanding, reduced ignorance, and prejudice.

Thirdly, museums were key for education, which used to be seen as individual learning and public teaching and propaganda. Hein (1998) felt that "museums are extraordinary places where visitors have an incredible range of experiences. It is not unusual for visitors to describe epiphanies in recounting their own history with museums" (p. 2). His saying evidentially pointed to the educational role of museums in daily interaction and its range of education reached everyone from the young to the old. Museum education was crucial and happened normally in public museums. As a part of public goods, museums were responsible for knowledge spreading beyond the limited class and separating cultural differences. Gradually, Hein considered museum professionals had become increasingly aware of the educational role of museums and begun to re-evaluate the relationship between museums and their visitors. His study suggested visitors the best way of learning knowledge was actively constructed in their minds since visitors were physically, socially, and intellectually accessible to the exhibitions. This was how museums can adapt to create a unique education and learning environment that he called the "Constructivist Museum." Likely, Gray and McCall (2020) argued museums were

> reconceptualized as governmental instruments as they were harnessed as an "apparatus of power," consolidating royal power for a "civilizing mission." This power was characterized by the multiplicity of objectives regarding its authority, rationality, and interests that the state was pursuing at this time. (p. 84)

This quote showed museums as collective and social institutions applied their educational role in communities and societies even though the civilizing and reshaping purpose had been increasingly questioned. Researchers pointed defining the social role of museums around the role of education was established in the first half of the 19th century, which had an emphasis and a recognition of the diverse social characteristics and differentiated cultural attitudes toward visitors, alongside the emergence of a consumer or user's relationship between museums and the public. The usage of museums in education enlisted high cultural and collective insinuations and used a selected culture or kinds of cultures as a governmental tool which was a subtle approach to manage social behavior and impose social norms and values. This educated "methodology" is seen as being appropriate in rapidly changing societies.

3. Negative Roles of Museums

While museums are actively serving and involving individual's livelihoods, their negative roles and impacts are reviewed by researchers and scholars who are upgrading a better way the development of museums and public education. First, Mayo (2013) wrote museums were sites of critical pedagogical practices, which indicated museums' role in education. Under the democratic notion of education, museums were part of sites of "public pedagogy." These public pedagogical sites covered a range "from schools, to cinemas and youth entertainment areas, and are often deeply informed by consumerist and militarized ideologies" (p. 144). This showed the sites of public pedagogy had been intervened by politics and the military, which was equal to what had happened to the arguments for universities, schools, and other learning agencies to serve as the democratic public spheres. When people argued for educational institutions and curricular democratization, economic, political, and ideological processes enabled specific groups' knowledge to become "official" and popular, which would influence from individual to a community even a nation. Likely, Dorfman (2012) noticed "scientific collecting expeditions" in which many museums in Europe had "the power of collections-based knowledge extended into the realm of geopolitics and policy. Collections were even sometimes used as instruments of the state" (p. 24). Dorfman referred to a few examples to relate his concept. For instance, to share culture and knowledge about the known world and "assembled exclusively to enlighten and enhance the status of Russian sailors and other official travelers representing Russian interests," the Russian imperial authorities created a collection of exotic taxidermy in St Petersburg in the 19th century (2012, p. 24). Whereas, in the 20th century, the "foundation of museums by colonizing and occupying powers often left a positive legacy of institutions that have subsequently flourished post-independence and have helped to build new identities" (2012, p. 25). Examples include the United Kingdom, United States and Canada collected "treasures" from their colonies in Southeast Asia and the native Americans (Indigenous people) and continued to "collect" by buying and donating from NGOs, institutions and wealthy individuals. In different, some museums even presented collections and objects from the royal families or the empires to indicate the power and the past "history." So, the museum education and exhibitions had been politicized and controlled by power dynamics to share and display their interests and ideologies.

Secondly, Mayo (2013) pointed museum education and collections were problematic. "Issues of cultural democracy and notions of culture and power, particularly in conditions of learning in disenfranchised and racialized communities" (p. 145). This quote recognized museums as sites of public pedagogical agencies were studied an approach to education known as critical pedagogy, which is "fundamentally concerned with the centrality of politics and power in our understanding of education and learning" (p. 145). This quote focuses on the power dynamics in charge of knowledge spreading and the interests of selected culture and collective work through museum education and shared public sphere, which viewed museum education and exhibition existed the issue of racism and social stereotypes. On the other hand, museums consistently engage in knowledge production and sharing of cultural values. Researchers insisted that "museums, like curricula, depict selections from the cultures of society" (p. 145). This led the critical viewers and visitors even museum educators to challenge the "culture arbitrariness" of museum exhibitions, which refrained from visiting museums as "repositories of so-called neutral knowledge" (p. 145). By doing research, the researcher noted some of the international museums such as Maltese museums, the Museum of Modern Art (MOMA), and the American Crafts Museum had the Eurocentric and class bias of museums, which reflected in their exhibition and culture sharing. "The juxtaposition of what strikes us as highbrow cultural production, on the one hand, and subaltern and popular production, on the other, renders questions and considerations associated with critical pedagogy most relevant" (p. 145). Differently, some of the museums had a particular interest in displaying objects related to race and gender. The studies show (Janes, 2016; Mayo, 2013) that Canadian museums recognize "other voices" from indigenous people. Because of cultural diversity, climate change, and isolation, a surprising challenge was new to museum management, scholars, and mainstream museum traditions. That was the exclusion of the voice from First Nations who were defined as "the native." The northern Aboriginal peoples stated the perception that museums were a colonial legacy and never addressed the needs of living cultures. The Canadian museums and the construction of new buildings and museums in Aboriginal people's land and territories were examples of derecognizing the rights of the indigenous people and lacking the acknowledgment of their relationship toward the environment and landscape. Based on their perception, the heart of the museum, the collections and the way of exhibiting needed to preserve and highlight their cultural traditions rather than operating and exhibiting in a manner that conformed to professional museum practice. In other words, it was not "a permanent collection in an environmentally controlled building. The preservation and celebration of intangible cultural heritage – music, dance, and storytelling – were often the focus of concern" (Janes, 2016, p. 18). The wish of the Aboriginal people was to have an inclusive practice along with the development of Canadian museums and the young generation. The group of museum workers would ensure that museums and Aboriginal peoples continued to collaborate and flourish, in a manner that would confirm the human rights of all Indigenous peoples in museums and the correct way of land construction. At the same, the idea was that the protection of natural heritage and the development of museums were exclusively built for storing and educating visitors, but the focus was concentrated on tangible heritage (Dorfman, 2012; Janes, 2016; Mayo, 2013). The suggestion was made by the International Council of Museums (ICOM), whose committee had drawn attention to the intangible heritage produced by culture in different forms including oral traditions, performing arts, social practices, rituals and so on. As a matter of fact, Dorfman posted that intangible heritage was the root of tangible heritage. Communicating the intangible, a large part of museums was responsible to exhibit and explain a culturally diverse audience the tangible natural heritage rather than intangible cultural heritage. As museums developed, they did not fulfill the purpose of a method of "inclusiveness." It was very important the "poor collecting and documentation practices, as well as a lack of understanding of other cultures' beliefs and knowledge, sometimes placed barriers between people's differing cultural values" (Dorfman, 2012, p. 16). At the same time, for museums, financial constraints limited the development of these growing collections, while rising costs sometimes lead to the destruction and dispersion of collected material without regard to its worldwide cultural value.

Are museum education and museums forgotten when the concept of smart cities emerges? What's it like now? Museums and museum education are of great significance as they reflect the social and scientific empirical knowledge accumulated by the public throughout their lives, provide the opportunity to enjoy renowned culture, and at the same time serve as an extension of education (Kang & Lee, 2014; Yang, 2017). The digitization of museum education is to increase public accessibility through the development of online museum projects as a service. Unlimited possibilities for the development of a new type of education that expands the content of cultural and artistic appreciation and innovatively develops. In detail, today we live in a "smart" world. Our daily lives are surrounded by smart technologies, and we spend most of our time with them. We can search for the information we want to know anytime, anywhere, and access the Internet. Because of the development of smart technologies, life has become more digitalized, convenient, and intelligent. These digital technologies have had an impact not only on the social and economic sectors but also on the education sector, resulting in the adoption of various digital technologies in education. Museums are also making efforts to use digital technology for museum education, and in particular mobile technology for a mobile museum learning environment. Therefore, the rebirth of museum education is an important component of revitalizing museum education. It makes full use of the material and human resources of the museum to achieve a qualitative leap in the educational function and role of the museum. It is a powerful combination of learning and mobile technology. It enables museum education to overcome the limitations of time and space, thereby expanding the scope and disciplines of education.

4. The Innovation – The Rebirth of Museum Education

> How to improve and expand museum education in the context of developing smart cities? How do smart education and adult education inspire and support museum education?

4.1. Adult Education

Whether you like to admit it or not, notions of adult education can often inform our learning process and study eye-captured objects in our daily lives. Museums were not a passive "script" determined by outside forces to study and conceptualize. These independent institutions and buildings offered such great visiting experience, and individual engagement with collections and specimens. Rather, museums can be understood as storage of both tangible heritages and intangible cultures, which serve as a continuous process that fits people into a cultural hierarchy and learning environment. In particular, adult education has been involved in the museums' system and the learning environment, which altered and advanced the roles of museums. At first, learned from Taylor (2010), that museums were set as cultural institutions where hundreds and thousands of individuals, groups, families, and communities could visit and have a joyful time with accompanies. Museums were linked together as institutions devoted to collecting, preserving, and presenting the body of knowledge. Examples include artifacts, cultural heritages, documents, animals, plants, and other types of meaningful pieces. Museums were socially and culturally valued by learning communities in many disciplines.

When these institutions were gradually investigated more thoughtfully, it became easily apparent that as sites of adult education they fall within an inclusive frame and a series of connected learning places. "They are also dynamic and contested repositories of knowledge where the lifelong education of adults is inherent to their mission" (Taylor, 2010, p. 6). Readily, museums were viewed as places of cognitive change since "cultural institutions are not passive repositories. They are structures that promote and facilitate cognitive change and reflection" (Taylor, 2010, p. 6).

Secondly, Taylor (2010) believed understanding the changing nature of what visitors experience in cultural institutions (new ways of looking at the world) required understanding the unique context and relevant teaching methods within these institutions. One way to understand this setting was to frame an analysis through the lens of informal education, which is usually defined as any organized, intentional, and explicit effort to promote learning and improve the quality of life through non-school settings. In the same, the foundations and purposes of adult education demonstrated the range of provision of adult education by examining different ways of learning (Rogers, 2014; Spencer & Lange, 2014). Such as formal, non-formal, and informal education. These different kinds of learning experiences took place in institutions and other learning environments. Visiting and studying in museums gave me opportunities to experience different ways of learning. Such as non-formal and informal learning happened frequently. The definitions given by Rogers (2014) that non-formal learning was not provided by an education or training institution and typically did not lead to certification. Then, the non-formal learning was still intentional from the learner's perspective. For instance, others who love cooking may voluntarily help and cook for dinner after school. Their passion led them to spend more time cooking dishes. Differently, some classmates may have an informal learning experience. Rogers (2014) defined informal learning as either incidental or accidental learning. An example might be a person who saw an Instagram posting of cooking. Accidentally, he could be finding a new flavor by trying a new brand of Chinese hot pots. So did the museum learning experience. Taylor (2010) emphasized non-formal teaching that "teaching in cultural institutions offers diversity and complexity not found in more formal institutions." It includes "the educator (e.g., librarian, curator, docent, naturalist, zookeeper, exhibit), the content (exhibits, books, animals), and for some institutions the physical setting itself" (p. 7). This saying elaborated within socio-cultural and educational institutions, exhibits may include static and interactive displays in museums and science centers, as well as lived animals and plants in zooms and gardens. These institutions provided information and generated meaning through the visitor's experience, especially in an environment where the exhibition was not interpreted by any lecturer, allowing for shared authority in creating meaning.

Moreover, Clover (2015) explored adult education focused on social and environmental change within public institutions, such as museums and galleries in Canada, Scotland, and the United Kingdom. She noted these cultural spaces often faced criticism for elitism, colonialism, and xenophobia, leading to their exclusion from adult education discussions. However, the pressing social and ecological challenges of the 21st century necessitated that these institutions adapt and respond meaningfully. Her research highlighted that public art galleries and cultural institutions are dynamic yet controversial spaces that play a crucial role in addressing social and environmental issues. Through various initiatives like exhibitions, workshops, and seminars, these institutions engage with identity, advocate for decolonization, and promote alternative narratives. This approach seeks to challenge authority, privilege, and foster critical discussions through art. Clover's analysis included a Canadian case study on colonialism, emphasizing how the nation's identity as a settler state is closely linked to Eurocentrism, often distorting Indigenous narratives. In her subsequent work, Clover et al. (2016) observed that while contemporary art has become more socially engaged, adult education policies often prioritize measurable outcomes over deeper engagement. This gap encouraged a shift towards relational and participatory practices in galleries, enhancing social inclusion for all community members. Clover (2018) examined the role of public museums in tackling current social issues through non-formal education. She referenced a six-week course at Tate Modern and Tate Britain, which facilitated discussion on the ecological and social impacts of climate change, highlighting the responsibility of public institutions to promote critical understanding and action regarding these challenges.

Overall, Taylor, Rogers, and Clover's definitions, researches, and findings underscore the transformative potential of education including formal education, non-formal education, informal education and adult education with different settings such as public educational institutions and other typical learning spaces. This illustrates how creative pedagogical practices can foster significant learning experiences that resonate with contemporary societal needs.

4.2. Smart Education in Smart Cities

Smart technology has significantly involved in everyday life and changed faster than ever. Technical innovations impact the traditional education vision and onsite learning environment. Smart education has been defined as the effective and consistent use of information and communication technologies to achieve personalized learning outcomes and to develop talents of wisdom through appropriate teaching methods in the digital age (Demir, 2021; Zhu & He, 2012; Zhu et al., 2016). Smart education helps to deal with the shortcomings of the current education methods and systems. To develop new or improved teaching and learning spheres, the involvement of smart education and its frameworks in museum education will become a new structure of teaching and learning methods and systems. The adding layer with smart technologies structures indicates the importance of implementation of educational goals and learning outcomes. Here, the notion of smart education is not just about technologies but also about new teaching and learning approaches in a way of learning. The combination of smart education and museum education is a shift from traditional education and learning activities to more advanced approaches and practices in line with the digital age. Leaners of the digital age evolve smart education as one of the essential education in 21st-century new ways of teaching and learning. In a smart education environment, the context of smart education is the connectivity featured by digital educational technologies supporting education in a better assessment and quality of learning. Essentially, a smart learning environment can be considered as the new adaptive learning methods and systems under the ongoing changing society in the curriculums of developing smart cities with innovative and features and capabilities.

4.3. The Rebirth of Museum Education

The ongoing debate on smart cities research is challenged by the new expectations with innovative technologies and ideas focusing on smart development. To communicate other debating topics, museum education can enrich and interconnect with smart education and adult education collectively. In March 2021, Chinese Government Network (2021) reported China issued the Outline of the 14th Five-Year Plan (2021–2025) for National Economic and Social Development and Vision 2035 of the People's Republic of China. In "Part V An Initiative to Build a Digital China," it was proposed to build a digital economy, accelerate the building of a digital society, raise the level of digital government, and foster a healthy digital ecosystem. Accordingly, digitalizing museum education was born at the right time, which was described as "the rebirth" and different from traditional museum education.

Are museum education and museums forgotten when the concept of smart cities emerges? What's it like now? Museums and museum education are of great significance as they reflect the social and scientific empirical knowledge accumulated by the public throughout their lives, provide the opportunity to enjoy renowned culture, and at the same time serve as an extension of education (Kang & Lee, 2014; Yang, 2017). The digitization of museum education is to increase public accessibility through the development of online museum projects as a service. Unlimited possibilities for the development of a new type of education that expands the content of cultural and artistic appreciation and innovatively develops. In detail, today we live in a "smart" world. Our daily lives are surrounded by smart technologies, and we spend most of our time with them. We can search for the information we want to know anytime, anywhere, and access the Internet. Because of the development of smart technologies, life has become more digitalized, convenient, and intelligent. These digital technologies have had an impact not only on the social and economic sectors but also on the education sector, resulting in the adoption of various digital technologies in education. Museums are also making efforts to use digital technology for museum education, and in particular mobile technology for a mobile museum learning environment. Therefore, the rebirth of museum education is an important component of revitalizing museum education. It makes full use of the material and human resources of the museum to achieve a qualitative leap in the educational function and role of the museum. It is a powerful combination of learning and mobile technology. It enables museum education to overcome the limitations of time and space, thereby expanding the scope and disciplines of education.

5. The Prosperity and Development – The China's Case Study

5.1. Museums in China

The development of museums in China has grown so fast. Mostly, many of China's museums are public institutions, and their roles significantly influence in many aspects. First, museums were popular public sites. Mayo (2013) thought besides providing exhibitions, museums could develop into a group of popular public places since the space usually shared the "universe of knowledge" and people could engage in recreational pursuits and specific forms of creative cultural production. Meanwhile, museums had been used for storing and reserving purposes, but the change of the collections was from protecting tangible natural heritage to an extension adding intangible cultural heritages. Differently, Taylor (2010) felt museums were public institutions, but they were cultural and physical places situated in specific settings with unique stories to tell.

> The physical place or location of a cultural institution historically shapes the focus, content, and message that the institution shares with the large community (local, regional, national, or international). These locations, both physical and digital, celebrate the unique culture and history of a place, in contrast to a twenty-first-century American landscape marked by homogeneity and commercial/material ubiquity. (p. 8)

In contrast, Ji et al. (2014) conducted a study of sociocultural perspective and visit motivation. The finding displayed "human behavior is mediated by tools and artifacts that have been constructed through social interactions" (Ji et al., 2014, p. 82). Museums as public institutions help visitors interact with culture and "history." An individual's life experience was situated in a sociocultural context within a public educational system. Individual's motivation was to these institutions seeking a greater understanding of specific narratives, arts, heritages, and local knowledge, which indirectly influenced cognitive and emotional outcomes. Museums served the increasingly important function of community building and searched for a better way of organizing and grouping people who explored and revised their values and perceptions by visiting museums. For example, libraries and other cultural institutions offered places and shared values. Such as the intellectual commons were modeled and approached through events, programs, and chance encounters in a space that was hospitable, comfortable, and accessible to all citizens. Referring to Wang et al. (2020)'s research on the mode of public cultural participation and its influencing factors was of great significance to the sustainable development of regional culture. They identified which factors influenced the cultural participation of the Chinese public. It was found that education, occupation, discipline, income, distance, age, gender, and other factors affected public participation in museums.

Second, both Hein (1998) and Lu (2015) reviewed the significance of education in museums. While the educational function of museums was respectable, the definition of education and its relative importance in museums and the museum industry had changed over the past three decades. The modern world has changed the social and cultural fabric in which the "tradition" functions of museums. The very nature of education, including our definition and expectations of educational institutions, has changed. For instance, museum education was shifting to placing identity and maintaining traditions. Studied by tourism academia, Dimache et al. (2017) found a new model for understanding the visiting experience, museum education and narratives at the Hong Kong Museum of History. Learning history in local and national museums, personal exploration and visiting experiences can make individuals become storytellers and narrators who are able to pass on their own experiences in the place of memory. Having a tour of the museum could learn better Chinese culture and the development of China since Hong Kong was controlled once by the British and taken back with a special political agreement "One Country, Two System" in 1997. As same as Dimache et al. (2017), Vickers (2007) concentrated on the relationship between museums and nationalism in contemporary China. In particular, the study explored the impact on the museum of the shift in national ideology from socialism to patriotism, which had been particularly pronounced since the early 1990s. In contemporary China, museums were officially designated as "bases for patriotic education," but the "patriotism" they were meant to promote remains vague or problematic in many ways.

Thirdly, on account of the previous learning of Chinese museums and differentiated museum education, it is a suitable time to reconsider what is museum education, how will adult education participate and influence the development of museums, and the roles of museums. Hein (1998) talked about the definitions of education in a museum. People started to think critically and rationally because a major factor in this change in the definition of education was our current more sophisticated approach to education. Learning was now seen as the learner's active engagement with the environment. This concept of learning elevated experience (as opposed to the information contained in books) to a more important place in educational efforts. Museums focused on the "stuff" of the world. They specialized in objects that represented culture and nature and thus became central to any educational effort as the emphasis shifted from the written word to the learner's active participation through interaction with objects.

In sum, the implications for the field of adult education and museums were seen there were abundant possibilities to explore the cultural system from the eclectic perspective of adult education. As places of learning museums and art galleries can be found everywhere in the cultural landscape. They were easily accessible, dispersed, and regularly attracted a highly motivated and diverse group of adult learners. Therefore, the influence in the field of adult education was numerous. In a future planning, the field has a responsibility to make both scholars and adult educators aware of the untapped resources that exist in these institutions. Many were ideal places for practice and research in adult learning and teaching. It had long been believed that non-formal and informal education, and more specifically, cultural institutions, despite their unique challenges (e.g., limited resources and time constraints) provided much of the practice of "formal education" that taught by educators and outreach assistants. Such as some scheduled lectures with weekly or monthly special topics. The different learning experiences in museums and art galleries became mixed up together. Visitors and scholars may assume the boundary between formal learning to non-formal and informal learning was blurred.

5.2. The Transformation of Smart Education – The Danxiashan Geo-Museum in the Danxiashan UNESCO Global Geopark

Danxiashan, named after China Red Stone Park and Danxia landform, is located in Shaoguan City, Guangdong Province, with a total area of 292 km^2. It is a scenic spot with the Danxia landform and the landscape of socio-cultural of local people living in the mountain area. Because of the clam in color of red with a various degree, Danxiashan is acknowledged as the world's low-altitude mountain scenic area outstanding representative. It was opened to the public in 1980, approved as the world's first global geopark in 2004, and listed as a World Natural Heritage by UNESCO in August 2010 (The International Centre on Space Technologies for Natural & Cultural Heritage under the Auspices of UNESCO, 2016).

Moving forward to the museum, Danxiashan Geo-Museum is one of the five provincial education bases named by the Guangdong Provincial Department of Land and Resources as the Guangdong Provincial Land and Resources Science education base. Danxiashan Geo-Museum is located inside the UNESCO Global Geopark. It is the first stop for visitors to visit the Geopark. It covers a total area of more than 5,000 square meters. There are exhibition halls such as a panorama simulation sand table of Danxia Mountain, comprehensive Hall, multimedia Hall, Earth Science Hall, Geology Hall, Biology Hall, etc. Through the graphic display board, multimedia playback, physical specimens and other forms, it vividly shows the basic knowledge of geology, the formation and evolution of landforms, and the characteristics of regional biodiversity.

In the first quarter of 2023, the Danxiashan UNESCO Global Geopark received 316,600 tourists, which was an increase of 370% year-on-year (Ministry of Culture and Tourism of the People's Republic of China, 2023). With the huge volume of visitors coming over the geopark, Danxiashan tried to catch the attention and educate people along with its designed models. Starting from the idea of education tourism, it was a rapidly growing market in China. Danxiashan has been selected to be part of the scientific popularization campaign and become the flagship tourism product of the local area. This made Danxiashan a useful case study for understanding the development of smart education collaborating with museum education, natural engagement, and the power of social media.

As introduced, Danxiashan Geo-Museum played an important role in public education and the popularization of natural knowledge locally. In the past, Danxiashan Geo-Museum operated and managed in a very traditional way to fulfill the needs of museum education and knowledge spreading. With the new demand for knowledge-driven visiting, Danxiashan Geo-Museum inserted and introduced a new plan to achieve the educational purpose under the development of smart cities.

Firstly, introduced by the previous concept of smart education and the PRC 14th Five-Year Plan (2021–2025), informatization can be described as the second evolution of human beings. China proposed to build itself a digital country, and a smart society. China has it believes that digitalized China is a change of forming and digitalizing sectors and networks in various fields. So the smart society should always reflect on the value of people-oriented. Danxiashan Geo-Museum and the UNESCO Geopark support the digitalizing ecosystem and re-develop itself by inserting more layers into its learning environment and the ways of learning. According to the Danxiashan Management Committee of Shaoguan City, during the year 2020, the hardest situation (the COVID era) has never stopped Danxiashan UNESCO Global Geopark from making its effort to spread knowledge and develop a better digital platform for smart education. Examples were found that it held 200 science popularization activities independently and organized 20+ online science popularization training courses which added up together, more than 100,000 people participated in online learning (Summary of the Work of Danxiashan Global Geopark, 2020). Truly, COVID abandoned people to be mobile, but smart education emerged and launched with coping and adapting to the unexpected situation and global health emergency. The online platforms set for course learning and digital touring at the Danxiashan Geo-Museum and UNESCO Global Geopark gradually become popular and operatable for other geoparks and heritage sites to refer and learn from.

Secondly, since the year of 2018, The Danxiashan Geo-Museum and the Danxiashan Nature School promoted the public, especially young people, to respect, love and protect nature, and to recognize the organic links between culture, geology, biology, and other disciplines. Courses with different themes are set up to give the audience a deep understanding of the geology, ecology, culture and history of Danxia Mountain by means of lectures, investigations, production and experiments (Summary of the Work of Danxiashan Global Geopark, 2018). In the year 2020, UNESCO Global Geopark carried out 16 science popularization lectures on Danxia landform research practice and nature education online classes. From a visual perspective, the online promotion activities along with digital touring Danxiashan Geo-Museum enrich the learning experience and knowledge gaining of students and its target audiences. Besides the online platforms of course learning, digital touring is a new hotpot to advance smart education in China and the educational purpose brought by Danxiashan UNESCO Global Geopark. In detail, online tourism activities have been carried out in museums such as Danxiashan Geo-Museum. The form of digital touring has entered the public eye since the COVID. The form of digital touring has entered the public eye since the COVID (Xinhua News, 2022). The rise of the digital cultural tourism industry represented by digital touring has promoted the optimization and upgrading of smart education and re-structuring the learning environment under museum education. The form of digital touring – "cloud tourism" has entered the public eye since the COVID. The rise of digital tourism represented by "cloud tourism" and digital touring has promoted the optimization and upgrading of the learning environment and re-structuring the heritage education and museum education under the form of smart education. It also helped to integrate online and offline resources and realized the coordinated development of online and offline. At the same time, digital touring has a wide audience, is not limited in time and space, and is more convenient for communication. The public can enjoy heritage relics, knowledge learning, and beautiful scenery without leaving their homes and spending no money, and truly realize the sharing of resources by all.

To conclude, the case of the Danxiashan Geo-Museum in the Danxiashan UNESCO Global Geopark and the successful stories of its museum education under the processing of smart cities were well-matched with the ongoing socio-innovative development and the changes in such smart city in Shaoguan, Guangdong, China. Excitingly, in 2021, Danxiashan UNESCO Global Geopark has rewarded the World Heritage Institute of Training and Research for the Asia and the Pacific Region under the auspice of the UNESCO (WHITRAP) Award for Innovative Case Studies on World Heritage Education.

6. Discussion

In summary, the contribution of this research contribution proves that in the digitalizing age, everything is connected. Education is changing with the times. Especially in the context of smart education, museum education and world heritage education are changing rapidly. The rebirth of museum education is not only one of the most important ways of connecting people and world heritages but also a new way of development and innovation toward the general direction of heritage education and education sustainability. With the introduction of smart education, museum education is constantly updating its educational content and educational methods. This provides new ways of learning and more possibilities for future education and public education. Possibly, this educational approach and the new blending of ways of learning can be applied in the Gulf Region. As one of the successful cases, Danxiashan Geo-Museum promotes a new type of museum education in a cross-border innovative way. It combines and improves the new ideas and methods of museum education and good heritage education with the platform of creativity and technology. Finally, it uses digitalization to carry out popular science education and knowledge sharing. The combination of online and offline allows more people to experience the new type of museum education.

Interpreting the key findings in this chapter for diverse sectors and stakeholders in smart cities, the education sector and educational institutions such as public museums shall work along with the local, provincial, and central government to search and survey the feedback of the new changes on museum education and smart education. This is one of the ways to ensure that future education on heritages in the public learning sphere is on the right track fulfilling the needs of the citizens in smart cities worldwide. Urban planners and policymakers in smart societies shall use educational institutions and public resources wisely by envisioning the blueprint of ongoing smart innovations and societal changes. All in all, museum education and smart education under the construction of smart cities essentially shall blend to innovate and explore new ways of learning which will benefit citizens living in such smart cities for all toward lifelong learning.

7. Conclusion

In conclusion, the roles of museums significantly influenced and participated in both individual and collective life. Referencing and studying the previous research and existing knowledge, the frameworks studied the development of museums and the relationships of museums gave a view of how museums are involved in our livelihood. In detail, the roles of museums are split into two parts. Museums have positive roles in society because of their functionality and services. Museums were viewed as a group of public institutions that were functionalized to provide the public goods sharing knowledge, culture, and history. Museums were well-known for their services as storing and protecting both the tangible natural heritages and the intangible cultural heritages. Museums also fulfilled the educational roles that educated generation after generation to grow by committing to the construction of spiritual civilization and shared socio-cultural living space. Because of globalization, the global cooperation between museums has become closer and closer. Learning from Dorfman (2012), museums have evolved into global institutions from which new and important advances can emerge. If museums wanted to further develop their proper position in the global scientific community, they needed to do more work and ensure that all relevant parties could obtain tangible and intangible information gathered by them. Environmentalists, conservationists, education workers and researchers, and even academic naturalists can through the internet and other ways to communicate and share information. New frontiers of science, such as research into the possibility of cloning extinct species, may also help bring new life and resources to protect and develop the natural science collections in our museums. Nevertheless, museums have negative roles and impacts on our lives. For example, museums' exhibitions and the selected objects and specimens indicated the political intervention that ideologically shared social norms being interested in the leader of the nation and the powerful political party. The museum's collections and artifacts also display symbols of royalty. Consequently, a rising concern of "exclusion" and a debate on "othering" burst out at once. Museums had been critiqued on their exhibitions and information spreading. Researchers studied Canadian case studies showed the social and cultural exclusion toward to First Nation leading to a strong resonance and a public outcry, which related to the topic of colonialism and racism. As a citizen concerned about the construction and development of museums even a qualified global citizen, we must continue to expose and criticize traditions that perpetuate inequality and preserve the status quo, but in a world hungry for hope. Museums would offer new teaching possibilities. Like, it was important to protect the cultural rights of the public, promote sustainable development and reduce the worry of "otherness." It was necessary to coordinate education, planning and other aspects of cultural management and improve public cultural participation, rather than eliminate the economic threshold of public cultural participation only through public finance.

Furthermore, the importance of smart education makes a new contribution to museum education in China, inspiring visitors, educators, and scholars to rethink the roles of museums and the way of learning. Findings from research related to adult education and museum education indicated socially orientated expectations of museums, including the interests of all of society, heritage education and respecting cultural differences. The implications of smart education in the rebirth of museum education manifested the new online platforms setting for course learning and digital touring improve and extend museums' social, cultural, and educational roles through different ways of learning experiences including a large proportion of non-formal and informal learning in such digitalizing age.

Finally, studying the development of museums and their various roles helped to reinforce museum education and museum-related services. Indeed, future education and world heritage education in China will need a new evolution and a need layer inspired by the development of smart cities in China. The usage of digital technologies and the implications of smart education in museum education will create a better learning environment for sharing knowledge, appreciating heritage and innovating new educational methods. For future research, the study on museum education and smart education under the ongoing development of smart cities will be going to find ways to integrate the core contribution with other educational platforms and target particular groups of people for short-term and long-term experiments. In the future, the new era of the development of museums and their advanced roles in society will be valued on their importance of heritage exporting and soft power regionally and internationally.

References

Chinese Government Network. (2021). *Opinions on strengthening the construction and application of online teaching and learning resources in primary and secondary schools*. Retrieved February 10th, 2024, from http://www.gov.cn/xinwen/2021-02/08/content_5586111.htm

Clover, D. E. (2015). Adult education for social and environmental change in contemporary public art galleries and museums in Canada, Scotland, and England. *International Journal of Lifelong Education*. https://doi.org/10.1080/02601370.2014.993731

Clover, D. E. (2018). Critical adult education and the art gallery museum. *International Journal of Lifelong Education*, *37*(1), 88–102.

Clover, D. E., Sanford, K., Bell, L., & Johnson, K. (Eds.) (2016). *Adult education, museums and art galleries: Animating social, cultural and institutional change*. Sense Publishing.

Demir, K. A. (2021). Smart education framework. *Smart Learning Environments*, *8*(1). https://doi.org/10.1186/s40561-021-00170-x

Dimache, A., Wondirad, A., & Agyeiwaah, E. (2017). One museum, two stories: Place identity at the Hong Kong Museum of history. *Tourism Management*, *63*, 287–301. https://doi.org/10.1016/j.tourman.2017.06.020

Dorfman, E. (2012). *Intangible natural heritage new perspective on natural objects*. Routledge.

Gray, C., & McCall, V. (2020). *The role of today's museum*. Routledge.

Hein, G. E. (1998). *Learning in the museum*. Routledge.

Janes, R. R. (2016). *Museums without borders selected writings of Robert R. Janes*. Routledge, Taylor & Francis Group.

Ji, J., Anderson, D., Wu, X., & Kang, C. (2014). Chinese family groups' museum visit motivations: A comparative study of Beijing and Vancouver. *Curator: The Museum Journal*, *57*(1), 81–96. https://doi.org/10.1111/cura.12052

Kang, I., & Lee, S. (2014). A *conceptual framework for smart museum education*. MWA2014 Museums and the Web Asia 2014. Retrieved February 10th, 2024, from https://mwa2014.museumsandtheweb.com/paper/a-conceptual-framework-for-smart-museum-education/index.html

Lu, T. L. (2015). *Museums in China: Power, politics and identities*. Routledge, Taylor & Francis Group.

Mayo, P. (2013). Museums as sites of critical pedagogical practice. *Review of Education, Pedagogy, and Cultural Studies*, *35*(2), 144–153.

Rogers, A. (2014). *The Base of the Iceberg. Informal learning and its impact on formal and non-formal learning*. https://doi.org/10.3224/84740632

Spencer, B., & Lange, E. (2014). *The purposes of adult education: A short introduction* (3rd ed.). Thompson Educational Publishing Inc.

Summary of the Work of Danxiashan Global Geopark. (2018). Retrieved February 10th, 2024, from https://globalgeoparksnetwork.org/?page_id=1945

Summary of the Work of Danxiashan Global Geopark. (2020). Retrieved February 10th, 2024, from https://globalgeoparksnetwork.org/?page_id=4851

Taylor, E. W. (2010). Cultural institutions and adult education. *New Directions for Adult and Continuing Education*, *2010*(127), 5–14.

The International Centre on Space Technologies for Natural and Cultural Heritage under the Auspices of UNESCO. (2016). Retrieved February 10th, 2024, from http://www.unesco-hist.org/index.php?r=article/info&id=834

Vickers, E. (2007). Museums and nationalism in contemporary China. *Compare: A Journal of Comparative and International Education*, *37*(3), 365–382. https://doi.org/10.1080/03057920701330255

Wang, W., Fu, M., & Hu, Q. (2020). The behavioral pattern of Chinese PUBLIC cultural participation in museums. *Sustainability*, *12*(7), 2890. https://doi.org/10.3390/su12072890

Xinhua News. (2022). Retrieved February 10th, 2024, from http://www.xinhuanet.com/fortune/2022-06/04/c_1128712177.htm

Yang, Y. K. (2017). The case study of contents development for online museum art appreciation based on smart media. *Journal of Information Technology Services*. https://doi.org/10.9716/KITS.2017.16.1.139

Zhu, Z. T., & He, B. (2012). Smart education: New frontier of educational informatization. *E- Education Research*, *12*, 1–13.

Zhu, Z.-T., Yu, M.-H., & Riezebos, P. (2016). A research framework of smart education. *Smart Learning Environments*, *3*(1). https://doi.org/10.1186/s40561-016-0026-2

Chapter 7

The Synergy Between Advanced Innovative Water Management and Smart Cities Development: Enhancing Sustainability and Quality of Life

Hend Sameh Hafez Hassan[a,b], *Ahmed Abdelkader*[c,d] *and Oualid Abidi*[e]

[a]*Faculty of Business Administration, Arab Open University, Kuwait*
[b]*Faculty of Commerce, Mansoura University, Egypt*
[c]*Arab Open University, Kuwait*
[d]*Damietta University, Egypt*
eHuman Resources Management Department, College of Business, Australian University, Kuwait

Abstract

The concept of smart cities, driven by advancements in innovative information and communication technologies (ICTs), has gained significant attention in recent years. Smart cities aim to improve the quality of life for citizens by leveraging ICT to enhance the efficiency and effectiveness of urban services and infrastructure. One critical aspect of smart cities development is advanced innovations in water management, which play a vital role in achieving sustainability, prosperity of community and ensuring the availability of clean water resources. This chapter explores the relationship between advanced water management and smart cities development and highlights the synergies and benefits that arise from their integration. The chapter develops a framework for adopting innovative ICTs that support the gradual transformation toward next generation smart cities in the Gulf Cooperation Council (GCC) region. Such transformation aligns with the United Nations' sustainable development goals (SDGs) and the maintenance of various social, economic, and environmental developments.

The Emerald Handbook of Smart Cities in the Gulf Region:
Innovation, Development, Transformation, and Prosperity for Vision 2040, 117–134

doi:10.1108/978-1-83608-292-720241007

The chapter begins by discussing the fundamental principles of smart cities and the role of advanced sensing technologies in enabling efficient and automated processes within urban environments. It then delves into the concept of water-sensitive cities, the importance of urban water mass balance analysis in designing sustainable water management strategies, and the emerging trends in water management. Furthermore, the chapter explores the integration of smart program management and the role of citizen engagement in the design and development of smart cities in the GCC countries and finally challenges and concerns facing these programs.

Keywords: Smart cities; advanced water management; sustainable development; GCC countries; water scarcity management; water conservation strategies

Introduction

The fundamental principles of smart cities encompass a holistic approach to urban development, integrating advanced technologies, sustainability, and citizen-centric governance. Smart cities are characterized by their ability to leverage data and technology to enhance the efficiency of urban systems, improve the quality of life for residents, and promote sustainable development. These cities are built on principles such as accessibility, openness, and comfort of urban spaces, aiming to create conditions for interaction among inhabitants and foster a sense of community. Additionally, the three E's of sustainability – Economically Prosperous, Socially Equitable, and Environmentally Sustainable – are fundamental principles that guide the development of smart cities, ensuring that urban growth is balanced and sustainable.

Advanced sensing technologies play a pivotal role in enabling efficient and automated processes within smart cities. These technologies encompass a wide range of applications, including vehicular sensing networks, social opportunistic sensing, and IoT-based open platforms for urban farming services. The integration of advanced sensing technologies allows for the creation of ubiquitous sensing infrastructures, facilitating the collection of real-time data for analysis and decision-making. Moreover, the use of sensing technologies, such as optical fibers and lasers, enables structural health monitoring, electrical transmission line monitoring, and the automated detection of anthropogenic changes in urban development, contributing to the resilience and sustainability of urban infrastructure.

Furthermore, the role of advanced sensing technologies extends to the analysis of accessibility for disabled individuals in sustainable cities, where radio frequency technologies are employed to assess the efficiency and usability of urban spaces. These technologies provide new opportunities to analyze the adequacy of facilities and promote social integration capabilities, aligning with the principles of inclusivity and social equity within smart cities.

The principles of smart cities and the role of advanced sensing technologies are also closely linked to the concept of citizen-centric governance and the active participation of residents in urban decision-making processes. The citizen-centric smart city paradigm emphasizes the importance of engaging with public stakeholders and enhancing the sense of gain for citizens through the application of intelligent technologies. This approach shifts the focus from technology-centric development to citizen-centric development, ensuring that the needs and well-being of residents are at the forefront of urban planning and management.

Advanced Innovative Water Management and Smart Cities Development

The relationship between advanced water management innovations and the development and transformation of smart cities is a critical aspect of urban development. Water is a fundamental resource for any city, and its efficient management is essential for the sustainability and resilience of smart cities (Šulyová & Kubina, 2022). Advanced water management is closely linked to the concept of smart cities, as it plays a crucial role in ensuring the efficient use and distribution of limited water resources within urban environments (Hernaningsih et al., 2023). Smart water management (SWM) systems are being integrated into the infrastructure of smart cities to address water scarcity issues and ensure sustainable clean water supply (Gade, 2021). These systems utilize advanced technical solutions, such as the Internet of Things (IoT) and information and communication technology (ICT), to optimize water distribution and facilitate water recycling for various activities (Gade, 2021; Gonçalves et al., 2020).

Furthermore, the integration of advanced water management into smart cities aligns with the broader goals of urban sustainability and environmental resilience. It is emphasized that water management in smart cities is increasingly appreciated from financial and environmental sustainability perspectives, highlighting its significance in the water sector (Ramos et al., 2019). Additionally, the application of smart water protocols for irrigation in urban gardens within smart cities demonstrates the multifaceted importance of water management in enhancing the livability and sustainability of urban environments (Aldegheishem et al., 2022). This underscores the critical role of advanced water management in promoting sustainable practices within smart cities.

The relationship between advanced water management and smart cities is also interconnected with other aspects of urban development, such as energy sustainability and transportation. Studies have highlighted the contribution of smart cities to energy sustainability and the role of advanced information technology in realizing smart management and operation of cities, thereby promoting harmonious and sustainable urban growth (Badidi, 2022; Villa-Arrieta & Sumper, 2019). Moreover, the development of smart city traffic management and intelligent transportation systems is revolutionizing urban mobility and emergency response, further emphasizing the interconnected nature of advanced water management with broader urban management systems (Badidi, 2022; Mohamed & Al-Shalfan, 2021).

Water Management and Smart Cities Integration: Synergies and Benefits

The integration of advanced water management and smart cities yields numerous synergies and benefits that are pivotal for sustainable urban development. The convergence of SWM systems with the broader framework of smart cities results in multifaceted advantages that encompass environmental sustainability, resource efficiency, and enhanced quality of life for urban residents.

Firstly, the integration of advanced water management within smart cities facilitates the development of an integrated, sophisticated, and sustainable clean water supply system (Hernaningsih et al., 2023). This integration allows for the efficient distribution and control of water supply based on demand, as well as the facilitation of water recycling for various activities, addressing water scarcity issues and promoting sustainable water usage (Gade, 2021). Moreover, the incorporation of SWM contributes to the overall environmental sustainability of smart cities, aligning with the goals of financial and environmental sustainability within the water sector (Ramos et al., 2019).

Furthermore, the integration of advanced water management with smart cities leverages the interconnected nature of urban systems, leading to synergies among various domains such as energy, transportation, waste management, and e-governance (Calvillo et al., 2017; Lynggaard & Skouby, 2016). This interconnectedness enables the optimization and operation of systems, allowing smart cities to profit from potential synergies and enhance resource-efficient management and economy (Fokaides et al., 2018; Impedovo & Pirlo, 2020). Additionally, the integration of SWM contributes to the overall smart city vision by enriching the quality of life for urban citizens and promoting sustainable urban development (Angelidou et al., 2017).

The benefits of integrating advanced water management with smart cities extend to the realm of technological innovation and data-driven decision-making. The utilization of advanced sensing technologies and IoT-driven solutions for SWM enhances the automation of processes within cities, leading to more efficient and informed decision-making (Faisal et al., 2018; Hancke & Silva, 2012). Moreover, the integration of smart city solutions allows for real-time data collection and analysis, significantly enhancing decision-making capabilities and contributing to the overall efficiency of urban management (Faisal et al., 2018).

Water-Sensitive Cities

The concept of water-sensitive cities (WSCs) represents a paradigm shift in urban water management, emphasizing the integration of water into the fabric of urban development to achieve sustainability, resilience, and liveability. This approach envisions cities that effectively manage their water resources, minimize environmental impact, and enhance the overall quality of life for residents. The WSC concept encompasses a range of principles and practices aimed at transforming urban water management, including rainwater utilization, water conservation, source reduction, and ecological landscaping Yang et al. (2022). It represents a holistic approach to urban water management that integrates water supply, sanitation, flood protection, and environmental protection strategies to ensure long-term sustainability and prosperity (Wong et al., 2020).

The core principles of WSCs are rooted in the idea of harmonizing urban development with the natural water cycle, thereby minimizing the impact of urbanization on water resources and ecosystems. This involves the adoption of water-sensitive urban design (WSUD) principles, which integrate the urban water cycle, incorporating water supply, wastewater, stormwater, and groundwater management, along with urban design and environmental protection (Chandran & Gowda, 2014). The WSUD approach emphasizes the interconnectedness of water systems and urban infrastructure, promoting the sustainable use of water resources and the preservation of natural hydrological processes within urban environments (Lee, 2015).

Advanced sensing technologies play a crucial role in enabling the realization of WSCs by providing real-time data for informed decision-making and the efficient management of water resources. These technologies facilitate the monitoring of water quality, water flow, and environmental conditions, allowing for proactive management of urban water systems and the early detection of potential issues (Ari, 2023). Furthermore, the integration of sensing technologies with WSUD practices enables the development of innovative stormwater management solutions, such as rain gardens and leaky wells, which contribute to the sustainable control of stormwater quantity and quality within urban areas (Ahammed et al., 2012; Wanitchayapaisit et al., 2022).

The transition toward WSCs also involves the development of assessment frameworks and indicators to evaluate the progress and effectiveness of WSUD in urban planning. These frameworks aim to enhance city-to-city learning and provide a basis for evaluating the sustainability and resilience of urban water management practices (Koop & Leeuwen, 2015; Madonsela et al., 2019). Additionally, the concept of resilience to extreme climatic events and climate change plays a key role in the WSC paradigm, emphasizing the need for urban water management to adapt to changing environmental conditions and mitigate the impact of water-related challenges (Rodríguez et al., 2014).

The implementation of WSUD principles also extends to the social and human aspects of urban development, emphasizing the creation of water-sensitive communities and the integration of water into urban processes. This approach aims to enhance the sensory and recreational value of water within urban spaces, fostering a deeper connection between residents and their water environment (Vernon & Tiwari, 2009). Moreover, the WSC concept promotes an integrative approach to urban water management, involving interdisciplinary collaboration between experts in water management, architecture, engineering, urban planning, and landscape design to minimize the hydrological impact of urban development on the environment (Holubchak & Obynochna, 2022).

Urban Water Mass Balance Analysis

Urban water mass balance analysis plays a pivotal role in designing sustainable water management strategies by providing a comprehensive understanding of water flows, consumption, and distribution within urban environments. This analytical approach enables the quantification of all natural and anthropogenic flows and fluxes of water through a defined urban area, facilitating the identification of potential inefficiencies, losses, and opportunities for sustainable water management (Farooqui et al., 2016). The importance of urban water mass balance analysis lies in its ability to inform evidence-based decision-making, optimize water resource allocation, and enhance the resilience and sustainability of urban water systems.

The principles of mass balance analysis are instrumental in fostering robust water accounting, monitoring, and management within urban areas (Kenway et al., 2011). By quantifying all flows and fluxes of water into, through, and out of a defined urban boundary, mass balance analysis provides a comprehensive overview of the urban water cycle, including water supply, wastewater, stormwater, and groundwater management (Renouf et al., 2017). This holistic understanding is essential for identifying potential areas for improvement and implementing targeted interventions to enhance the efficiency and sustainability of urban water management.

Furthermore, urban water mass balance analysis serves as a fundamental tool for evaluating the performance of urban water systems and assessing the impact of various urban design options on the water balance. Through the analysis of evapotranspiration, stormwater, and other water balance components, this approach enables the assessment of the effects of urban development on the water cycle, guiding the design of sustainable urban infrastructure and landscape features (Mitchell et al., 2007). It also supports the evaluation of the local influence of residential infill on urban water flows, providing insights into the performance of different design typologies in managing urban water resources.

In addition, water mass balance analysis contributes to the development of integrated urban water management strategies by bridging the gap between visions and performance assessment of urban water resource management (Renouf et al., 2017). This approach quantifies all natural and anthropogenic water flows, providing a basis for informed decision-making and the prioritization of sustainable water management strategies (Banihabib et al., 2020). It also supports the evaluation of the impact of transitions from centralized to decentralized water solutions on existing infrastructures, facilitating the identification of opportunities for enhancing the sustainability and resilience of urban water systems (Sitzenfrei et al., 2013).

Moreover, the application of water mass balance analysis in urban water management aligns with the principles of sustainability and resilience, as it enables the identification of opportunities for sustainable water resource management and the development of strategies to address the challenges posed by climate change and urbanization (Worku, 2017). By quantifying all water flows and providing a comprehensive understanding of the urban water cycle, mass balance analysis supports the development of adaptive responses to climate change and the formulation of feasible strategies for sustainable urban water management (Kinyanjui, 2019).

The evolving trends in water management encompass a wide array of developments and innovations that are shaping the sustainable use and governance of water resources in urban and regional contexts. These trends reflect the evolving challenges, technological advancements, and policy shifts that are influencing the management and conservation of water resources.

Several scholars provide valuable insights into the emerging trends in water management, shedding light on the diverse factors and developments that are shaping the contemporary water management landscape.

In this regard, Brown and Farrelly (2009) highlight the hurdles and advancements in sustainable urban water management, emphasizing the need for mainstream adoption of sustainable practices in the water and development industries. This underscores the ongoing efforts to overcome barriers and promote the widespread implementation of sustainable water management strategies, reflecting an emerging trend toward greater adoption of sustainable practices in urban water management. Similarly, Meene et al. (2011) discuss the evolving governance arrangements for sustainable urban water management, indicating a trend toward the transition from hierarchical and centralized management structures to market governance structures. This shift in governance arrangements reflects an emerging trend toward more adaptive and market-oriented approaches to urban water management, signaling a departure from traditional management paradigms.

Liu et al. (2021) delve into the shifting entrepreneurial landscape and development performance of water startups in emerging water markets, highlighting the technological and commercial revolutions in the water sector. This points to an emerging trend of increased innovation and entrepreneurship in addressing water management challenges, reflecting a growing emphasis on technological and commercial advancements in the water sector.

The need for sustainable urban water management in response to rapid urbanization, population growth, and the effects of climate change has been emphasized in Barron et al. (2017) study. This underscores the emerging trend of heightened awareness and action toward sustainable urban water management in the face of urbanization and environmental challenges, reflecting a growing emphasis on sustainability in urban water management practices.

The key trends in water legislation, including the attraction of water resources into the public domain, limitations on governmental authority, controlled trading of water rights, and ensuring user participation in decision-making have all been addressed by Burchi (2012). This reflects an emerging trend toward more inclusive, participatory, and sustainable approaches to water legislation and governance, signaling a shift toward more collaborative and equitable water management practices.

The Role of Citizen Engagement in the Design and Development of Smart Cities

Citizen engagement plays a pivotal role in the design and development of smart cities, fostering inclusive governance, participatory decision-making, and the co-creation of urban environments. The involvement of citizens in the planning and implementation of smart city initiatives is essential for ensuring that urban development aligns with the needs, preferences, and aspirations of the community. The references provide valuable insights into the multifaceted role of citizen engagement in the design and development of smart cities, shedding light on the diverse factors and developments that are shaping the contemporary urban governance landscape.

The multidisciplinary understanding of smart city governance, as discussed by Webster and Leleux (2018), emphasizes the opportunities for citizen engagement in the co-production of service delivery and decision-making. This underscores the role of citizens as active participants in the governance processes, contributing to the co-creation of urban services and policies.

The study by Przeybilovicz et al. (2020) provides empirical insights into citizen engagement in smart city governance, demonstrating that citizens undertake dynamic roles and functions that evolve. This highlights the active and evolving nature of citizen participation in smart city initiatives, emphasizing the need for flexible and adaptive governance structures that accommodate diverse forms of citizen engagement. Additionally, Zhu et al. (2022) emphasize that citizen engagement serves as an effective measure of the service quality of smart cities, underscoring the role of citizens as key stakeholders in evaluating and improving urban services.

The concept of "smart citizens," as discussed by (Shin et al., 2021), underscores the significance of citizen engagement in smart cities, portraying citizens as active and empowered participants who leverage technology to express themselves, connect with others, and contribute to decision-making processes. This highlights the transformative role of citizens in shaping the digital and social fabric of smart cities, emphasizing their agency in driving urban innovation and governance. Moreover, Kusumastuti and Rouli (2021) emphasize that citizen engagement is a crucial element of smart city governance, underscoring the importance of citizen involvement in shaping the development and implementation of smart city initiatives.

The case study presented by Chantry (2022) utilizes citizen engagement documentation from the Quayside smart city project in Toronto to assess citizen participation in smart city planning. This approach aims to develop a new heuristic for evaluating citizen engagement, emphasizing the need for robust methodologies to assess and enhance citizen involvement in smart city initiatives. Furthermore, Khan et al. (2019) propose a novel blockchain and edge computing-based architecture to engage citizens in urban planning, highlighting the potential of technological innovations to facilitate citizen participation in smart city governance.

The study by Choo et al. (2023) identifies barriers to citizen engagement in smart city plans, emphasizing the need to overcome obstacles and enhance citizen involvement in the planning and development of smart cities. This underscores the importance of addressing challenges and creating enabling environments that promote inclusive and meaningful citizen engagement. Additionally, Lim et al. (2018) evaluate the level of citizen engagement in building citizen-centric smart cities, emphasizing the value of participatory approaches in shaping urban services and public value chains.

The URBiNAT project, as discussed by (Nunes et al., 2021), emphasizes the role of citizen engagement in the co-creation of nature-based solutions (NBS) for sustainable urban development. This highlights the potential of citizen involvement in shaping urban regeneration and the deployment of NBS, underscoring the transformative impact of citizen engagement on urban sustainability. Furthermore, Ansari (2021) emphasizes that citizen participation is instrumental in producing socially equitable urbanization, highlighting the role of citizens as active agents in shaping inclusive and equitable urban development.

Citizen engagement also plays a crucial role in SWM, fostering inclusive governance, participatory decision-making, and the co-creation of sustainable water management strategies. The involvement of citizens in the planning and implementation of smart water initiatives is essential for ensuring that urban water management aligns with the needs, preferences, and aspirations of the community. The references provide valuable insights into the multifaceted role of citizen engagement in SWM, shedding light on the diverse factors and developments that are shaping the contemporary urban water governance landscape.

The study by Bermejo-Martín and Rodríguez-Monroy (2020) emphasizes the role of citizen engagement in the design of urban water sustainability solutions, highlighting the importance of participatory approaches in shaping sustainable water management strategies. This underscores the transformative impact of citizen involvement on urban water sustainability, emphasizing the need for inclusive and meaningful citizen engagement in water management decision-making processes.

Moreover, Preston et al. (2020) present practical lessons on how citizen engagement can facilitate the co-creation of low-carbon SWM strategies, highlighting the importance of citizen involvement in shaping sustainable urban water development, particularly in the context of energy in the built environment. This underscores the pivotal role of citizens as active stakeholders in shaping the future of SWM, emphasizing the importance of fostering meaningful and inclusive citizen engagement in urban water governance and decision-making processes.

The concept of "smart citizens" represents an evolving paradigm in the context of smart cities, emphasizing the active role of citizens in shaping urban innovation, governance, and decision-making processes. Smart citizens are envisioned as empowered and engaged individuals who leverage technology, data, and participatory approaches to contribute to the co-creation of urban environments and the improvement of public services. The references provide valuable insights into the multifaceted nature of smart citizens, shedding light on the diverse factors and developments that are shaping the contemporary urban governance landscape. The concept of "smart citizens" underscores the significance of citizen engagement in SWM, portraying citizens as active and empowered participants who leverage technology to express themselves, connect with others, and contribute to decision-making processes. This highlights the transformative role of citizens in shaping the digital and social fabric of SWM, emphasizing their agency in driving urban innovation and governance.

The study by Przeybilovicz et al. (2020) identifies emerging patterns of contextually specific citizen interaction in SWM, highlighting the importance of citizen engagement in shaping the development and implementation of smart water initiatives. This underscores the pivotal role of citizens as active stakeholders in shaping the future of SWM, emphasizing the importance of fostering meaningful and inclusive citizen engagement in urban water governance and decision-making processes.

The URBiNAT project is a European Union Horizon 2020 initiative focused on promoting inclusive urban regeneration and the co-creation of NBS for sustainable urban development. The project aims to engage citizens and stakeholders in all stages of the co-creation process, emphasizing the importance of participatory approaches in shaping urban environments and enhancing the quality of life in cities. The URBiNAT project seeks to address urban challenges by integrating NBS into the fabric of cities, fostering social cohesion, environmental sustainability, and resilience. By involving citizens and stakeholders in the co-creation of NBS, the project aims to develop innovative approaches to urban regeneration that align with the needs and aspirations of the community.

One of the key objectives of the URBiNAT project is to develop guidelines for citizen engagement and the co-creation of NBS. These guidelines are designed to facilitate meaningful participation and collaboration among citizens, stakeholders, and local authorities in the development and implementation of NBS. By promoting inclusive urban regeneration, the project aims to create more sustainable and liveable urban environments that enhance the well-being of citizens. By engaging citizens in the co-creation of NBS, the project seeks to empower communities to contribute to the design and implementation of sustainable urban interventions. This participatory approach is intended to foster a sense of ownership and collective responsibility for the development of NBS, ultimately leading to more resilient and inclusive urban spaces.

Furthermore, the URBiNAT project aligns with the broader discourse on smart cities and citizen engagement. It emphasizes the importance of citizen-centric approaches to urban development, highlighting the transformative impact of citizen involvement on urban governance and decision-making processes. By promoting citizen engagement in the co-creation of NBS, the project aims to contribute to the development of more sustainable, inclusive, and citizen-centric smart cities. The URBiNAT project emphasizes the role of citizen engagement in the co-creation of NBS for sustainable urban water development. This highlights the potential of citizen involvement in shaping urban water regeneration and the deployment of NBS, underscoring the transformative impact of citizen engagement on urban water sustainability.

The Concept of Smart Cities in the GCC Countries

Smart cities in the Gulf Cooperation Council (GCC) countries represent a significant effort to leverage technology and innovation to enhance urban living, sustainability, and economic development. The GCC is a political and economic alliance between six Middle Eastern countries, namely, Bahrain, Kuwait, Oman, Qatar, Saudi Arabia, and the United Arab Emirates. The concept of smart cities in the GCC countries aligns with the broader global discourse on urban innovation, digital transformation, and sustainable development.

The GCC countries are keen on building upon their infrastructure to push their "smart city" agenda, which would go a long way in furthering the economic diversification objective of the region besides improving the quality of public services. The development of smart cities in the GCC countries is driven by the need to enhance urban infrastructure, improve public services, and foster economic diversification. These initiatives aim to leverage technology and innovation to create more sustainable, efficient, and liveable urban environments.

The development of smart cities in the GCC countries is closely linked to economic diversification efforts, as these countries seek to reduce their reliance on oil and gas revenues and transition toward knowledge-based economies. The implementation of smart city initiatives is intended to drive economic growth, enhance the quality of life for residents, and attract investment and talent to the region. These efforts are aligned with the broader goal of fostering innovation, entrepreneurship, and sustainable development in the GCC countries.

The concept of smart cities in the GCC countries encompasses a wide range of initiatives, including the deployment of advanced infrastructure, the adoption of digital technologies, and the implementation of sustainable urban development strategies. These initiatives aim to enhance urban mobility, improve energy efficiency, and promote environmental sustainability. The development of smart cities in the GCC countries reflects a commitment to leveraging technology and innovation to address urban challenges and enhance the well-being of citizens.

The implementation of smart city initiatives in the GCC countries is underpinned by a commitment to citizen-centric approaches to urban development. These initiatives emphasize the active involvement of citizens in shaping the future of urban environments, fostering inclusive governance, and participatory decision-making. The concept of smart cities in the GCC countries underscores the transformative role of citizens as active stakeholders in shaping the future of urban development, emphasizing the importance of fostering meaningful and inclusive citizen engagement in urban governance and decision-making processes.

SWM in the GCC countries represents a significant effort to leverage technology and innovation to enhance water conservation, distribution, and quality. The GCC countries, which include Bahrain, Kuwait, Oman, Qatar, Saudi Arabia, and the United Arab Emirates, are investing in SWM initiatives to address water scarcity, improve water infrastructure, and promote sustainable water use.

The concept of SWM in the GCC countries encompasses a wide range of initiatives, including the deployment of advanced water infrastructure, the adoption of digital technologies for water monitoring and management, and the implementation of sustainable water conservation strategies. These initiatives aim to enhance water efficiency, reduce water loss, and ensure the sustainable use of water resources in the region.

The GCC countries are facing significant water challenges, including scarcity, high demand, and the need for efficient water distribution. SWM initiatives in the region focus on leveraging technology, data analytics, and real-time monitoring to optimize water distribution networks, detect leaks, and improve water quality. These initiatives aim to enhance the resilience and sustainability of water infrastructure in the GCC countries.

The implementation of SWM initiatives in the GCC countries is underpinned by a commitment to citizen-centric approaches to water management. These initiatives emphasize the active involvement of citizens in shaping the future of water infrastructure, fostering inclusive governance, and participatory decision-making. The concept of SWM in the GCC countries underscores the transformative role of citizens as active stakeholders in shaping the future of water management, emphasizing the importance of fostering meaningful and inclusive citizen engagement in water governance and decision-making processes.

The development of SWM in the GCC countries also aligns with broader efforts to promote sustainable development and environmental conservation. These initiatives aim to enhance water quality, reduce water waste, and promote the efficient use of water resources. By leveraging technology and innovation, the GCC countries are working toward creating more sustainable, efficient, and resilient water management systems.

SWM in the GCC countries faces several challenges that are crucial to address for sustainable water resource management. The GCC countries are characterized by arid and semi-arid climates, limited freshwater resources, and high water demand due to rapid urbanization and economic development. The following are some of the key challenges in SWM in the GCC countries:

1. Water Scarcity: The GCC countries face significant water scarcity due to limited freshwater resources and high water demand, driven by population growth, urbanization, and industrial development. Addressing water scarcity through sustainable management practices, conservation efforts, and the development of alternative water sources is a critical challenge.
2. Energy-Intensive Desalination: Desalination is a primary source of freshwater in the GCC countries. However, desalination is energy-intensive and places a significant burden on the region's energy resources. Balancing the energy requirements of desalination with sustainable water production is a key challenge in SWM.
3. Wastewater Management: Effective management of wastewater is essential for water conservation and environmental protection. The GCC countries face challenges in treating and reusing wastewater to alleviate pressure on freshwater sources and minimize environmental impact.
4. Climate Change Impacts: Climate change poses a threat to water resources in the GCC countries, leading to changes in precipitation patterns, increased temperatures, and heightened water stress. Adapting to the impacts of

climate change and implementing resilient water management strategies are critical challenges.

5. Groundwater Depletion: Over-extraction of groundwater for agricultural, industrial, and domestic use has led to groundwater depletion in some GCC countries. Sustainable groundwater management and aquifer recharge initiatives are essential to address this challenge.
6. Infrastructure Development: The rapid urbanization and population growth in the GCC countries require significant investment in water infrastructure, including distribution networks, storage facilities, and treatment plants. Developing and maintaining robust water infrastructure is a key challenge for SWM.
7. Governance and Regulation: Effective governance, policy frameworks, and regulatory mechanisms are essential for sustainable water management. Strengthening water governance, enhancing regulatory frameworks, and promoting integrated water resource management are critical challenges in the GCC countries.
8. Public Awareness and Participation: Engaging the public in water conservation efforts and promoting awareness about the importance of sustainable water management are crucial challenges. Encouraging public participation and behavioral changes to reduce water consumption are essential for SWM.

Addressing these challenges requires a multi-faceted approach, including the adoption of advanced technologies for water monitoring and management, the implementation of water conservation and reuse practices, investment in sustainable infrastructure, and the development of effective policies and regulations. Collaboration among stakeholders, public awareness campaigns, and regional cooperation are also essential for addressing the complex challenges of SWM in the GCC countries.

To overcome the challenges of SWM in the GCC countries, a multi-faceted approach is required. This includes the integration of smart technologies, policy and strategy development, renewable energy applications, rainwater harvesting, and the development of new irrigation methods. These strategies can contribute to addressing water scarcity, ensuring sustainable economic growth, and enhancing food security in the region.

The scarcity of water resources in the GCC countries is a major challenge, with an annual per capita water availability of 500 m^3 compared to the world average of 6,000 m^3 (Qureshi, 2020). This scarcity is largely compensated through high-energy reserves and economic abundance (Al-Saidi & Saliba, 2019). To address this, SWM that integrates ICT to monitor water resources, diagnose problems, improve efficiency, and coordinate management can help overcome supply challenges and strengthen public participation in governance (Heland et al., 2015). Additionally, the use of renewable energy for desalination in urban agriculture presents possibilities and challenges that can contribute to sustainable water management in the GCC countries (Al-Jabri & Ahmed, 2018).

Furthermore, the GCC Supreme Council has issued a directive toward a long-term comprehensive Gulf water strategy, emphasizing the need for serious and speedy steps to be taken to address water challenges in the region (Al-Zubari et al., 2017). This highlights the importance of policy and strategy development to address water management challenges. Additionally, the application of renewable energies for water desalination can contribute to addressing the limited water and energy resources in the region (Goosen et al., 2011). Moreover, the use of rainwater harvesting and management practices, particularly rooftop rainwater harvesting, can be considered a soft-path water management strategy to mitigate short-term flooding problems in urban areas and contribute to sustainable water resource management (Amin, 2013).

In the context of smart city development, challenges such as traffic and transportation management, energy and water distribution and management, air quality, and waste management monitoring need to be addressed (Wang et al., 2021). Smart grid technologies in the GCC countries also present opportunities for addressing challenges in the electricity market, which is closely linked to water management and sustainability (Okedu & ALSalmani, 2019). Additionally, managed aquifer recharge and the development of new technologies and methods of irrigation are essential to utilize water resources effectively in the GCC countries (Hassen & Bilali, 2019; Parimalarenganayaki, 2020).

Conclusion

The concept of smart cities development, driven by innovations and advancements in ICTs, has gained significant attention in recent years. Smart cities aim to improve the quality of life for citizens by leveraging ICT to enhance the efficiency and effectiveness of urban services and infrastructure. One critical aspect of smart cities is advanced water management, which plays a vital role in achieving sustainability and ensuring the availability of clean water resources. The framework proposed in this chapter assists academic, practitioner, and industrial stakeholders in understanding the relationship between advanced innovations in water management and the development of smart cities by highlighting the synergies, benefits, and challenges that arise from their integration. The chapter discussed the fundamental principles of smart cities development and the role of innovative sensing technologies in enabling efficient and automated processes within urban environments. The concepts of WSCs, the importance of urban water mass balance analysis in designing sustainable water management strategies, and the emerging trends in water management have been discussed. Furthermore, the chapter explored the integration of smart program management and the role of citizen engagement in the design and development of smart cities in the GCC countries. Finally, the challenges and concerns facing the application and execution of smart cities principles.

Advanced sensing technologies and data analytics play a pivotal role in monitoring and managing water resources, enabling efficient and automated processes within cities. The concept of WSCs emphasizes the need for sustainable water management strategies that consider the urban water mass balance, thereby promoting the efficient use of water resources and minimizing wastage. Moreover, citizen engagement and smart program management are essential components in the development of smart cities, as they foster public participation and awareness, leading to more effective water management practices.

In the context of the GCC countries, the challenges and concerns facing SWM programs are significant. These include the physical scarcity of water resources, the need for sustainable economic growth, and the importance of policy and strategy development to address water management challenges. Additionally, the integration of renewable energy for water desalination and the implementation of rainwater harvesting practices present opportunities and challenges that can contribute to sustainable water management in the GCC countries. Furthermore, the development of new technologies and methods of irrigation, as well as the exploration of managed aquifer recharge, are essential for utilizing water resources effectively in the region.

Future researchers are encouraged to investigate the obstacles that policy makers, legislators, practitioners, and industrial stakeholders face in the integration of advanced water management innovations in smart cities. This is crucial for addressing the challenges of water scarcity, ensuring sustainable water supply, and enhancing the overall resilience of urban environments.

References

Ahammed, F., Hewa, G., & Argue, J. (2012). Introducing leaky-well concept for stormwater quantity control in Dhaka, Bangladesh. *Applied Water Science*, *3*(1), 115–123.

Aldegheishem, A., Alrajeh, N., García, L., & Lloret, J. (2022). Swap: Smart water protocol for the irrigation of urban gardens in smart cities. *IEEE Access*, *10*, 39239–39247.

Al-Jabri, S., & Ahmed, M. (2018). Use of renewable energy for desalination in urban agriculture in the GCC countries: Possibilities and challenges. *Journal of Agricultural and Marine Sciences [JAMS]*, *22*(1), 48.

Al-Saidi, M., & Saliba, S. (2019). Water, energy and food supply security in the Gulf Cooperation Council (GCC) countries – A risk perspective. *Water*, *11*(3), 455.

Al-Zubari, W., Al-Turbak, A., Zahid, W., Al-Ruwis, K., Al-Tkhais, A., Al-Muataz, I., … Al-Sulaymani, Z. (2017). An overview of the GCC unified water strategy (2016–2035). *DWT*, *81*, 1–18.

Amin, M. (2013). Soft path water management in dry regions of the Arabian Peninsula by rainwater harvesting. *American Journal of Environmental Sciences*, *9*(2), 156–163.

Angelidou, M., Psaltoglou, A., Komninos, N., Kakderi, C., Tsarchopoulos, P., & Panori, A. (2017). Enhancing sustainable urban development through smart city applications. *Journal of Science and Technology Policy Management*, *9*(2), 146–169.

Ansari, S. (2021). Participatory urban development in India: A tale of two townships. *Environment and Urbanization Asia*, *12*(1), 136–147.

Ari, I. (2023). A scenario of water-sensitive urban design for Kimchi City, Indonesia. *IOP Conference Series Earth and Environmental Science*, *1199*(1), 012028.

Badidi, E. (2022). Edge AI and blockchain for smart sustainable cities: Promise and potential. *Sustainability*, *14*(13), 7609.

Banihabib, M., Noori, A., Jurík, L., Gacko, I., & Mirzaie, N. (2020). Prioritization of sustainable water management strategies in arid and semi-arid regions using swot coupled the technique in addressing SDGs. *Acta Scientiarum Polonorum Formatio Circumiectus*, *19*(2), 35–52.

Barron, N., Kuller, M., Yasmin, T., Castonguay, A., Copa, V., Duncan-Horner, E., & Deletic, A. (2017). Towards water sensitive cities in Asia: An interdisciplinary journey. *Water Science & Technology*, *76*(5), 1150–1157.

Bermejo-Martín, G., & Rodríguez-Monroy, C. (2020). Design thinking methodology to achieve household engagement in urban water sustainability in the city of Huelva (Andalusia). *Water*, *12*(7), 1943.

Brown, R., & Farrelly, M. (2009). Delivering sustainable urban water management: A review of the hurdles we face. *Water Science & Technology*, *59*(5), 839–846.

Burchi, S. (2012). A comparative review of contemporary water resources legislation: Trends, developments and an agenda for reform. *Water International*, *37*(6), 613–627.

Calvillo, C., Sánchez-Miralles, Á., Villar, J., & Martín, F. (2017). Impact of EV penetration in the interconnected urban environment of a smart city. *Energy*, *141*, 2218–2233.

Chandran, C., & Gowda, K. (2014). Water sensitive urban design: Investigating opportunities for Thiruvananthapuram. *Environmental Quality Management*, *24*(1), 1–13.

Chantry, W. (2022). 'built from the Internet up': Assessing citizen participation in smart city planning through the case study of Quayside, Toronto. *Geojournal*, *88*(2), 1619–1637.

Choo, M., Choi, Y., Yoon, H., Bae, S., & Yoon, D. (2023). Citizen engagement in smart city planning: The case of living labs in South Korea. *Urban Planning*, *8*(2), 32–43.

Faisal, H., Comsats, L., Usman, S., & Zahid, S. (2018). In what ways smart cities will get assistance from the Internet of Things (IoT)? *International Journal of Education and Management Engineering*, *8*(2), 41–47.

Farooqui, T., Renouf, M., & Kenway, S. (2016). A metabolism perspective on alternative urban water servicing options using water mass balance. *Water Research*, *106*, 415–428.

Fokaides, P., Apanaviciene, R., & Klumbyte, E. (2018). *5.12 energy management in smart cities* In Comprehensive energy systems (Vol. 5, pp. 457–473). Academic Press. https://www.sciencedirect.com/science/article/abs/pii/B9780128095973005265 (pp. 457–473).

Gade, D. (2021). Reinventing smart water management systems through ICT and IoT-driven solutions for smart cities. *International Journal of Applied Engineering and Management Letters*, 132–151.

Gonçalves, R., Soares, J., & Lima, R. (2020). An IoT-based framework for smart water supply systems management. *Future Internet*, *12*(7), 114.

Goosen, M., Mahmoudi, H., Ghaffour, N., & Sablani, S. (2011). *Application of renewable energies for water desalination*. InTech. doi: 10.5772/14011

Hancke, G., & Silva, B. (2012). The role of advanced sensing in smart cities. *Sensors*, *13*(1), 393–425.

Hassen, T., & Bilali, H. (2019). Food security in the Gulf Cooperation Council countries: Challenges and prospects. *Journal of Food Security*, *7*(5), 159–169.

Heland, F., Nyberg, M., Bondesson, A., & Westerberg, P. (2015). *The citizen field engineer: Crowdsourced maintenance of connected water infrastructure. scenarios for smart and sustainable water futures in Nairobi, Kenya.*

Hernaningsih, T., Said, N., Yudo, S., Wahyono, H., Widayat, W., & Rifa'i, A. (2023). Application of the concept of smart city and smart water management for the new capital city. *IOP Conference Series Earth and Environmental Science*, *1201*(1), 012103.

Holubchak, K., & Obynochna, Z. (2022). Architectural & urban planning strategy "sponge city" as a means of fighting the consequences of global climate change. *Urban Development and Spatial Planning*, (80), 149–158.

Impedovo, D., & Pirlo, G. (2020). Artificial intelligence applications to smart city and smart enterprise. *Applied Sciences*, *10*(8), 2944.

Kenway, S., Gregory, A., & McMahon, J. (2011). Urban water mass balance analysis. *Journal of Industrial Ecology*, *15*(5), 693–706.

Khan, Z., Abbasi, A., & Pervez, Z. (2019). Blockchain and edge computing–based architecture for participatory smart city applications. *Concurrency and Computation*

Practice and Experience, *32*(12)., 1–20, (e5566). https://onlinelibrary.wiley.com/doi/full/10.1002/cpe.5566

Kinyanjui, D. (2019). Adaptive responses to climate change: Evaluating the implementation and effectiveness of water resource management strategies in selected hotels within the environs of Lake Naivasha. *International Journal of Environment and Climate Change*, 287–296.

Koop, S., & Leeuwen, C. (2015). Assessment of the sustainability of water resources management: A critical review of the city blueprint approach. *Water Resources Management*, *29*(15), 5649–5670.

Kusumastuti, R., & Rouli, J. (2021). Smart city implementation and citizen engagement in Indonesia. *IOP Conference Series Earth and Environmental Science*, *940*(1), 012076.

Lee, J. (2015). *Urban resilience: Principles for the water sensitive cities. In Proceedings of the 8th conference of the international forum on Urbanism (IFoU)*, Incheon, 22–24 June 2015, MDPI: Basel, Switzerland, doi:10.3390/ifou-D021

Lim, S., Malek, J., Hussain, M., & Tahir, Z. (2018). Citizen participation in building citizen-centric smart cities. *Malaysian Journal of Society and Space*, *14*(4), 42–53.

Liu, P., Huang, Y., & Hermanowicz, S. (2021). Shifting entrepreneurial landscape and development performance of water startups in emerging water markets. *Plos One*, *16*(2), e0246282.

Lynggaard, P., & Skouby, K. (2016). Complex IoT systems as enablers for smart homes in a smart city vision. *Sensors*, *16*(11), 1840.

Madonsela, B., Koop, S., Leeuwen, K., & Carden, K. (2019). Evaluation of water governance processes required to transition towards water-sensitive urban design – An indicator assessment approach for the city of Cape Town. *Water*, *11*(2), 292.

Meene, S., Brown, R., & Farrelly, M. (2011). Towards understanding governance for sustainable urban water management. *Global Environmental Change*, *21*(3), 1117–1127.

Mitchell, V., Cleugh, H., Grimmond, S., & Xu, J. (2007). Linking urban water balance and energy balance models to analyse urban design options. *Hydrological Processes*, *22*(16), 2891–2900.

Mohamed, S., & Al-Shalfan, K. (2021). Intelligent traffic management system based on the Internet of Vehicles (IoV). *Journal of Advanced Transportation*, *2021*, 1–23.

Nunes, N., Björner, E., & Hilding-Hamann, K. (2021). Guidelines for citizen engagement and the co-creation of nature-based solutions: Living knowledge in the URBiNAT project. *Sustainability*, *13*(23), 13378.

Okedu, K., & ALSalmani, W. (2019). Smart grid technologies in Gulf Cooperation Council (GCC) countries: Challenges and opportunities. *International Journal of Smart Grid*.

Parimalarenganayaki, S. (2020). Managed aquifer recharge in the gulf countries: A review and selection criteria. *Arabian Journal for Science and Engineering*, *46*(1), 1–15.

Preston, S., Mazhar, M., & Bull, R. (2020). Citizen engagement for co-creating low carbon smart cities: Practical lessons from Nottingham city council in the UK. *Energies*, *13*(24), 6615.

Przeybilovicz, E., Cunha, M., Geertman, S., Leleux, C., Michels, A., Tomor, Z., Webster C. W. R ., & Meijer A. (2020). Citizen participation in the smart city: findings from an international comparative study. *Local Government Studies*, *48*(1), 23–47.

Qureshi, A. (2020). Challenges and prospects of using treated wastewater to manage water scarcity crises in the Gulf Cooperation Council (GCC) countries. *Water*, *12*(7), 1971.

Ramos, H., McNabola, A., Jiménez, P., & Sánchez, M. (2019). Smart water management towards future water sustainable networks. *Water*, *12*(1), 58.

Renouf, M., Serrao-Neumann, S., Kenway, S., Morgan, E., & Choy, D. (2017). Urban water metabolism indicators derived from a water mass balance – Bridging the gap between visions and performance assessment of urban water resource management. *Water Research*, *122*, 669–677.

Rodríguez, C., Ashley, R., Gersonius, B., Rijke, J., Pathirana, A., & Zevenbergen, C. (2014). Incorporation and application of resilience in the context of water-sensitive urban design: Linking European and Australian perspectives. *Wiley Interdisciplinary Reviews Water*, *1*(2), 173–186.

Shin, S., Kim, D., & Chun, S. (2021). The digital divide in advanced smart city innovations. *Sustainability*, *13*(7), 4076.

Sitzenfrei, R., Möderl, M., & Rauch, W. (2013). Assessing the impact of transitions from centralised to decentralised water solutions on existing infrastructures – Integrated city-scale analysis with vibe. *Water Research*, *47*(20), 7251–7263.

Šulyová, D., & Kubina, M. (2022). Integrated management of limited water resources in smart cities. *IOP Conference Series Earth and Environmental Science*, *1077*(1), 012003.

Vernon, B., & Tiwari, R. (2009). Place-making through water-sensitive urban design. *Sustainability*, *1*(4), 789–814.

Villa-Arrieta, M., & Sumper, A. (2019). Contribution of smart cities to the energy sustainability of the binomial between city and country. *Applied Sciences*, *9*(16), 3247.

Wang, K., Zhao, Y., Gangadhari, R., & Li, Z. (2021). Analyzing the adoption challenges of the Internet of Things (IoT) and artificial intelligence (AI) for smart cities in China. *Sustainability*, *13*(19), 10983.

Wanitchayapaisit, C., Suppakittpaisarn, P., Charoenlertthanakit, N., Surinseng, V., Yaipimol, E., & Rinchumphu, D. (2022). Rain garden design for stormwater management in Chiang Mai, Thailand: A research-through-design study. *Nakhara Journal of Environmental Design and Planning*, *21*(3), 222.

Webster, C., & Leleux, C. (2018). Smart governance: Opportunities for technologically-mediated citizen co-production. *Information Polity*, *23*(1), 95–110.

Wong, T., Rogers, B., & Brown, R. (2020). Transforming cities through water-sensitive principles and practices. *One Earth*, *3*(4), 436–447.

Worku, H. (2017). Rethinking urban water management in Addis Ababa in the face of climate change: An urgent need to transform from traditional to a sustainable system. *Environmental Quality Management*, *27*(1), 103–119.

Yang, J., Xue, F., Sun, X., Luo, Y., & Wang, J. (2022). Developing comprehensive resilient water systems and their driving factors for the 21st-century society in China. *IOP Conference Series Earth and Environmental Science*, *1087*(1), 012037.

Zhu, W., Yan, R., & Song, Y. (2022). Analysing the impact of smart city service quality on citizen engagement in a public emergency. *Cities*, *120*, 103439.

Chapter 8

AI's Role in the Development and Transformation of Smart Cities

Andreia de Bem Machado[a], *Gabriel Osório de Barros*[b], *João Rodrigues dos Santos*[c], *Silvana Secinaro*[d], *Davide Calandra*[d] *and Maria José Sousa*[e]

[a]*Federal University of Santa Catarina, Brazil*
[b]*Office for Strategy and Studies (GEE) of the Portuguese Ministry of Economy and Maritime Affairs, Portugal*
[c]*ISEC Lisboa e CETRAD/UTAD, Portugal*
[d]*University of Torino, Italy*
[e]*Iscte, Portugal*

Abstract

Humans now enjoy a better life because of Artificial Intelligence (AI). AI has a significant impact on the creation of smart cities. Modern applications based on big data, Internet of Things (IoT) systems, and deep learning require extensive use of complex computational solutions. Thus, the following problems arise: (1) what are smart cities? (2) what is AI? (3) How is AI used in smart cities? To respond to this problem, the following objective was set: to map how AI is used in smart cities. For this purpose, a qualitative methodology based on a narrative analysis of the literature was used. It is concluded that AI and smart cities are complementary technologies that can assist cities in tackling difficult issues including public safety, transportation, energy management, environmental monitoring, and predictive maintenance. This chapter's findings, while broadly applicable, offer valuable insights into the Gulf region's unique context, where rapid urbanization and technological adoption intersect with cultural and environmental considerations. The integration of AI in smart cities presents a promising avenue for the Gulf region to address its specific challenges and leverage its economic and infrastructural strengths, thereby contributing to the

The Emerald Handbook of Smart Cities in the Gulf Region: Innovation, Development, Transformation, and Prosperity for Vision 2040, 135–169

doi:10.1108/978-1-83608-292-720241008

broader goals of innovation, development, prosperity, and well-being as envisioned in the region's Vision 2040 initiatives.

Keywords: Smart cities; technology; digital cities; governance; digital cities; innovation

1. Introduction

Smart cities are planned through information and communication technologies (ICTs) with the aim of improving all dimensions of society, creating accessible spaces, and providing more security in the lifestyle and well-being of citizens residing in that city, in addition to having companies with responsibility focused on green in sustainable practices.

At their core, smart cities utilize data, gathered through sensors and analytics, to optimize resources, reduce waste, and efficiently address urban challenges such as transportation, energy consumption, and public safety. This concept not only embodies the digital transformation of urban management but also fosters a participatory approach to city governance, where citizens actively engage in shaping their living environments.

Thus, this study proposed an analysis of the different dimensions of smart cities to map how artificial intelligence (AI) is used in smart cities. Furthermore, this study is focused on the governance of smart cities, technology, and AI.

This chapter will use a qualitative exploratory methodology, targeting policy makers, politicians, business managers, citizens, and researchers. The chapter was organized as follows: the first section explains the literature review on smart cities and AI, the second section explains the methodology, followed by the third section which was the example of cases of the seven most relevant smart cities according to the "Digital Cities Index (DCI), 2022" published by Economist which were: Rank 1 – Copenhagen (Denmark); Rank 2 – Amsterdam (Holland); Rank 3 – Beijing (China); Rank 4 – London, (United Kingdom); Rank 5 – Seoul (South Korea); Rank 6 – New York (USA); Rank 7 – Sydney (Australia). Finally, the fourth section presents the final considerations.

While this chapter broadly explores the integration of AI in the development of smart cities, its insights and strategic frameworks hold particular relevance for the Gulf region. The region's rapid urbanization, coupled with a strong commitment to technological innovation and sustainability, aligns with the principles and applications discussed herein. This research, therefore, serves as a valuable resource for stakeholders across the Gulf region seeking to harness the transformative potential of smart cities to achieve their visionary goals.

2. Smart Cities

Smart cities are urban areas that utilize advanced technology and data analytics to manage resources and services efficiently (Caragliu et al., 2011a) to improve the quality of life for its citizens while also promoting sustainable development.

This improves various aspects of urban life by implementing interconnected infrastructure systems, such as transportation, energy, communication, health, education, and security, which are designed to optimize efficiency, reduce costs, and enhance overall urban services (Chourabi et al., 2012; Neirotti et al., 2014).

According to Markets and Markets (2022), the global smart cities market is estimated to be worth \$511.6 billion in 2022, and it is projected to reach \$1024.4 billion by 2027, growing at a compound annual growth rate (CAGR) of 14.9% from 2022 to 2027. The growing mobile subscriber base, broadband penetration, and adoption of advanced technologies across businesses drive the market. The report identifies AI and Machine Learning as technologies that accelerate the development of smart cities, empowering systems to be self-sufficient and futuristic (Borges et al., 2020).

However, privacy and security concerns are identified as restraints, as the vulnerability of the ecosystem is high, and stakeholders must address these challenges holistically. Advancements in technologies such as AI, cloud, Internet of Things (IoT), and analytics are identified as opportunities, enabling better connectivity and resulting in a complete remodeling of the smart city ecosystem. The report also identifies challenges, such as the need for more appropriate infrastructure in developing and underdeveloped economies, making it difficult to implement smart city solutions. The smart healthcare services are expected to register higher CAGR during the forecast period, while smart utilities solution and advanced metering infrastructure account for the largest market share. North America is expected to hold the largest share of the overall smart cities market. The US and Canada are leading smart city usage due to sustainable and well-established economies that invest strongly in R&D activities.

The process of creating smart cities has accelerated in recent years, as more people move to urban areas (Thales. 2018). This has put pressure on city officials and planners to find innovative solutions to manage resources efficiently and create more sustainable and liveable cities. Smart city initiatives have emerged as a way to harness the power of technology and data to optimize urban infrastructure, improve public services, and enhance the quality of life for residents (Mohanty & Kumar, 2021; Zigurat. 2019). With the rapid pace of urbanization, smart cities have become a critical focus for urban planners and policymakers worldwide.

The European Union (EU, 2019) recognized the potential of smart urban technologies to contribute to the sustainable development of European cities. In the Europe 2020 strategy, innovation was placed at the heart of the plan to drive competitiveness and create new jobs, while tackling major societal challenges such as climate change and energy efficiency. The EU has developed a shared European vision of sustainable urban and territorial development, where cities should be places of advanced social progress and environmental regeneration, as well as engines of economic growth based on a holistic integrated approach.

The European Commission (2019) identifies several key challenges and opportunities that cities face. The challenges include providing affordable housing, addressing mobility issues, improving access to services, dealing with ageing populations, promoting urban health, reducing social segregation, mitigating environmental impacts, and taking climate action. The report suggests that cities are uniquely equipped to address these challenges and can harness the power of technology and innovation, citizen engagement, good governance, and community resilience to achieve sustainable development. The report highlights the importance of appropriate management of new technologies and data, the need for affordable housing, the role of cities in implementing global agendas and engaging citizens, the fight for sustainability, the collective power of cities and city networks, the risk of polarization, and the importance of a holistic approach to service provision. The report emphasizes that city communities will play a substantial role in reshaping their own futures, and greater efforts will be required to anticipate the impacts of future trends and to build urban resilience to new challenges.

This is an area where different stakeholders participate in order to be able to create living and functional smart cities. The International Telecommunication Union (ITU) Smart Sustainable Cities report (2021) provides valuable insights into the current state of smart city initiatives and the opportunities and challenges that lie ahead. The report highlights the importance of collaboration between different stakeholders, including governments, industry, academia, and citizens, in achieving sustainable and inclusive smart cities. It also emphasizes the need for a data-driven approach and the use of emerging technologies such as 5G, AI, and the IoT to enhance urban services and address social and environmental issues. Additionally, the report discusses the role of standardization, policy frameworks, and financing mechanisms in supporting the development of smart cities.

The "Smart Cities Council Readiness Guide" (Smart Cities Council. 2015) emphasizes the importance of an integrated approach to smart city development, which considers the needs and perspectives of all stakeholders, including citizens, businesses, and government, and identifies several important uses of smart cities, including:

- Improving transportation: Smart cities can optimize traffic flow, reduce congestion, and improve public transportation through the use of sensors, data analytics, and other technologies;
- Enhancing public safety: Smart cities can use real-time data to monitor crime and other public safety issues, allowing for faster response times and more effective emergency services;
- Improving energy efficiency: Smart cities can use data and analytics to optimize energy use, reduce waste, and increase the use of renewable energy sources;
- Enhancing sustainability: Smart cities can promote sustainable practices by reducing greenhouse gas emissions, promoting public transportation, and encouraging the use of renewable energy sources;
- Improving citizen engagement: Smart cities can use technology to enhance citizen engagement, allowing for greater transparency and collaboration between citizens and government.

According to IoT Analytics' report (2020), the top 10 smart city use cases that are being prioritized now are: Connected Public Transport, Traffic Monitoring and Management, Water level/Flood Monitoring, Video Surveillance and

Analytics, Waste Management, Smart Parking, Air Quality Monitoring, Energy Management, Noise/Pollution Monitoring, and Smart Lighting. These use cases are intended to help cities become more efficient and sustainable while improving the quality of life for their citizens. The top four uses identified are:

- Connected Public Transport: This use case involves real-time monitoring of public transportation vehicles' locations and routes combined with notifications and personalized travel news to passengers, improving the efficiency of public transport systems – connected public bus system may allow for smart management and control of the public transport system;
- Traffic Monitoring and Management: Traffic monitoring and management help monitor and manage traffic flows efficiently to reduce congestion – intelligent traffic lights may help prioritize bikes and buses, and help clear congested streets after selected events such as soccer games;
- Water Level/Flood Monitoring: This use case involves monitoring the level of water in public water basins such as rivers, channels, or even lakes and seas, which can be crucial in densely built-up areas such as cities – systems can be implemented to offer real-time information about the condition of water levels;
- Video Surveillance and Analytics: Cities use cameras and analytics to detect specific situations, e.g., accidents, crimes, potential threats, or recognize specific features (face recognition, demographics, etc.) – solutions like this can help making the streets of the city much safer.

Also, according to IoT Analytics' research on Smart Cities (2021), there are seven things that successful smart cities do differently:

- Holistic smart city development:
  - Smart cities of the future require a comprehensive and integrated approach to city development that incorporates both new and existing infrastructure and city applications;
  - This approach should be facilitated by a centralized management software or smart city platform that offers a complete view of various smart city applications;
- Citizens-come-first mindset:
  - Successful smart cities prioritize the needs and preferences of their citizens, fostering an environment in which citizens are empowered to participate in shaping the future of their city;
  - This includes developing infrastructure that enables citizens to become co-creators of their future city;
- Alignment with government initiatives:
  - Smart cities should align their initiatives with larger municipality or government initiatives in a smart city program or plan, to ensure a coordinated and integrated approach to city development;

- Long-term vision:
  - Smart city programs of the future should have a long-term character, with a clear and defined vision for the future development of the city;
- Sustainability as top priority:
  - Environmental sustainability should be a top priority for smart cities of the future, with specific smart city solutions designed to improve environmental sustainability being prioritized within smart city programs;
- Public–Private-Partnership programs:
  - Public–Private-Partnerships are an essential component of successful smart city programs, encouraging and facilitating collaboration between the public and private sectors in the development and implementation of smart city projects;
- Open, city-wide databases and platforms:
  - One of the top priorities for future smart cities is to create open databases and platforms that are accessible to the public, which contain various datasets and information related to the city;
  - By doing so, citizens can have access to valuable information that can help them make informed decisions and participate more actively in the development of their city;
  - This approach promotes transparency and fosters collaboration between the city and its residents, and facilitates transparency and collaboration between citizens, businesses, and government, leading to more efficient and effective decision-making and problem-solving.

The following figure summarizes the most important stakeholders and areas involved in a smart sustainable city scheme (InsightsIAS. 2018). The figure encompasses six key pillars, namely Smart Mobility, Smart People, Smart Economy, Smart Environment, Smart Living, and Smart Government, and they represent a comprehensive approach to the development of urban areas that integrate advanced technologies, sustainable practices, and community engagement to enhance the quality of life for all citizens:

- Smart Mobility involves creating mixed-modal access to transportation, offering clean and non-motorized options, and integrating ICTs to optimize traffic flow and reduce congestion;
- The Smart People pillar focuses on 21st-century education, inclusivity, and creativity, aiming to create an engaged and knowledgeable citizenry;

- Smart Economy fosters entrepreneurship, innovation, productivity, and local and global interconnectedness to support economic growth and prosperity;
- Smart Environment prioritizes green buildings, renewable energy, and sustainable urban planning to reduce environmental impact and improve resource management;
- The Smart Living pillar seeks to create culturally vibrant and happy communities that are safe and healthy for all citizens;
- Finally, Smart Government aims to enable supply and demand-side policies that promote transparency and open data, foster innovation, and leverage ICT and eGovernment solutions to improve citizen services and engagement.

Overall, the Smart Sustainable City scheme provides a holistic framework for urban development that prioritizes sustainability, innovation, and community engagement, as shown in Fig. 1.

Smart cities represent an important opportunity for enterprises. According to a sponsored article by Pictet.(2018), published in *The Economist*, smart cities have potential to create investment opportunities for companies involved in urban development. With more than half of the world's population living in cities and this number set to increase by an additional 2 billion by 2045, the article highlights the pressure this will put on infrastructure, resources, and the environment. However, authorities worldwide are working with the private sector to make cities safer, more sustainable, and better connected. The article splits emerging investment opportunities into three areas: building the city, running the

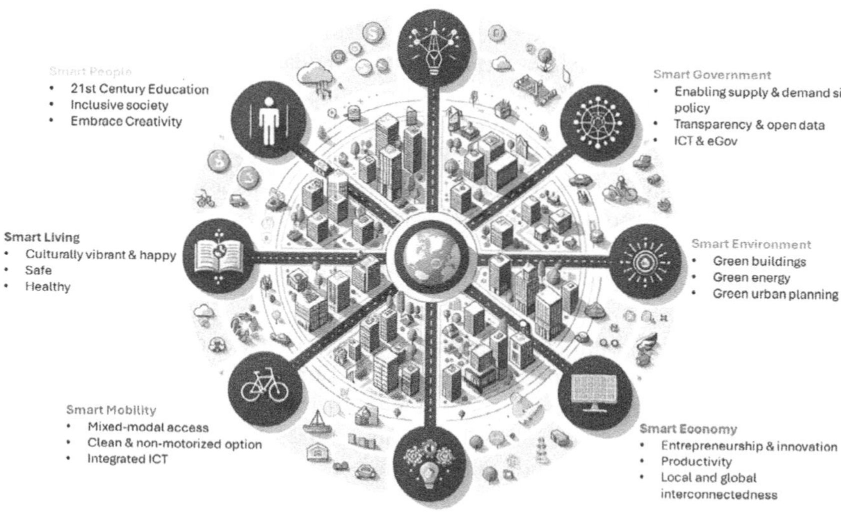

Fig. 1. Smart Sustainable City Scheme. *Source*: Image generated by AI through OpenAI's DALL-E and edited by the authors based on InsightsIAS.(2018).

city, and living in the city. It notes that businesses that help cities run smoothly and sustainably should do well and discusses how changing consumer tastes are shaping the demands of city dwellers, creating opportunities for companies to provide services such as healthier convenience food, flexible offices, and childcare. The article emphasizes the potential benefits of investing in sustainable and eco-friendly solutions such as low-emission public transportation, renewable energy sources, improving energy efficiency in commercial buildings, and effective waste management. According to the article, adopting such measures could help reduce substantially global energy costs.

Giffinger et al. (2007) consider that urban development is a complex process, and its perception and evaluation depend on the actors themselves and their objectives:

- The study's ranking features innovative aspects and results, such as focusing on medium-sized cities, analyzing characteristics and factors decisive for a successful forward-looking city development, using a transparent and comprehensive catalogue of indicators, and elaborating detailed city profiles based on 74 indicators providing applicable statements on comparative strengths and weaknesses.
- The study suggests that truly smart cities use this city ranking as a tool to benchmark with other cities, draw lessons from better-performing cities, and transfer policies.

Because data are essential for smart cities, Databroker.(2021) conducted a survey of smart city stakeholders in 11 countries to identify the most valuable data sets for driving smart city initiatives. With urbanization on the rise and more than half of people living in cities, the volume and quality of data are expanding exponentially, providing valuable insights for managing infrastructure, mobility, the environment, and health. The predefined data sets were ranked by importance under the categories of Geographics, Environment, Transport, People, Energy, and Economics, and the survey aimed to identify which data sets had the potential to drive insights and open up business opportunities to support smart cities:

Geographical data sets:

- Relate to locations, human-environment interactions, and human mobility movements, are crucial in understanding the Earth's physical characteristics and the state of human development at a granular level;
- The survey revealed that the list of companies and types is very important as understanding business trends in a region is essential for attracting business growth and ensuring economic activity has a net positive impact on citizens' lives and the environment;
- Another data set of high interest was parking locations and parking usage because real-time parking availability data can vastly improve efficiencies in managing parking and reduce negative impacts on cities such as increased traffic and reduced economic activity.

Environment data sets:

- These include air quality, biodiversity, climate change, noise pollution, and waste management, and can be leveraged to improve quality of life while minimizing environmental impact;
- Climate change data received the highest importance among the respondents, followed by air quality, waste management and recycling, and noise pollution;
- Climate change data can help cities understand the extent of the challenge they are facing, track the effectiveness of new methods for cooling the city, and plan infrastructure upgrades for the future;
- Air quality data sets help smart cities identify factors that contribute to air quality and deploy data-driven methods for improving it;
- Waste management and recycling data sets are an essential starting point for designing effective waste management and recycling systems, and noise pollution data is crucial for solving noise pollution problems in most cities.

Transport data sets:

- Include traffic volumes, bike and car usage, shared vehicle usage, navigation services, traffic violations, and public transportation usage, and are used primarily to optimize the movement of people within cities;
- Smart city representatives are highly interested in car traffic data, followed by public transportation hours and usage, and shared bike/car usage;
- Traffic data are essential for driving insights relating to the effectiveness and optimization of existing infrastructure and designing new infrastructure;
- Public transportation usage data are necessary for optimizing new and existing public transportation infrastructure, and shared vehicle usage data are important for promoting the sharing economy in cities and reducing the strain on public transportation and road infrastructure;
- With the right data, smart cities can leverage AI to improve outcomes for bike sharing systems and make optimal infrastructure design decisions based on real-time or close to real-time data on public transportation usage and human mobility data.

People data sets:

- This category includes social media and app usage, health, demographics, education, and human mobility data;
- Health data were indicated as the most important data set in this category, followed by demographics data;
- Smart cities can use health and demographics data to optimize services and identify the need for additional health facilities;
- Data-driven approaches improve healthcare effectiveness and lower healthcare costs for smart city residents;
- Human mobility data were also used during the Covid-19 pandemic to gauge the effectiveness of lockdowns, bolster attempts at contact tracing and exposure notification.

Energy and water consumption data sets:

- This category includes gas, fuel, electricity, water consumption, energy performance in buildings, and lighting usage;
- Smart cities can use data in this category to reduce costs, lower carbon footprints and environmental impact;
- Survey respondents indicated that energy performance in buildings is highly valuable, followed by water consumption data;
- Building energy performance data can identify efficient building designs and drive building owners and operators to reduce their energy consumption;
- Water consumption data can identify potential shortfalls and be used to optimize strategies for reducing consumption;
- IoT-enabled sensors can be used to collect real-time data to detect leaks, monitor water distribution, and develop data-driven policies that incentivize optimal water consumption decisions.

Economy data set:

- The category includes data related to in-city payments, retail potential, goods consumption, tax records, and housing and building prices;
- Among these, retail potential is the most valuable dataset, followed by in-city payments;
- In-city payments are essential for smart cities as they indicate the level of digital maturity and contribute an estimated 3% to the city's GDP;
- Smart payments generate positive net benefits for citizens such as time savings, improved budgeting, and more personalized customer service, while the government benefits from better data on citizens' needs and increased revenue;
- House and building prices were rated the lowest, but they are important indicators of the real estate market's impact on the economy and the city's development, which can be sensitive to economic conditions.

5G is also a new trend with great importance for smart cities. According to Osório de Barros (2021), 5G technology will revolutionize telecommunications, hastening digital transformation across various sectors and enhancing citizens' leisure experiences through applications such as smart cities. Deloitte (2020) explores the potential impact of 5G on smart cities and identifies the key areas where this technology can be leveraged to enhance city services, operations, and sustainability:

- The whitepaper explores the potential impact of 5G on smart cities, identifying the key areas where 5G can be leveraged to enhance city services, operations, and sustainability;
- 5G can be a key enabler for smart city initiatives, providing high-speed connectivity, low latency, and massive device connectivity, which can support new applications and services;

- The whitepaper identifies four categories of smart city use cases that can be enabled or enhanced by 5G:

  - o Smart mobility,
  - o Public safety and emergency services,
  - o Environmental sustainability, and
  - o Digital citizen services;

- In the smart mobility category, 5G can support autonomous vehicles, traffic management, and public transportation systems, enabling real-time data sharing and communication between vehicles, infrastructure, and passengers;
- In the public safety and emergency services category, 5G can provide faster and more reliable communication and data transfer between emergency responders, citizens, and public safety agencies, enabling better emergency response and disaster management;
- In the environmental sustainability category, 5G can enable smart energy grids, environmental monitoring, and waste management systems, supporting sustainability initiatives and reducing the carbon footprint of cities;
- In the digital citizen services category, 5G can support telemedicine, smart education, and e-government services, enabling better access to healthcare, education, and public services;
- The whitepaper also highlights the importance of collaboration between stakeholders in the smart city ecosystem, including cities, telecommunications operators, technology providers, and citizens, to ensure successful implementation and adoption of 5G-enabled smart city initiatives;
- Finally, the whitepaper also identifies several key challenges that need to be addressed for successful 5G-enabled smart city implementation, including regulatory issues, cybersecurity, privacy, and data management, and the need for significant investment in infrastructure and technology.

Recently, AI gained attention in many areas, namely regarding its use in smart cities. Pellegrin et al. (2021) explore the role of AI in urban areas and its impact on socio-economic and territorial cohesion. The study finds that AI has the potential to bring about deep transformations in people's daily lives, patterns of economic functionality, and the way governments operate. While AI is expected to have a positive impact on smart city initiatives, its actual benefits are yet to be fully assessed, and there are potential risks associated with its application:

- AI can provide numerous solutions in different areas, including improved urban management, support for decision-making, and the release of new or improved services for citizens;
- However, its application in urban development is fraught with risks such as security and privacy risks, performance risks, and economic risks, which can jeopardize socio-economic and territorial cohesion;
- Local and urban authorities need to ensure that necessary conditions are met, including data access, interoperability, legal frameworks, appropriate

governance structure, administrative capacity, relevant skills, and active citizen participation;

- The EU's AI Policy and its Cohesion Policy can help mitigate these risks, but they need to address the territorial dimension of AI more explicitly;
- The study also recommends mobilizing Cohesion Policy effectively and explicitly to address AI in an urban context, consolidating a knowledge base concerning the benefits and risks of AI for urban development, mobilizing Digital Innovation Hubs, and fostering awareness among citizens regarding the potential and risks associated with AI and city development.

Overall, the technological infrastructure required for a city to become "smart" is multifaceted and comprises a range of interconnected systems and technologies. These technologies include the IoT (Microsoft, 2023; Thales. 2018), which enables cities to collect and analyze data from various sources to manage assets and resources more efficiently. Cloud computing is another crucial component (Microsoft, 2023; Thales, 2018), providing on-demand access to shared computing resources, such as servers and software, that can store, process, and analyze large amounts of data without requiring costly hardware or software investments. Data analytics is also essential (McKinsey, 2018), transforming raw data into meaningful insights that inform decision-making and problem-solving. Additionally, AI and Machine Learning (Microsoft, 2023) have emerged as key enablers of smart cities, allowing tasks that typically require human intelligence, such as learning, reasoning, and planning, to be performed more efficiently. With these technologies in place, smart cities can improve infrastructure, modernize government services, drive sustainability, and accelerate economic development.

In conclusion, smart cities present a significant opportunity for enterprises as urbanization continues to rise and the volume and quality of data expands exponentially, providing valuable insights for managing infrastructure, mobility, the environment, and health. Investments in lower-emission public transport, renewable energy, and increasing efficiency in commercial buildings and waste management may be able cities to cut energy costs. The EU has recognized the potential of smart urban technologies to contribute to the sustainable development of European cities. The European Commission highlights the importance of technology and innovation, citizen engagement, good governance, and community resilience to achieve sustainable development. Furthermore, smart cities may benefit from valuable data sets, with geographical, environmental, transport, people, energy, and economic data sets all playing a critical role in optimizing and improving city services. Finally, the role of AI in urban areas and its impact on socio-economic and territorial cohesion may be important shortly. It is essential to highlight the potential benefits and risks associated with its application. Overall, the multifaceted technological infrastructure required for a city to become "smart" includes interconnected systems and technologies such as IoT, cloud computing, data analytics, and AI, enabling smart cities to improve infrastructure, modernize government services, drive sustainability, and accelerate economic development.

3. Artificial Intelligence

In a two-month workshop at Dartmouth College in the United States, John McCarthy was the first to define AI, even though the term has been the subject of numerous definitions in recent years. AI is a branch of study that focuses on artificially imitating human intelligence's cognitive capacities to build computer programs or other devices that can carry out tasks traditionally handled by people. Russell and Norvig (2016) state that "a machine simulates the cognitive functions that humans associate with other human minds, such as learning and problem-solving," that is when the phrase "AI" is employed. McCarthy originally introduced the term "AI" in the workshop proposal in 1956 (Russell & Norvig, 2016). They explicitly state in their definition of the term AI that it refers to the engineering and science of creating intelligent devices, particularly intelligent computer programs. It is connected to the goal of using computers to study human intelligence, although AI need not be constrained to techniques that may be observed biologically.

The history of AI has evolved along two axes: human-centric, which emphasizes empirical methods like hypothesis testing and experimentation, and rationalistic, which combines engineering and mathematical methods. Early studies on AI that relied on learning skills showed promising improvements in tasks requiring judgment and problem-solving. Although the development of AI has progressed more slowly than anticipated, governments and businesses have continued to invest heavily in research and development throughout the 1990s. The expansion of cloud service options, the improvement of IT's ability to collect and store task-related data across organizations, the development of methods and technologies, and the ease with which it is possible to acquire new technology tools are just a few of the factors that have contributed to the widespread adoption of AI in businesses (Russell & Norvig, 2016; Von Krogh, 2018).

IoT, big data, cloud computing, and the adoption of sensors (cyber-physical systems – CPS), among other technologies, give systems the ability to make decisions in a highly volatile and dynamic environment in a more effective manner, going beyond the simple automation of processes (Ahuett-Garza & Kurfess, 2018). In the manufacturing industries, AI advancements are related to strengthening Industry 4.0. Machines now play a significant role in decision-making and forming new production models, such as networked manufacturing, thanks to their capacity for machine learning (Li et al., 2017; Yang et al., 2017). IT management (Pandl et al., 2020), decision-making (Ding et al., 2020), sustainable performance evaluation (Souza et al., 2019), and the future of work have been the primary business sectors where AI has been adopted (Wang & Siau, 2019).

AI should be based on the hypothesis that, in theory, every part of learning or any other aspect of intelligence can be characterized precisely enough for a computer to imitate it. There will be an attempt to figure out how to make robots understand language, create abstractions and concepts, solve human-only issues, and advance themselves (Zawacki-Richter et al., 2019).

AI is defined by Baker et al. (2019) as machines that carry out cognitive processes that are usually similar to those of the human mind, especially learning and problem-solving. Additionally, they clarify that AI does not refer to a specific technology. It is a catch-all phrase for a wide range of tools and techniques, including genetic algorithms, neural networks, machine learning, and natural language processing.

A different way to define AI is as a collection of models, procedures, and tools (such as knowledge search, reasoning and representation, decision mechanisms, perception, planning, natural language processing, uncertainty handling, and machine learning) that, individually or collectively, address issues requiring logical reasoning (Sichman, 2021).

As a result, the term "AI" has a wide range of applications that include knowledge, reasoning, problem-solving, perception, learning, planning, and the ability to operate and move objects (Guerreiro & Barros, 2019). This scenario takes into account the presence of AI in a number of scenarios, including the context of smart cities.

Hence, in urban areas, AI can be utilized to implement technology and intelligent solutions for energy management, environmental monitoring, public safety, transportation, and predictive maintenance (Koumetio Tekouabou et al., 2022). In recent years, the climate for urban research has shifted from being data-scarce to being data-rich. Digital twin research has significantly increased as a result of the ongoing integration of big data and computational algorithms into the built environment and human daily life.

4. Methodology

This chapter employs a qualitative methodology. According to Miles et al. (2014) and Jackson et al. (2007), such methodologies enable researchers to comprehend and create an analytical framework from primary or secondary data. In addition, using qualitative research is possible derives inductive investigation of phenomena (Secinaro et al., 2021). In addition, these procedures permit interpretive approaches that are extremely valuable in the social sciences for determining how a phenomenon unfolds and for establishing linkages and epistemological virtuosity (Dal Mas et al., 2020). Lastly, qualitative approaches allow for contextual negotiations, i.e., unanticipated modifications in the study design that add value to the initial contextual assumptions, as contrasted to quantitative methods (Damgaard et al., 2000).

Furthermore, this methodological design also uses content analysis. Concerning the use of questionnaires and focus groups, it is an alternative way of analyzing data from multiple sources systematically and objectively (Krippendorff, 2004). Therefore, qualitative analysis through the content lens will allow researchers to select, interpret, and theorize on elements related to smart cities.

The analysis was rooted in a narrative analysis of literature (Massaro et al., 2016), which allows the main sources of dialogue in literature to be identified non-systematic; in this sense, a Scopus analysis using the keywords "smart cit*" is performed.

In the second stage, the following research questions were investigated experimentally (1) What are smart cities? (2) What is AI? (3) How is AI used in smart cities? and their factors using an exploratory case study. According to Yin (2017), the case study technique is appropriate when the subject includes one or more representative cases. In this instance, employing a triangulation strategy, real-world situations, and researchers' observations might be advantageous for deriving theoretical methods and logical replications in various contexts. Finally, the third step in this chapter is to select the most appropriate case studies (Calandra et al., 2022). In particular, the "DCI 2022," published by *The Economist*, has been selected as the world's most relevant smart cities. Economist Impact of the Economist Group develops it "considers four key pillars of digital connectivity, services, culture and sustainability to assess the extent and impact of digitisation in 30 global cities." The top seven cities in the 2022 ranking were: Rank 1 – Copenhagen (Denmark); Rank 2 – Amsterdam (Netherlands); Rank 3 – Beijing (China); Rank 4 – London (United Kingdom); Rank 5 – Seoul (South Korea); Rank 6 – New York (USA); Rank 7 – Sydney (Australia). Romme et al. (2015) suggest that using practitioners' sources can enhance the debate between academics and practitioners. Therefore, such sources could be interesting for developing dialogue skills by drawing from functional pragmatism and reporting success stories in recent literature.

Finally, after selecting the seven most relevant smart cities in the world, the researchers started a process of codifying the information based on certain elements typical of this field. These included: climate actions in place (Huang-Lachmann, 2019), technologies adopted (Secinaro et al., 2021), educational measures (Molnar, 2021), and health strategies (Pacheco Rocha et al., 2019).

5. Examples of Smart Cities That Use Artificial Intelligence

"Smart cities" cover various ICTs. The main objective of a smart city is to efficiently and effectively face the major issues of the 21st century, all related to sustainability and also to safety and security. Currently, the central issues are: the energy efficiency of buildings; the increasing mobility of people and goods; the potential scarcity of potable water; waste and health management in urban centers; increased road traffic, especially in city centers; and greenhouse gasses.

de Bem Machado et al. (2021) underline that a "smart city" is "a concept that is still under construction, not very clear, without a standardized nomenclature that can be effectively described" (p. 3).

Pandiyan et al. (2023) state that a "smart city is an efficient and sustainable urban center that maximizes the use of its resources to provide a high-quality living environment to its citizens."

Considering the principle of scarcity, which states that Man's desires are unlimited, but the available resources are finite, then it is possible to assess an indisputably defining characteristic of the concept of "Smart City": "the role of this entity in the efficient management of the scarce terrestrial resources available to Man" (de Bem Machado et al., 2021).

It can be assumed that sustainable development corresponds to the most important challenge of a smart city. "A Smart City places people at the center of development, incorporates information and communication technologies into urban management and uses these elements as tools for efficient management" (Bouskela et al., 2016).

Environmental concerns are the order of the day. The decarbonization of cities is one of the main goals. Even before we have achieved this objective, there are already those who say that we are currently experiencing the first decarbonization crisis, as everything seems to be worth replacing Russia's "fossil energy resources" with "fossil energy resources" produced in other geographies, when the objective should be to replace fossil fuels with renewable energy.

Nowadays, we can identify two major global initiatives in the field of combating global warming and climate change: the "Net Zero Carbon Cities" program (a program of the World Economic Forum, whose objective is to provide cities with greater resilience in terms of environmental issues); and the 2030 Agenda of the United Nations, which, in addition to indicating global development goals, also includes 17 "Sustainable Development Goals" that break down into specific actions.

In fact, "resilience" is one of the most important issues nowadays regarding urban areas and managing scarce resources. Decision-makers and the various entities with decision-making power in a city must today, more than ever, "have the capability to plan for and adjust to any change through forecasting, safeguarding, responsive capability and reviving and recovering strategy" (Rakshit & Paul, 2020, p. 54).

On a global scale, public administrations and governments have been able to recognize the importance of environmental sustainability in urban spaces and have been encouraging options for organizing spaces and urban occupations that, more than not harming, increase the utility obtained from the scarce resources available. In this regard, Bouskela et al. (2016) point out that "integration and cohesion are at the heart of every Smart City. Breaking down the barriers that usually divide Public Administration departments is fundamental. Teams must have complementary skills and work together."

And, considering the highly troubled and warlike period the world is going through, efficient resource management has never been so essential. "A smarter community is composed of not only a more integrated, collaborative, and inclusive "whole" but also of multiple neighborhoods and communities of interest and of kind. These are not in conflict" (Kanter & Litow, 2009, p. 3).

Eventually, it is because of this reason that, despite technological innovation every day, the existing urban centers continue to grow. According to Caragliu et al. (2011b), this triggers additional difficulties in resource management.

In fact, large urban centers entail many management difficulties. Even so, the world's population has been increasingly concentrated in cities. The UN's World Cities Report 2022 points out that the world's population will be 68% urban by 2050 and that urban areas already house 55% of the world's population. This trend is understandable, as a smart city "improves the competitiveness of urban contexts in the global knowledge economy." (Nam & Pardo, 2011). In fact, it is the capitalist economic model that continues to dictate the rules at this level, and, of course, the concentration of economic resources is a crucial element in dense urban spaces. In this context, Santos (2020) underlines that "the accumulation of specialized capital goods is fundamental for economic growth and development." (p. 23).

An environment of "excessive" urbanization will result in insufficiency of the most basic urban functions. This undermines the citizens' standard of living. Problems such as transversal insufficiency in managing the scarce resources available (natural resources, failures in the collection and treatment of urban waste, poor management of water resources, among many others). Civil protection, risk and disaster prevention, education, security, the railroad system, the mobility system, health, and even culture, all these areas are subject to many limitations in terms of efficient resource management.

According to Cohen (2011, p. 27):

> A lesson provided to us by cities that have put in motion highly advanced programs is that the solution to current problems must be to build the future. This can be achieved by creating new jobs, building modern, sustainable, and technologically advanced infrastructure, and by promoting innovation, human capital formation, and improving urban management.

Considering the issues mentioned previously, it is important to underline that the problem of managing the environment and scarce resources is an increasingly important priority on the agenda of countries all over the world.

In Portugal, for example, the issue of environmental sustainability assumes legislative prominence. It is a good example of this importance the Law n.°. 31/2014, of May 30, which establishes the general bases for public policy on land, land use planning, and urban planning. It is possible to read in paragraph b) of article 2.° that one of the legislative objectives is "to guarantee sustainable development" and, for example, in article 3.° that the law must ensure the "environmental and financial sustainability of the options adopted by the programs and territorial plans." Article 2.° d) is even clearer, stressing that the purpose of land use planning should be "to increase the resilience of the territory to the effects of extreme weather phenomena, combat the effects of erosion, minimize the emission of greenhouse gasses and increase energy and carbon efficiency."

In the United States, it is possible to highlight President Biden's Executive Order n.° 14057, which focuses on boosting "green" industries and jobs through increased federal sustainability, being materialized through the "Federal Sustainability Plan," which sets out an ambitious set of targets to reduce US greenhouse gas emissions by 50%–52% from 2005 levels by 2030 and limit global warming to 1.5 °C, as required, for example, by the so-called "Goal of Paris."

In section n.° 209 of the referred Executive Order, on d), it is possible to confirm that one of the targets of this Order is to "develop and enhance tools that assess climate change impacts and support climate adaptation planning and implementation."

In England, Part 1 (Environmental Governance) Schedule 7A of the UK Environment Act, 2021 "makes provision for grants of planning permission in England to be subject to a condition to secure that the biodiversity gain objective is met."

In France, for example, in the 2022 Report "Public planning establishments," the Ministry of Ecological Transition and Territorial Cohesion (2022),[1] regarding the role of the state and actors in urban planning and regional development, underlines the major importance of facilitating "the emergence shared and innovative strategies for sustainable development reconciling development and preservation" (p. 40).

As a federal executive order, it assumes a legally binding nature for the various levels of the North American Public Administration. For this reason, the targets to reduce US greenhouse gas emissions must necessarily be translated into the practice of territorial planning in the United States.

To address just one more case, in the Parliament of Australia's official webpage (Parliament of Australia, 2023), it is possible to confirm that one of the major concerns of the Australian Government was always "the impacts of urban development for an expanding population gives rise to issues of energy use, land availability, transport systems and infrastructure planning." Issues like: Future transport systems; Energy use in buildings, and Planning infrastructure are still some of the main focuses of urban planning in Australia.

Today, few government decision-makers need to incorporate the issue of environmental sustainability into their institutional and legal systems. This articulation between preoccupations related to the efficient management of resources and the economic performance of territories is an exercise that appears to be increasingly complex. The selection of the most suitable investments is an increasingly complex process. Smart cities, producing performance indicators in real time, offer essential assistance at this level. In this sense, reinforcing the importance of the "digital city" concept, Yovanof and Hazapis (2009) highlight that "the digital city is primarily characterized by the ability to implement communication technologies, promoting broad access to tools, contents and management systems, in order to meet the needs of public authorities and their servants, citizens and organizations."

Compliance with current legislation (and its revision), at present, is only likely to be achieved with the use of ICTs. And only with the integration of all the information collected by the sensory and detection equipment available today in a smart city will it be possible to efficiently and effectively manage the various thematic layers of an urban space.

In fact, "the goal of smart city architecture is to come up with a solution that may integrate all the real time response applications" (Usman et al., 2023, p. 2). And, of course, the logic of these authors is the same about legislation in the field of sustainability. Only the use of ICTs makes it possible to fully achieve the objectives of the legislation.

---

[1]Ministère de la Transition Écologique et de la Cohésion des Territoiresles.

On a global scale, there are several cases of very successful smart cities. Below are some of these examples.

For this chapter, the first seven cases of the "DCI 2022" were selected. The DCI 2022, developed by Economist Impact of the Economist Group "considers four key pillars of digital connectivity, services, culture and sustainability in order to assess the extent and impact of digitisation in 30 global cities." The top seven cities in the 2022 ranking were: Rank 1 – Copenhagen (Denmark); Rank 2 – Amsterdam (Netherlands); Rank 3 – Beijing (China); Rank 4 – London (United Kingdom); Rank 5 – Seoul (South Korea); Rank 6 – New York (USA); Rank 7 – Sydney (Australia).

Rank 1 – Copenhagen

According to the Smart-I portal, which operates in the area of providing solutions for smart cities using AI to solve constraints in the public sphere and to improve the quality of life in cities, both for the resident population and for the present population, "Copenhagen's smart city approach is based on three main goals: achieving carbon neutrality by 2025; creating a greener, more sustainable capital city; supporting economic growth."

To achieve its objectives, Copenhagen is developing, "on the ground," a set of actions.

The Strategic Climate Action Plan is one of Copenhagen's most important projects. This project includes a set of actions and initiatives aimed at gradually reducing Carbon Dioxide emissions to zero. For example, in Denmark, the substantial decrease in road traffic is one of the main changes to increase environmental sustainability and improve citizens' quality of life.

For example, the Copenhagen Strategic Climate Action Plan predicts that 75% of trips in the city will be made by bicycle, using public transport or simply walking. It should be noted that Denmark has long been known as the "bicycle nation." The change in the behavioral paradigm of Danish citizens in relation to their displacements was "motivated" by the high taxes on motor vehicles, ending up with a transfer to the so-called "bicycle culture."

In this regard, Ulpiani et al. (2023) refer that "cities are hotspots of climate impacts, but also a crucial part of the solution." Copenhagen is the best example of all in this sense.

Smart parking is also an integral part of smart culture in Copenhagen. This project also contributes to the reduction of car traffic and, therefore, to a lower carbon footprint in the atmosphere. Smart parking consists of an online platform that informs, in real-time, about the availability of parking in the various car parks in the city.

Finally, reference is made to the "City Data Exchange" project, which consists of a public–private collaboration to assess the possibilities for interaction and exchange of information between public entities and private parties. "The goal of this project is to create a platform that allows the citizens to buy, sell, and share data among themselves, institutions, and organizations/companies" (World Smart Cities Forum, 2019).

Rank 2 – Amsterdam

On the Amsterdam Smart City portal, it is possible to consult a wide range of smart projects running in Amsterdam. The Amsterdam Smart City project is a partnership "of twenty permanent partners who make Amsterdam Smart City possible: governments, knowledge institutions, social organizations and innovative companies active in the Amsterdam Metropolitan Area." The aim of the project is to combine synergies that lead to the emergence of smarter and greener solutions that guarantee sustainability for the future of the Amsterdam Metropolitan Area.

Amsterdam Smart City organizes smart projects into five groups: Circular City; Energy Mobility; Citizens and Living; Digital City; Smart City Academy.

Of the various open projects, the following stand out:

- The "Amsterdam entrepreneurs and SMEs invited to join Impact Deals and become more sustainable," which encourages Amsterdam entrepreneurs to be more sustainable, challenging them to present proposals with impact, in order to receive support from the City of Amsterdam;
- The "InnoBuyer: Looking for public and private organizations with challenges to solve," a "project that brings together public and private organizations (Challengers) and innovative companies (Solvers) to jointly co-create innovative solutions, and speed-up the leading process to a concrete innovation procurement";
- The "Dialoog/game – Climate Privilege Walk," which aims to promote environmental education through a game that allows participants from different socioeconomic conditions to conclude how unequal the energy conditions of families are and how important the principles of energy efficiency are.

In fact, the increase in education and knowledge is also an integral part of a "Smart" city.

The article "Smart Cities: Explaining the Relationship Between City Growth and Human Capital" has a very suggestive title. In this article, Shapiro (2003) deeply develops the virtues that result from knowledge and education to improve the quality of life in cities. The author concludes that "areas with more educated populations experience more rapid growth in the quality of life" (p. 2).

Rank 3 – Beijing

From Beijing, the "Beijing City Lab" project stands out within the scope of this chapter. This platform is dedicated to the study of the Chinese capital, but is not limited exclusively to Beijing, presenting projects from other regions of China, which can serve as an example or "inspiration" toward intelligence in the management of urban spaces in their various layers. "The Lab focuses on employing interdisciplinary methods to quantify urban dynamics, generating new insights for urban planning and governance, and ultimately producing the science of cities required for sustainable urban development." The core of a smart city can be found in this project, which combines urban planning, urban geography, Geographic Information Technologies, economics, computing, and digital education.

In the "Beijing City Lab," it is possible to consult on some of the most important projects and works in the scope of the application of available resources in a smart city.

One of the works presented at the "Beijing City Lab" is the "Smart technologies for fighting against pandemics: Observations from China during COVID-19" (next figure), as shown in Fig. 2.

This work focuses on China's performance in the face of COVID-19, making a systematic review of the application of smart technologies in the context of fighting the pandemic. The aim is "to provide a reference for applying smart technologies against pandemics in the future" (Li & Long, 2022).

Reinforcing the importance of works such as the one presented above, Zhu et al. (2019) underline that "Smart technologies are the critical tools for increasing urban smartness as well as the ability to fight against pandemics."

Another very "smart" example from Beijing is the virtual card held by all resident citizens. This virtual card is responsible for managing all citizens' identity and identification documents. This added value facilitates and streamlines the daily interaction of many millions of citizens with public and private institutions. For example, all public transport can be paid for by mobile phone.

Rank 4 – London

First of all, the city of London has officially a Chief Digital Officer.

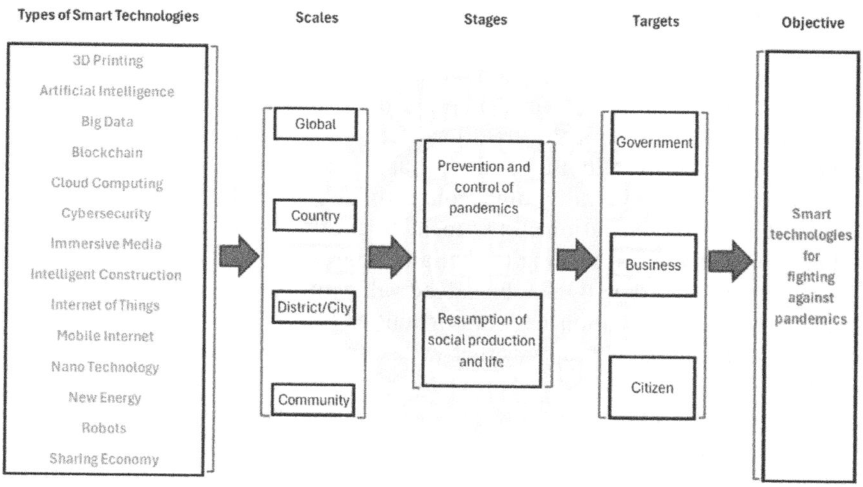

Fig. 2. The Conceptual Framework of Haw Smart Technologies Fight Against Pandemics. *Source*: Elaborated by the authors based on Beijing City Lab (https://www.beijingcitylab.com/projects-1/43-smart-cities-review/), with image generated by AI through OpenAI's DALL-E.

In 2018, London launched the "Smarter London Together" project, a public program that aims to modernize and transform the city, emphasizing data sharing, smart infrastructure and technological innovation, considering these as engines for sustainable development.

Even before it was a smart city, London was already recognized as the "European hub of technology companies." In this regard, according to the We Build Value Digital Magazine (2022):

> One-third of all Europe's technology "unicorn companies" (one-person-led start-ups worth at least $1 billion (€888 million) are in the London area; 46,000 tech companies are based there, employing 240,000 people and generating an income of £44 billion (€51 billion). In the decade 2006-2016, employment in the digital sector grew by 77%, while the number of companies increased by 90%.

One of London's several "smart" initiatives is the Retrofit Accelerator – Homes. Retrofitting is a program partially funded by the European Regional Development Fund, which consists of carrying out works on existing buildings in order to "make them more efficient, sustainable and better for the environment."

The Retrofit Accelerator – Homes program is part of the strategy to combat climate change, through interventions that guarantee the insulation of homes and the adoption of heat and energy options with zero/reduced carbon emissions. This project is central to the fulfillment of the City of London's strategy to achieve zero carbon emissions by 2030.

The London Datastore is another high-impact "smart" project in London. It consists of an open and free portal for sharing information about the city and is accessible to anyone. "Whether you're a citizen, business owner, researcher, or developer, the site provides over a thousand datasets to help you understand the city and develop solutions to London's problems."

Finally, the "Connected London" project and the "Adding to fiber connectivity" intervention. About the first, it can be read on the official Transport for London portal, of the City of London, that "by the end of 2024, tube customers will be able to access full mobile connectivity and the Internet within every London Underground station and tunnel." Regarding the intervention "Adding to fiber connectivity," according to the same source, the expansion of the high-capacity fiber network will take this added value directly to the most peripheral areas of London, "creating new opportunities for commercial and domestic usage with gigabit-capable speeds that will support digital inclusion and future growth."

This increased capacity for interconnectivity in London is a central element within the "smart city" concept. In this sense, Vershinina and Volkova (2020) state that "in smart cities, all problems are approached as complex urban problems, since all components of the city are closely interconnected and are parts of the same networks."

Rank 5 – Seoul

Time Magazine (2022) states that Seoul "is the first city government to integrate with the Metaverse." "The metaverse is a digital space that blends virtual and/or augmented reality with permanent virtual spaces" (Bibi et al., 2022). Users navigate and "inhabit" this virtual, digital, and alternative world. Surprisingly, or perhaps not, users seek, as in real and physical life, to interact, "live" and even own some dimensions and aspects of the metaverse. Through Metaverse Seoul, citizens and tourists can access all dimensions of the city and experience them virtually, even interacting with various dimensions of the city (administrative, business, and tourist). Expectedly, "Metaverse Seoul" intends to be a technological and multilayer ecosystem covering all areas of the Seoul Metropolitan Government, covering cultural, economic, educational, tourism and other public services.

Even before the Metaverse Seoul project, Seoul already had the Smart City Platform (next figure). The Smart City Platform corresponds to a screen with tactile technology installed in the municipality of Seoul's main building, allowing you to see all 605.2 km^2 of Seoul at the distance of a click. And the platform's advantages are not limited to visualization. Featuring tactile technology, it allows interaction through touch and voice, intending to access a huge amount of data and information related to the various layers of the city of Seoul. It is a powerful tool to support the decision-making process. "This state-of-the-art system was created for work efficiency and city management transparency for the Mayor's use." (United Nations, 2022), as shown in Fig. 3.

The Smart City Platform was designed to allow real-time monitoring and tracking of daily maintenance operations, incidents and contingencies, and all types of events susceptible to the response by the authorities. Since the platform offers the possibility of visualizing the territory in its various dimensions (via image and alphanumeric data), it allows decision-makers to support their decisions based on reliable information from the city, obtained in real time and consulted in a very intuitive way. These features, in addition to streamlining the process, undoubtedly increase decision-makers' efficiency.

Incheon International Airport (Seoul) is also on the verge of being one of the smartest and most environmentally sustainable airports globally. It operates a set of self-sufficient buses in terms of energy and with state-of-the-art autonomous mechanisms that allow airport users' registration and automatic processing. These buses are equipped with disinfectant robots.

Korean Air also signed a cooperation agreement to acquire hydrogen-based fuel to stimulate commercial aircraft development with zero carbon emissions by 2035 at Airbus.

Seoul's public transport system also stands out worldwide in terms of innovation. Some of the recent initiatives include the installation of mechanisms in passenger buses that renew the indoor air quality; the possibility of free charging of smartphones, and free and unlimited use of 5G Wi-Fi.

Finally, another highly innovative project announced by Samsung for Seoul should be highlighted. "Hyundai Motor Group recently announced it will

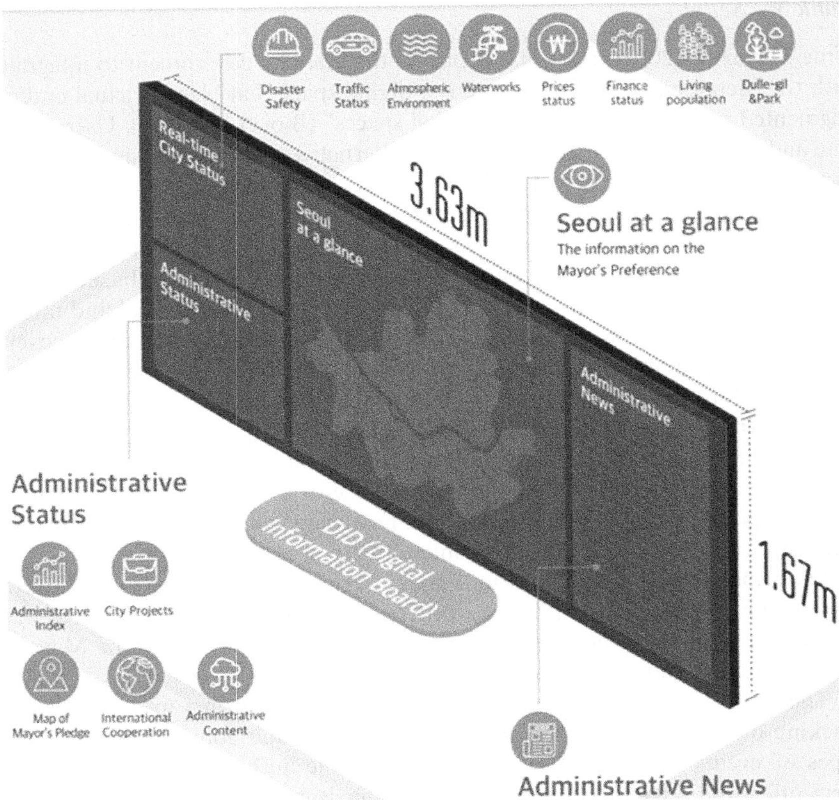

Fig. 3. Seoul: Smart City Platform. *Source*: United Nations – Department of Economic and Social Affairs (2022).

pilot a RoboRide, the first car-hailing service with autonomous driving vehicles, in Gangnam in Seoul, with plans to add more than 300 by 2026" (Time Magazine, 2022).

Rank 6 – New York

> From LED light projects and water leakage detection systems to smart garbage cans, New York smart city initiatives are paving the way for other major cities to address growing populations, carbon emissions and other environmental impacts. (Earth.Org, 2022)

New York has the objective of achieving carbon neutrality by 2050. To this end, it has not only to decarbonize its productive sectors, but also to implement solutions that enhance efficiency in the daily management of functions and resources. The bet on cutting-edge technology and sensors has been a strong bet in the administration of the city of New York. The objective is to transition as quickly as possible to the absolute condition of "Smart City."

One of the first initiatives carried out in New York dates back to 2013. Accelerated Conservation and Efficiency enabled the installation of LED lighting in around 650 buildings owned by 16 agencies in the city. The project has greatly reduced carbon emissions into the atmosphere and has also allowed savings of around SD$800,000 a year in electricity.

Drinking water waste is also a top concern for the NYC Department of Environmental Protection to tackle this problem, a large-scale program "Automated Meter Reading" was launched. This program, in addition to providing an in-depth understanding of how water is used and how it is applied, has also ensured significant savings for families and companies in New York.

In 2020, a massive program was launched to install intelligent technological sensors, aimed at different areas of analysis, from measuring road traffic, detecting water leaks, measuring air quality, among others. The registered information has allowed to improve the efficiency of the public services in the management of the different dimensions of the city.

Finally, regarding solid urban waste, the New York sanitation department is the largest in the world. This suggests well the massive amount of urban waste that must be collected daily. That is about 10,500 tons a day. To reduce the likelihood of excessive accumulation of waste and to facilitate the logistical process associated with collection, "smart" waste collection equipment (BigBelly) is installed throughout the city (next image), as shown in Fig. 4.

This equipment runs on solar energy and has filling sensors installed that help, in real time, the authorities to define "optimal" routes for waste collection

Fig. 4. Smart Waste Collection Solutions. *Source*: Image for illustrative purposes only, generated by AI through OpenAI's DALL-Ey.

around the city. It also has a highly effective automatic compression system, which extends the time of equipment availability between collections carried out by public services. The "BigBelly" has allowed increasing the carbonic efficiency and allocation of resources immensely in the context of the collection of urban solid waste in New York.

Rank 7 – Sydney

In 2020, Sydney published the "Smart City Strategic Framework." The objective of the document was to systematize the advantages provided by the digital disruption so that it would be less complex to transform them into sustainable development opportunities for Sydney, and that this would translate into an increase in the competitiveness and attractiveness of the city on a global scale.

The Smart City Strategic Framework defines three priorities (City of Sydney, 2020, p. 17):

1. Data-driven monitoring, prediction and management of city conditions and impacts of shocks and stresses
2. New technologies propelling a greener city and a carbon-neutral future, powered by the circular economy and affordable renewable energy
3. Informed and prepared communities actively participating to strengthen the local area's sustainability and resilience.

In this sense, Rakshit and Paul (2020) emphasize that *"resilience capabilities of the entities must be reinforced to respond to the aftermath of unforeseen events"* (p. 54). The conceptual essence of smart cities (a sensory and digital essence) is essential to achieve the desideratum of resilience in urban spaces in the 21st century.

Increasing the resilience of territories currently presupposes the adoption of "zero carbon" strategies. Reducing greenhouse gas emissions is a priority for Sydney.

Technology Magazine (2022) states that *"from July 2020, the City of Sydney began using 100% renewable power to satisfy its demands. This will reduce the city's emissions by about 18,000 tonnes per year, comparable to the electricity use of 4,000 residents."*

According to the About Smart Cities portal (2021), Digital Infrastructure, Sustainable Energy, Smart Lighting, Smart Waste Management, Open Data, Innovation, are other major areas of activity in the city of Sydney. It is certainly this scope of investment in multiple areas that allowed the city of Sydney to become one of the seven most important smart cities in the world in 2022, in the ranking established by the " DCI 2022."

For example, the "Smart Lighting" project is important not just because it allows reducing energy consumption. It "can simultaneously be a mechanism for enhancing the sense of safety, as well as a channel for creative art and placemaking" (City of Sydney, 2020, p. 27).

Another innovative project in Sydney, related to "Smart Waste Management," stands out. The objective of this project is *"to stop recycling ending up as waste"*

(Startup Daily, 2022). This waste collection equipment is closed and blocked by default. Opening is automatic and only when "qualified" items are detected for each of the deposits. Thus, it significantly reduces the risk of contamination of each of the selective deposits by unadjusted residues. This technology makes it possible to greatly leverage the efficiency of the recovery and reuse process of recyclable waste.

The target of the "Charopy bins" (next figure) are, mainly, *"schools, stadiums, councils, shopping centers and major companies."*

For example, a study conducted by Asio (2021), about *"environmental consumption and waste recycling schemes in school,"* reveals that students *"throw the plastic, metal, and paper to separate recycling boxes or bins" "often,"* but not *"every time"* (p. 39). This evidence shows that human beings will always need some support to be perfect. And that is the purpose of Charopy bins, in particular, and smart cities, in general, as shown in Fig. 5.

Fig. 5. Smart Bins. *Source*: Image for illustrative purposes only, generated by AI through OpenAI's DALL-E.

6. Conclusion and Discussion

Summary of Contribution

This chapter delineates a strategic framework that leverages AI to foster the evolution of smart cities. By integrating AI with big data, IoT systems, and deep learning, the framework addresses critical urban challenges such as public safety, transportation, energy management, environmental monitoring, and predictive maintenance. While the framework is globally applicable, its principles hold particular promise for the Gulf region, aligning with its ambitious Vision 2040 goals. The intent is to catalyze urban innovation, development and transformation, contributing to the overarching prosperity and well-being of Gulf societies.

Key Findings

The key findings of the chapter revolve around the pivotal role of AI in the development, transformation and sustainable management of smart cities. The chapter outlines how AI, in conjunction with big data, IoT systems, and deep learning, addresses critical urban challenges like public safety, transportation, energy management, environmental monitoring, and predictive maintenance. It presents a comprehensive overview of smart cities, including their definition, the integration of AI technologies and the resulting benefits and challenges.

The discussion extends to the practical applications of AI in enhancing city operations, improving residents' quality of life and contributing to the broader goals of sustainable urban development. The chapter also underscores the importance of multi-stakeholder collaboration in realizing the full potential of smart cities, highlighting the role of policymakers, urban planners and the community in fostering innovation and inclusivity in urban development.

Furthermore, the chapter emphasizes the global applicability of these findings while acknowledging the unique opportunities and challenges faced by the Gulf region. By aligning with the Vision 2040 goals, the insights from this chapter offer a strategic roadmap for the Gulf cities to harness AI and smart technologies in pursuit of innovation, prosperity, and enhanced urban well-being.

Implications for the Smart Cities Agenda

The implications of this research are far-reaching, providing actionable insights for a broad spectrum of stakeholders involved in the smart cities' agenda. For urban planners and architects, the integration of AI presents new avenues for designing adaptable, resilient urban spaces. Policymakers and government officers can leverage these insights to drive regulatory frameworks and governance models that support sustainable urban development. Furthermore, the educational sphere, including faculty and students, gains a resource for understanding and innovating on smart city technologies and their societal impacts. In the context of the Gulf region, this framework can serve as a blueprint for harnessing digital transformation to meet the unique environmental, economic and social challenges of its rapidly evolving urban centers.

Propositions for Smart Cities Initiatives

Based on the findings of this chapter, several propositions emerge for smart cities strategies and initiatives. First, the development of AI-powered urban operational centers could enhance real-time decision-making and urban service delivery. Secondly, fostering public–private partnerships can accelerate the deployment of smart infrastructure, from sustainable buildings to intelligent transportation systems. These initiatives, underpinned by the research presented, offer a roadmap for cities in the Gulf region to navigate their smart city journeys effectively.

Policy Recommendations

To promote the impact of smart cities, the following policy recommendations are proposed:

- Incentivize AI integration: Governments should offer incentives for the adoption of AI technologies in urban development projects, encouraging innovation and sustainable practices;
- Foster collaborative ecosystems: Establish multi-stakeholder platforms that bring together government, industry, academia, and civil society to co-create smart city solutions, ensuring they are inclusive and equitable;
- Invest in digital literacy: Enhance educational curricula to include digital literacy and smart technologies, preparing citizens for the demands and opportunities of AI-enhanced urban environments.

This book chapter has explored the intersection of smart cities and AI, highlighting the potential of these technologies to drive innovation and transform the urban environment. The rapid growth of AI and the increasing need for sustainable, efficient, and livable cities provide a fertile ground for exploring the use of AI in urban contexts. We have examined various aspects of smart cities and AI, including their definitions, history, benefits, challenges, and potential applications.

Through this chapter, we have shown that smart cities and AI are complementary technologies that can help cities address complex challenges, such as energy management, environmental monitoring, public safety, transportation, and predictive maintenance. Additionally, integrating big data and computational algorithms into the built environment and human daily life has created opportunities for digital twin research.

We have also demonstrated the importance of collaboration and interdisciplinary approaches in designing and implementing smart cities and AI solutions. It is crucial to involve stakeholders from various domains and backgrounds, including government, academia, industry, and the public, to ensure the success and sustainability of these initiatives.

As we have seen in the examples of the top smart cities, adopting smart solutions has been a multi-faceted approach, covering areas such as energy management, environmental monitoring, public safety, transportation, and waste management. The cities that have successfully implemented smart solutions have done so by focusing on the needs of their citizens and creating partnerships with the private sector, academia, and other stakeholders to ensure the best outcomes. With the continued development of technology and the increasing availability of data, the potential for smart cities to continue improving citizens' lives is immense.

In sum, this chapter contributes significantly to the smart cities discourse by providing a theoretical and practical framework that underscores the symbiotic relationship between AI and urban development. Theoretically, it enriches the understanding of smart cities by illustrating how AI can be a pivotal force in addressing urban challenges, thus offering a nuanced perspective to the ongoing academic dialogue. Practically, the framework serves as a guide for urban planners, policymakers and industry practitioners, illustrating the implementation of AI-driven solutions in smart cities, from conceptualization to governance. Future research could explore the integration of this framework with emerging technologies such as blockchain and the IoT to further enhance the resilience and efficiency of smart cities. Additionally, empirical studies assessing the long-term impacts of AI applications in diverse urban contexts could provide deeper insights.

Overall, this chapter contributes to advancing knowledge and understanding of Smart Cities and AI, providing a valuable resource for researchers, practitioners, policymakers, and anyone interested in the future of cities. The potential of these technologies is vast, and we look forward to seeing their continued development and deployment in the urban context. For future work, it is proposed to discuss the sustainability of smart cities using AI and blockchain.

7. Teaching Orientation

Suggested Teaching Assignments

Case study analysis: Select and analyze a successful implementation of AI in a smart city, focusing on aspects such as the challenges faced, the solutions implemented and the outcomes achieved.

- Design a smart city solution: Design an AI-based solution for a common urban issue, such as traffic congestion, waste management, or energy efficiency. This assignment may involve drafting a proposal that outlines the problem, the proposed AI solution, the expected impact, and any potential challenges.

Suggested Titles for Future Research Essays

"Evaluating the impact of AI on urban sustainability"

"The role of AI in enhancing public safety in urban environments"

"Challenges and opportunities in integrating AI technologies in smart city infrastructure"

"AI and citizen engagement: How technology is reshaping urban governance"

Recommended Projects URLs

Amsterdam Smart City – https://amsterdamsmartcity.com/
Beijing City Lab – https://www.beijingcitylab.com/
Digital Dubai – https://www.digitaldubai.ae/
EIT Digital City – https://www.eitdigital.eu/
New York City Economic Development Corporation – https://edc.nyc/industry/smart-cities
Seoul Metaverse – https://www.tomorrow.city/seoul-metaverse/
Sidewalk Labs – https://www.sidewalklabs.com/
Sidewalk Toronto – https://sidewalktoronto.ca/
Smart City Copenhagen – https://www.beesmart.city/en/city-portraits/copenhagen
Smart City London – https://www.beesmart.city/en/city-portraits/london-europes-smartest-city

References

About Smart Cities. (2021). *Sydney*. https://www.aboutsmartcities.com/smart-city-sydney

Asio, J. M. R. (2021). Disaster awareness and level of compliance to disaster programs in a highly urbanized city. *Aquademia*, *5*(1), ep21003

Ahuett-Garza, H., & Kurfess, T. (2018). A brief discussion on the trends of habilitating technologies for Industry 4.0 and smart manufacturing. *Manufacturing Letters*, *15*, 60–63.

Baker, T., Smith, L., & Anissa, N. (2019). *Educ-AI-tion rebooted? Exploring the future of artificial intelligence in schools and colleges*. NESTA. Retrieved June 23, 2022, from www.nesta.org.uk/ documents/1190/Future_of_AI_and_education_v5_WEB.pdf

Beijing City Lab. https://www.beijingcitylab.com/projects-1/43-smart-cities-review

Bibi, B., Lee-Ah Mat, V., Gomez, A., & Pirovich, J. (2022, March 16). *Discussion paper: First nations' culture in the metaverse*. http://dx.doi.org/10.2139/ssrn.4058777

Borges, A. F., Laurindo, F. J., Spínola, M. M., Gonçalves, R. F., & Mattos, C. A. (2020). The strategic use of artificial intelligence in the digital era: Systematic literature review and future research directions. *International Journal of Information Management*, *57*, 102225.

Bouskela, M., Casseb, M., Bassi, S., De Luca, C., & Facchina, M. (2016). *The road toward smart cities: Migrating from traditional city management to the smart city*. Inter-American Development Bank. https://publications.iadb.org/bitstream/handle/11319/7743/La-ruta-hacia-las-smart-citiesMigrando-de-una-gestion-tradicional-a-la-ciudad-inteligente.pdf

Calandra, D., Secinaro, S., Massaro, M., Dal Mas, F., & Bagnoli, C. (2022). The link between sustainable business models and Blockchain: A multiple case study approach. *Business Strategy and the Environment*.

Caragliu, A., Del Bo, C., & Nijkamp, P. (2011a). Smart cities in Europe. In B. Murgante, G. Borruso, & A. Lapucci (Eds.), *Computational science and its applications – ICCSA 2011. ICCSA 2011. Lecture Notes in Computer Science* (Vol. 6782). Springer. https://www.semanticscholar.org/paper/Smart-Cities-in-Europe-Caragliu-Bo/0565b38859f76df988248e6054f0ed72156371ab

Caragliu, A., Del Bo, C., & Nijkamp, P. (2011b). Smart cities in Europe. *Journal of Urban Technology*, *18*(2), 65–82. http://dx.doi.org/10.1080/10630732.2011.601117

Chourabi, H., Nam, T., Walker, S., Gil-Garcia, J. R., Mellouli, S., Nahon, K., Pardo, T. A., & Scholl, H. J. (2012). Understanding smart cities: An integrative framework.

In A. Hameurlain, J. Küng, R. Wagner, S. Kirn, & E. J. Neuhold (Eds.), *Database and expert systems applications. DEXA 2012*. Lecture Notes in Computer Science (Vol. 7446). Springer. https://asu.pure.elsevier.com/en/publications/understanding-smart-cities-an-integrative-framework

City of Sydney. (2020). *Smart city strategic framework*. https://www.cityofsydney.nsw.gov.au/strategies-action-plans/smart-city-strategic-framework

Cohen, M. P. (2011). *Cities in times of crisis: The response of local government in light of the global economic crisis – The role of the formation of human capital, urban innovation and strategic planning*. Institute of Urban and Regional Development. https://www.econstor.eu/bitstream/10419/59396/1/647147017.pdf

Dal Mas, F., Dicuonzo, G., Massaro, M., & Dell'Atti, V. (2020). Smart contracts to enable sustainable business models. A case study. *Management Decision*, *58*(8), 1601–1619. https://doi.org/10.1108/MD-09-2019-1266

Damgaard, T., Freytag, P. v., & Darmer, P. (2000). Qualitative methods in business studies. *Advances in Business Marketing and Purchasing*, *9*, 143–186. https://doi.org/10.1016/S1069-0964(00)09005-0/FULL/XML

Databroker. (2021). *Most valuable data to smart cities*. https://www.databroker.global/about/updates/most-valuable-data-to-smart-cities-122

de Bem Machado, A., Santos, R. J., Richter, M. F., & Sousa, M. J. (2021). *Smart cities: Building sustainable cities. Green technological innovation for sustainable smart societies – Post pandemic era*. Springer International Publishing. https://doi.org/10.1007/978-3-030-73295-0

Deloitte. (2020). *5G smart cities whitepaper*. https://www2.deloitte.com/content/dam/Deloitte/cn/Documents/technology-media-telecommunications/deloitte-cn-tmt-empowering-smart-cities-with-5g-white-paper-en-200702.pdf

Digital Cities Index (DCI). (2022). Digital Cities Index 2022. *The Economist*. Retrieved March 20, 2023, from https://impact.economist.com/projects/digital-cities/2022-executive-summary

Ding, R.-X., Palomares, I., Wang, X., Yang, G.-R., Liu, B., Dong, Y., Herrera-Viedma, E., & Herrera, F. (2020). Large-scale decision-making: Characterization, taxonomy, challenges and future directions from an artificial intelligence and applications perspective. *Information Fusion*, *59*, 84–102.

Earth.Org. (2022). *How New York smart city projects are leading the way*. https://earth.org/new-york-smart-city

European Commission. (2019). *The future of cities: Opportunities, challenges and the way forward*. EUR 29752 EN. ISBN 978-92-76-03847-4. https://doi.org/10.2760/375209

Executive Order n.° 14057. (2021, December 8). *Catalyzing clean energy industries and jobs through federal sustainability. A presidential document by the executive office of the President of the USA*. https://www.federalregister.gov/d/2021-27114

Giffinger, R., Fertner, C., Kramar, H., Kalasek, R., Milanović, N., & Meijers, E. (2007). *Smart cities – Ranking of European medium-sized cities*. https://www.researchgate.net/publication/261367640_Smart_cities_-_Ranking_of_European_medium-sized_cities

Guerreiro, A., & Barros, D. M. V. (2019). *Novos desafios da educação a distância: programação e uso de Chatbots*.

Huang-Lachmann, J. T. (2019). Systematic review of smart cities and climate change adaptation. *Sustainability Accounting, Management and Policy Journal*, *10*(4), 745–772.

InsightsIAS. (2018). *Insights into Editorial: The smart cities project must promote diversity*. https://www.insightsonindia.com/2018/05/24/insights-into-editorial-the-smart-cities-project-must-promote-diversity

International Telecommunication Union (ITU). (2021). *ITU smart sustainable cities*. https://www.itu.int/en/ITU-T/ssc/united/Documents/Reports-on-SSC/ITU_smart_sustainable_cities_brochure.pdf

IoT Analytics. (2020). *The top 10 smart city use cases that are being prioritized now*. https://iot-analytics.com/top-10-smart-city-use-cases-prioritized-now

IoT Analytics. (2021). *Smart cities of the future: 7 things that successful cities do*. https://iot-analytics.com/smart-cities-future-7-things-successful-cities-do

Jackson, R. L., Drummond, D. K., & Camara, S. (2007). What is qualitative research? *Qualitative Research Reports in Communication*, *8*(1), 21–28.

Kanter, R. M., & Litow, S. S. (2009). Informed and interconnected a manifesto for smarter cities. *Harvard Business School General Management Unit Working Paper*, *9*(141), 1–27. https://www.hbs.edu/faculty/Publication%20Files/09-141.pdf

Koumetio Tekouabou, S. C., Diop, E. B., Azmi, R., & Chenal, J. (2022). Artificial Intelligence based methods for smart and sustainable urban planning: A systematic survey. *Archives of Computational Methods in Engineering*, 1–18.

Lei n.° 31/2014, de 30 de maio. *Lei de bases gerais da política pública de solos, de ordenamento do território e de urbanismo*. https://www.pgdlisboa.pt/leis/lei_mostra_articulado.php?nid=2123&tabela=leis&so_miolo

Li, B.-h., Hou, B.-c., Yu, W.-t., Lu, X.-b., & Yang, C.-w. (2017). Applications of artificial intelligence in intelligent manufacturing: A review. *Frontiers of Information Technology & Electronic Engineering*, *18*(1), 86–96.

Li, W., & Long, Y. (2022). Smart technologies for fighting against pandemics: Observations from China during COVID-19. *Transactions in Urban Data, Science, and Technology*, *1*(3–4), 105–120. https://doi.org/10.1177/27541231221127152

Markets and Markets. (2022, November). *Smart cities market by focus area, smart transportation, smart buildings, smart utilities, smart citizen services (public safety, smart healthcare, smart education, smart street lighting, and E-Governance) and Region – Global Forecast to 2027*. Report Code: TC 3071. https://www.marketsandmarkets.com/Market-Reports/smart-cities-market-542.html

Massaro, M., Dumay, J., & Guthrie, J. (2016). On the shoulders of giants: Undertaking a structured literature review in accounting. *Accounting, Auditing & Accountability Journal*, 29(5), 767–801.

McKinsey. (2018). *Smart cities: Digital solutions for a more liveable future*. https://www.mckinsey.com/capabilities/operations/our-insights/smart-cities-digital-solutions-for-a-more-livable-future

Microsoft. (2023). *Smart cities: The cities of the future*. https://www.microsoft.com/en-us/industry/government/resources/smart-cities

Miles, M., Huberman, A., & Saldaña, J. (2014). Sampling: Bounding the collection of data. In *Qualitative data analysis: A methods sourcebook* (pp. 26–30). Sage.

Ministère de la Transition Écologique et de la Cohésion des Territoiresles. (2022). *Établissements Publics D'aménagement (édition 2022)*. https://www.ecologie.gouv.fr/sites/default/files/les_etablissements_publics_d_amenagement_edition_2022.pdf

Mohanty, R., & Kumar, B. P. (2021). Urbanization and smart cities. In *Smart city technologies, handbook on planning, design, development, and regulation* (Chapter 7, pp. 143–158). https://www.sciencedirect.com/science/article/pii/B9780128168165000073

Molnar, A. (2021). Smart cities education: An insight into existing drawbacks. *Telematics and Informatics*, *57*, 101509.

Nam, T., & Pardo, T. A. (2011). Conceptualizing smart city with dimensions of technology, people and institutions. In *Proceedings of the 12th annual international conference on digital government research* (pp. 282–291). Association for Computing Machinery. https://www.ctg.albany.edu/media/pubs/pdfs/dgo_2011_smartcity.pdf

Neirotti, P., De Marco, A., Cagliano, A. C., Mangano, G., & Scorrano, F. (2014). Current trends in Smart City initiatives: Some stylised facts. *Cities*, *38*, 25–36. https://www.researchgate.net/publication/260015335_Current_trends_in_Smart_City_initiatives_Some_stylised_facts

Osório de Barros, G. (2021). *Digitalisation, skills and cybersecurity in Portugal – Critical factors in a digital economy driven by Covid-19*. TemaEconómico 89, GEE – Office for

Strategy and Studies of the Ministry of Economy and Digital Transition, Portugal. https://www.gee.gov.pt/pt/estudos-e-seminarios/estudos-de-temas-economicos-category/31552-te-89-digitalisation-skills-and-cybersecurity-in-portugal-critical-factors-in-a-digital-economy-driven-by-covid-19

Pacheco Rocha, N., Dias, A., Santinha, G., Rodrigues, M., Queirós, A., & Rodrigues, C. (2019). Smart cities and healthcare: A systematic review. *Technologies*, *7*(3), 58.

Pandiyan, P., Saravanan, S., Usa, K., Kannadasan, R., Alsharif, M. H., & Kim, M.-K. (2023). *Technological Advancements Toward Smart Energy Management in Smart Cities*. http://dx.doi.org/10.2139/ssrn.4350031

Pandl, K. D., Thiebes, S., Schmidt-Kraepelin, M., & Sunyaev, A. (2020). On the convergence of artificial intelligence and distributed ledger technology: A scoping review and future research agenda. *IEEE Access*, *8*, 57075–57095.

Pellegrin, J., Colnot, L., & Delponte, L. (2021). *Research for REGI Committee – Artificial intelligence and urban development*. European Parliament, Policy Department for Structural and Cohesion Policies. http://www.europarl.europa.eu/RegData/etudes/STUD/2021/690882/IPOL_STU(2021)690882_EN.pdf

Pictet. (2018). The future of cities. *The Economist*. https://innovationmatters.economist.com/the-future-of-cities

Rakshit, D., & Paul, A. (2020, June 6). *Impact of COVID-19 on Sectors of Indian Economy and business survival strategies*. http://dx.doi.org/10.2139/ssrn.3620727

Romme, A. G. L., Avenier, M. J., Denyer, D., Hodgkinson, G. P., Pandza, K., Starkey, K., & Worren, N. (2015). Towards common ground and trading zones in management research and practice. *British Journal of Management*, *26*(3), 544–559.

Russell, S., & Norvig, P. (2016). *Artificial intelligence: A modern approach*. Pearson.

Santos, J. R. (2020). *Economia Indispensável*. LisbonInternationalPress

Secinaro, S., Brescia, V., Calandra, D., & Biancone, P. (2021). Towards a hybrid model for the management of smart city initiatives. *Cities*, *116*, 103278. https://doi.org/10.1016/J.CITIES.2021.103278

Shapiro, J. (2003). *"Smart Cities: Explaining the relationship between city growth and human capital"*. Urban/Regional 0309001, University Library of Munich, Germany. http://dx.doi.org/10.2139/ssrn.480172

Sichman, J. S. (2021). Inteligência Artificial e sociedade: avanços e riscos. *EstudosAvançados*, *35*(101), 37–50. https://doi.org/10.1590/s0103-4014.2021.35101.004

Smart Cities Council. (2015). *The smart cities readiness guide – A handbook for planning action*. https://www.slideshare.net/mmudassar/smartcitiescouncilreadiness-guidev1571714

Souza, J. T. de, Francisco, A. C. de, Piekarski, C. M., Prado, G. F. do, & Oliveira, L. G. de (2019). Data mining and machine learning in the context of sustainable evaluation: A literature review. *IEEE Latin America Transactions*, *17*(03), 372–382.

Startup Daily. (2022). *Sydney 'smart bin' startup Charopy ramps up to stop recycling ending up as waste*. https://www.startupdaily.net/topic/climate-tech/sydney-smartbin-startup-charopy-ramps-up-to-stop-recycling-ending-up-as-waste

Technology Magazine. (2022). *Top 10 sustainable smart cities helping make a better world*. https://technologymagazine.com/top10/top-10-sustainable-smart-cities-helping-make-a-better-world

Thales. (2018). *Secure, sustainable smart cities and the IoT*. https://www.thalesgroup.com/en/markets/digital-identity-and-security/iot/inspired/smart-cities

Time Magazine. (2022). *Seoul – The smartest city*. https://time.com/collection/worlds-greatest-places-2022/6194601/seoul

UK Environment Act. (2021). https://www.legislation.gov.uk/ukpga/2021/30/contents/enacted

Ulpiani, G., Vetters, N., Shtjefni, D., Kakoulaki, G., & Taylor, N. (2023). *Let's hear it from the cities: On the role of renewable energy in reaching climate neutrality in Urban Europe*. http://dx.doi.org/10.2139/ssrn.4358085

United Nations. (2022). *World Cities Report 2022 – Envisaging the future of cities*. United Nations Human Settlements Programme. https://unhabitat.org/sites/default/files/2022/06/wcr_2022.pdf

United Nations – Department of Economic and Social Affairs. (2022). *Seoul Smart City Platform White Paper*. https://publicadministration.un.org/Unpsa/Portals/0/UNPSA_Submitted_Docs/2019/4f4c5d55-C9e6-4e61-Bc76-F8fc93bfdcdf/2020%20UNPSA_SCPM_Whitepaper_27112019_011429_0bc8129d-6963-46a7-A4a2-0d29c203ca4c.Pdf?Ver=1441-03-30-011430-090

Usman, M., Balsalobre-Lorente, D., Jahanger, A., & Ahmad, P. (2023). Are Mercosur economies going green or going away? An empirical investigation of the association between technological innovations, energy use, natural resources and GHG emissions. *Gondwana Research*, *113*, 53–70.

Vershinina, I. A., & Volkova, L. (2020). Smart cities: Challenges and opportunities. *Revista Espacios*, *41*(15), 23–24. https://www.revistaespacios.com/a20v41n15/a20v41n15p23.pdf

Von Krogh, G. (2018). Artificial intelligence in organizations: New opportunities for phenomenon-based theorizing. *Academy of Management Discoveries*, *4*(i.4), 404–409.

Wang, W., & Siau, K. (2019). Artificial intelligence, machine learning, automation, robotics, future of work and future of humanity: A review and research agenda. *Journal of Database Management*, *30*(1), 61–79.

We Build Value Digital Magazine. (2022). *Hi-tech London, or the making of a smart city*. https://www.webuildvalue.com/en/megatrends/smart-city-london.html

World Smart Cities Forum. (2019). *Copenhagen*. https://worldsmartcities.org/copenhagen

Yang, J., Chen, Y., Huang, W., & Li, Y. (2017). *Survey on artificial intelligence for additive manufacturing* [Paper presentation]. 2017 23rd International Conference on Automation and Computing (ICAC).

Yin, R. K. (2017). *Case study research and applications: Design and methods*. Sage Publications.

Yovanof, G. S., & Hazapis, G. N. (2009). An architectural framework and enabling wireless technologies for digital cities & intelligent urban environments. *Wireless Personal Communications*, *49*, 445–463. https://doi.org/10.1007/s11277-009-9693-4

Zawacki-Richter, O., Marín, V. I., Bond, M., & Gouverneur, F. (2019). Systematic review of research on artificial intelligence applications in higher education – Where are the educators? *International Journal of Educational Technology in Higher Education*, *16*(1). https://doi.org/10.1186/s41239-019-0171-0

Zhu, Q., Loke, S. W., Trujillo-Rasua, R., Jiang, F., & Xiang, Y. (2019). Applications of distributed ledger technologies to the Internet of Things: A survey. *ACM Computing Surveys (CSUR)*, *52*(6), 1–34.

Zigurat. (2019). *Smart city and urbanization challenges*. https://www.e-zigurat.com/blog/en/smart-cities-urbanization-challenges

Chapter 9

The Impact of Innovation and Sustainable Development in Smart Cities

Khider Hamid Khider

Lecturer at Arab Open University -Kuwait Branch

Abstract

This chapter is concerned about the effect of innovation and sustainable development on smart cities, since smart cities focus on the improvement of society, by providing fine and good services, so it's a must to use new ideas and solve arising problems in a clear and nice way that will allow smart cities to provide excellent type of work. Innovation and sustainability will robust the capability of smart cities to tackle uncertain issues in the environment of our urban area. The usage of new technology will increase competitive advantages in the market of technology, because each organization wants to satisfy the need for smart cities, by presenting the latest technology in those new and smart cities, which ultimately enhances the services applied in these improved and developed cities.

Keywords: Smart cities; innovation; sustainable development; digital transformation; urban; waste management

Introduction

Smart cities are urban areas that leverage technology and innovation to enhance the quality of life for residents while promoting sustainability and efficient resource management. Innovation and sustainable development play crucial roles in shaping the evolution of smart cities, offering numerous benefits and opportunities. In this chapter, we are going to discuss about innovation which means tackling problem

The Emerald Handbook of Smart Cities in the Gulf Region:
Innovation, Development, Transformation, and Prosperity for Vision 2040, 171–180

doi:10.1108/978-1-83608-292-720241009

situations in a unique and new way to find solutions and a good understanding of that situation also, we will demonstrate how sustainable development could emphasize the clear and fine application of smart cities; the chapter will address how principles and factors of innovation and sustainable development could improve smart cities and how this progress could affect the life of the people and improve it.

What Is Innovation, Sustainable, and Smart Cities

No doubt that innovation is core stone for any development nowadays. Whatever happened was due to digital transformation built on new ideas and creativity. Most projects and organizations take advantages of innovation to compete with other organizations as well as survive in the market and provide their services. In this chapter, we are going to discuss the impact of innovation on smart cities. Smart cities' ideas were built on innovation and creativity meaning. So first we will define smart cities, second we will move to innovation and then finally link them to each other. If smart cities use innovation to improve their services, this will lead to sustainable development which means improvement in all parts of society in the economy, community, and environment.

Smart cities are urban areas that leverage technology and data-driven solutions to enhance efficiency, sustainability, and quality of life for residents. These cities integrate various Internet of Things (IoT) devices, sensors, and data analytics to optimize infrastructure, transportation, energy usage, public services, and more. The goal is to create cities that are more liveable, resilient, and responsive to the needs of citizens while minimizing resource consumption and environmental impact. Smart city initiatives often involve the deployment of smart grids, intelligent transportation systems, digital governance platforms, and innovative urban planning strategies (Caragliu et al., 2011).

Innovation refers to the process of introducing new ideas, methods, products, or services that create value or improve existing ones. It involves creativity, problem-solving, and the application of novel approaches to address challenges or opportunities. Innovation can occur in various domains, including technology, business, science, art, and social systems. It often leads to advancements, efficiency gains, and improvements in quality of life by fostering change and adaptation to evolving needs and circumstances (Rogers, 2010).

Sustainable development is a concept that aims to meet the needs of the present generation without compromising the ability of future generations to meet their own needs. It involves balancing economic growth, social progress, and environmental protection to ensure a harmonious and equitable development path. Sustainable development recognizes the interdependence of economic, social, and environmental factors and seeks to address them in an integrated and holistic manner (Kelly, 2012).

Key Principles of Sustainable Development

Environmental Sustainability: Ensuring that natural resources are used efficiently and responsibly, minimizing pollution and waste, and preserving biodiversity and ecosystems for future generations.

Economic Sustainability: Promoting inclusive and equitable economic growth that generates prosperity and opportunities for all while minimizing disparities and ensuring the long-term viability of economic activities.

Social Sustainability: Fostering social inclusion, equity, and justice by addressing poverty, inequality, and social exclusion, promoting access to education, healthcare, and basic services, and empowering communities to participate in decision-making processes.

Sustainable development requires cooperation and collaboration among governments, businesses, civil society organizations, and individuals at local, national, and global levels. It involves adopting policies, practices, and technologies that promote sustainable consumption and production, promote resilience to environmental and social challenges, and foster a more equitable and prosperous future for all.

Innovation plays a pivotal role in shaping the development and functionality of smart cities, enhancing efficiency, sustainability, and quality of life.

How Innovation Impacts Smart Cities

Here are some ways innovation impacts smart cities, along with examples:

Infrastructure Management: Innovations such as IoT sensors, data analytics, and predictive maintenance algorithms help in efficiently managing infrastructure like roads, bridges, and utilities. For instance, Barcelona uses IoT sensors to monitor traffic flow, optimize bus routes, and manage parking spaces, reducing congestion and improving mobility.

Energy Efficiency: Smart cities leverage innovations like smart grids, renewable energy sources, and energy-efficient technologies to reduce energy consumption and carbon footprint. Copenhagen, for example, aims to become carbon-neutral by 2025 through initiatives such as wind energy integration, district heating systems, and smart lighting.

Transportation: Innovations such as autonomous vehicles, ride-sharing platforms, and smart public transportation systems revolutionize urban mobility. Singapore's autonomous vehicle trials and comprehensive public transit network, combined with congestion pricing schemes, demonstrate a holistic approach to sustainable transportation.

Public Services: Innovations in e-governance, digital platforms, and artificial intelligence (AI)-driven service delivery to enhance the efficiency and accessibility of public services. Tallinn, Estonia, is renowned for its digital governance model, offering e-residency, digital voting, and streamlined online services for citizens.

Sustainability: Innovations in waste management, water conservation, and green building technologies contribute to environmental sustainability. Amsterdam's circular economy initiatives involve recycling organic waste into biogas, retrofitting buildings for energy efficiency, and promoting sustainable urban agriculture.

Safety and Security: Innovations such as surveillance systems, predictive policing algorithms, and emergency response technologies enhance public safety. Dubai's smart city initiatives include advanced surveillance systems, AI-powered crime prediction, and smart emergency response systems to ensure the safety of residents and visitors.

Community Engagement: Innovations like social media platforms, mobile apps, and digital community hubs facilitate citizen participation and engagement in urban governance. The "My Neighbourhood" initiative in Seoul, South Korea, empowers residents to propose and prioritize local improvement projects through a digital platform, fostering community cohesion and civic involvement.

Innovation is fundamental to the development of smart cities, enabling them to address urban challenges effectively and improve the overall quality of life for residents (Sachs, 2015).

Innovation and Capacity of Smart Cities

People in the present age are in need of advanced services, which is distinguished by innovation, smart cities need more technological innovation which will lead society to sustainable development, and this will bring about more comfort among nations. The development that could happen in all life sectors due to innovation is in need for people awareness about the role of innovation in smart cities, so education and other social services should be in high standard to match the growth that could happen in humans life. This will show the impact of innovation on smart cities.

Innovation has the potential to significantly increase the capabilities of smart cities in various ways:

Technological Advancements: Continuous innovation leads to the development of more advanced technologies, such as IoT sensors, AI, and data analytics tools. These technologies enhance the ability of smart cities to collect, analyze, and act upon vast amounts of data in real-time, enabling more efficient and effective management of urban systems.

Interconnected Systems: Innovations in connectivity and interoperability allow different urban systems to communicate and work together seamlessly. For example, integrating transportation systems with energy management systems can optimize traffic flow based on energy consumption patterns, leading to reduced congestion and lower emissions.

Customization and Adaptability: Innovative solutions can be tailored to the specific needs and challenges of individual cities. This customization ensures that smart city initiatives are aligned with local context, culture, and priorities, maximizing their impact and relevance.

User-Centric Design: Innovation in design methodologies and human-centered approaches ensures that smart city solutions are user-friendly, accessible, and inclusive. By prioritizing the needs and preferences of residents, businesses, and visitors, cities can create more engaging and responsive urban environments.

Sustainability and Resilience: Innovative technologies and strategies promote sustainability and resilience in smart cities. For example, advancements in renewable energy generation, energy storage, and grid management enhance the ability of cities to reduce their carbon footprint and withstand environmental challenges such as climate change and natural disasters.

Collaboration and Partnership: Innovation fosters collaboration and partnership among various stakeholders, including government agencies, private companies, academic institutions, and community organizations. By working together, these stakeholders can leverage their respective expertise and resources to address complex urban challenges more effectively.

Economic Growth and Competitiveness: Smart cities that embrace innovation attract investment, talent, and businesses, driving economic growth and enhancing their competitiveness on a global scale. By creating a conducive environment for innovation and entrepreneurship, cities can stimulate job creation, innovation-led growth, and prosperity for their residents.

Continuous Improvement: Innovation is an ongoing process that encourages continuous improvement and learning. Smart cities that embrace a culture of innovation are better equipped to adapt to changing circumstances, anticipate future challenges, and continuously evolve to meet the needs of their citizens.

Innovation plays a crucial role in enhancing the capabilities of smart cities by driving technological advancements, promoting collaboration and customization, fostering sustainability and resilience, and stimulating economic growth and competitiveness.

Innovation plays a pivotal role in driving sustainable development in smart cities by introducing new technologies, processes, and approaches that address environmental, social, and economic challenges.

Innovation and Sustainable Development in Smart Cities

Here's how innovation contributes to sustainable development in smart cities:

Renewable Energy: Innovation in renewable energy technologies, such as solar, wind, and hydropower, enables smart cities to reduce reliance on fossil fuels and mitigate greenhouse gas emissions. Smart grids and energy storage systems optimize the integration of renewable energy sources into the urban energy infrastructure, promoting a transition toward cleaner and more sustainable energy systems.

Energy Efficiency: Innovative solutions, such as smart meters, building automation systems, and energy-efficient appliances, help smart cities optimize energy consumption and reduce waste. Real-time monitoring and data analytics enable more efficient use of energy resources, leading to cost savings and environmental benefits.

Green Building Design: Innovations in green building materials, construction techniques, and design principles promote sustainable development in smart cities. Buildings with features like energy-efficient insulation, passive heating and cooling, and rooftop gardens reduce energy consumption, enhance indoor air quality, and minimize environmental impact.

Smart Transportation: Innovation in transportation technologies, such as electric vehicles, autonomous vehicles, and intelligent transportation systems, contributes to sustainable mobility in smart cities. By reducing reliance on fossil fuel-powered vehicles, optimizing traffic flow, and promoting alternative modes of transportation like cycling and public transit, smart transportation solutions help mitigate traffic congestion, air pollution, and carbon emissions.

Waste Management: Innovative waste management technologies, such as waste-to-energy facilities, recycling robots, and smart bins equipped with sensors, enable smart cities to optimize waste collection, processing, and disposal. By promoting recycling, composting, and resource recovery, these innovations minimize landfill waste and promote a circular economy.

Water Management: Innovation in water management technologies, such as smart irrigation systems, leak detection sensors, and water recycling facilities, helps smart cities optimize water use and conserve precious resources. Real-time monitoring and data analytics enable more efficient water distribution, reduce water loss due to leaks and inefficiencies, and promote sustainable water practices.

Urban Agriculture: Innovative urban agriculture techniques, such as vertical farming, aquaponics, and hydroponics, enable smart cities to produce fresh, nutritious food locally and sustainably. By utilizing underutilized urban spaces and reducing the carbon footprint associated with food production and transportation, urban agriculture contributes to food security, resilience, and sustainability.

Community Engagement: Innovation in digital technologies and social platforms facilitates community engagement and participation in sustainable development initiatives. By empowering citizens to contribute ideas, feedback, and resources, smart cities can foster a sense of ownership, collaboration, and collective action toward common sustainability goals (United Nations, 2015).

Overall, innovation plays a crucial role in driving sustainable development in smart cities by introducing new technologies, processes, and approaches that promote energy efficiency, renewable energy, green building design, smart transportation, waste management, water conservation, urban agriculture, and community engagement. By harnessing the power of innovation, smart cities can create more resilient, liveable, and sustainable urban environments for current and future generations (Sen, 1999).

When we are looking at the growth of smart cities due to innovation we should not forget, the development of society and how the capability of people could increase to match the improvement that happened in cities (smart cities) and could bring about sustainability.

Role of Information Technology

The success of smart cities in achieving innovation and sustainable development depends not only on technological advancements but also on the capability of people to embrace and adapt to these changes. Here's how the capability of people influences the development of smart cities:

Digital Literacy and Skills: As smart cities rely heavily on digital technologies and data-driven solutions, the capability of people to use and understand these technologies is crucial. Digital literacy programs and training initiatives help residents develop the skills needed to navigate digital platforms, access online services, and participate in smart city initiatives effectively.

Innovation Mindset: Fostering an innovation mindset among residents encourages creativity, experimentation, and collaboration in addressing urban challenges. Smart cities promote a culture of innovation through educational programs, entrepreneurship initiatives, and platforms for idea-sharing and co-creation.

Community Engagement: The capability of people to engage actively in community decision-making processes is essential for the success of smart city initiatives. Citizens' participation in public consultations, neighborhood meetings, and online forums enables them to provide feedback, share insights, and contribute to the design and implementation of sustainable development projects.

Behavioral Change: Sustainable development in smart cities requires changes in behavior and lifestyle patterns to reduce resource consumption and environmental impact. Public awareness campaigns, incentives for sustainable practices, and social norms interventions encourage residents to adopt eco-friendly behaviors such as energy conservation, waste reduction, and alternative transportation modes.

Education and Awareness: Educating residents about the benefits of innovation and sustainable development fosters a sense of ownership and responsibility toward the future of their city. Schools, universities, and community organizations play a vital role in promoting environmental stewardship, social equity, and economic resilience through curriculum integration, outreach programs, and awareness-raising campaigns.

Accessibility and Inclusivity: Smart cities must ensure that innovation and sustainable development initiatives are accessible and inclusive for all residents, regardless of age, income, or ability. Designing inclusive digital platforms, providing assistive technologies, and offering multilingual support services enable broader participation and engagement among diverse populations.

Empowerment and Participation: Empowering residents to take an active role in shaping the future of their city strengthens community cohesion and resilience. Participatory budgeting, citizen science projects, and grassroots initiatives empower residents to propose, implement, and evaluate sustainable development projects that address their unique needs and priorities.

Resilience and Adaptation: Building resilience to environmental hazards and socioeconomic challenges requires the capability of people to adapt to changing circumstances and mitigate risks effectively. Community-based disaster preparedness, climate adaptation strategies, and social support networks enhance residents' capacity to cope with and recover from adverse events (Hawken et al., 1999).

In summary, the capability of people to stay abreast of innovation and actively contribute to sustainable development is essential for the success of smart cities. By fostering digital literacy, promoting an innovation mindset, encouraging community engagement, facilitating behavioral change, promoting education and awareness, ensuring accessibility and inclusivity, empowering participation, and building resilience, smart cities can harness the collective potential of their residents to create more liveable, equitable, and sustainable urban environments.

So according to what we discussed above, we want to know what importance of innovation is to smart cities.

Importance of Innovation in Smart Cities

Innovation plays a critical role in the development and sustainability of smart cities. Here are several reasons why innovation is important in the context of smart cities:

Efficiency and Optimization: Innovation enables the implementation of new technologies and processes that enhance efficiency in various aspects of city management, including transportation, energy usage, waste management, and public services.

Sustainability: Smart cities aim to minimize their environmental footprint by adopting sustainable practices. Innovation drives the development of clean energy solutions, green infrastructure, and resource-efficient systems, helping cities reduce emissions and mitigate environmental impact.

Quality of Life: Innovative technologies and services contribute to improving the quality of life for residents. For example, smart transportation systems reduce traffic congestion and improve air quality, while smart healthcare solutions enhance access to healthcare services and promote public health.

Economic Growth: Innovation fosters economic growth by creating opportunities for businesses, entrepreneurs, and startups. Smart cities attract investment and talent by offering an environment conducive to innovation and technological advancement.

Resilience and Adaptability: Smart cities are better equipped to respond to challenges such as natural disasters, pandemics, and infrastructure failures through innovative solutions. Technologies like IoT sensors, data analytics, and AI enable real-time monitoring and rapid response, enhancing the city's resilience and adaptability.

Citizen Engagement and Participation: Innovation facilitates citizen engagement and participation in decision-making processes. Digital platforms, mobile apps, and IoT-enabled devices empower residents to provide feedback, report issues, and participate in community initiatives, fostering a sense of ownership and accountability.

Data-Driven Decision-Making: Innovation generates vast amounts of data that can be leveraged to make informed decisions. By harnessing data analytics, AI, and machine learning, smart cities gain valuable insights into urban trends, patterns, and behaviors, enabling policymakers to implement evidence-based policies and strategies.

Inclusive Development: Innovation promotes inclusive development by ensuring that smart city initiatives benefit all segments of the population, including marginalized communities. By addressing issues of accessibility, affordability, and equity, smart cities strive to create a more inclusive and equitable urban environment (Albino et al., 2015).

Overall, innovation serves as a catalyst for transforming traditional cities into dynamic, resilient, and sustainable smart cities that prioritize the well-being and prosperity of their residents.

In essence, digital transformation, fueled by innovation, is paving the way for a future where technology not only enhances our capabilities but also creates a more sustainable and equitable world. Smart cities are a prime example of how this transformation can lead to a better quality of life, demonstrating the potential of technology to solve complex challenges in ways that benefit everyone.

Conclusion

The impact of innovation on smart cities is profound and multifaceted. Innovation drives efficiency, sustainability, and economic growth while enhancing the quality of life for residents. It empowers cities to tackle complex challenges, adapt to change, and foster inclusive development. By leveraging technology, data, and creativity, smart cities can create more resilient, connected, and livable urban environments. However, to fully realize the potential of innovation, it is essential for stakeholders to collaborate, invest in infrastructure, and prioritize the needs of communities. Innovation is not just about deploying new technologies; it is about creating smarter, more equitable cities that improve the well-being and prosperity of everyone who calls them home. These cities are the focal point of urban evolution, utilizing technology and progressive policies to address the complex challenges of the 21st century. By prioritizing efficiency, connectivity, inclusivity, and quality of life, smart city initiatives are reshaping the urban landscape in ways that promote environmental stewardship, economic prosperity, and social equity. As smart cities continue to evolve, it is essential to maintain a focus on collaboration, adaptability, and citizen-centric design. By engaging stakeholders across sectors and embracing emerging technologies and best practices, cities can ensure that their development remains responsive to evolving needs and aspirations. Ultimately, the success of smart cities lies in their ability to balance innovation with sustainability, creating vibrant, resilient communities that thrive in an ever-changing world.

References

Albino, V., Berardi, U., & Dangelico, R. M. (2015). Smart cities: Definitions, dimensions, performance, and initiatives. *Journal of Urban Technology*, *22*(1), 3–21.

Caragliu, A., Del Bo, C., & Nijkamp, P. (2011). Smart cities in Europe. *Journal of Urban Technology*, *18*(2), 65–82.

Hawken, P., Lovins, A., & Lovins, L. H. (1999). *Natural capitalism: Creating the next industrial revolution*. Little, Brown and Company.

Kelly, K. (2012). *What technology wants*. Penguin Books.

Rogers, E. M. (2010). *Diffusion of innovations*. Simon and Schuster.

Sachs, J. D. (2015). *The age of sustainable development*. Columbia University Press.

Sen, A. (1999). *Development as freedom*. Oxford University Press.

United Nations. (2015). *Transforming our world: The 2030 Agenda for Sustainable Development*. United Nations.

Chapter 10

Digital Intersections: Synergies Between the Gig Economy and Evolution of the Smart City

Sawsan Malik[a], *Afnan Alkhaldi*[a], *Jarrah Fahad Al-Mansour*[a] *and Miltiadis D. Lytras*[b]

[a]*Arab Open University, Kuwait*
[b]*College of Engineering, Efrat University, Saudi Arabia*

Abstract

Technology and a gig economy have affected how people are employed, with shorter-term, more flexible, and technologically mediated work arrangements increasingly becoming the norm. The digital platform, which acts both as a marketplace and a "shadow employer," makes this possible, with workers provided with unprecedented autonomy and flexibility. The authors examine the interplay between the gig economy and the concept of the smart city – an urban environment reliant upon information and communication technologies – and focus on how technological integrations such as blockchain, the collective network of connected devices, and the technology that facilitates communication between devices and the cloud, as well as between the devices themselves (Internet of Things [IoT]), and 5G, bolster opportunities for gig workers to carve out an existence in a perpetually changing landscape. Examples are provided for the global giant Amazon and the regional businesses Talabat and Raha in the Gulf Cooperation Council (GCC). Application of real-time tracking, precision-driven management, and decentralized transactions, provides gig workers new opportunities for employment in their evolving digital smart-city environment.

Keywords: Gig economy; gig worker; smart city management; shadow employer; Gulf Cooperation Council; Internet of Things; digitalization

---

The Emerald Handbook of Smart Cities in the Gulf Region:
Innovation, Development, Transformation, and Prosperity for Vision 2040, 181–187

doi:10.1108/978-1-83608-292-720241010

Introduction

Rapid technological advancements are transforming global landscapes. The gig economy – a labor market characterized by the prevalence of short-term, flexible, and freelance work arrangements (James, 2020; Wood et al., 2018) – relies heavily on technology and digital platforms to connect workers with consumers or clients (Wood et al., 2018). Work can be more flexible, autonomous, and the nature of tasks performed is more technology-mediated. Smart cities, the peak in urban development, combine information and communication technologies (ICT) with everyday urban functions to create responsive, efficient, and sustainable urban environments. The smart-city digital platform connects independent workers to dynamic job opportunities, and smart-city infrastructure amplifies their reach and efficiency.

Herein we describe the symbiotic relationship between the gig economy and smart cities, the opportunities and challenges gig workers encounter, the technological catalysts that drive this change, and the broader implications for urban living and work in the 21st century.

Gig Economy: Benefits and Challenges

The gig economy is often associated with a rise in the sharing economy and establishment of a capital-labor relationship between workers and digital platforms (Gandini, 2018). In this economy, workers engage in small tasks or "gigs" mediated by these digital platforms, which act as market intermediaries and "shadow employers" (Gandini, 2018). The use of technology and digitization is integral to this economy, with digital platforms and smart applications playing important roles in facilitating the supply and demand for gig work (Wood et al., 2018).

Gig workers can access a range of job opportunities, and because they can choose when and where they work, the potential exists to improve their work balance with other commitments, and increase their control over their schedules (James, 2020; Wood et al., 2018), and ability to earn additional income through short-term gigs or freelance work, especially developing countries (Wood et al., 2018). Gig workers are better positioned to establish their own businesses, engage in entrepreneurial activities, pursue their passions, increase their earnings (Wood et al., 2018), and improve their job satisfaction (Cropanzano et al., 2023). However, they might also experience volatile income because of fluctuations in gig availability and demand, a lack of job security, and limited access to social protections (Anwar & Graham, 2019; Ayentimi et al., 2023). Gig workers are often classified as independent contractors, who typically do not receive benefits such as health insurance, paid leave, or retirement plans (Anwar & Graham, 2019; James, 2020) and are vulnerable to exploitation, making it difficult for them to address workplace issues.

The gig economy and concept of a smart city (which leverages technology and data to improve urban services and the quality of life of residents) are closely linked. In a smart city, technology and smart applications can assist gig workers to locate job opportunities more easily and efficiently and to access resources and services to support their work. For example, gig workers in transportation can benefit from real-time traffic information and route optimization algorithms, enhancing their efficiency and earning potential (Barrios et al., 2020).

The car washing system in both traditional and smart gig economies has been influenced by the emergence of the gig economy and the utilization of digital platforms. In the traditional gig economy, car washing services have been a common form of gig work, where individuals offer their services as independent contractors through various platforms (Duggan et al., 2019). These platforms facilitate the connection between car owners and gig workers, allowing for flexible and on-demand car washing services.

The gig economy in India has emerged as a significant force, reshaping the traditional employment landscape, and offering unique opportunities for workers seeking flexible work arrangements (Singh & Bhushan, 2023). The surge of internet users and the demand for food delivery services have created broad opportunities for India's gig workers to work as per their convenience (Behl et al., 2022). There is a positive sentiment and optimism around the gig economy in India (Pant & Majumder, 2022).

Gig Economy and Smart Cities: A Symbiotic Connection

Smart cities use innovative technologies and data-driven approaches to address urban challenges and promote sustainable development (Treude, 2021). The combination of a gig economy and smart city has transformed traditional ways of work and urban living into more technologically advanced and efficient systems. Because a smart city uses ICT to enhance the quality of life for its citizens and to improve the efficiency of urban services and infrastructure (Albino et al., 2015; Meijer & Bolívar, 2015), technological systems such as sensors, data analytics, and connectivity are integrated to enable real-time monitoring, analysis, and management of urban processes (Kitchin, 2013).

In smart cities, the efficiency and effectiveness of the urban system (e.g., transportation, energy, waste management, and public safety) are enhanced by the integrating of ICT (Neirotti et al., 2014). Resource allocation can be optimized, environmental impacts can be reduced, and the delivery of public services can be improved in smart cities through the collection and analysis of a range of data (Nooringsih & Susanti, 2022). Smart cities also foster citizen engagement and participation by providing platforms for collaboration, feedback, and co-creation of urban solutions (Prevelianaki et al., 2022).

Innovative transportation solutions such as ridesharing and bike-sharing services that are common in smart cities are typically facilitated by gig economy platforms. Individuals seeking transportation services are connected with independent drivers or bike providers, providing more efficient and convenient mobility options (Kaine & Josserand, 2019). The gig economy also contributes to the economic vitality of smart cities, with the flexible nature of gig work allowing individuals to engage in the labor market on their own terms, increasing employment opportunities and income generation, and ultimately contributing to economic growth and development within the city (Barrios et al., 2020). Additionally, the gig economy provides opportunities to supplement income for residents, improving their overall well-being and financial stability (Chen et al., 2023).

Gig workers play an important role in meeting the increased demand for food delivery and home services. Smart city platforms can help gig workers to provide efficient and timely delivery of goods and services, improving the overall convenience and accessibility for residents (Au & Tsang, 2022). Because individuals can engage in entrepreneurial activities and launch their own businesses (Barrios et al., 2020), the gig economy further contributes to the smart-city entrepreneurial ecosystem.

Unlocking Gig Worker Potential: Tech Advances in Smart Cities

The ongoing integration of blockchain, the collective network of connected devices and the technology that facilitates communication between devices and the cloud, as well as between the devices themselves (Internet of Things [IoT]), and 5G into smart city frameworks benefits municipal infrastructures and broader urban mechanisms and extends tangible advantages to individual gig workers. These workers, who often navigate flexible employment patterns and disparate task environments, benefit immensely from the technological advancements of smart cities (Aldribi & Singh, 2022). For example, gig workers, particularly those in delivery or ride-hailing services, use sophisticated GPS and tracking systems powered by 5G's rapid communication channels to enhance efficiency and accuracy (Rahman et al., 2019; Yin, 2022). Access to these instantaneous data allows workers to optimize routes, avoid congested areas, improve efficiency, and maximize earnings within a limited timeframe.

The extensive array of interconnected devices now available in smart cities provides gig workers with a nuanced platform for real-time monitoring and precision-driven management (Rahman et al., 2019). A home-based craftsman can use IoT-enabled tools to send automated inventory notifications when supplies are low (Yin, 2022). The decentralized nature of blockchain technology can also improve the quality of services (Alkhaldi et al., n.d.) and enable secure and instantaneous financial transactions for gig workers (Aldribi & Singh, 2022).

The Global Gig Economy

Amazon epitomizes the interplay of technology and flexible employment structures. Its Mechanical Turk platform – a "clickwork" model (Heeks, 2017) – democratizes access to digital tasks. Its Amazon Flex initiative, akin to models seen in companies such as Uber, uses technological solutions to enable gig-style package delivery (Burtch et al., 2018). The profound influence of technology in reshaping work experiences in the gig landscape is also apparent in Amazon's embrace of advanced monitoring systems designed to optimize worker movement in its warehouses (Linder et al., 2022). In the broader context, Amazon exemplifies the multifaceted platform economy delineated by Hoang et al. (2020), merging marketplace offerings with platform-based workforce management.

Gig Economy in the Gulf Cooperation Council (GCC)

The online food delivery platform Talabat originated in Kuwait in 2004 and, within about two decades, had extended its reach to offer online food delivery services throughout the GCC (Sharma, & Waheed, 2018). Talabat, now one of the leading food delivery platforms in GCC countries, offers a range of cuisines, and partners with numerous restaurants (Alsumait & Alenezi, 2019). This platform aims to provide convenient and efficient food delivery services to customers so that they may enjoy their favorite meals from the comfort of their own homes (Alsumait & Alenezi, 2019).

Talabat's adoption of smart technology has played a significant role in its growth and popularity. The company has developed a user-friendly mobile application for customers to place orders within a few taps on their smartphone screens. This convenience, coupled with features like real-time order tracking and in-app communication with delivery riders, has made the food delivery experience seamless and efficient for customers (Safieddine & Nakhoul, 2018), improving the overall customer experience. Features such as real-time order tracking, in-app communication with delivery riders, and personalized recommendations further contribute to customer satisfaction and engagement. Talabat's use of smart technology aligns with the broader concept of smart cities, which aim to leverage technology to provide convenient services to residents.

Gig Economy in Kuwait

Raha. an innovative e-grocery platform based in Kuwait, is redefining the traditional shopping experience. This digital platform represents a convenient way for consumers to access a wide range of groceries, fresh produce, and household essentials (Raha, 2023). As a testament to its commitment to modernization and efficiency, Raha employs Kuwait's first fully automated robotics-fulfillment center, ensuring consistent quality and product freshness. However, beyond being an e-grocery service, a "Raha Recipes" feature provides a unique blend of shopping and interactive cooking, mirroring the gig economy's ethos of offering tailored and unique experiences. Through its tech-centric and user-focused strategies, Raha exemplifies the shift toward personalized, flexible services, aligning with global trends in the gig economy landscape.

Conclusion

The convergence of the gig economy and smart city frameworks has transformed both the nature of employment and urban environments. Platforms such as Amazon, Talabat, and Raha epitomize this dynamic union, combining technology with adaptable work modalities. Gig workers operating within this realm are granted unprecedented flexibility, digital enablement, and access to many opportunities. Smart cities benefit from this synergy, with their efficiency being augmented, service delivery improved, and resident engagement bolstered. However, gig workers must also navigate potential pitfalls of precarious employment, and care must be taken to ensure that rapid technological advancements do not aggravate pre-existing inequalities or compromise sustainability. Stakeholders, from policymakers to platform developers, must collaborate to ensure that the digital revolution underpinning the gig economy in smart cities is inclusive and progressive.

References

Albino, V., Berardi, U., & Dangelico, R. (2015). Smart cities: Definitions, dimensions, performance, and initiatives. *Journal of Urban Technology*, *22*(1), 3–21. https://doi.org/10.1080/10630732.2014.942092

Aldribi, A., & Singh, A. (2022). Blockchain empowered smart home: A scalable architecture for sustainable smart cities. *Mathematics*, *10*(14), 2378. https://doi.org/10.3390/math10142378

Alkhaldi, A., Alrashidi, H., Alhasan, K., Alsadeeqi, A., & Alshami, A. (n.d.). The use of blockchain technology to build smart cities: Creating public value in Kuwait. *Global Knowledge, Memory and Communication.*

Alsumait, S., & Alenezi, R. (2019). Online food industry (A Study Case: Talabat). *International Journal of Tourism and Hospitality in Asia Pasific (IJTHAP)*, *2*(2).

Anwar, M. A. & Graham, M., (2019). The global gig economy: Toward a planetary labor market. In *The digital transformation of labor* (pp. 213–234). Routledge.

Au, W., & Tsang, N. (2022). Gig workers' self-protective behaviour against legal risks: An application of protection motivation theory. *International Journal of Contemporary Hospitality Management*, *35*(4), 1376–1397. https://doi.org/10.1108/ijchm-12-2021-1537

Ayentimi, D. T., Abadi, H. A., & Burgess, J. (2023). Decent gig work in Sub Sahara Africa?. *Journal of Industrial Relations*, *65*(1), 112–125.

Barrios, J., Hochberg, Y., & Yi, H. (2020). Launching with a parachute: The gig economy and entrepreneurial entry. *SSRN Electronic Journal*. https://doi.org/10.2139/ssrn.3557279

Behl, A., Rajagopal, K., Sheorey, P., & Mahendra, A. (2022). Barriers to entry of gig workers in the gig platforms: Exploring the dark side of the gig economy. *Aslib Journal of Information Management*, *74*(5), 818–839. https://doi.org/10.1108/ajim-08-2021-0235

Burtch, G., Carnahan, S., & Greenwood, B. N. (2018). Can you gig it? An empirical examination of the gig economy and entrepreneurial activity. *Management Science*, *64*(12), 5497–5520. https://doi.org/10.1287/mnsc.2017.2916

Chen, J., Luo, J., Hu, W., & Ma, J. (2023). Fit into work! From formalizing governance of gig platform ecosystems to helping gig workers craft their platform work. *Decision Support Systems*, 174, 114016. https://doi.org/10.1016/j.dss.2023.114016.

Cropanzano, R., Keplinger, K., Lambert, B. K., Caza, B., & Ashford, S. J. (2023). The organizational psychology of gig work: An integrative conceptual review. *Journal of Applied Psychology*, *108*(3), 492.

Duggan, J., Sherman, U., Carbery, R., & McDonnell, A. (2019). Algorithmic management and app-work in the gig economy: A research agenda for employment relations and HRM. *Human Resource Management Journal*, *30*(1), 114–132. https://doi.org/10.1111/1748-8583.12258

Gandini, A. (2018). Labour process theory and the gig economy. *Human Relations*, *72*(6), 1039–1056. https://doi.org/10.1177/0018726718790002

Heeks, R. (2017). Decent work and the digital gig economy: A developing country perspective on employment impacts and standards in online outsourcing, crowdwork, etc. *Development Informatics Working Paper*, (71).

Hoang, L., Blank, G., & Quan-Haase, A. (2020). The winners and the losers of the platform economy: Who participates? *Information, Communication &Amp; Society*, *23*(5), 681–700. https://doi.org/10.1080/1369118x.2020.1720771

James, A. (2020). The gig economy: A critical introduction. *Economic Geography*, *97*(1), 113–114. https://doi.org/10.1080/00130095.2020.1831908

Kaine, S., & Josserand, E. (2019). The organisation and experience of work in the gig economy. *Journal of Industrial Relations*, *61*(4), 479–501. https://doi.org/10.1177/0022185619865480

Kitchin, R. (2013). The real-time city? big data and smart urbanism. *Geojournal*, *79*(1), 1–14. https://doi.org/10.1007/s10708-013-9516-8

Linder, R., Hunter, C., McLemore, J., Dutta, S., Akbar, F., Grover, T., … Williams, A. C. (2022). Characterizing work-life for information work on Mars. *Proceedings of the ACM on Human-Computer Interaction*, *6*(GROUP), 1–27. https://doi.org/10.1145/3492859

Meijer, A., & Bolívar, M. (2015). Governing the smart city: A review of the literature on smart urban governance. *International Review of Administrative Sciences*, *82*(2), 392–408. https://doi.org/10.1177/0020852314564308

Neirotti, P., Marco, A., Cagliano, A., Mangano, G., & Scorrano, F. (2014). Current trends in smart city initiatives: Some stylised facts. *Cities*, *38*, 25–36. https://doi.org/10.1016/j.cities.2013.12.010

Nooringsih, K., & Susanti, R. (2022). Implementation of smart city concept for sustainable development in Semarang old town area. *IOP Conference Series Earth and Environmental Science*, *1082*(1), 012034. https://doi.org/10.1088/1755-1315/1082/1/012034

Pant, J. J., & Majumder, M. G. (2022). Themes and narratives of gig economy: An Indian HR perspective. *NHRD Network Journal*, *15*(1), 83–99. https://doi.org/10.1177/26314541211064751

Prevelianaki, K., Sherratt, F., & Henjewele, C. (2022). Iso standards or global indices: Who decides if a city is smart? *IOP Conference Series Earth and Environmental Science*, *1101*(2), 022045. https://doi.org/10.1088/1755-1315/1101/2/022045

Raha. (2023). *About us*. Retrieved October 7, 2023, fromhttps://shopraha.zendesk.com/hc/en-us/articles/5587312471825-About-Raha

Rahman, M., Rashid, M., Hossain, M., Hassanain, E., Alhamid, M., & Guizani, M. (2019). Blockchain and IoT-based cognitive edge framework for sharing economy services in a smart city. *IEEE Access*, *7*, 18611–18621. https://doi.org/10.1109/access.2019.2896065

Safieddine, F. & Nakhoul, I. (2018). Generic business process model for smes in m-commerce based on talabat's case study. *Lecture Notes in Computer Science*, 264–278. https://doi.org/10.1007/978-3-030-02131-3_24

Sharma, K., & Waheed, K. A. (2018). Consumption of online food app services: An exploratory study among college students in Dubai. *Middle East Journal of Business*, *13*(4), 4–11.

Singh, R., & Bhushan, V. (2023). Understanding the fundamentals and dynamics of the gig employment landscape in India. *International Journal of Advanced Research*, *11*(07), 300–315. https://doi.org/10.21474/ijar01/17232

Treude, M. (2021). Sustainable smart city – Opening a black box. *Sustainability*, *13*(2), 769. https://doi.org/10.3390/su13020769

Wood, A., Graham, M., Lehdonvirta, V., & Hjorth, I. (2018). Good gig, bad gig: Autonomy and algorithmic control in the global gig economy. *Work Employment and Society*, *33*(1), 56–75. https://doi.org/10.1177/0950017018785616

Yin, Y. (2022). Machine learning computing migration and management based on edge computing of multiple data sources in the Internet of Things. *Computational Intelligence and Neuroscience*, *2022*, 1–11. https://doi.org/10.1155/2022/8065767

Chapter 11

Enhancing Prosperity Through Smart Tourism: A Competitive Analysis of Gulf Cities Using STCCI

Hyunae Lee[a] *and Namho Chung*[b]

[a]*Smart Tourism Research Center, Kyung Hee University, South Korea*
[b]*College of Hotel and Tourism Management, Kyung Hee University, South Korea*

Abstract

This chapter introduces the Smart Tourism Cities Competitiveness Index (STCCI), an innovative benchmarking tool developed in collaboration with the Seoul Tourism Organization and the Smart Tourism Research Center at Kyung Hee University. The STCCI evaluates cities on smart tourism using a comprehensive framework that covers five key dimensions: attractiveness, accessibility, digitalization readiness, sustainability, and co-partnership, further detailed into 17 categories and 51 specific indicators. The application of STCCI to the Gulf cities of Dubai, Abu Dhabi, and Doha reveals their strengths and areas for enhancement in the realm of smart tourism. Dubai shines in attracting visitors and embracing digital technology, while Abu Dhabi leads in online visibility and the accessibility of tourism information. Doha is recognized for its superior accessibility. Despite these strengths, each city faces unique challenges that need addressing. The STCCI not only provides a deep dive into the current state and potential growth of smart tourism infrastructures in these cities post-pandemic but also positions them on a global scale in terms of competitiveness in smart tourism. The index suggests strategic improvements, supporting the evolution of these cities into advanced smart cities in the Gulf region, characterized by innovation, development, transformation, and prosperity. This chapter contributes valuable perspectives for ongoing smart cities research and development across the Gulf, emphasizing the

The Emerald Handbook of Smart Cities in the Gulf Region: Innovation, Development, Transformation, and Prosperity for Vision 2040, 189–204

doi:10.1108/978-1-83608-292-720241011

significance of this approach in guiding the future of urban tourism in a smarter, more connected world.

Keywords: Smart tourism; smart tourism cities; Smart Tourism Cities Competitiveness Index; Gulf cities (Dubai; Abu Dhabi and Doha); diagnosis of cities' competitiveness; city assessment

1. Introduction

The concept of smart cities transcends mere technological advancement, embodying an integrated approach to urban development that leverages digital technology, data analytics, and the Internet of Things (IoT) to enhance the quality of life, efficiency of urban operation and services, and competitiveness of cities. At its core, a smart city harnesses information and communication technologies to improve decision-making, sustainability, and connectivity. In the tourism sector, this paradigm shift toward smart tourism cities is pivotal for augmenting visitor experiences through personalized services, optimizing resource management, and ensuring health and safety, thereby significantly contributing to a city's global allure and economic vitality (Boes et al., 2016; Buhalis & Amaranggana, 2013; Gretzel et al., 2015).

In the wake of the pandemic, the transition toward smart tourism destinations has become even more critical. Cities are now focusing on adapting to new health and safety norms, enhancing visitor experiences through technology, and efficiently managing resources to revitalize and expand their tourism sectors (Buhalis et al., 2023; Gretzel & Koo, 2021; Mehraliyev et al., 2020). Since 2020, the collaboration between the Seoul Tourism Organization (STO) and the Smart Tourism Research Center (STRC) at Kyung Hee University in refining the Smart Tourism Cities Competitiveness Index (STCCI) reflects a systematic effort to align tourism practices with the evolving societal and technological landscape, particularly through the inclusion of emerging trends identified via text mining and comprehensive analyses.

The Gulf region, represented by dynamic cities such as Dubai, Abu Dhabi, and Doha, serves as a fascinating case study for this research. These cities, known for their rapid economic growth, luxury tourism, and ambitious smart city projects, are at the forefront of integrating smart tourism initiatives within their broader urban development strategies. Dubai and Abu Dhabi, in particular, have become prime destinations, drawing in 16.73 million and 11.35 million international visitors respectively in 2019 (Department of Culture & Tourism of Abu Dhabi, 2020; Government of Dubai, 2020). Meanwhile, Doha has recently witnessed remarkable growth and diversification in its tourism sector, especially after hosting prestigious international events, including the 2022 FIFA World Cup. This growth has positioned Qatar's tourism industry as a cornerstone of economic development and sustainability (The Median Line, 2024). Furthermore, as part of a larger initiative for economic diversification in the Gulf region, they are strategically aiming to decrease their dependence on oil-based revenues. This approach involves fostering economies that are sustainable, diversified, and rooted in knowledge, extending beyond the traditional oil sector (Ben Hassen, 2022). Such a shift is crucial for the long-term economic resilience and growth of these cities in a rapidly changing global landscape. This strategic redirection underscores a commitment to innovation and sustainable development within the region. Therefore, assessing their smart tourism cities' competitiveness level is integral to understanding the evolving tourism dynamics in the Middle East, a region actively diversifying its economy with a considerable emphasis on smart tourism. This chapter aims to apply the updated STCCI to evaluate the smart tourism competitiveness of these key Gulf cities.

This chapter is organized into six main sections to comprehensively address our research objectives. The second section presents the methodology behind the development and recent updates to the STCCI. The third section delves into the specifics of the 2023 edition, detailing its subindices, pillars, and indicators. The second and third sections summarise the STO's reports titled "Mid- to Long-Term Consulting Strategies for Becoming a Global Tourism City in 2023 (provisional title)." The fourth section offers an analysis of the assessment results, wherein the performance of the Gulf region is compared with other regions and the global average. This section also includes a detailed assessment of each of the three Gulf cities, highlighting their common strengths and weaknesses within the framework of a smart tourism city. In the fifth section, we provide some implications for the Gulf region and place these findings in the global smart tourism context. Finally, the sixth section wraps up by summarizing the theoretical, practical, and political contributions of STCCI to smart tourism and urban development in the Gulf cities, emphasizing its role in shaping future strategies and policies for enhancing smart city competitiveness and sustainability.

2. Development and Updating of STCCI

In 2023, we refined the STCCI to better understand the changing landscape of tourism in the aftermath of the pandemic. This update involved a detailed look into how societies and their tourism sectors have evolved, using modern techniques like text mining of online news and in-depth analysis of recent studies and reports.

We collected data from 50 cities worldwide, recognized for their tourism appeal or smart city initiatives, including Seoul, Tokyo, Shanghai, Singapore, Bangkok, Ho Chi Minh City, Jakarta, Istanbul, Kuala Lumpur, Hong Kong, Busan, Osaka, Beijing, Taipei, London, Barcelona, Copenhagen, Amsterdam, Milan, Munich, Dublin, Zurich, Oslo, Prague, Vienna, Lisbon, Brussels, Helsinki, Paris, Budapest, Stockholm, Rome, Athens, New York, San Francisco, Los Angeles, Vancouver, Toronto, Rio de Janeiro, Buenos Aires, Mexico City, Sydney, Melbourne, Auckland, Dubai, Abu Dhabi, Doha, Cape Town, Johannesburg, and Cairo. This wide selection ensures our index reflects global trends and practices.

The data were meticulously gathered from a variety of primary and secondary sources. For primary data, we employed "Web Scraper," a Google Chrome Extension, to systematically extract a substantial amount of data from platforms including Google, Google Maps, Airbnb, YouTube, official DMO website and

SNS channels, and others. Following the data collection, we eliminated the data deemed non-essential, such as online content with zero views. Secondary data were rigorously compiled from authoritative reports and statistical publications issued by government bodies, international organizations, and esteemed consulting companies. To compare cities fairly, we normalized all data on a scale from 1 to 7, adjusting for positive or negative impacts on smart tourism competitiveness.

Drawing from existing literature (Hair et al., 2017; Kim et al., 2010; Petter et al., 2007), we reconfigured the STCCI into a formative structural model. The STCCI is conceptualized to influence three key performances in tourism and urban contexts: inbound tourism revenue, quality of life, and happiness level. These measures act as reflective indicators, representing the tangible outcomes of the overall competitiveness fostered by the STCCI. Subsequently, we executed the verification process for four kinds of validity: (1) internal consistency, (2) construct validity, (3) external validity, and (4) empirical validity.

3. 2023 STCCI

As a result, the STCCI version 2023 was proposed, featuring 5 subindices, 17 pillars, and 44 indicators. This comprehensive structure is designed to provide a nuanced and in-depth assessment of smart tourism competitiveness in cities. The index is visually represented in Fig. 1, which illustrates the structure of the subindices and various pillars.

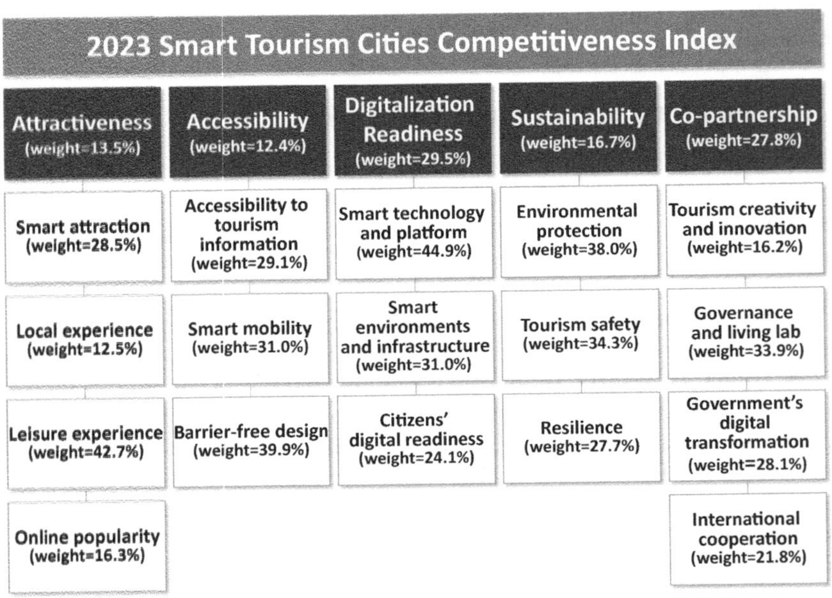

Fig. 1. Smart Tourism Cities Competitiveness Index (*Source*: Seoul Tourism Organization, 2023).

The five subindices are defined as follows (Seoul Tourism Organization. 2023):

- *Attractiveness* looks at how well a city's natural beauty, culture, and unique experiences are promoted online, not just by official bodies but also by individuals like residents and tourists.
- *Accessibility* measures how easily tourists can access information online and physically navigate the city, emphasizing smart mobility solutions and inclusive design.
- *Digitalization Readiness* assesses the city's technological infrastructure and the community's adaptability to digital innovations for enhancing the tourism experience.
- *Sustainability* focuses on a city's commitment to environmental protection, safety, and resilience in its tourism sector.
- *Co-partnership* evaluates the level of collaboration among city stakeholders in fostering innovation, governance, and digital transformation in tourism.

Weights for both the 5 subindices and the 17 pillars were determined based on the loadings of the pillars on the derived subindices and the path coefficients of the subindices on the whole STCCI. The weights were normalized to ensure that the total sum of weights for each dimension equaled 1 (Wetzels et al., 2009). The scores of the subindices were calculated by summing the products of each indicator value and its corresponding pillar weight. These subindex scores were then multiplied by their respective subindex weights to compute the total score for each city. The weight values are shown in Fig. 1.

4. City Assessment

Fig. 2 illustrates the individual subindex scores for the three cities in the Gulf region alongside the average subindex scores for all 50 cities in the study. Dubai excels notably in attractiveness and digitalization readiness. Abu Dhabi demonstrates a well-rounded performance across all five subindices. While Doha shows strength in accessibility, it requires additional focus in the other subindices.

Regarding attractiveness, Dubai stands out with the highest score among the three cities, exceeding the mean values. In contrast, Doha records the lowest score in this subindex within the three cities. When assessing accessibility, Doha leads with the highest scores among the three cities, whereas Dubai and Abu Dhabi present relatively lower scores than the average. In the aspect of digitalization readiness, both Dubai and Abu Dhabi demonstrate a notably high level of digital readiness, while Doha displays a comparatively lower level. Concerning sustainability, each of the three cities scores below the average observed across the 50 cities. Among the three cities, Abu Dhabi has the highest score in this subindex. In terms of co-partnership, Dubai, Abu Dhabi, and Doha all register scores that are lower than the overall average of the cities evaluated.

In Fig. 3, the blue line depicts the average values for the three cities in the Gulf region (Dubai, Abu Dhabi, and Doha), while the black line represents the average values across the 50 cities in each pillar (referred to as the global average in

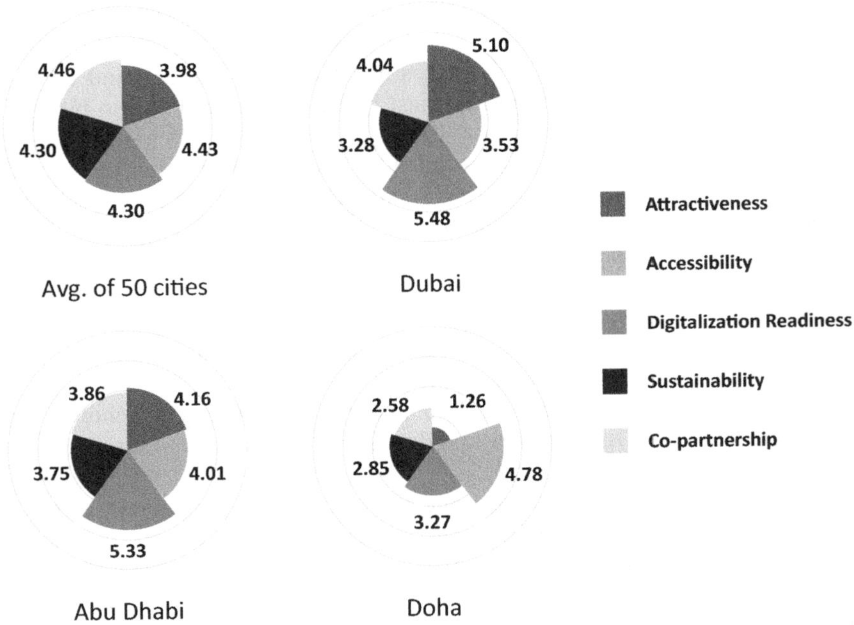

Fig. 2. Subindex Scores Overview (*Source*: Authors).

this chapter). The Gulf region's average scores are notably higher than those of the 50-city cohort in specific areas such as online popularity, accessibility of tourism information, barrier-free design, and smart environment and infrastructure. However, in many other pillars, the Gulf region's mean values fall below those of the 50 cities. This gap is especially evident in tourism safety, and governance and living lab, where the Gulf region's averages are significantly lower. These findings may be ascribed to the Gulf region's inferior performance in areas of surveillance data accessibility and transparency, racial equity, and political participation.

4.1. Dubai

Fig. 4 shows the scores of Dubai in the pillars of STCCI. In terms of attractiveness, Dubai achieved notably high scores in the two pillars, smart attraction and online popularity. This exceptional performance is partly attributed to Dubai's substantial amount of information about natural and cultural resources, which are frequently searched on search engines. Furthermore, the strong social media presence of Dubai's DMO, as evidenced by a high follower count on its official SNS account, plays a significant role in enhancing its attractiveness score. Regarding the leisure experience pillar, Dubai's score falls below average. The lower rating in leisure experience is primarily linked to the limited availability of outdoor activities like trails and rock-climbing routes.

In terms of accessibility, while Dubai scores above the average in the pillar of accessibility to tourism information, it falls below the average in two other

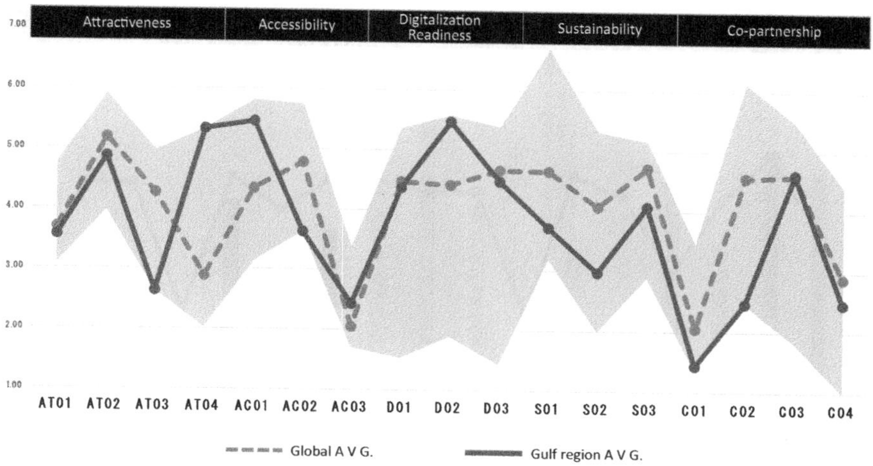

Fig. 3. Gulf Region's Pillar Performance Overview (*Source*: Authors).

Note: AT01: Smart attraction; AT02: Local experience; AT03: Leisure experience; AT04: Online popularity; AC01: Accessibility to tourism information; AC02: Smart mobility AC03: Barrier-free design; D01: Smart technology and platform; D02: Smart environments and infrastructure; D03: Citizens' digital readiness; S01: Environmental protection; S02: Tourism safety; S03: Resilience; C01: Tourism creativity and innovation; C02: Governance and living lab; C03: Government's digital transformation; C04: International cooperation.

pillars within this category: smart mobility and barrier-free design. Specifically, the low number of shared bicycles per 100,000 population and the limited variety of car-sharing services available in the city contribute to Dubai's lower level of smart mobility. Furthermore, Dubai encounters difficulties in the barrier-free design pillar, as evidenced by the low percentages in all three indicators that assess accessibility for disabled guests in hotels and shared accommodation facilities. Considering that the average score for the barrier-free design pillar is generally low, this finding is notably worrisome.

In the area of digitalization readiness, Dubai showcases a relatively impressive performance. The city particularly excels in the pillars of smart technology and platform, and smart environments and infrastructure, where it registers high scores. These high scores reflect Dubai's strong commitment to integrating cutting-edge technology and digital innovation in its urban infrastructure. However, despite leading among the 50 cities in mobile Internet speed, Dubai's 5G experience is relatively lower. In the citizens' digital readiness pillar, Dubai's performance aligns closely with the average.

In terms of sustainability, Dubai's performance falls below average across all three pillars. While the city shows favorable results in aspects such as cleanliness, security, and urbanization rate, it faces challenges in other critical areas. Notable issues include poor air quality, limited accessibility and transparency of

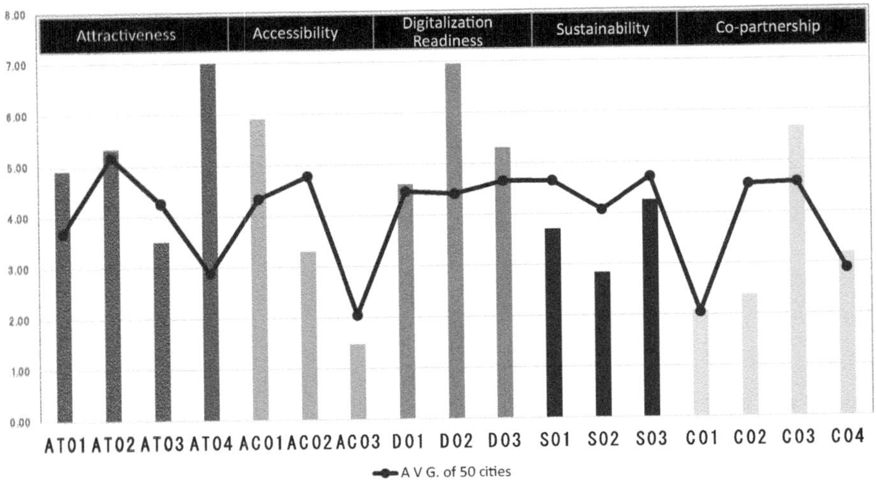

Fig. 4. Scores of Dubai in the Pillars of STCCI (*Source*: Authors).

Note: AT01: Smart attraction; AT02: Local experience; AT03: Leisure experience; AT04: Online popularity; AC01: Accessibility to tourism information; AC02: Smart mobility AC03: Barrier-free design; D01: Smart technology and platform; D02: Smart environments and infrastructure; D03: Citizens' digital readiness; S01: Environmental protection; S02: Tourism safety; S03: Resilience; C01: Tourism creativity and innovation; C02: Governance and living lab; C03: Government's digital transformation; C04: International cooperation.

surveillance data, challenges in achieving racial equality, and a low prevalence of low carbon driving technologies. These factors collectively contribute to Dubai's lower sustainability scores.

Regarding Co-partnership, Dubai demonstrates above-average performance in the government's digital transformation pillar. However, its scores in the governance and living lab pillar are significantly lower. This is primarily due to the very low level of political participation. As for the other two pillars, (1) tourism creativity and innovation and (2) international cooperation, Dubai's performance is not distinguished. Specifically, the limited number of tourism startups and a low number of prestigious universities specializing in hotel and tourism management are areas that contribute to these moderate scores.

4.2. Abu Dhabi

Fig. 5 shows the scores of Abu Dhabi in the pillars of STCCI. In terms of attractiveness, Abu Dhabi achieves notably high scores in the online popularity pillar. This high score is largely due to the substantial number of total views of videos related to tourism information on a popular video-sharing platform. This reflects Abu Dhabi's significant online presence and appeal in the digital tourism landscape. The city's performance in both the smart attraction and local experience

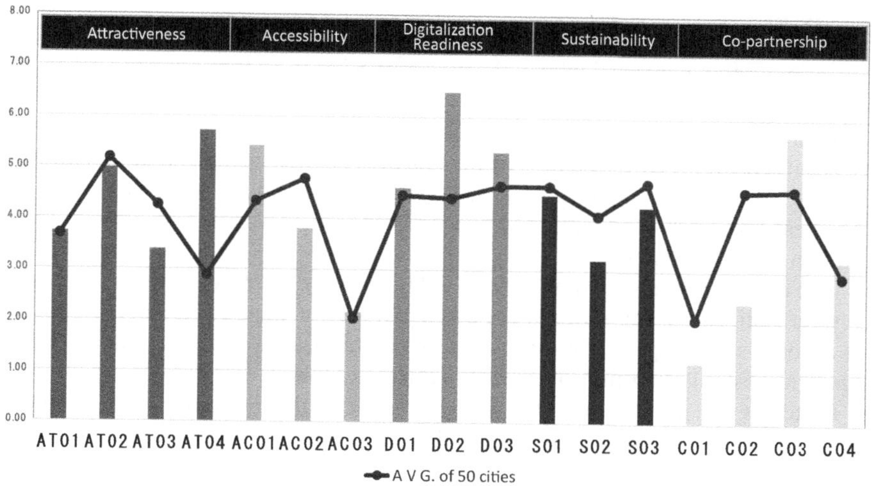

Fig. 5. Scores of Abu Dhabi in the Pillars of STCCI (*Source*: Authors).

Note: AT01: Smart attraction; AT02: Local experience; AT03: Leisure experience; AT04: Online popularity; AC01: Accessibility to tourism information; AC02: Smart mobility AC03: Barrier-free design; D01: Smart technology and platform; D02: Smart environments and infrastructure; D03: Citizens' digital readiness; S01: Environmental protection; S02: Tourism safety; S03: Resilience; C01: Tourism creativity and innovation; C02: Governance and living lab; C03: Government's digital transformation; C04: International cooperation.

pillars is at an average level. However, its score in the leisure experience pillar falls below average, primarily due to the limited availability of outdoor activities, including trails and rock-climbing routes.

Regarding accessibility, Abu Dhabi performs well in the accessibility to tourism information pillar. This achievement is largely due to the diverse range of online channels operated by the DMO, which effectively disseminate tourism information. However, in the smart mobility pillar, Abu Dhabi's score is below average. A contributing factor to this lower score is the relatively low number of shared bicycles available per 100,000 population. In the barrier-free design pillar, Abu Dhabi's performance is average. However, it is important to note that the mean score for this pillar is generally low, suggesting an industry-wide need for improvement. Specifically, there is a need for more hotels and shared accommodations to incorporate facilities for disabled guests.

Since most indicators of digitalization readiness, sustainability and co-partnership were measured using the same data for the United Arab Emirates, Dubai and Abu Dhabi received very similar evaluations in this aspect. Nevertheless, compared to Dubai, Abu Dhabi excels in air quality (environmental protection), average one-way transportation time, and crime index (tourism safety), but falls behind in mobile Internet download speed (smart environments and infrastructure) and number of tourism startups (tourism creativity and innovation).

4.3. Doha

Fig. 6 shows the scores of Doha in the pillars of STCCI. In the context of STCCI, Doha's performance, aside from the two pillars in accessibility (accessibility to tourism information and barrier-free design), is below average in all other pillars. This outcome is partly influenced by the frequent absence of data in authoritative reports and statistical publications from government bodies and international organizations. Such data gaps can impede a comprehensive assessment of a city's tourism competitiveness.

One of the most concerning aspects is Doha's very low level of attractiveness. Despite achieving a top score in the average rating of local experience products on a home-sharing platform (an indicator of the local experience pillar), Doha lags in almost all other indicators within attractiveness.

In terms of accessibility, Doha demonstrates above-average performance in the two pillars: access to tourism information and barrier-free design. This is largely attributed to the diverse range of online channels operated by Doha's DMO, which significantly enhances access to tourism information. Furthermore, Doha achieves the highest scores in the indicator of "Percentage of hotels with facilities for disabled guests," a metric that substantially contributes to its strength in barrier-free design. Nevertheless, it is important to note that, as previously

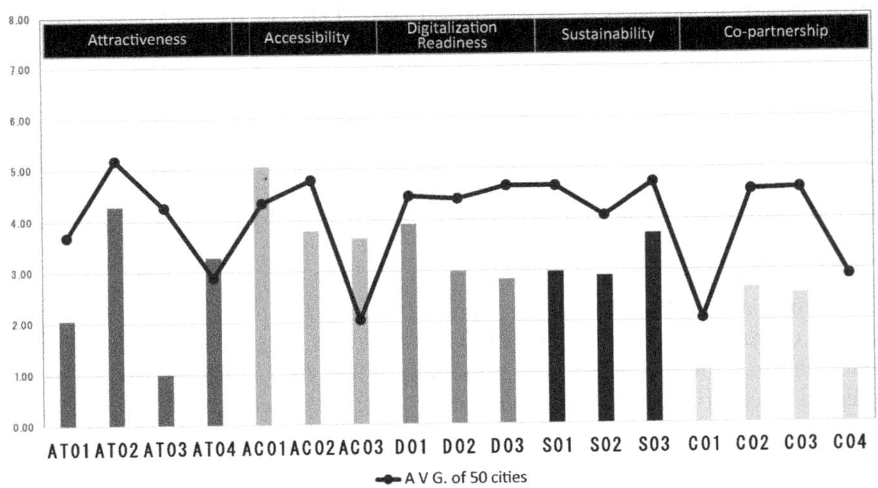

Fig. 6. Scores of Doha in the Pillars of STCCI (*Source*: Authors).

Note: AT01: Smart attraction; AT02: Local experience; AT03: Leisure experience; AT04: Online popularity; AC01: Accessibility to tourism information; AC02: Smart mobility AC03: Barrier-free design; D01: Smart technology and platform; D02: Smart environments and infrastructure; D03: Citizens' digital readiness; S01: Environmental protection; S02: Tourism safety; S03: Resilience; C01: Tourism creativity and innovation; C02: Governance and living lab; C03: Government's digital transformation; C04: International cooperation.

mentioned, the average score for the barrier-free design pillar is comparatively low across all evaluated cities. This indicates that while Doha is performing well relative to other cities, there is an overall need for improvement in barrier-free design within the broader context of the STCCI.

Within the realm of digitalization readiness, Doha's overall performance falls below the average, marked by notably low scores in the domain of data openness. This aspect distinctly sets it apart from its regional counterparts, Dubai and Abu Dhabi, both of which exhibit superior performance in this area.

In the area of sustainability, Doha's performance is below average across all three pillars. The city has particularly low scores in two indicators: surveillance data accessibility and transparency, and racial equity. These deficiencies lead to a significant decrease in the level of tourism safety. Meanwhile, when compared with Dubai and Abu Dhabi, Doha's urbanization rate is slightly higher.

Regarding Co-partnership, Doha scores below average across all four pillars. A key area where Doha lags is in the pillar of E-government development, setting it distinctly apart from Dubai and Abu Dhabi, both of which show superior performance in this respect. Additionally, Doha's performance in two indicators of tourism innovation – number of tourism startups and number of prestigious universities specializing in hotel and tourism management – is notably lower. This is not only in comparison to a broader set of 50 cities but also specifically against Dubai and Abu Dhabi, which are in the same Gulf region.

4.4. Comparative Analysis of the Competitiveness of the Cities

Table 1 shows the color-coded according to the cities' position relative to global scores. We used three quartiles of each indicator. Performance in the top 25% was coded in dark blue; for the next quartile (top 25% to 50%), it was coded in blue; for the third quartile (top 50% to 75%), the coding was light blue; and for the lowest quartile (bottom 25%), the coding was in light sky blue. This color-coding effectively visualizes the range of performance across different quartiles, providing a clear and hierarchical representation of the city's standing in each indicator.

The comparative analysis of the competitiveness of Dubai, Abu Dhabi, and Doha as smart tourism cities based on the STCCI reveals distinct strengths, weaknesses, and opportunities for each city.

Dubai emerges as a leader in the attractiveness, demonstrating exemplary performance in smart attraction and online popularity. The city's rich tapestry of natural and cultural offerings, combined with a robust social media presence, underpins its superior attractiveness. Furthermore, Dubai's commitment to digitalization is evident in its outstanding digitalization readiness, particularly in smart technology and infrastructure, showcasing its forefront position in integrating cutting-edge technologies and digital innovation into its urban landscape. However, Dubai faces challenges in accessibility, with specific limitations in smart mobility and barrier-free design, indicating a need for enhanced inclusivity and mobility solutions. The city also encounters sustainability issues, notably in air quality and surveillance data accessibility, pointing to areas requiring strategic focus to bolster its environmental stewardship and transparency. In terms of

Table 1. Cities Position Relative to Global Scores (*Source*: Authors).

Subindex	Pillar	Dubai	Abu Dhabi	Doha
Attractiveness	Smart attraction			
	Local experience			
	Leisure experience			
	Online popularity			
Accessibility	Accessibility to tourism information			
	Smart mobility			
	Barrier-free design			
Digitalization readiness	Smart technology and platform			
	Smart environments and infrastructure			
	Citizens' digital readiness			
Sustainability	Environmental protection			
	Tourism safety			
	Resilience			
Co-partnership	Tourism creativity and innovation			
	Governance and living lab			
	Government's digital transformation			
	International cooperation			

Bottom 25% Top 25%

Co-partnership, while Dubai shows strength in the government's digital transformation efforts, it requires further enhancement in governance and living lab, primarily due to low levels of political participation and tourism innovation.

Abu Dhabi showcases a well-rounded performance, particularly excelling in online popularity, reflecting its significant digital footprint in the tourism landscape. The city also performs commendably in accessibility to tourism information, facilitating easy access to tourism-related data for visitors. Nonetheless, Abu Dhabi shares similar challenges with Dubai in leisure experience and smart mobility, highlighting opportunities for diversification in outdoor activities and improvement in smart transportation options. While Abu Dhabi's performance in sustainability shows slight advantages, particularly in air quality, it still faces challenges in environmental protection and tourism safety, underscoring the need for concerted efforts in these domains. Co-partnership in Abu Dhabi, similar to Dubai, suggests potential areas for further development, especially in fostering innovation and international collaboration in the tourism sector.

Doha distinguishes itself in the accessibility subindex, particularly in access to tourism information and barrier-free design, showcasing its commitment to inclusivity and information dissemination. However, the city's attractiveness and digitalization readiness lag behind, with notable deficiencies in several indicators, signaling critical areas for improvement to enhance its competitive edge in the smart tourism domain. Sustainability and co-partnership emerge as significant areas of concern for Doha, with challenges in surveillance data accessibility, racial equity, and E-government development. These issues highlight the need for strategic initiatives aimed at enhancing sustainability practices, promoting racial equity, and bolstering government involvement and innovation in the tourism sector.

5. Discussion

5.1. Implications of Common Strengths

- *High Ratings on Local Experience*: The high level of average ratings for local experience products on home-sharing platforms indicates a strong tourist satisfaction with the cultural and experiential offerings of these cities. This is crucial for smart tourism, as it reflects effective utilization of digital platforms to enhance visitor experiences.
- *Digital Competitiveness and Mobile Speed*: The forefront position in adopting digital technologies and fast mobile internet speed signifies a robust digital infrastructure. This is a key component of smart tourism, enabling efficient information dissemination and enhancing the overall tourist experience.
- *Cleanliness, Security, and Urbanization*: High standards of cleanliness and security, along with a significant urbanization rate, indicate well-developed urban environments conducive to tourism. These factors contribute to creating a safe and appealing destination for tourists.
- *Cyber Risk Management*: Effective cyber risk management, indicated by strong legal, technical, and organizational measures, is essential in protecting digital data and infrastructure, which are integral to smart tourism cities.

5.2. Implications of Common Weaknesses

- *Limited Culinary and Leisure Options*: The limited number of Michelin restaurants and leisure experiences point to a need for diversifying the tourism offerings. Expanding these aspects can enhance the cities' appeal to a wider tourist demographic.
- *Accessibility Challenges*: The small percentage of accommodations with Braille facilities and wheelchair accessibility highlights a gap in inclusivity, an important aspect of smart tourism cities aiming to cater to all visitor needs.
- *Lower 5G Speed and Air Quality*: Despite good mobile internet speeds, the low level of 5G speed and poor air quality are areas for improvement, as they can impact visitor experiences and the cities' reputations as advanced, sustainable tourist destinations.
- *Racial Equity and Data Transparency*: The low levels of racial equity and surveillance data accessibility and transparency suggest a need for more inclusive policies and transparent governance, which are key for sustainable and responsible tourism.
- *Limited Support for Tourism Innovation*: The lack of investment in tourism startups and related research indicates a potential area for growth, which can foster innovation and sustainable development in the tourism sector.

5.3. Positioning Within Global Smart Tourism

The strengths of these cities align well with the global trends in smart tourism, particularly in leveraging digital technologies and ensuring high-quality visitor experiences. However, the identified weaknesses highlight areas where these cities lag behind global standards, particularly in sustainability, inclusivity, and innovation. Addressing these areas is crucial for these cities to maintain their competitive edge in the rapidly evolving global tourism landscape. Continuous investment in technology, sustainability, and innovation, along with a focus on inclusivity and environmental concerns, will be key in positioning these cities as leading global smart tourism destinations.

6. Conclusions

This study makes several theoretical contributions to the understanding of smart cities and smart tourism. First, STCCI provides a comprehensive framework that encapsulates essential dimensions of smart tourism, including attractiveness, accessibility, digitalization readiness, sustainability, and co-partnership. This framework enriches the theoretical landscape by offering a structured approach to evaluating and enhancing smart tourism initiatives within the broader context of smart city development. It emphasizes the multifaceted nature of smart tourism, integrating economic, social, technological, and environmental considerations. Future studies can build on this holistic approach to explore the complex interdependencies within smart cities and the role of tourism in urban innovation and sustainability. Second, by breaking down the broad dimensions of smart tourism into specific subindices, categories, and indicators, the STCCI operationalizes abstract smart tourism concepts into measurable components. This contribution is crucial for theory as it provides a methodological basis for quantitatively assessing and comparing the smart tourism performance of cities. It enables a more nuanced understanding of how various factors contribute to the competitiveness of tourism destinations in the digital age. Future research can further refine these measures or adapt them to different contexts, facilitating empirical studies that contribute to the body of knowledge on smart tourism and smart cities. Third, the research provides a foundation for exploring future directions in the intersection of smart cities and tourism. It suggests areas for innovation and improvement, such as digital literacy, sustainable mobility, and public-private partnerships, that are critical for the evolution of smart cities. These propositions open new avenues for theoretical exploration and practical application, encouraging future studies to investigate the evolving landscape of smart urban development and its implications for residents, visitors, and the global community.

This research makes significant practical and political contributions to the Smart Cities Agenda by introducing the STCCI, a comprehensive tool designed to evaluate and enhance the smart tourism capabilities of cities. Specifically, in the context of the Gulf region, this contribution is timely and crucial. The Gulf cities of Dubai, Abu Dhabi, and Doha have been examined through the STCCI, uncovering insights into their smart tourism practices. This evaluation aligns with the Gulf region's ambitious Smart Cities Agenda, aiming to foster innovation, sustainability, and digital transformation in urban tourism. By identifying strengths and areas for improvement, the STCCI offers a roadmap for these cities to advance their smart tourism infrastructure, an essential component of their broader smart city development goals. Furthermore, the findings from this research are highly beneficial for a broad range of stakeholders involved in smart cities, such as urban planners, architects, policymakers, academics, students, and government officials. For diverse communities and stakeholders in smart cities, including urban planners, architects, policymakers, faculty, students, and government officials, the key findings of this research offer valuable insights. The STCCI framework demonstrates the importance of integrating digital readiness, accessibility, sustainability, and co-partnership in urban planning and development. For urban planners and architects, this means designing cities that are not only physically accessible but also digitally inclusive. Policymakers and government officials can leverage these insights to prioritize investments in technology and infrastructure that support smart tourism, enhancing the overall attractiveness and competitiveness of their cities.

Acknowledgment

This work was supported by the Ministry of Education of the Republic of Korea and the National Research Foundation of Korea (NRF-2023S1A5C2A03095253) and Seoul Tourism Organization.

References

Ben Hassen, T. (2022). The GCC economies in the wake of COVID-19: Toward post-oil sustainable knowledge-based economies? *Sustainability*, *14*(18), 11251.

Boes, K., Buhalis, D., & Inversini, A. (2016). Smart tourism destinations: Ecosystems for tourism destination competitiveness. *International Journal of Tourism Cities*, *2*(2), 108–124.

Buhalis, D., & Amaranggana, A. (2013). Smart tourism destinations. In *Information and communication technologies in tourism 2014: Proceedings of the international conference in Dublin, Ireland*, January 21–24, 2014 (pp. 553–564). Springer International Publishing.

Buhalis, D., O'Connor, P., & Leung, R. (2023). Smart hospitality: From smart cities and smart tourism towards agile business ecosystems in networked destinations. *International Journal of Contemporary Hospitality Management*, *35*(1), 369–393.

Department of Culture and Tourism of Abu Dhabi. (2020, February 24). *Abu Dhabi welcomes record-breaking 11.35 million international visitors in 2019*. https://dct.gov.ae/en/media.centre/news/abu.dhabi.welcomes.record.breaking.11.35.million.international.visitors.in.2019.aspx

Government of Dubai. (2020, July 1). *Annual visitor report 2019*. https://www.dubaitourism.gov.ae/en/research-and-insights/annual-visitor-report-2019

Gretzel, U., & Koo, C. (2021). Smart tourism cities: A duality of place where technology supports the convergence of touristic and residential experiences. *Asia Pacific Journal of Tourism Research*, *26*(4), 352–364.

Gretzel, U., Sigala, M., Xiang, Z., & Koo, C. (2015). Smart tourism: Foundations and developments. *Electronic Markets*, *25*, 179–188.

Hair Jr, J., Hair Jr, J. F., Hult, G. T. M., Ringle, C. M., & Sarstedt, M. (2017). *A primer on partial least squares structural equation modeling (PLS-SEM)*. Sage Publications.

Kim, G., Shin, B., & Grover, V. (2010). Research note: Investigating two contradictory views of formative measurement in information systems research. *MIS Quarterly*, 345–365.

Mehraliyev, F., Chan, I. C. C., Choi, Y., Koseoglu, M. A., & Law, R. (2020). A state-of-the-art review of smart tourism research. *Journal of Travel & Tourism Marketing*, *37*(1), 78–91.

Petter, S., Straub, D., & Rai, A. (2007). Specifying formative constructs in information systems research. *MIS Quarterly*, 623–656.

Seoul Tourism Organization. (2023). *Mid- to long-term consulting strategies for becoming a global tourism city in 2023 (provisional title)*.

The Median Line. (2024, January 29). *Qatar's tourism industry thrives post-world cup: A 347% increase in arrivals expected*. https://themedialine.org/life-lines/qatars-tourism-industry-thrives-post-world-cup-a-347-increase-in-arrivals-expected/

Wetzels, M., Odekerken-Schröder, G., & Van Oppen, C. (2009). Using PLS path modeling for assessing hierarchical construct models: Guidelines and empirical illustration. *MIS Quarterly*, 177–195.

Chapter 12

Innovative Digital Transformation of Issuing Final Clearance Certificates for Infrastructure Projects in Dubai

Anmar AlObaidi[a] *and Wael A. Samad*[b]

[a]*Right of Way Department, Roads and Transportation Authority, Dubai, UAE*
[b]*Department of Mechanical & Industrial Engineering, Rochester Institute of Technology, Dubai, UAE*

Abstract

This chapter investigates the essential practices to digitally transform the process of issuing final clearance certificates. In Dubai, a final clearance system enhancement was introduced recently to combine all stakeholders as it is part of Electronic No Objection Certificate (eNOC) portal to issue the certificate. The eNOC platform acts as a centralized hub, bringing together multiple service entities from both government and private organizations to issue unified NOCs for infrastructure projects within the Right-Of-Way. Employing the DMAIC (Define, Measure, Analyze, Improve, Control) approach to engineering management, the chapter systematically evaluates the project's aspects. Real-life case studies illustrate challenges encountered in the manual final clearance process for construction projects such as electric cable lines, water lines, and sewerage networks. The proposed transformation seeks to automate application reception and bolster departmental ownership within the eNOC system, incorporating all relevant entities and departments involved in base construction NOC approvals. Departmental assignments for final clearance applications based on project type aim to optimize the process. Interface upgrades will include alert messages guiding applicants to select the correct application type, reducing submission errors on a centralized web portal. Comprehensive reports will provide an overview of final clearance application statuses for effective monitoring. The chapter presents a comprehensive framework

The Emerald Handbook of Smart Cities in the Gulf Region:
Innovation, Development, Transformation, and Prosperity for Vision 2040, 205–218

doi:10.1108/978-1-83608-292-720241012

tailored for infrastructure projects, seamlessly integrating with the eNOC system. Anticipated outcomes encompass improved project management, streamlined processes, heightened customer satisfaction, and sustainable infrastructural growth, particularly pertinent to Smart Cities' development within the Gulf Cooperation Council (GCC) region, aligning with aspirations for technological innovation and urban progress.

Keywords: Final clearance application; smart cities; infrastructure projects; engineering management; automation; digital transformation

1. Introduction

The concept of a smart city has rapidly gained prominence in contemporary discussions about sustainability, and understandably so. While the term "Smart City" first appeared in the literature in the late 1990s, there remains some ambiguity surrounding its precise definition, contextual relevance, and methods for assessment and quantification (Samad & Azar, 2019). The International Telecommunication Union Focus Group on Smart Sustainable Cities examined nearly one hundred definitions of smart cities, sustainable cities, and smart sustainable cities, ultimately proposing a novel definition for a smart sustainable city. This new definition describes an innovative city that harnesses information and communication technologies (ICTs) and other tools to enhance the quality of life, efficiency of urban operations and services, and overall competitiveness. Crucially, it ensures that present and future generations' economic, social, environmental, and cultural needs are met (FG-SSC, ITU-T, 2014). In simpler terms, a smart city can be characterized as an urban area that utilizes digital technology to optimize performance and enhance the well-being of its inhabitants (Dameri, 2013). This transformation is achieved by leveraging data and ICT to optimize infrastructure and services (Geray, 2018).

In the context of infrastructure projects, the transition toward smart cities represents a significant shift, particularly in the issuance of final clearance certificates (NOC). This pivotal step in project management serves as the gateway to ensuring that infrastructure projects comply with regulatory standards, secure necessary approvals from relevant authorities, and guarantee safety for public use (Fewings & Henjewele, 2019). Historically, the process of issuing final clearance certificates has been labor-intensive and paper-bound, entailing copious paperwork, in-person interactions, and convoluted approval workflows (Koseoglu et al., 2019). However, this transformation signifies a comprehensive change in how governments and organizations oversee infrastructure project delivery (Baldini et al., 2019; Manny et al., 2021).

Within the framework of this chapter's analysis, our perspective on smart cities emphasizes their role in mega-construction projects and digital transformation. Smart cities utilize advanced technologies and data-driven strategies to improve sustainability, efficiency, and quality of life in large-scale urban developments. These cities strategically implement sophisticated infrastructure, including smart buildings, intelligent transportation systems, energy-efficient utilities, and digital governance platforms, to realize their objectives. In this chapter, we focus on Dubai's initiative to digitize the final clearance system, with significant enhancements embodied in the Electronic No Objection Certificate (eNOC) portal. The eNOC portal functions as a centralized platform, integrating various service entities from diverse governmental and private organizations. Its primary objective is to issue unified NOCs for infrastructure, thus contributing to Dubai's evolution into a smart city. The approach adopted to drive this transformation leverages the DMAIC (Define, Measure, Analyze, Improve, Control) methodology within the field of engineering management (Srinivasan et al., 2016). This method ensures a systematic evaluation of various aspects of the project. Throughout this comprehensive case study, we delve into the intricacies of the manual final clearance process, with a particular focus on prominent projects such as electric cable lines, water lines, drainage systems, sewerage networks, and other essential services.

The presented transformation aims to automate application processes and enhance departmental ownership within the eNOC system. Key aspects of automation involve incorporating all relevant departments responsible for reviewing basic construction NOC applications into the final clearance process. Furthermore, different departments will be assigned the task of handling final clearance applications based on the project type, optimizing the overall process. Additionally, the user interface of the eNOC system will undergo upgrades to include alert messages that guide applicants in selecting the appropriate final clearance application type, thus effectively preventing submission errors. Furthermore, the introduction of comprehensive reports is poised to provide an overview of final clearance application statuses, allowing for more effective monitoring and evaluation.

This research contributes insights into engineering management practices, particularly within the context of infrastructure projects. By enhancing the process through the eNOC system, we anticipate that the findings of this study will significantly impact project management practices in the region. Expected outcomes include streamlined processes, increased customer satisfaction, and sustainable infrastructural growth in the pursuit of a smarter and more efficient city. Moreover, and along similar lines, Dubai's Roads Authority has unveiled its Digital Strategy 2023–2030, a substantial AED1.6 billion investment comprising 82 projects and initiatives. The strategy's core goals are to enhance Dubai's global leadership in digital transformation, achieve 100% agility and scalability in digital infrastructure, enable 100% fintech-driven mobility, attain 95% digital service adoption, fully digitized government employee skills, and create 50 artificial intelligence use cases (RTA, 2023).

The digital transformation of infrastructure project management offers benefits that extend beyond smart cities to various sectors. It enhances government service delivery by streamlining processes and improving access to information. In healthcare, it optimizes patient care and record management, while in education, it facilitates personalized learning and administrative efficiency. Financial services benefit from improved access and security through online banking and digital payments. Transportation systems become more efficient and safer with real-time management, and manufacturing sees increased productivity and agility through automation and smart technologies. Overall, digital transformation drives innovation, efficiency, and sustainability across sectors, fostering economic growth and improving quality of life.

2. Description of the Process

As mentioned earlier, the process of issuing final clearance certificates is a crucial step in infrastructure project management. The customer, in this case, is the contractor who executed the project. To ensure the process is transparent, adhering to regulatory standards, having obtained all necessary approvals, and ensuring safety for public use, significant measures are taken. Traditionally, this process has been cumbersome, involving extensive paperwork, face-to-face interactions, and complex approval chains (Zhang et al., 2008). However, with the advent of digital transformation, the organization is responsible for embracing technology to streamline this process. This entails the digitization of documents, the automation of workflows, and the integration of various stakeholders into a unified digital platform. The process of issuing the final clearance certificate is outlined in Fig. 1, where the customer (aka contractor) and the various departments are linked during the overall process. The goal is to reduce processing times, enhance transparency, minimize errors, and ultimately provide a more efficient service to contractors and project managers.

Table 1 provides a detailed depiction of the process, spanning from application submission to the receipt of the final clearance certificate passing through the various stages in between.

3. Proposed Approach

The proposed approach for digital transformation in the issuance of final clearance certificates is a structured and systematic strategy aimed at leveraging technology to optimize the existing manual process. This approach follows a well-defined path that involves several key stages, each contributing to the overarching goal of enhancing efficiency, transparency, and accuracy in the final clearance certificate issuance process.

In Table 2, the approach is outlined with a central focus on enhancing the efficiency, transparency, and accuracy of the final clearance certificate issuance process. The primary objectives are aimed at reducing processing times, minimizing errors, and improving overall transparency, all aligned with the broader goals of smart cities. To measure our success, we will establish Key Performance Indicators (KPIs), including a significant reduction in processing times and a decrease in error rates. Identifying stakeholders is of paramount importance for project success. Key players in this endeavor include applicants (customers), inspectors, engineers, regulatory authorities, and various departments within the Dubai eNOC portal. It is crucial to incorporate the needs and concerns of each stakeholder group to ensure that the digital transformation aligns with their expectations. Vendor selection plays a pivotal role in this process, with a strong emphasis on technology solutions that align with

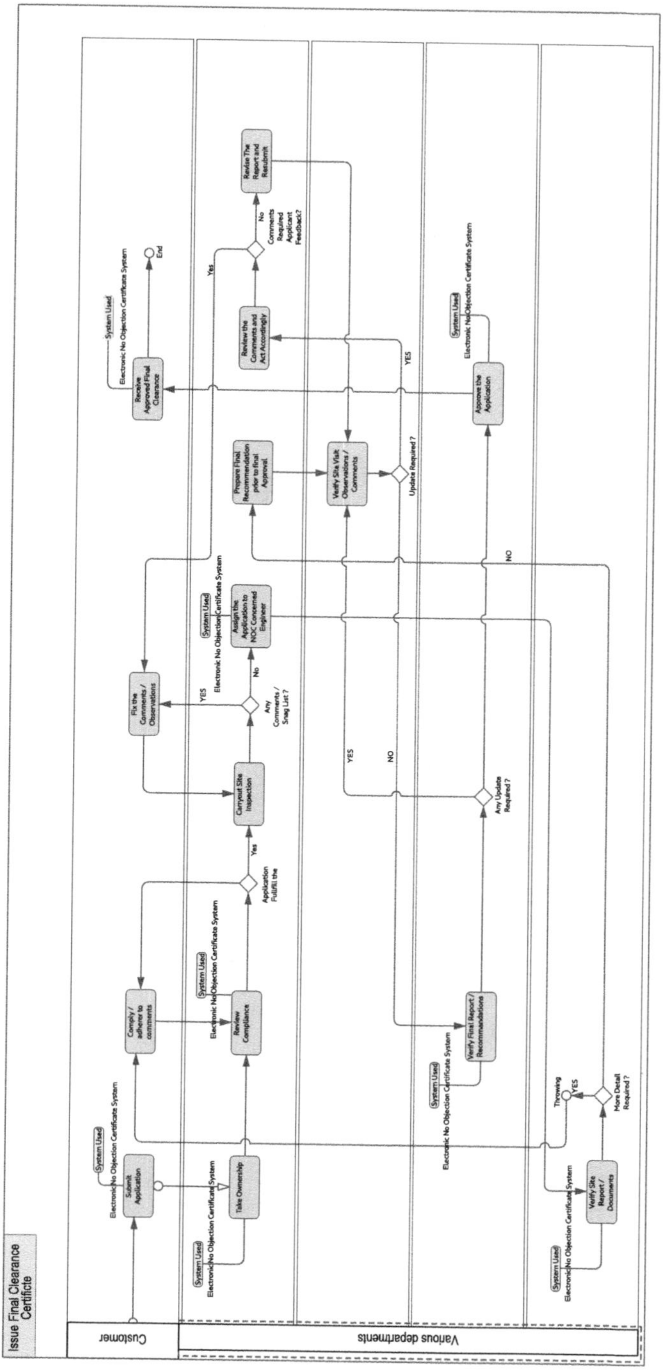

Fig. 1. *Final Clearance System Process.*

Table 1. *Final Clearance System Process Details.*

S.R	Activity	Description	Performed By
1.0	Submit application	Applicant submits Final Clearance request/application through the system	Applicant
1.1	Take ownership	The inspector owns the application based on the assigned projects/utility	Inspector
1.2	Review Compliance	The inspector reviews the application from the perspective of document compliance	Inspector
1.3	Carryout Site Inspection	The inspector will arrange for a site inspection in coordination with the applicant to verify work compliance according to issued NOCs	Inspector
1.4	Fix the Comments / Observations	The applicant will arrange to Fix the Comments / Observations provide evidence that action has been taken place and submit them through the system	Applicant
1.5	Assign the Application to NOC concerned Engineer	The inspector will Assign the Application to the NOC concerned Engineer for review and feedback	Inspector
1.6	Verify Site Report / Documents	NOC Engineer will review the application and site visit report from perspectives concerning the compliance with undertaking letters (if any) and integration aspects with future projects.	Concerned Engineer
1.7	Comply / Adherer to comments	The applicant will receive the comments raised by the inspector for compliance and adherence before proceeding further by the inspector	Applicant
1.8	Prepare Final Recommendation before final Approval	The inspector will prepare the Final Recommendation for the system including the recommendations and forward to the Team Leader	Inspector

1.9	Verify Site Visit Observations / Comments	Team Leader will verify the site visit observations/recommendations as well as overall attachments and act accordingly	Team Leader
2.0	Review the Comments and Act Accordingly	The inspector will Review the Comments and Act Accordingly	Inspector
2.1	Revise the Report and resubmit	The inspector will revise the report and resubmit to the team leader in a way that incorporates all comments	Inspector
2.2	Verify Final Report / Recommendations	The Inspection Manager will receive the final report/recommendations and act accordingly	Manager
2.3	Approve the Application	The Inspection Manager will approve the application	Manager
2.4	Receive Approved Final Clearance	Applicant will Receive Approved Final Clearance through the system	Applicant

Table 2. *Proposed Approach Stages.*

Stage No.	Topic	Description	Goal
1	Assessment & Analysis	Current State Assessment: Understand the existing manual process, identify inefficiencies, and assess regulatory requirements Data Gathering: Collect and analyze data on processing times, error rates, and user feedback	Gain a comprehensive understanding of the current process and establish a baseline for improvement
2	Stakeholder Engagement	Identify Stakeholders: Identify all involved stakeholders, including applicants, inspectors, engineers, and regulatory authorities. Gather Input: Collect feedback and input from stakeholders to consider their needs and concerns	Ensure that the digital transformation aligns with stakeholder expectations and requirements
3	Define Objectives	Set Clear Goals: Define transformation objectives, such as reducing processing times, minimizing errors, and enhancing transparency KPIs: Establish measurable KPIs to evaluate success, e.g., processing time reduction	Clearly define what the digital transformation aims to achieve and how success will be measured
4	Technology Selection	Evaluate Technology Solutions: Explore available tech solutions like EDMS, Workflow Automation Tools, and eNOC systems Customization: Decide whether to customize off-the-shelf software or build a bespoke solution	Select the technology that best aligns with transformation objectives and organizational needs
5	Process Redesign	Workflow Mapping: Redesign the process workflow to align with the chosen technology, simplifying steps and automating tasks Integration: Ensure seamless integration with existing systems like eNOC platforms	Streamline and optimize the workflow to maximize the benefits of digital tools and integration

6	Pilot Implementation	Select Pilot Projects: Choose a subset of projects for testing Implement Technology: Introduce digital tools and workflows in the pilot phase for real-world testing and refinement	Test the digital solution in a controlled environment and gather feedback for adjustments and improvements
7	Training and Change Management	Training Programs: Provide training to stakeholders on how to use digital tools effectively Change Management: Implement strategies for a smooth transition from manual to digital processes	Ensure that all stakeholders are equipped with the knowledge and resources to embrace the digital transformation
8	Monitoring & Optimization	Continuous Monitoring: Monitor the digital process, track KPIs, and gather user feedback Feedback Loop: Use feedback to make iterative improvements and ensure scalability.	Continuously refine the digital process based on feedback and ensure its ability to handle increased volumes
9	Evaluation & Documentation	Performance Evaluation: Compare pre- and post-implementation KPIsDocumentation: Maintain detailed records of the transformation process, including lessons learned and best practices	Assess the impact of the transformation and create a knowledge base for future improvements and expansions
10	Scaling up	Gradual Expansion: Based on pilot success, expand digital tools to cover more applications and projects Replication: Apply lessons learned to replicate successful strategies in other departments or agencies	Extend the benefits of digital transformation to a wider range of applications and replicate successful approaches in other contexts

our transformation objectives. Our evaluation criteria encompass assessing the Electronic Document Management System (EDMS), Workflow Automation Tools, and compatibility with the existing eNOC platform. A customized software is preferred to ensure the long-term sustainability of the project. Effective integration with the eNOC platform is crucial, facilitating a seamless operation of digital tools within the existing system. This integration ensures a unified and streamlined process, allowing real-time access to data and progress for all relevant departments. Scalability considerations, such as additional manpower and database capacity, are essential when accommodating other departments/entities to serve a larger customer base. Before full implementation, we will conduct User Acceptance Testing (UAT), involving stakeholders in real-world testing of digital tools and workflows. This phase is instrumental in identifying potential issues, gathering feedback, and refining the system to meet practical needs and transformation goals. The Go-Live phase represents the transition to full implementation, deploying digital tools and workflows across the entire final clearance application process. Stakeholders will shift from manual to digital processes, supported by a robust change management strategy that includes comprehensive training programs. Continuous monitoring will ensure prompt resolution of challenges, marking a significant milestone in our digital transformation journey.

4. Case Study

This section of the chapter presents a case study focusing on the digital transformation for the final clearance certificate application in Dubai, UAE. More specifically, the case study addresses the comprehensive supply, installation, testing, and commissioning of 132kV cable works for 10 new 132/11kV substations and their associated infrastructure. The project underscores the critical importance of streamlined project management in large-scale infrastructure development.

Stage 1: Railway Network Coordination

A notification was received that the proposed work falls within the protection zone and is under technical review. Awaiting the completion of the technical review for NOC issuance.

Stage 2: Maintenance Department Site Visit

Supporting documents and a site visit were requested. The subsequent exchange included the submission of site photos and a formal request for approval.

Stage 3: Request for Joint Site Visit

Coordination for a joint site visit within the Railway Protection Zone was urged. Joint site visits were identified as crucial for ensuring compliance with railway regulations.

Stage 4: Collaboration with Electricity Authority

Concerns about missing data were raised, and a list of crossings was requested. Subsequently, a revised drawing was submitted, addressing the raised comments. This exemplified the collaborative effort required between various utility providers.

Stage 5: Coordination with (Telecom Services)

A signed report for telecommunication services identification and site verification was requested. In response, a telecommunication company representative-verified drawing was provided for review and approval.

Stage 6: Inspection by Traffic Department

The NOC was forwarded for verification of existing traffic services. Following this, a site inspection was completed, and approval was sought. This step illustrated the integration of various departments and the importance of verifying existing services for seamless project execution.

Stage 7: Detailed Documentation and Submission

Various documents were requested, including revised key plans, as-built details for specific circuits, soil test reports, and clearances. The subsequent submissions included a thorough set of documents such as key plans, circuit details, soil test reports, and clearances from relevant entities.

Stage 8: Stakeholder Coordination

Specific details related to the project's location and boundaries were provided. This action underscored the critical role of stakeholder engagement with developers, regulatory bodies, and internal departments in defining project boundaries and ensuring alignment with authority standards.

Stage 9: Final Inspection and Project Entry

The inspector initiated a request for a revised overall as-built key plan, and clearance from the project consultant and the final clearance certificate was

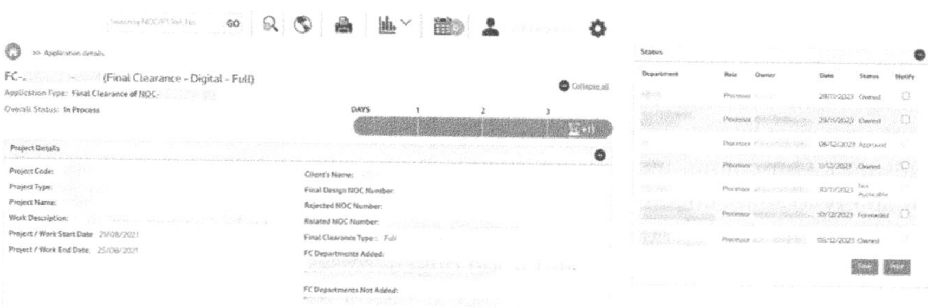

Fig. 2. *Final Clearance Dashboard (Example).*

issued. In response, a comprehensive set of documentation, including the revised key plan and clearances, was meticulously submitted. This exchange highlighted the importance of clearances from both the project consultant (Resident Engineer) and the area developer.

Fig. 2 provides an example of what the Final Clearance process dashboard looks like, highlighting a streamlined status and notification system between the various stakeholders and approval layers involved.

Case Study Conclusion

This case study, presented in a sequence, provides a comprehensive understanding of the steps undertaken in the digital transformation of infrastructure project management. It underscores the significance of stakeholder coordination, detailed documentation, regulatory clearances, and ongoing communication in ensuring the success of large-scale infrastructure projects. Each interaction demonstrates the complexity and interdependence of various entities involved in the project, emphasizing the need for a seamless and integrated approach to project management. Moreover, to ensure time management and meeting deadlines, a KPI is set based on the number of working days, in this case, to ensure the application does not exceed 90 working days.

5. Results and Discussion

In this section, we delve into the outcomes and implications arising from the digital transformation of the final clearance certificate issuance process, particularly within the context of infrastructure projects. This transformation, rooted in the principles of smart cities and the ongoing digital revolution, has been meticulously crafted to enhance efficiency, transparency, and accuracy. The key stages of this approach, ranging from initial assessment to final documentation, have played a pivotal role in ensuring its success. Each step is a crucial building block in achieving a seamless alignment with stakeholders' expectations and requirements, supported by clear objectives and measurable KPIs.

Through precise technology selection and process redesign, this transformation has streamlined workflows, resulting in tangible benefits such as reduced processing times and minimized errors. The pilot phase has allowed for real-world testing and fine-tuning, while the implementation of training and change management strategies has ensured that all stakeholders are well-prepared for the digital transition. Our commitment to continuous monitoring and optimization ensures our ability to adapt effectively to evolving needs. Furthermore, the comprehensive evaluation and documentation efforts create a valuable knowledge base for future enhancements, enabling us to build upon our successes.

The research outlined in this chapter makes a substantial contribution to advancing smart cities agendas, especially within the Gulf Region, by showcasing the effectiveness of digital transformation in infrastructure management. It provides valuable insights for urban planners, architects, policymakers, and government officials, offering avenues to optimize projects and improve city management practices. Key recommendations include investing in digital infrastructure, promoting collaborative platforms, and standardizing digital project management procedures. Additionally, policy suggestions entail the establishment of supportive regulatory frameworks, incentivizing private sector involvement, and fostering partnerships between government and technology sectors to drive urban innovation.

In summary, this digital transformation signifies a significant advancement in infrastructure project management, promising to foster greater efficiency, transparency, and accuracy while perfectly aligning with the principles of smart cities and contributing meaningfully to the sustainability of urban environments. Scaling up this transformation will extend its advantages to a broader range of projects, marking a significant milestone in our journey.

6. Conclusion

The digital transformation of the final clearance certificate issuance process has profoundly impacted smart cities, significantly enhancing efficiency, transparency, and accuracy. This evolution from a manual, paper-based system to a streamlined, technology-driven model has led to remarkable improvements in efficiency, enabling faster clearance issuance and project implementation. Moreover, digitization and workflow automation have minimized errors, ensuring greater precision in regulatory compliance and approvals.

Furthermore, the digitalization of infrastructure project management has introduced heightened transparency, providing stakeholders with real-time access to information. This increased transparency fosters trust and accountability in project management, benefiting stakeholders and facilitating sustainable growth within the region. By expediting regulatory compliance, this transformation aligns infrastructure projects with environmental and social sustainability goals.

The success of this digital transformation serves as a guiding example for future urban infrastructure management. It showcases the potential of technology-driven approaches to enhance efficiency, transparency, and sustainability within smart cities. Through the strategic use of digital tools and automation, cities can effectively manage infrastructure projects, fostering more efficient, transparent, and sustainable urban environments.

7. Teaching Orientation (Optional)

- Suggested teaching assignments: N/A
- Recommended readings:
  - Samad, W. A., & Azar, E. (2019). *Smart cities in the Gulf*. Springer.
  - Salem, F. (2016). *A smart city for public value: Digital transformation through agile governance-the case of 'Smart Dubai'*. World Government Summit Publications.

- El Khatib, M. M., Alzoubi, H. M., Ahmed, G., Kazim, H. H., Al Falasi, S. A. A., Mohammed, F., & Al Mulla, M. (2022, February). Digital transformation and SMART-The analytics factor. In *2022 International Conference on Business Analytics for Technology and Security (ICBATS)* (pp. 1–11). IEEE.
- Batty, M., Axhausen, K. W., Giannotti, F., Pozdnoukhov, A., Bazzani, A., Wachowicz, M., … Portugali, Y. (2012). Smart cities of the future. *The European Physical Journal Special Topics, 214*, 481–518.
- Silva, B. N., Khan, M., & Han, K. (2018). Towards sustainable smart cities: A review of trends, architectures, components, and open challenges in smart cities. *Sustainable Cities and Society, 38*, 697–713.

Case study: N/A (the chapter presented here includes one to a certain extent)

- Titles for research essays: N/A
- Recommended Projects URL: https://noc.rta.ae/

References

Baldini, G., Barboni, M., Bono, F., Delipetrev, B., Duch Brown, N., Fernandez Macias, E., Gkoumas, K., Joossens, E., Kalpaka, A., & Nepelski, D. (2019). *Digital transformation in transport, construction, energy, government and public administration.* Publications Office of the European Union.

Dameri, R. P. (2013). Searching for smart city definition: A comprehensive proposal. *International Journal of Computers & Technology*, *11*(5), 2544–2551.

Fewings, P., & Henjewele, C. (2019). *Construction project management: An integrated approach*. Routledge.

FG-SSC, ITU-T. (2014). *An overview of smart sustainable cities and the role of information and communication technologies*. International Telecommunication Union (ITU-T) Focus Group on Smart Sustainable Cities (FG-SSC).

Geray, O. (2018). An impact-driven smart sustainable city framework to address urban challenges: Smart Dubai Experience. In W. A. Samad & E. Azar (Eds.), *Smart cities in the Gulf: Current state, opportunities, and challenges* (pp. 13–39). Springer Nature Singapore.

Koseoglu, O., Keskin, B., & Ozorhon, B. (2019). Challenges and enablers in BIM-enabled digital transformation in mega projects: The Istanbul new airport project case study. *Buildings*, *9*(5), 115.

Manny, L., Duygan, M., Fischer, M., & Rieckermann, J. (2021). Barriers to the digital transformation of infrastructure sectors. *Policy Sciences*, *54*, 943–983.

Roads and Transport Authority. (2023). www.rta.ae

Samad, W. A., & Azar, E. (2019). *Smart cities in the Gulf*. Springer.

Srinivasan, K., Muthu, S., Devadasan, S. R., & Sugumaran, C. (2016). Six Sigma through DMAIC phases: A literature review. *International Journal of Productivity and Quality Management*, *17*(2), 236–257.

Zhang, W., Hill, A. V., Schroeder, R. G., & Linderman, K. W. (2008). Project management infrastructure: The key to operational performance improvement. *Operations Management Research*, *1*, 40–52.

Chapter 13

Innovating Urban Policy Development: Lusail Smart City's Role in Qatar's Transition to a Knowledge-Based Economy

Ahmed Badran

Department of International Affairs, Qatar University, Qatar

Abstract

In recent years, a shift in economic focus has occurred, recognizing humans as the primary resource and indispensable element in economic relationships. This shift has given rise to a new economic model known as the knowledge-based economy, which places emphasis on human resources, experiences, skills, ideas, and knowledge generation. The knowledge-based economy relies heavily on information as a strategic commodity to enhance services and national income while increasing employment rates. This economic model has found application across various fields, including education, commerce, medicine, and urban planning. Particularly, the concept of smart cities has emerged as a modern notion linking the knowledge-based economy with urban development. Despite global interest in these concepts, the literature suggests a lack of attention to the knowledge-based economy and smart cities in the Arab Gulf region, including Qatar. This chapter aims to explore the relationship between smart city initiatives and the transformational processes of the knowledge-based economy, especially in Gulf Cooperation Council (GCC) countries like Qatar. It argues that smart cities and the knowledge-based economy are mutually interdependent. Smart city initiatives rely on the information and knowledge produced by the knowledge-based economy, while the latter requires better utilization of knowledge by smart applications to flourish. Using the case of Lusail smart city in Qatar, the chapter examines how the integration of smart city

The Emerald Handbook of Smart Cities in the Gulf Region:
Innovation, Development, Transformation, and Prosperity for Vision 2040, 219–238

doi:10.1108/978-1-83608-292-720241013

technologies contributes to Qatar's progression toward a knowledge-based economy. It emphasizes the strategic significance, facilitation of innovation, and practical implementation of sustainable urban development strategies in Qatar's development trajectory.

Keywords: Smart cities; knowledge-based economy; urban policy; the State of Qatar; Lusail City

1. Introduction

Since the dawn of time, man has been driven by his nature and instincts to advance both his own position in society and that of others around him. This is something that we can see in all facets of a person's life since we are always noticing the evolution of his attire, his gait, the setting in which he lives and is protected, among many other things. Humans' thoughts, ideas, and knowledge originate from their interactions with nature and with technology. Different ideas and various methods which allow humans to operate and live in a more contemporary and cutting-edge manner can be developed as human knowledge aids in observing and making sense of the social phenomena.

The study of economics is arguably one of the most significant areas where the idea of human nature driving growth has arisen. For instance, when discussing the agricultural revolution, it becomes clear that there were only two basic components to economic connections at the time: man as a producer and crops as a commodity. However, since human requirements are continuously changing and because man is constantly thinking of ways to make his life easier, more efficient, and more effective, man started to turn to industry and mechanization. He created tools and employed them to cultivate the soil and harvest the crops. Then, factories and people started having economic ties. Man has just recently come to understand that he himself is the fundamental resource and component that no business connection can exist without. Man also understood that true capital – knowledge – is created when human experiences intersect with actual life events (Salman, 2009). As a result, a novel economic theory was developed that focused on human resources, including their knowledge, abilities, and perspectives. It focuses on enhancing capabilities and competencies while also attempting to strengthen human relationships and cooperation. This is what the knowledge-based economy is referred to today. Information is the primary strategic good and source of income in this sort of economy (At-Tahir, 2015, p. 271). As such, the knowledge-based economy can manifest itself in a variety of industries, including education, trade, medical, and urban planning (Zygiaris, 2013). In this environment, a number of contemporary concepts have emerged that combine urbanism and the knowledge-based economy, including the concept of smart cities.

Smart cities are underpinned by three primary pillars: technical, environmental, and social. From a technical perspective, smart cities leverage a diverse array of smart information and communication technologies (ICTs), including wireless networks, virtual reality, and sensor networks, to establish a multi-layered digital infrastructure essential for their operation. As such, a smart city is characterized by the integration of infrastructure, capital, behaviors, and cultures within a smart system, emphasizing the interconnectedness of these elements. Moreover, smart cities are designed to prioritize environmental sustainability by minimizing negative environmental impacts. This entails the incorporation of energy-efficient production and distribution networks, environmental technologies, and the utilization of renewable energy sources as foundational components of smart city projects. At the social level, smart cities harness the creativity and knowledge of their inhabitants and research institutions to enhance digital communications and optimize knowledge management processes. Active citizen participation is recognized as pivotal in the realization of smart city objectives, distinguishing them from previous digitization initiatives (Badran, 2019, 2021).

Furthermore, smart cities transcend mere technological utilization, emphasizing collective problem-solving and active citizen engagement as fundamental principles. They serve as urban hubs facilitated by networks and digital technologies, offering interactive e-services across various domains. Smart cities leverage the intelligence of individuals, institutions, and technologies to address societal challenges while promoting social and environmental sustainability alongside knowledge-based economic growth for enhanced competitiveness. In that sense, smart city leverages ICT to enhance liveability, workability, and sustainability, relying on converged ICT systems, high-speed broadband access, sensing infrastructure, analytics, and innovative services to achieve these goals. Additionally, investments in human capital are deemed essential for fostering innovation and problem-solving capabilities within smart city contexts. Consequently, smart cities serve as urban laboratories, innovation ecosystems, and agents of change (Badran, 2021).

Consequently, the aim of smart cities projects is to make use of modern technologies/information in order to establish more communicative and responsive governance structures. That in turn should enable the societal actors including citizens, the private sector companies, and non-governmental organizations to become a part of the governance mechanism including knowledge production and sharing (At-Tahir, 2015). For this purpose, the policy and decision-makers in the Gulf Cooperation Council (GCC) countries have to develop an integrated strategy that facilitates the process of transformation to smart cities. In this context, the development of smart technologies and solutions should be done taking into account the nature of the community locally in order to ensure that these applications meet the needs of customers (Badran, 2021).

To examine the relationship between the knowledge-based economy and smart cities, the discussion in this chapter will focus on analyzing these two concepts. The main goal is to understand the nature of this concept and the requirements for its application in practice. The chapter also aims to shed light on the relationship between smart cities and the knowledge-based economy, which the author believes is a relationship of mutual dependence, where both concepts depend on each other and feed each other. This discussion at the theoretical and conceptual level will be reflected in one of the smart city initiatives in the State of Qatar, which is the Lusail City project. Hence, this applied part of the chapter aims to identify the application requirements as well as the difficulties facing the process of implementing smart city initiatives and the impact of this on the shift toward a knowledge-based economy.

Following on from the above, this chapter holds significant relevance to smart cities in the Gulf Region due to several key factors. Firstly, the Gulf Region, including countries like Qatar, Saudi Arabia, and the United Arab Emirates, has been at the forefront of smart city development initiatives in recent years. These nations have recognized the importance of leveraging technological advancements to address urban challenges and enhance the quality of life for their citizens. The pillars of smart cities highlighted in the research – technical, environmental, and social – are particularly pertinent to the Gulf Region. With rapid urbanization and population growth, there is a pressing need for efficient infrastructure, sustainable environmental practices, and social inclusion initiatives in Gulf cities. The deployment of smart ICT technologies, renewable energy solutions, and citizen engagement programs aligns closely with the goals of Gulf countries' smart city agendas. Furthermore, the emphasis on knowledge-based economies and human capital development resonates strongly with the Gulf Region's aspirations for economic diversification and talent development. Countries in the Gulf are actively investing in education, research, and innovation to transition from resource-based economies to knowledge-driven ones. The research underscores the importance of integrating human intelligence with technological advancements to drive innovation and problem-solving in smart city contexts, a principle that aligns with the Gulf Region's strategic objectives. Additionally, the challenges and opportunities identified in the research, such as legal and legislative barriers, information security concerns, and the need for comprehensive city management, are highly relevant to the Gulf context. Gulf countries are navigating similar challenges as they seek to implement large-scale smart city projects. By drawing lessons from the research findings and best practices, policymakers and stakeholders in the Gulf Region can develop tailored strategies to address these challenges effectively and maximize the benefits of smart city initiatives.

2. Smart Cities and Knowledge-Based Economy: A Theoretical Framework

This section of the chapter aims to build a theoretical and conceptual framework for smart cities and knowledge-based economy. As is the case with many social science terms, there is no agreement among researchers on a single definition for these terms. Then the definitions vary according to the perspective from which the researcher approaches the study of these topics. That in turn necessitates clarifying the key elements that link both concepts, as well as the way in which they will be used in this chapter.

2.1. Smart Cities in Concept and Practice

Having a thorough grasp of smart cities will make it easier to pinpoint the motivations behind their conception, as well as their numerous applications, and evaluation techniques. At the European Digital City Conference in 1994, the phrase "smart city" was first used (Toli & Murtagh, 2020). Following this innovation, European authorities launched initiatives like the "European Digital City," which sped up the spread of this contemporary idea. Many Arab nations have accepted comparable projects, including Saudi Arabia, Egypt, Kuwait, Qatar, and the United Arab Emirates. The idea of "urban conglomerates that encompass three essential elements: technological, social, and environmental" is frequently used to characterize smart cities. Thus, smart cities combine virtual, cognitive, and environmental characteristics (Albino et al., 2015; Mlaih, 2017). In order to support sustainable economic growth, and to ensure social and environmental sustainability, smart cities are defined as urban areas that stand out by their reliance on contemporary technology networks (Sadik, 2013). Such networks are primarily based on information and communication infrastructure. Smart cities also depend on a knowledge-based economy to stay competitive (At-Tajaddat Magazine, 2019).

In the context of this chapter, the idea of "smart cities" can be broadly conceived from an urban development/knowledge-based economy perspective as cities that rely heavily on thoughts, creativity, scientific research, human skills, knowledge, and technology in all domains. That extend to include varied residential, service, recreational, environmental, and other sectors. Urban development has been among the fields that traditional economies have largely ignored due to their focus solely on economic growth. In contrast, the notion of the knowledge-based economy, which emphasize the comprehensive integration of social, environmental, economic, and cultural components, has changed this idea. It emphasizes how these elements are linked to one another. According to this viewpoint, the knowledge-based economy makes use of the commodities and services generated and provided in metropolitan areas in a number of ways, the most recent of which is the creation of smart cities. Thus, the following inquiries are made: How do smart cities work? What do they do, and how do they fit into the knowledge-based economy?

Smart cities are metropolitan places that provide a high quality of life for their citizens by using cutting-edge techniques to solve problems and meet their requirements across many different fields. Modern ICTs are used in conjunction with new approaches and concepts in these creative strategies (Ben Hassen, 2020). The fundamental concept underlying smart cities is to apply cutting-edge technology to many sociological, economic, and environmental spheres in order to provide high-quality services. Smart cities also emphasize the sustainability of growth for both people and the environment (Al-Buali, 2013; Omar, 2010).

Several countries make large financial investments in building smart cities. To ensure that these initiatives are successfully carried out, they also engage with significant real estate firms. Nevertheless, a frequent query is, "What is the point of smart cities?" the main motivation for creating smart cities is a global trend toward urbanization. The movement of people from rural to urban regions is known as urbanization. This tendency is noticeable in countries such as Egypt where people are increasingly choosing to live and work in cities like Cairo and Alexandria as opposed to choosing to pursue agricultural or non-urban jobs. The management of urban areas and the provision of services (such as transportation, airports, education, healthcare, water, and electricity) become more challenging as cities grow in both population and size. In larger, more complex metropolitan contexts, conventional techniques of service delivery may prove ineffective and inefficient (Sadik, 2013).

Understanding the concept of smart cities is a crucial starting point. However, the real value lies in the practical application of this concept, which requires the adoption of appropriate strategies and the active involvement of governments, institutions, and society as a whole. By effectively implementing these strategies, smart cities can achieve their objectives and contribute to the progress of countries and their citizens. Consequently, at the implementation level, there are several factors to consider by the entities responsible for smart city initiatives. Chief among those considerations is the importance of providing the appropriate legislative and regulatory frameworks for the implementation process (Badran, 2021; Keshvardoost et al., 2018). During this step, it is necessary to ensure the availability of regulatory frameworks and legislations that control the functions of various aspects in smart cities in order to protect the privacy of users' information (Badran, 2024). If there are any legislative gaps, implementing agencies must fill these gaps by developing appropriate legislation, such as those related to obtaining, circulating, and accessing information by a third party. The second factor to consider is ensuring the availability of the appropriate infrastructure to implement smart city initiatives. Implementing such initiatives requires a modern technological structure, advanced information networks, and high-speed Internet services. In the absence of such infrastructure, smart city initiatives cannot be implemented successfully, and it becomes difficult to produce and circulate knowledge, which forms the backbone of the idea of a knowledge-based economy (Salman, 2009).

Added to the above, in order to secure the success of smart city projects, collaboration between various sectors is crucial. The growth and improvement of smart cities depend heavily on cooperation between the public and private sectors. The private sector may offer a vibrant market with cutting-edge technological goods and services. Additionally, it can build the fundamental infrastructure for information technology. While this is happening, the public sector concentrates on setting legal guidelines for these technologies and information as well as on fostering innovation through the creation of company incubators (Meijer & Bolívar, 2016). At the societal level, individuals should prioritize and put into practice supporting activities for smart city initiatives. No matter how much effort a nation and its various institutions put into developing technological networks or offering financial and moral support to individuals, these efforts will not be successful unless the society as a whole actively seeks, makes, uses, and advances knowledge. The community should actively work to learn new things, use them, and advance their application. Beyond that, it ought to inspire those with talents to contribute to the government or its associated organizations, which will be to everyone's advantage (Harrison et al., 2010).

It is clear from the requirements for successful implementation of smart city initiatives that the implementation process is not an easy task. The ability of these initiatives to achieve their goals can be compromised by a number of difficulties.

One of these difficulties is the issue of protecting privacy and users' data. This issue is gaining great importance in light of the increase in cyber-attacks on the websites of many governmental and private institutions, which results in some entities and parties not being authorized to access user data obtaining that data. Additionally, the absence of state or governmental backing, knowledge and skill limitations brought on by poor education, or a lack of institutions dedicated to research and development are all challenges that might face the smart city initiatives (Knight, 2014). Challenges could also stem from the brain drain that occurs because of negative circumstances in some countries (Mlaih, 2017).

2.2. Definition and Motives for Knowledge-Based Economy

Human society has undergone major and observable changes over the millennia, which has caused people's needs to change. In contemporary, convoluted society, simple needs such as food, drink, and shelter have been replaced by ones that are more complicated. The desire for knowledge, education, the right to use ICT, and access to the Internet and social networks are only a few examples of these modern requirements (Asheim & Hansen, 2009). The human mind has become more significant because of these changes. The reason for this is that the human mind is the primary source for producing goods and services. In terms of importance, the human intellect has overtaken work and wealth (Acs & Sanders, 2012). As a result, the knowledge-based economy, which has supplanted previous economic theories from the Natural Economy to the Industrial Economy, came into being. The information revolution that took place in the latter part of the twentieth century is what gave rise to this economy. Human civilizations underwent a profound and obvious transformation because of this revolution as they started to rely heavily on information, data, research findings, information technology, and communication for a variety of economic operations (Montgomery et al., 2013).

The idea of the knowledge-based economy is relatively new and has garnered substantial popularity (Bendickson et al., 2021). The knowledge-based economy is known by many different names, including the Digital Economy, Electronic Economy, and Virtual Economy. There is typically no agreement on a single description of its notion among academics and specialists in this discipline. The knowledge-based economy is viewed from many angles by each researcher or professional, focusing on the features they feel vital. That is why there is not a clear description. Accordingly, a knowledge-based economy is one that relies heavily on knowledge, innovation, cutting-edge technology, and leverages information and internet networks in a variety of economic activities (Abdullah & Noor, 2016).

Although definitions of the knowledge-based economy vary depending on the perspective and area of study of each researcher, it can be said that information, data, research findings, information technology, and communication technologies play a major role in the knowledge-based economy. These components assist in developing and producing items, services, or technologies that are more effective, sophisticated, and able to satisfy the complicated needs of society. In the knowledge-based economy, knowledge, skills, and information are crucial because they can be converted into goods or services that can be sold and generate income (Ciacci & Ivaldi, 2023).

The shift toward a knowledge-based economy has changed the standards of countries' power and success. A nation's success and progress are no longer determined by the volume of manual output or the amount of capital it has. The information revolution the globe experienced in the last century and the subsequent transition toward the knowledge-based economy have put more emphasis on knowledge (Audretsch et al., 2021). The capacity to generate information and then apply it to diverse economic, social, and technological elements have become the primary indicator for power and success. The nations with the strongest economies worldwide are those that have made investments in the creation and use of knowledge (Komninos, 2011). For instance, when we examine nations like China, Japan, and Singapore, we discover that they have a lot in common when it comes to investing in research facilities, educational systems, and development.

One of the main forces behind the shift to the knowledge-based economy is the belief that relying on smart technology means performing tasks, creating commodities, and providing services with higher efficiency and effectiveness while requiring less work, time, and expenses. Computers and the Internet are now producing data and knowledge that are used in almost all industries and institutions as inputs in products and operations deigns. Additionally, the new technological products have the capacity to precisely transfer, store, and process enormous amounts of data. They can quickly deliver appropriate services to consumers or clients by retrieving this information (Asheim & Hansen, 2009).

The returns that the knowledge-based economy generates is another important factor to take into account. Such returns frequently outpace that of any other economy. The reason for this is that goods and services that incorporate knowledge, including cutting-edge technologies, fetch high prices. When we look at the biggest corporations in the world, we frequently discover that many of them are built on innovative human minds and cutting-edge technologies. For instance, Apple, a market leader in the smartphone sector, is thought to be worth about $605 billion, with Alphabet, which includes Google, coming in second and Microsoft in third. The knowledge-based economy therefore gives nations the chance to generate sizable economic returns, which in turn aids in the advancement of their numerous development strategies, including social, economic, environmental, and other ones.

2.3. Smart Cities and Knowledge-Based Economy

In examining the relationship between smart cities and the knowledge-based economy, it is clear from the previous conceptual discussion that both concepts have an organic integrated relationship (Chang et al., 2018). The knowledge-based economy focuses on methods of generating and using various information and knowledge to improve many processes, whether those related to products, services, or decision-making. At the same time, generating this type of knowledge and information requires the engagement of many smart technologies that are considered an integral part of any smart cities' initiative. These smart technologies generate a huge amount of knowledge and information that different parties can use to improve their products in addition to improving economic growth rates in general. Furthermore, adopting the concept of the knowledge-based economy results in the production of goods and services with high economic returns due to their degree of complexity and the knowledge component invested in them (Yigitcanlar, 2011).

As such, knowledge and expertise must be acquired, shared, exchanged, and used in order for the knowledge-based economy to function. It places emphasis on the use and flow of ideas. By definition, smart cities support the following principles, which in turn reflect in the functionality of the knowledge-based economy (El Khatib et al., 2023):

- **Data and Information Production and Sharing**: Sensors, IoT devices, and other sources are used by smart cities to gather a lot of data. This data can be examined to glean insightful and useful information that can be applied to resource allocation and city administration decisions.
- **Research and Innovation**: To create and execute cutting-edge solutions, smart cities usually work with technology firms, academic institutions, and creative start-ups. These collaborations support creativity and advance the knowledge-based economy.
- **Skills Development**: People who live in smart cities regularly engage with cutting-edge services and technology. Individuals benefit personally and the information-based economy as a result of this exposure to learning new skills and knowledge.
- **Technology Transfer**: To install and maintain cutting-edge systems, many smart cities work with technology firms. The transmission of knowledge and technological developments is aided by this technology transfer.

Hence, smart cities are crucial in developing a knowledge-based economy because they shape people's knowledge, encourage innovation, and facilitate the sharing of knowledge and skills. They make use of technology and data to improve urban living, encourage sustainable growth, and adhere to the tenets of the knowledge-based economy (Ivaldi & Ciacci, 2023). Added to this, smart cities use the Internet and modern technology to deliver their electronic services, which include things like education, commerce, government transactions, healthcare records, and more (Gangi, 2017). This indicates that highly sensitive user data is there. Therefore, it is crucial to create procedures that safeguard sensitive data. These services must be continued by smart cities while maintaining the confidentiality and security of the data. In order to do this, a special system must be created by accountable organizations in charge of managing smart city initiatives. The following elements should be part of the data protection system (Mlaih, 2017):

- Making sure the security product meets all necessary security standards.
- Implementing penetration testing to check the security system's sturdiness. The system or the system containing the data should be able to quickly detect any threats and respond by shutting down or defending themselves.

- Product authorities that offer certification can help evaluate a product and the solutions it offers to accountable institutions.
- Verifying the security and quality of data, as well as guaranteeing accurate operational procedures that can differentiate between authorized and suspect operations and stop the latter.

3. Qatar's Efforts Toward the Knowledge-Based Economy and Smart Cities

Qatar has adopted a number of policies, strategies, and projects to achieve these goals as a result of the leadership's vision and dedication to enhancing its economy, achieving high levels of human development, promoting environmental sustainability, and offering a quality of life for its citizens. The National Vision 2030, published by the former emir, Sheikh Hamad bin Khalifa Al Thani, in 2008, was one of the most important actions the government of Qatar took. This vision acted as a catalyst for the creation of strategies, plans, and programs that were used to carry out projects that aimed to advance the nation's economy, people, society, and environment. The knowledge-based economy and smart cities were two of these objectives, and we shall go into more detail about them in this part.

3.1. The Knowledge-Based Economy and Key Smart Cities Initiatives in Qatar

The Qatar National Vision 2030 is one of the most effective plans and policies for the development of Qatar's knowledge-based economy. It is supported by a number of pillars, the economic one being one of the most significant. This pillar's main objective is to establish a knowledge-based economy through advanced research, invention, and education as well as the creation of an informational and physical infrastructure (General Secretariat for Development Planning. 2008). The inclusion of this direction in the vision demonstrates the government's understanding of the significance of this trend and its significant implications for Qatar's societal and economic development. By making this a pillar, it inspires other pertinent organizations to work tirelessly toward achieving these goals. This in turn guarantees the continuation of significant initiatives to change Qatar's traditional economy into one based on knowledge (Ben Hassen, 2021). The human development pillar, which focuses on building a knowledgeable and competent society by offering equal educational and training opportunities for everyone and establishing cutting-edge, effective educational institutions, is also a crucial pillar that the vision pursues (Gangi, 2017; Rottleb & Kleibert, 2022). Additionally, it intends to promote cooperation between the public and commercial sectors, the government, and foreign research institutions in order to finance scientific research. As education, training, knowledge acquisition, and scientific research are all crucial components of creating a knowledge-based economy in Qatar, these goals are crucial and important (Al-Mansoori & Koç, 2019; Al-Qahtani & Shirazi, 2023).

As was already established, knowledge, innovation, and rejuvenation produced by the results of scientific research are one of the key components of the knowledge-based economy. The Qatar National Research Fund (QNRF) was founded in 2006 in order to fulfill the duty of developing a knowledge-based economy in Qatar, which is supported by the Qatar Foundation for Education, Science, and Community Development. The QNRF's main goals are to promote a keen interest in science among Qatari society's citizens by supporting and hosting competitions for scientific research and other endeavors. As a result, the state gains a lot and is better able to develop a knowledge-based economy (Baker et al., 2015).

There are different levels of projects related to smart cities in Qatar (Adalbi et al., 2022). At the national level, Smart Qatar Initiative "Tasmu" stands as a major project. This initiative aims to transform Qatar into a fully smart nation and serves as the umbrella project from which other smart city projects like Lusail, Msheireb, and The Pearl originate. It provides a unified platform for data exchange among these projects (Al Sharif & Pokharel, 2022). At the system level, Kahramaa (Qatar General Electricity & Water Corporation) has made progress in this area by implementing what is known as the smart grid to enhance service delivery. Additionally, Ashghal (Public Works Authority) has also introduced a smart transportation system. At the service level, Qatar offers various smart services related to water networks, electricity, traffic signals, and smart parking, which are available in public spaces and roads throughout the country. At the Government Level, Qatar has a "smart government" where services are provided efficiently and electronically to individuals through services like "Metrash 2". While not yet at 100% implementation, there has been significant progress in this area over the past decade.

3.2. Challenges Facing Qatar Regarding Smart City Projects

Security and Privacy Challenges: As highlighted in the literature, security and privacy challenges are among the most significant obstacles facing the establishment and operation of smart cities. It is essential to address these challenges, as cyberattacks have become a growing concern, potentially affecting critical infrastructure such as gas, water, and electricity networks. This can lead to catastrophic consequences. Therefore, cybersecurity is a crucial aspect that requires both technological and legislative attention. Qatar has taken steps to address this issue by establishing laws to protect personal data and by creating entities like the Qatar Computer Emergency Response Team (Q-CERT) to respond to cyberattacks. However, there is room for improvement in implementing and enforcing these cybersecurity measures (Laufs et al., 2020).

Integrated Management of Smart Cities: Integrated management is essential for the success and sustainability of smart cities. It involves coordinating different departments, entities, and sectors to ensure precise and efficient service delivery. Integrated management helps tailor services to individual needs and facilitates the exchange of information between different entities, resulting in more accurate and effective decision-making. While the importance of integrated management is clear, it is often overlooked. Many believe that smart systems can manage themselves independently. However, the challenge lies in achieving a seamless integration of various systems and ensuring data privacy and security. Questions arise about whether users should have a single identity or number for all systems and services or if there should be consent and ethical considerations regarding the sharing of user data between different sectors. Finding optimal solutions to these challenges requires further research and exploration (Nam & Pardo, 2011; Shareh et al., 2019).

4. Case Study: Lusail Smart City Project

The Lusail Smart City project is sponsored by the Qatari company, Diar, and is considered the largest in the state of Qatar. Lusail City spans an area of 38 square kilometers and consists of 19 zones with diverse uses, including residential, commercial, and recreational areas. Additionally, the city features four islands known as the "Qetaifan Islands." Lusail City has been designed to accommodate over 200,000 residents and approximately 170,000 daily visitors. Smart cities possess several characteristics that set them apart from traditional cities, making their services more efficient and effective. Smart cities literature made it clear that, in contrast to traditional cities that typically rely on conventional methods, smart cities leverage modern knowledge and technology to provide services across various sectors. Therefore, it is important to examine the smart features in project like Lusail City. It is equally important know the mechanism via which such projects deliver smart services in contrast to other traditional cities in Qatar. Several smart features of Lusail Smart City were identified (Adalbi et al., 2022):

- **District Cooling System**: This system is unique to Qatar and is implemented in both Lusail and Al-Daayen cities. It involves replacing conventional rooftop air conditioning units with a centralized cooling district. The cooling energy is then distributed from this district to various parts, buildings, and houses in the city.
- **Gas Networks**: Most areas in Lusail City are served by gas networks, with the exception of residential villa areas.
- **Waste Disposal Systems**: Lusail Smart City has advanced waste disposal systems in three areas. Unlike traditional tower buildings where waste is disposed of through openings in the towers, Lusail City uses specific systems to transport waste to a central location after disposal. At this central location, the waste is analyzed, recycled, or incinerated based on its nature, ensuring appropriate disposal methods for each element.
- **Programmable Irrigation Systems**: These systems help conserve water by programming irrigation based on specific schedules and needs.
- **Smart Traffic Signals**: Although these systems are still in the implementation phase, they are designed to monitor all four directions simultaneously. They can automatically detect the direction with the highest traffic volume and

adjust traffic signals accordingly, reducing congestion. This technology highlights the real-time decision-making capability present in smart cities, which traditional cities often lack.

- **Metro Project**: Lusail City originally initiated the metro project within its boundaries, later transitioning it into a separate company covering all regions of Qatar. The city is fully serviced by the metro network.
- **Water Taxis and the Waterfront City**: Lusail Smart City includes water taxis and a waterfront city featuring a hotel and a water-based entertainment center.
- **Sustainability**: All buildings in Lusail City are sustainable and must obtain a Gulf Organization for Research and Development (GORD) Green Building certification. This certification evaluates the level of sustainability through the "GSAS" (Global Sustainability Assessment System) rating system. Buildings are rated on a scale from one star to six stars, each star representing specific sustainability criteria. A minimum of two stars is required for any building in Lusail City, ensuring energy and water conservation, among other factors. Additionally, local contractors, materials, and architectural styles are encouraged to be employed in the design and construction of buildings.

These smart features collectively contribute to making Lusail Smart City a cutting-edge and sustainable urban development in Qatar, with an emphasis on efficiency, resource conservation, and advanced technology integration.

4.1. The Role of Research and Society in the Development of the Lusail Smart City Project

Research and the community play vital roles in the establishment of smart cities and knowledge economies. They are sources that generate the necessary knowledge for the development of a smart city and the growth of a knowledge-based economy. As discussed in the first chapter when examining the relationship between the concept of smart cities and the knowledge-based economy, knowledge is the link that makes the relationship between these two concepts interdependent. Therefore, it is essential to consider that both scientific research and community members are necessary for the development and sustainability of any smart city project (Al-Qahtani & Shirazi, 2023).

It became clear that Lusail City has benefited significantly from research conducted by the GORD. Initially focused solely on sustainability research, GORD shifted its main focus to the Lusail project. This relationship has strengthened through mutual support, where the Qatari company, Diar, responsible for the Lusail project, financially supports GORD through projects and marketing activities. In return, GORD provides specialized research on sustainability and how to achieve it effectively in Lusail City (Al-Mansoori & Koç, 2019).

Regarding the role of the community in the development of the Lusail project, the analysis of previous studies led to an important point related to the necessary strategies for establishing smart cities. It is essential to have an aware community capable of creating and utilizing knowledge, and supporting individuals with expertise and knowledge. An aware community is crucial because the state expects awareness and an understanding of the importance of the project. Providing sustainable networks and smart services throughout Lusail City is not enough if there is no awareness among the Qatari community about their significance and how to employ them correctly. Personal experiences and daily interactions with these services play a significant role in spreading awareness. People realize that these smart services save them time and effort, thus increasing their interest in experiencing and utilizing them. Therefore, personal experiences are the most effective way to transfer and disseminate knowledge in the community (Gangi, 2017).

4.2. The Partnership Between the Public and Private Sectors for the Establishment of Lusail City

Coordination between the public and private sectors is a key element when discussing highly efficient smart cities. Such cooperation is required for the establishment of any smart city, as concluded from previous literature and the analysis of the reciprocal relationship between Lusail City and the GORD. The analysis confirmed the presence of a genuine cooperative relationship, and various entities contributed to the support and establishment of the Lusail project (Adalbi et al., 2022).

1. **Government Institutions**: Many government agencies participated in the development of the project, such as the Ministry of Interior, Kahramaa (Qatar General Electricity and Water Corporation), Public Works Authority (Ashghal), and Ooredoo (a telecommunications company). These entities contributed to the development of infrastructure networks and the provision of optical fibers, which are essential for ICT.
2. **Private Companies**: Companies like Hyundai played a significant role in supporting infrastructure projects within Lusail City. Additionally, the GORD had a prominent role in the project's development. Infrastructure projects in Lusail City are mainly handled by foreign companies, with financial contributions from Diar to develop and work on these projects, whether in terms of design or actual execution.
3. **External Institutions**: Lusail City has also enlisted the help of real estate developers from different countries.

The partnership between the public and private sectors has been instrumental in the successful establishment and ongoing development of Lusail City, highlighting the importance of collaboration for the growth of smart cities.

4.3. Challenges Facing the Implementation of the "Lusail" Smart City Project

Smart city projects face several challenges that need to be identified and addressed to ensure the successful implementation of high-quality smart cities (Shareh et al., 2019; Sharif & Pokharel, 2022). According to what was discussed in, the challenges encountered in implementing the "Lusail" smart city project are as follows:

1. **Real Estate Slump**: One of the most significant challenges faced during the implementation of the "Lusail" project is the real estate slump. This coincided with a drop-in oil prices, and the project management is currently working to address this challenge by adapting to new real estate laws and making changes to some projects to make them more beneficial for property owners, the Qatar Diar Company, or the "Lusail" project itself.
2. **Political Challenges**: During the period of the blockade, challenges arose due to shortages of some building materials imported from certain countries. Additionally, some real estate developers from those countries were involved in projects in the city. However, the company managed to overcome this challenge and provide quick alternatives.
3. **Legal and Legislative Framework**: It has been emphasized that a robust legal and legislative framework is essential to support the sustainability and effectiveness of smart city projects. These regulations serve as the foundation for smart cities. Interestingly, these regulations can sometimes be a result of smart city projects themselves. For example, Qatar did not have an effective sustainability system in place, but the "Lusail" project introduced this system. Similarly, there were no laws regulating relationships between property owners in residential apartments, but due to the significant number of residential units in "Lusail," the company and government institutions worked on developing laws to address this issue. Furthermore, the "Lusail" project introduced the idea of GORD (GSAS) certificates, which became mandatory for all projects in Qatar. This was not the case before.
4. **Security and Privacy**: Security and privacy are major challenges in discussions about smart cities, as acknowledged in various research. While Qatar has taken some measures and enacted laws to mitigate this challenge, the efforts of the Qatar Diar Company responsible for the "Lusail" project are still in the early stages. Currently, there is ongoing work to ensure the security of information exchanged in the real estate market's dedicated application, which includes residential and commercial units.

4.4. The Role of the "Lusail" Smart City Project in Achieving Knowledge-Based Economy in Qatar

The primary goal of this study is to determine the role that the "Lusail" smart city project plays in achieving a Knowledge-based economy in Qatar. The knowledge-based economy, as discussed, is a key objective outlined in Qatar's Vision 2030. Through the analysis, it was found that one aspect in which the "Lusail" smart city project contributes to enhancing the knowledge-based economy in Qatar is through tourism.

At the outset of the project in 2005, the vision and goals behind "Lusail" were to create a city that could accommodate the urban and population growth Qatar was experiencing during that period. Being a smart and sustainable city, "Lusail" was envisioned as a unique and civilized destination that would represent Qatar to the rest of the world. Thus, "Lusail," built on the principles of knowledge and modern technology, is expected to become a major tourist attraction in Qatar. For example, some projects within the city, such as the "Place Vendome Mall," the largest shopping complex in Qatar, and the Qetaifan Islands, which will include a water city, a hotel, a sports club, and the largest water theme park in Qatar, are expected to draw the attention of tourists.

In conclusion, the role of the "Lusail" smart city project in achieving a knowledge-based economy in Qatar is exemplified by its application of one of the key pillars of the knowledge-based economy: harnessing knowledge and technology to achieve economic diversification. This is expected to generate significant economic returns for the country and enhance its position in the knowledge-based economy landscape.

The aforementioned discussion of Lusail City's role in Qatar's transition to a knowledge-based economy and smart cities has significant implications for the broader smart cities agenda, particularly within the Gulf Region. By highlighting the strategic importance of Lusail City as a model for sustainable urban development and innovation, the research underscores the potential for similar initiatives to drive economic diversification, enhance the quality of life, and promote environmental sustainability in other urban centers across the region. In that sense, such a treatment offers valuable insights into the practical implementation of smart city technologies and strategies, providing a roadmap for policymakers and urban planners seeking to replicate Lusail City's success in their own contexts.

Added to the above, the key findings of this research have important implications for diverse stakeholders involved in smart city initiatives, including urban planners, architects, policymakers, educators, and government officials. For urban planners and architects, the research highlights the importance of integrating smart technologies and sustainable design principles into urban development projects to create more efficient, livable, and resilient cities. Policymakers can use these findings to inform policy decisions aimed at promoting innovation, fostering collaboration between public and private sectors, and improving infrastructure to support smart city initiatives. Additionally, educators and students can benefit from the insights provided by this research to develop curricula and programs focused on smart city technologies and sustainable urban development.

Based on the research findings, several propositions for initiatives related to smart cities strategies and projects can be justified. These include initiatives aimed at enhancing digital connectivity and infrastructure, promoting innovation hubs and research centers, incentivizing investment in sustainable technologies and renewable energy, and fostering public–private partnerships to support smart city initiatives. Furthermore, initiatives focused on promoting citizen engagement and participation, improving public transportation and mobility solutions, and enhancing environmental conservation efforts can also be justified based on the research findings.

In terms of policy recommendations, three key recommendations emerge from the research. Firstly, policymakers should prioritize investment in smart infrastructure and technologies to support knowledge-based economy and sustainable development goals. This includes investments in digital infrastructure, innovation initiatives, and smart technological systems. Secondly, there is a need for comprehensive urban planning strategies that prioritize sustainability, resilience, and inclusivity in smart city development projects. Finally, policymakers should prioritize collaboration and knowledge sharing between government agencies, private sector stakeholders, and academic institutions to foster innovation and ensure the effective implementation of smart city initiatives.

5. Conclusion

In conclusion, this chapter not only underscores the pivotal role of smart cities in shaping the future of urban development but also makes significant contributions to the theoretical understanding, practical implementation, and applied industry of smart city initiatives. Through an in-depth analysis of the relationship between smart cities and knowledge-based economies, this study contributes to the theoretical framework surrounding smart city concepts. By emphasizing the interconnectedness of these two concepts and their potential for driving economic transformation, the study advances our theoretical understanding of how smart cities can serve as catalysts for knowledge creation, innovation, and sustainable development. In terms of practical implications, the discussion offers valuable insights for policymakers, urban planners, and industry stakeholders involved in smart city projects. By highlighting the challenges and opportunities associated with implementing smart city initiatives, the study provides practical guidance on overcoming barriers and maximizing the benefits of smart city development. Moreover, by examining the case of Lusail City in Qatar, the study offers real-world examples of successful smart city projects, serving as a roadmap for other cities seeking to embark on similar initiatives.

The propositions and recommendations put forth in this study have significant implications for the design, implementation, and governance of smart cities. By emphasizing the importance of investing in human capital, fostering innovation ecosystems, and addressing challenges such as information security and privacy concerns, the study provides practical strategies for designing and managing smart city projects effectively. Furthermore, by advocating for collaborative approaches and interdisciplinary partnerships, the study encourages the integration of diverse perspectives and expertise in the development of smart city solutions. Looking ahead, there are several avenues for future research that could build upon the findings of this study. One potential direction is to explore the socio-economic impacts of smart city initiatives, including their effects on employment, social equity, and quality of life. Additionally, future research could focus on integrating the core contributions of this study with other works in related fields, such as urban planning, sustainability, and technology innovation. By fostering interdisciplinary collaboration and knowledge exchange, future research endeavors can further advance our understanding of smart cities and their role in shaping the future of urban development.

References

Abdullah, K., & Noor, S. (2016). Iraq's integration into the knowledge-based economy: Reality and challenges. *Tikrit Journal of Administrative and Economic Sciences, 12*(36), 220–243.

Acs, Z. J., & Sanders, M. (2012). Patents, knowledge spillovers, and entrepreneurship. *Small Business Economics, 39*(4), 801–817. https://doi.org/10.1007/s11187-011-9322-y

Adalbi, M., Salim Ferwati, M., Mohammad Ahmad, A., & Cheol Lee, Y. (2022). Smart cities strategies in developing countries: The case of Lusail City, Qatar. In *Intelligent Human Systems Integration (IHSI 2022) Integrating People and Intelligent Systems*. https://doi.org/10.54941/ahfe1001015

Al Sharif, R., & Pokharel, S. (2022). Smart city perspectives in the context of Qatar. In J. M. Corchado & S. Trabelsi (Eds.), *Sustainable smart cities and territories* (Vol. 253, pp. 103–113). Springer International Publishing. https://doi.org/10.1007/978-3-030-78901-5_10

Albino, V., Berardi, U., & Dangelico, R. M. (2015). Smart cities: Definitions, dimensions, performance, and initiatives. *Journal of Urban Technology, 22*(1), 3–21. https://doi.org/10.1080/10630732.2014.942092

Al-Buali, Y. (2013). The reality of knowledge-based economy in the Gulf Cooperation Council Countries according to digital content indicators. *Gulf Economic Journal, 10*(2019), 39–74.

Al-Mansoori, R. S., & Koç, M. (2019). Toward knowledge-based economy: Innovation and transformational leadership in public universities in Texas and Qatar. *Sustainability, 11*(23), 6721. https://doi.org/10.3390/su11236721

Al-Qahtani, A. S. S. A., & Shirazi, N. S. (2023). What are the binding constraints for a knowledge-based economy in Qatar? *Sustainability, 15*(5), 3871. https://doi.org/10.3390/su15053871

Asheim, B., & Hansen, H. K. (2009). Knowledge bases, talents, and contexts: On the usefulness of the creative class approach in Sweden: Economic geography. *Economic Geography, 85*(4), 425–442. https://doi.org/10.1111/j.1944-8287.2009.01051.x

At-Tahir, A. M. (2015). *Knowledge management* (2nd ed.). Dar Wael for Publishing and Distribution.

At-Tajaddat Magazine, T. (2019). *Smart cities: Features and transformation indicators in the world and the Arab Region. 1*(5), 1–11.

Audretsch, D. B., Belitski, M., & Korosteleva, J. (2021). Cultural diversity and knowledge in explaining entrepreneurship in European cities. *Small Business Economics, 56*(2), 593–611. https://doi.org/10.1007/s11187-019-00191-4

Badran, A. (2019). Smart-governments for smart cities: The case of Dubai Smart-Government. In W. A. Samad & E. Azar (Eds.), *Smart cities in the Gulf* (pp. 59–82). Springer Nature Singapore. https://doi.org/10.1007/978-981-13-2011-8_5

Badran, A. (2021). Developing smart cities: Regulatory and policy implications for the state of Qatar. *International Journal of Public Administration*, 1–14. https://doi.org/10.1080/01900692.2021.2003811

Badran, A. (2024). Regulating cyberspace for children: Reflections on the Qatari Case. *British Journal of Cyber Criminology, 3*(1). https://www.pubtexto.com/pdf/?regulating-cyberspace-for-children-reflections-on-the-qatari-case

Baker, D. P., Crist, J. T., Zhang, L., Powell, J. J. W., Shima, K., & Stock, M. (2015). *Science productivity, higher education development and the knowledge society (SPHERE Project) Final Report*. Georgetown University School of Foreign Service in Qatar.

Ben Hassen, T. (2020). The entrepreneurship ecosystem in the ICT sector in Qatar: Local advantages and constraints. *Journal of Small Business and Enterprise Development, 27*(2), 177–195. https://doi.org/10.1108/JSBED-04-2019-0119

Ben Hassen, T. (2021). The state of the knowledge-based economy in the Arab world: Cases of Qatar and Lebanon. *EuroMed Journal of Business, 16*(2), 129–153. https://doi.org/10.1108/EMJB-03-2020-0026

Bendickson, J. S., Irwin, J. G., Cowden, B. J., & McDowell, W. C. (2021). Entrepreneurial ecosystem knowledge spillover in the face of institutional voids: Groups, issues, and actions. *Knowledge Management Research & Practice, 19*(1), 117–126. https://doi.org/10.1080/14778238.2020.1768810

Chang, D. L., Sabatini-Marques, J., Da Costa, E. M., Selig, P. M., & Yigitcanlar, T. (2018). Knowledge-based, smart and sustainable cities: A provocation for a conceptual framework. *Journal of Open Innovation: Technology, Market, and Complexity, 4*(1), 1–17. https://doi.org/10.1186/s40852-018-0087-2

Ciacci, A., & Ivaldi, E. (2023). Smart sustainable cities and knowledge-based economy for people, workers, and enterprises: Mutually reinforcing dynamics. In E. Ivaldi & A. Ciacci (Eds.), *Smart sustainable cities and knowledge-based economy* (pp. 19–51). Springer International Publishing. https://doi.org/10.1007/978-3-031-25038-5_2

El Khatib, M. M., Abidi, N., Al-Nakeeb, A., Alshurideh, M., & Ahmed, G. (2023). Dubai smart city as a knowledge based economy. In M. Alshurideh, B. H. Al Kurdi, R. Masa'deh, H. M. Alzoubi, & S. Salloum (Eds.), *The effect of information technology on business and marketing intelligence systems* (Vol. 1056, pp. 1657–1672). Springer International Publishing. https://doi.org/10.1007/978-3-031-12382-5_90

Gangi, Y. A. (2017). The role of entrepreneurship education and training on creation of the knowledge economy: Qatar leap to the future. *World Journal of Entrepreneurship, Management and Sustainable Development, 13*(4), 375–388. https://doi.org/10.1108/WJEMSD-06-2017-0032

General Secretariat for Development Planning. (2008). *Qatar National Vision 2030*. https://www.mdps.gov.qa/en/knowledge/HomePagePublications/QNV2030_Arabic_v2.pdf

Harrison, C., Eckman, B., Hamilton, R., Hartswick, P., Kalagnanam, J., Paraszczak, J., & Williams, P. (2010). Foundations for smarter cities. *IBM Journal of Research and Development, 54*(4), 1–16. https://doi.org/10.1147/JRD.2010.2048257

Ivaldi, E., & Ciacci, A. (2023). *Smart sustainable cities and knowledge-based economy: Policy implications*. Springer.

Keshvardoost, S., Renukappa, S., & Suresh, S. (2018). Developments of policies related to smart cities: A critical review. *2018 IEEE/ACM international conference on utility and cloud computing companion (UCC Companion)*, 370–375. https://doi.org/10.1109/UCC-Companion.2018.00083

Knight, J. (Ed.). (2014). *International education hubs: Student, talent, knowledge-innovation models*. Springer Netherlands. https://doi.org/10.1007/978-94-007-7025-6

Komninos, N. (2011). Intelligent cities: Variable geometries of spatial intelligence. *Intelligent Buildings International, 3*(3), 172–188. https://doi.org/10.1080/17508975.2011.579339

Laufs, J., Borrion, H., & Bradford, B. (2020). Security and the smart city: A systematic review. *Sustainable Cities and Society, 55*, 102023. https://doi.org/10.1016/j.scs.2020.102023

Meijer, A., & Bolívar, M. P. R. (2016). Governing the smart city: A review of the literature on smart urban governance. *International Review of Administrative Sciences, 82*(2), 392–408. https://doi.org/10.1177/0020852314564308

Mlaih, Y. (2017). Smart cities in Morocco: Foundations, experiences, and implementation potential. *Masalik: Journal of Thought, Politics, and Economy, 28*(2019), 189–208.

Montgomery, M. R., Stren, R., Cohen, B., & Reed, H. E. (2013). *Cities transformed* (0 ed.). Routledge. https://doi.org/10.4324/9781315065700

Nam, T., & Pardo, T. A. (2011). Smart city as urban innovation: Focusing on management, policy, and context. In *Proceedings of the 5th International conference on theory and practice of electronic governance* (pp. 185–194). https://doi.org/10.1145/2072069.2072100

Omar, G. (2010). The role of information and communication technology in the knowledge-based economy. *Journal of Economy and Human Development*, *2*(2019), 1–11.

Rottleb, T., & Kleibert, J. M. (2022). Circulation and containment in the knowledge-based economy: Transnational education zones in Dubai and Qatar. *Environment and Planning A: Economy and Space*, *54*(5), 930–948. https://doi.org/10.1177/0308518X221077105

Sadik, K. (2013). *Approaches to smart city planning: A case study of Damascus* [Unpublished master's thesis]. Department of Planning and Environment, Faculty of Architecture, University of Damascus, Syria.

Salman, J. D. (2009). *Knowledge-based economy*. Dar Al-Yazouri Scientific Publishing and Distribution.

Shareh, M. B., Navidi, H., Javadi, H. H. S., & HosseinZadeh, M. (2019). Preventing Sybil attacks in P2P file sharing networks based on the evolutionary game model. *Information Sciences*, *470*, 94–108. https://doi.org/10.1016/j.ins.2018.08.054

Sharif, R. A., & Pokharel, S. (2022). Smart city dimensions and associated risks: Review of literature. *Sustainable Cities and Society*, *77*, 103542. https://doi.org/10.1016/j.scs.2021.103542

Toli, A. M., & Murtagh, N. (2020). The concept of sustainability in smart city definitions. *Frontiers in Built Environment*, *6*, 77. https://doi.org/10.3389/fbuil.2020.00077

Yigitcanlar, T. (2011). Position paper: Redefining knowledge-based urban development. *International Journal of Knowledge-Based Development*, *2*(4), 340. https://doi.org/10.1504/IJKBD.2011.044343

Zygiaris, S. (2013). Smart city reference model: Assisting planners to conceptualize the building of smart city innovation ecosystems. *Journal of the Knowledge Economy*, *4*(2), 217–231. https://doi.org/10.1007/s13132-012-0089-4

Chapter 14

The Role of Government in Ensuring Sustainability in Smart Urban Development: A Case Study of Kuwait

Afnan Alkhaldi[a], *Sawsan Malik*[a], *Salah Alhammadi*[a] *and Miltiadis D. Lytras*[b]

[a]*Arab Open University, Kuwait*
[b]*College of Engineering, Efrat University, Saudi Arabia*

Abstract

The emergence of smart cities, metropolises that integrate physical infrastructure, digital technology, and data analytics, and that focus on urban sustainability, have profoundly changed urban development. In the modern digital era, robust infrastructure has become an indispensable catalyst for urban advancement. Kuwait is dedicated to the integration of diverse renewable energy technologies in the development of smart cities that enhance energy security, promote innovation, and contribute to global climate change mitigation efforts. Focusing on smart cities within Gulf Cooperation Council (GCC) countries, a review is presented of how successfully they have effectively combined technology, infrastructure, and sustainability to serve as models for new global and regional developments. Insights into what makes a city smart are provided in different settings.

keywords: Urban development; sustainable infrastructure; smart cities governance; urban resilience; Gulf Cooperation Council (GCC)

The Emerald Handbook of Smart Cities in the Gulf Region:
Innovation, Development, Transformation, and Prosperity for Vision 2040, 239–249

doi:10.1108/978-1-83608-292-720241014

Introduction

The concept of the "smart city" has become increasingly important as countries strive to enhance the quality of life for their citizens (Neirotti et al., 2014). A "smart city" uses smart computing technologies to make components of its critical infrastructure and services more intelligent, interconnected, and efficient (Meijer & Bolívar, 2016). It is a city in which the community has learned to learn, adapt, and innovate (Caragliu et al., 2011). The use of information and technology is combined to increase efficiency, transparency, and service quality (Sitna et al., 2021).

Despite the importance of designing smart cities, there is no universally accepted definition. As such, the concept of the "smart city" is open to interpretation and subject to the mindset or agenda of those referring to them (Prevelianaki et al., 2022). They have been described as, for example, a "smart sustainable city," an "eco-city," "cyberville," and a "digital city" (Ohtsuki, 2017). The lack of a universally accepted definition for what constitutes a smart city highlights the evolving nature of the concept, and the different approaches taken by different stakeholders. For example, smart cities can be seen through technology-centric lenses, emphasizing advanced tech integration; citizen-centric perspectives, focusing on citizen engagement and quality of life; sustainability-centric viewpoints, emphasizing environmental responsibility; data-driven orientations, highlighting data usage for urban optimization; infrastructure-centric strategies, promoting modern and resilient city structures; economic development outlooks, fostering innovation and growth; or customized approaches, combining various elements as per the unique context and priorities of each city (Alhammadi, 2024; Alkhaldi et al., 2024).

The success of a smart city depends on the appropriate application of technology, governance, and urban planning (Meijer & Bolívar, 2016). Local and municipal governments apply the concept of the smart city to deliver benefits to people, the government and economy, mobility and living, and the environment (Alkhaldi et al., 2023). A successful smart city requires effective urban management systems, and introduction of safe and comfortable living conditions (Parubochaya et al., 2020).

United Nations Sustainable Development Goals (SDGs) shape the initiatives and objectives of smart cities. These SDGs provide a comprehensive framework to address global challenges and to promote sustainable development (Bibri & Krogstie, 2017), especially numbers 7 (Affordable and Clean Energy), 9 (Industry, Innovation, and Infrastructure), 11 (Sustainable Cities and Communities), and 13 (Climate Action). Successful smart city development requires sustainability to be considered, because these cities are designed and intended to operate in a way that satisfies the needs of current and future generations (Khan et al., 2020). The development and resilience of these cities are aided by incorporation of sustainable infrastructure. Examples of successful smart cities worldwide are provided, with a focus on those within the Gulf Cooperation Council (GCC) region, especially Kuwait, as exemplified by analysis of the Abdali economic zone.

Key Components of Sustainability Infrastructure in Smart Cities

Urban communities are increasingly using innovative solutions to tackle complex issues. For example, smart grid technologies have been implemented in American cities such as San Diego and Austin (Nagasawa et al., 2012). The use of cutting-edge sensors and communication tools in these grids contributes to improved energy distribution, fewer power interruptions, and a seamless integration of renewable energy sources (Owunna & Obeagu, 2022). Singapore has become a leader in the transportation industry by implementing cutting-edge technology such as digital road tolling, real-time transit data, and intelligent traffic control (Hin & Subramaniam, 2012) to reduce traffic congestion and boost overall effectiveness. Amsterdam's Edge, the so-called "smartest building in the world" (Araszkiewicz, 2017), exemplifies the potential of thoughtful building design in a world of architectural masterpieces; the building has cutting-edge sensors and data-driven algorithms that efficiently control energy usage, lighting, and indoor surroundings, significantly reducing energy consumption and improving user comfort and satisfaction (Komninos, 2011).

Masdar City, Abu Dhabi, is an example of sustainable development in the GCC. The city integrates renewable energy, cutting-edge waste management technology, and sustainable transportation, and creates a carbon-neutral and waste-free model (Reiche, 2010). Similar ideas and characteristics are demonstrated by Lusail City in Qatar, with the ideals of sustainability and energy efficiency, supported by modern technology, at the heart of its goal; using environmentally friendly building methods, cutting-edge smart water practices, and the promotion of sustainable transportation, Lusail embodies the future of urban development (Shaaban & Adalbi, 2021).

"Madinat Al Hareer" project in Kuwait is a pioneering endeavor, initially launched but temporarily halted before resuming. This city blends traditional and cutting-edge elements, poised to incorporate state-of-the-art sustainable building practices, energy-saving strategies, and automated transportation systems (Saprykina, 2018). The prioritization of renewable energy sources and development of an eco-friendly infrastructure are central to minimizing adverse environmental impacts. According to the specialized German market and consumer data website Statista, "Madinat Al Hareer" ranks fifth among the largest ongoing projects globally, with a total estimated value of $132 billion. This city is set to become home to the world's tallest building (Al-Anba Newspaer, 2023).

Governments both globally and in regions such as the GCC are actively pursuing intelligent and environmentally friendly infrastructure to align with United Nations SDGs. The overall aim is to create an urban environment that utilizes technology and allows for sustainable growth.

Energy Management and Efficiency

Smart Grid Implementation: Technologies and Strategies

The use of smart grids is inevitable as urban areas increasingly adopt the concept of smart cities. These grids guarantee effective, sustainable, and dependable energy systems by combining conventional power infrastructure with cutting-edge digital communication and control technologies (Fang et al., 2012). Smart grids encourage bidirectional connectivity, enabling real-time communication between electricity suppliers and customers, encouraging adaptive energy use. Advanced metering systems provide real-time consumption information, enabling residents and administrations to make informed decisions regarding energy usage. When connected to a city's digital infrastructure, this link optimizes energy distribution across various urban utilities, and improves monitoring and systematic enhancement of energy supply (Lee et al., 2023).

Smart grids allow urban planners and energy experts to estimate more precisely current and project future energy needs, match energy supply plans with urban needs, and identify ways to improve efficiency. The ability of smart grids to include renewable energy sources is particularly important for smart cities. A resilient and sustainable energy matrix is provided by integrating resources like solar panels and wind turbines into urban areas (Ahmed et al., 2022). Real-time data enables people to plan energy consumption and promote sustainability, and by reducing waste, provide a financial incentive to adopt smart city technologies (Quitzow & Rohde, 2021).

Japan's Energy Evolution: The Pivotal Role of Smart Grids

Japan's global leadership in delivering sustainable energy can be attributed to its steadfast commitment to smart grid development. A need for resilient and sustainable energy infrastructure became a priority following the earthquake and tsunami of 2011 (Joo, 2019). The Fujizawa Sustainable Smart Town, an example of urban development among many developed initiatives, combines modern smart grid technology, renewable energy integration, and energy-efficient architectural design (Kanyowa & Mahoso, 2018); by providing residents with real-time access to information regarding their energy consumption, they can make more-informed decisions, exert control over their energy usage, and realize significant energy savings.

Saudi Arabia's Urban Energy Renaissance

Smart grid technology has become increasingly important in Riyadh, the capital city of Saudi Arabia, with the integration of smart grid technologies, particularly the introduction of smart meters and demand response mechanisms (Salam & Khan, 2017). These meters and mechanisms facilitate the city's ability to streamline electricity distribution in real-time. By providing residents with real-time consumption data, these smart systems also incentivize energy-efficient behavior, promoting energy consciousness in Saudi Arabia's urban centers (Imam et al., 2019).

The Role of Renewable Energy Integration in Smart City Development

Renewable energy integration is essential if sustainability goals are to be achieved. The integration of renewable energy sources into existing energy systems contribute to the sustainability of technologies and help mitigate risks (Barykina et al., 2023). For instance, in areas prone to hurricanes or severe storms, traditional power grids are susceptible to damage and can experience widespread outages. However, when renewable energy sources such as solar panels and wind turbines are integrated into the energy system, they can generate electricity even during adverse weather conditions. This redundancy in the energy system contributes to ensuring that critical infrastructure (e.g., hospitals and emergency response centers) can remain operational, reduces the vulnerability of the energy supply to external factors, and enhances the overall resilience of the community. Renewable energy also creates opportunities for renewable energy deployment, overcoming existing obstacles, and supporting sustainable development (Li et al., 2022).

The integration of renewable energies into electrical networks is important because of increasing energy demand, depletion of fossil fuel reserves, and environmental impacts (Bencs et al., 2020). Strategies for sustainable development rely heavily on resources such as wind, solar and wave power, and biomass (e.g., wood, crop residues, agricultural waste, algae, and various organic materials) (Lund, 2007). Large-scale implementation plans for renewable energy must include strategies that integrate renewable sources into coherent energy systems, and consider energy savings and efficiency measures (Lund & Mathiesen, 2009). Additionally, assessing sustainability indicators for renewable energy technologies is needed to ensure their long-term viability and contribution to sustainability (Evans et al., 2009).

Integrating distributed power-generation systems into the grid is necessary for sustainable and resilient power system development, particularly with highly intermittent renewable energy resources (Blaabjerg et al., 2017). The nexus between tourism, renewable energy, income, and environmental quality is an important aspect of sustainability. Renewable energy consumption plays a significant role in enriching tourism, accelerating economic growth, and combating CO_2 emissions (Bano et al., 2021).

Challenges of Integrating Renewable Energy

The integration of renewable energy sources is not without obstacles, particularly with the potentially intermittent nature of energy generation from the wind and sun (Blaabjerg et al., 2017). The GCC, and Kuwait in particular, faces unique challenges because of extreme weather conditions. In Kuwait, high temperatures and sandstorms increase solar energy maintenance costs, requiring the frequent cleaning and maintenance of panels and systems to ensure optimal performance (Al-Dousari et al., 2019). These conditions require advanced management solutions to maintain grid stability (Riaz et al., 2021). Additionally, infrastructure adjustments can be costly and complex, and require collaboration among stakeholders, i.e., the government, citizens, and public–private partnerships (Barykina & Song, 2023). Despite the high initial investment, the long-term economic, environmental, and social benefits of renewable energy integration are significant (Riaz et al., 2021). Thus, a holistic approach that encompasses technology, policy, public engagement, and maintenance is essential for successful renewable energy adoption in regions such as Kuwait.

Benefits of Integrating Renewable Energy

Harnessing renewable energy is a holistic approach to sustainability. Such energy alternatives curtail greenhouse gas emissions that contribute to climate change, and substantially reduce harmful pollutants derived from fossil fuels (Güney, 2019). Transitioning to renewables also bolsters energy security by minimizing reliance on imported and finite fossil resources (Rani et al., 2022). Because these energy sources are both abundant and widespread, their use also fosters decentralized energy generation, rendering energy infrastructure less susceptible to outage (Güney, 2019).

Economically, the renewable sector could spur growth and generate jobs, especially roles in manufacturing, installation, and maintenance (Berényi & Deutsch, 2020). With technological advancements, renewable systems are also becoming increasingly affordable and competitive (Rani et al., 2022), facilitating the establishment of a robust and resilient energy network. This shift also underscores sustainable development by addressing its multifaceted nature – encompassing environmental, social, and economic considerations – and aligns with the broader United Nations SDGs involving the use of clean energy, health improvement, poverty eradication, and sustainable growth (Blanc, 2015).

Shagaya Renewable Energy Project: Kuwait's Sustainable Vision

Kuwait's Shagaya Renewable Energy Project is an example of the nation's visionary approach toward a sustainable future (Kuwait Institute for Scientific Research. 2019). Its location in western Kuwait leverages natural assets, namely constant sunshine and periodic strong winds. By combining solar panels, wind turbines, and state-of-the-art concentrated solar power systems, this energy project aims to produce a substantial portion of the nation's electricity demand from renewable sources (Hassaan et al., 2021). Concentrated solar power system technology allows energy storage in the form of heat, ensuring a stable power supply during periods of cloud cover and at night.

The use of multiple renewable technologies is an example of Kuwait's goals of achieving energy resilience and reducing its dependency on fossil fuels (Steensma et al., 2019). Kuwait is reducing its carbon footprint and mitigating the effects of climate change by transitioning to cleaner energy sources. The impact extends beyond environmental benefits in that it also provides economic benefits, drives innovation in green technologies, and potentially establishes Kuwait as a regional renewable energy leader (Majed et al., 2020). Moreover, the Shagaya Renewable Energy Project serves as an educational and research hub, promoting knowledge exchange and fostering innovation in renewable energy technologies, and functions as a blueprint for other nations regarding the benefits and feasibility of shifting toward renewable energy (Steensma et al., 2019).

Energy-Efficient Buildings and Infrastructure

Energy-efficient buildings and infrastructure play pivotal roles in bolstering sustainability and diminishing the environmental effects of the built environment. Researchers have delved into sustainable design methodologies, such as the incorporation of green infrastructure, nature-driven solutions, and renewable energy technologies (Anderson & Gough, 2021; Harrington & Hsu, 2018). These studies underscore the necessity of embedding energy-efficient components from the nascent phases of building design to enhance energy output and minimize environmental repercussions.

Another area of research has focused on evaluating and optimizing energy efficiency within buildings. Frameworks and methodologies have been developed to assess a building's energy performance, considering elements like daylight utilization, energy consumption, and the quality of the view (Ezema & Maha, 2022; Pilechiha et al., 2020). For example, "Leadership in Energy and Environmental Design," a widely recognized framework used to assess building sustainability and energy efficiency, evaluates aspects of building design and operation, such as energy efficiency, water usage, materials, and indoor environmental quality (Wu et al., 2017). The "Europe Building Research Establishment Environmental Assessment Method" is also widely used to evaluate the sustainability and environmental performance of buildings, particularly in the United Kingdom (Cordero et al., 2019). Because this methodology considers various factors such as energy usage, water consumption, and the ecological impact of the building, it provides a comprehensive assessment of a building's environmental performance and a rating system to indicate its level of sustainability. Approaches such as these steer decision-making protocols to meet elevated energy-efficiency benchmarks.

Green infrastructure, which encompasses urban green spaces and plant life, can bolster energy conservation by offering natural cooling, diminishing the urban heat island phenomenon, and refining air quality (Anderson & Gough, 2021; Vargas-Hernández & Zdunek-Wielgołaska, 2020). The infusion of green infrastructure into urban planning and design amplifies the resilience and sustainability of urban regions. These investigations have revealed the importance of robust policies and efficient governance to facilitate adoption of energy-efficient technologies (Anderson & Gough, 2021; Harrington & Hsu, 2018).

Energy Efficiency in Mushereib Smart City: A Sustainable Model

Doha's Mushereib Smart City, Qatar, blends modernity with sustainability, and is a classic example of how urban developments can harmoniously integrate energy efficiency and modern vibrancy. Transformation is rooted in the adoption of passive design strategies, incorporation of LED lighting, and integration of smart building automation systems (Al-Sulaiti 2021; Corchado et al., 2022). These features are reinforced by the city's commitment to renewable energy, with a specific focus on solar panel deployment. Smart grids are used to optimize energy distribution, coupled with district cooling systems to further streamline energy consumption. A comprehensive transportation network also promotes sustainable mobility, encompassing pedestrian pathways, cycling lanes, and efficient public transport.

Conclusions

While the concept of the smart city has evolved, and no universally accepted definition of such a city exists, their design is driven by a desire to enhance the quality of life for urban dwellers. These cities leverage smart computing technologies to create more intelligent, interconnected, and efficient infrastructure and services. However, the smart city concept extends beyond application of technology, in that it also encompasses governance, urban planning, and most importantly, sustainability. Sustainable infrastructure, energy efficiency, and renewable energy integration are all important components of cities that are technologically advanced, environmentally responsible, and resilient. The integration of renewable energy sources, such as solar and wind power, into smart grids ensures reliable and sustainable energy supplies, even in adverse conditions.

Japan has shown remarkable leadership in sustainable energy through its commitment to smart grid development, with projects like the Fujizawa Sustainable Smart Town exemplifying the successful integration of modern smart grid technology, renewable energy, and energy-efficient design. Kuwait has similarly made significant progress, with the Shagaya Renewable Energy Project being an example of its commitment to integrating renewable energies – the incorporation of which positions Kuwait as a proactive contributor to global climate change mitigation efforts. By combining modernity with energy efficiency through passive design, LED lighting, and smart building automation, Mushereib Smart City is also a model of sustainable urban development, blending modern vibrancy with sustainable practices.

Smart cities are at the forefront of global efforts to create urban environments that are technologically advanced and sustainable, resilient, that align with United Nations SDGs. Such cities are proof that sustainable practices and technology can be integrated, paving the way for a more prosperous and environmentally responsible future.

References

Ahmed, Z., Ahmad, M., Rjoub, H., Kalugina, O. A., & Hussain, N. (2022). Economic growth, renewable energy consumption, and ecological footprint: Exploring the role of environmental regulations and democracy in sustainable development. *Sustainable Development*, *30*(4), 595–605.

Al-Anba. (2023, July 20). *Madinat Al Hareer Among the World's 10 Largest Projects*. Al-Anba Newspaper. Retrieved from https://www.alanba.com.kw/1178900

Alhammadi, S., 2024. Islamic finance as a driver for enhancing economic sustainability and innovation in the GCC. *Journal of Science and Technology Policy Management.*

Alkhaldi, A., Alrashidi, H., Alhasan, K., Alsadeeqi, A. and Alshami, A. (2023). The use of blockchain technology to build smart cities: creating public value in Kuwait. *Global*

Knowledge, Memory and Communication, Vol. ahead-of-print No. ahead-of-print. https://doi.org/10.1108/GKMC-11-2022-0263

Alkhaldi, A., Malik, S. and Lytras, M.D. (2024). Transformative leadership in Kuwait Direct Investment Promotion Authority: Investing in Talent, Innovation and the Next Generation. In S. Baroudi, and M. D., Lytras, (Eds.) *Transformative Leadership and Sustainable Innovation in Education: Interdisciplinary Perspectives* (Emerald Studies in Higher Education, Innovation and Technology, pp. 27–38). Emerald Publishing Limited.

Al-Sulaiti, K. I., Abaalzamat, K. H., Khawaldah, H., & Alzboun, N. (2021). Evaluation of Katara Cultural Village events and services: a visitors' perspective. Event Management, 25(6), 653–664.

Al-Dousari, A., Al-Nassar, W., Al-Hemoud, A., Alsaleh, A., Ramadan, A., Al-Dousari, N., & Ahmed, M. (2019). Solar and wind energy: Challenges and solutions in desert regions. *Energy*, *176*, 184–194. https://doi.org/10.1016/j.energy.2019.03.180

Anderson, V., & Gough, W. (2021). Nature-based cooling potential: A multi-type green infrastructure evaluation in Toronto, Ontario, Canada. *International Journal of Biometeorology*, *66*(2), 397–410. https://doi.org/10.1007/s00484-021-02100-5

Araszkiewicz, K. (2017). Digital technologies in facility management – The state of practice and research challenges. *Procedia Engineering*, *196*, 1034–1042.

Barykina, Y. N., & Shao, Y. (2023, July). Integration of renewable energy sources into energy systems. *IOP Conference Series: Earth and Environmental Science*, *1212*(1), 012018.

Bano, S., Alam, M., Khan, A., & Liu, L. (2021). The nexus of tourism, renewable energy, income, and environmental quality: An empirical analysis of Pakistan. *Environment Development and Sustainability*, *23*(10), 14854–14877. https://doi.org/10.1007/s10668-021-01275-6

Bencs, P., Alktranee, M., & Mészáros, K. (2020). Effects of solar panels on electrical networks. *Analecta Technica Szegedinensia*, *14*(1), 50–60. https://doi.org/10.14232/analecta.2020.1.50-60

Berényi, L., & Deutsch, N. (2020). Assessment of the future role of some energy sources by higher education students in Hungary. *WSEAS Transactions on Environment and Development*, *16*, 689–698.

Bibri, S. E., & Krogstie, J. (2017). On the social shaping dimensions of smart sustainable cities: A study in science, technology, and society. *Sustainable Cities and Society*, *29*, 219-246.

Blaabjerg, F., Yang, Y., Yang, D., & Wang, X. (2017). Distributed power-generation systems and protection. *Proceedings of the IEEE*, *105*(7), 1311–1331. https://doi.org/10.1109/jproc.2017.2696878

Blanc, D. (2015). Towards integration at last? The sustainable development goals as a network of targets. *Sustainable Development*, *23*(3), 176–187.

Caragliu, A., Del Bo, C., & Nijkamp, P. (2011). Smart cities in Europe. *Journal of urban technology*, *18*(2), 65–82.

Corchado, J. M., Pinto-Santos, F., Aghmou, O., & Trabelsi, S. (2022). Intelligent development of smart cities: Deepint. net case studies. *Sustainable Smart Cities and Territories* (pp. 211–225). Springer International Publishing.

Cordero, A. S., Melgar, S. J. G., & Andújar, J. M. (2019). Green building rating systems and the new framework level(s): A critical review of sustainability certification within Europe. *Energies*, *13*(1), 66. https://doi.org/10.3390/en13010066

Fang, X., Misra, S., Xue, G., & Yang, D. (2012). Managing smart grid information in the cloud: Opportunities, model, and applications. *IEEE Network*, *26*(4), 32–38.

Evans, A., Strezov, V., & Evans, T. (2009). Assessment of sustainability indicators for renewable energy technologies. *Renewable and Sustainable Energy Reviews*, *13*(5), 1082–1088. https://doi.org/10.1016/j.rser.2008.03.008

Ezema, I., & Maha, S. (2022). Energy efficiency in high-rise office buildings: An appraisal of its adoption in Lagos, Nigeria. *IOP Conference Series Earth and Environmental Science, 1054*(1), 012037. https://doi.org/10.1088/1755-1315/1054/1/012037

Gough, W. A., & Anderson, V. (2022). *Changing Air Quality and the Ozone Weekend Effect during the COVID-19 Pandemic in Toronto*, Ontario, Canada. Climate 2022, 10, 41.

Güney, T. (2019). Renewable energy, non-renewable energy and sustainable development. *International Journal of Sustainable Development & World Ecology, 26*(5), 389–397.

Harrington, E., & Hsu, D. (2018). Roles for government and other sectors in the governance of green infrastructure in the U.S. *Environmental Science & Policy, 88*, 104–115. https://doi.org/10.1016/j.envsci.2018.06.003

Hassaan, M., Hassan, A., & Al-Dashti, H. (2021). GIS-based suitability analysis for siting solar power plants in Kuwait. *The Egyptian Journal of Remote Sensing and Space Science, 24*(3), 453–461. https://doi.org/10.1016/j.ejrs.2020.11.004

Hin, L. T. W., & Subramaniam, R. (2012). Creating smart cities with intelligent transportation solutions: Experiences from Singapore. In O. Y. Ercoskun (Ed.), *Green and ecological technologies for urban planning: Creating smart cities* (pp. 174–190). IGI Global.

Imam, A. A., Al-Turki, Y., & R., S. K. (2019). Techno-economic feasibility assessment of grid-connected PV systems for residential buildings in Saudi Arabia – A case study. *Sustainability, 12*(1), 262. https://doi.org/10.3390/su12010262

Joo, J. (2019). Infusion process of smart grid-related technology based on coping theory. *Sustainability, 11*(12), 3445. https://doi.org/10.3390/su11123445

Kanyowa, T., & Mahoso, R. (2018). Assessing the feasibility of smart grid technology on electrical distribution grids. *EAI Endorsed Transactions on Smart Cities, 166008*. https://doi.org/10.4108/eai.18-8-2020.166008

Khan, H., Malik, M., Zafar, R., Goni, F., Chofreh, A., Klemeš, J., & Alotaibi, Y. (2020). Challenges for sustainable smart city development: A conceptual framework. *Sustainable Development, 28*(5), 1507–1518. https://doi.org/10.1002/sd.2090

Komninos, N. (2011). Intelligent cities: Variable geometries of spatial intelligence. *Intelligent Buildings International, 3*(3), 172–188.

Lee, H., Chung, B., & Huh, S. (2023). Consumer Preferences for Smart Energy Services Based on AMI Data in the Power Sector. Energies. https://doi.org/10.3390/en16093961

Kuwait Institute for Scientific Research. (2019). *The Shagaya Solar Panels Project*. Retrieved October 7, 2023, from https://www.kisr.edu.kw/ar/gi/2/details/

Li, M., Lu, Y., & Xu, X. (2022). Mapping the scientific structure and evolution of renewable energy for sustainable development. *Environmental Science and Pollution Research, 29*(43), 64832–64845. https://doi.org/10.1007/s11356-022-20361-4

Lund, H. (2007). Renewable energy strategies for sustainable development. *Energy, 32*(6), 912–919. https://doi.org/10.1016/j.energy.2006.10.017

Lund, H., & Mathiesen, B. (2009). Energy system analysis of 100% renewable energy systems – the case of Denmark in years 2030 and 2050. *Energy, 34*(5), 524–531. https://doi.org/10.1016/j.energy.2008.04.003

Majed, A. R., Gueymard, C. A., Al-Khayat, M., Ismail, A., Lee, J. A., & Al-Duaj, H. (2020). Performance evaluation of a utility-scale dual-technology photovoltaic power plant at the Shagaya Renewable Energy Park in Kuwait. *Renewable and Sustainable Energy Reviews, 133*, 110139.

Neirotti, P., De Marco, A., Cagliano, A. C., Mangano, G., & Scorrano, F. (2014). Current trends in Smart City initiatives: Some stylised facts. *Cities, 38*, 25–36.

Nagasawa, K., Upshaw, C. R., Rhodes, J. D., Holcomb, C. L., Walling, D. A., & Webber, M. E. (2012, July). Data management for a large-scale smart grid demonstration project in Austin, Texas. In *Energy sustainability* (Vol. 44816, pp. 1027–1031). American Society of Mechanical Engineers.

Ohtsuki, T., & Ohtsuki, T. (2017). Deep learning the quantum phase transitions in random electron systems: Applications to three dimensions. *Journal of the Physical Society of Japan*, *86*(4), 044708.

Owunna, T. A., & Obeagu, E. I. (2022) Overview of smart grid technology as a renewable energy source. *Journal of Energy Research and Reviews*, 6–15. https://doi.org/10.9734/jenrr/2022/v12i3239

Parubochaya, E. F., Piskunov, N. V., & Drinova, E. M. (2020, March). The "Smart Cities" concept in the European Union and the Russian Federation: from project to practical implementation. In *Institute of Scientific Communications Conference* (pp. 976–986). Springer International Publishing, Cham.

Pilechiha, P., Mahdavinejad, M., Rahimian, F., Carnemolla, P., & Seyedzadeh, S. (2020). Multi-objective optimisation framework for designing office windows: Quality of view, daylight, and energy efficiency. *Applied Energy*, *261*, 114356. https://doi.org/10.1016/j.apenergy.2019.114356

Quitzow, L., & Rohde, F. (2022). Imagining the smart city through smart grids? Urban energy futures between technological experimentation and the imagined low-carbon city. *Urban Studies*, *59*(2), 341–359.

Prevelianaki, K., Sherratt, F., & Henjewele, C. (2022). ISO standards or global indices: Who decides if a city is smart? *IOP Conference Series Earth and Environmental Science*, *1101*(2), 022045. https://doi.org/10.1088/1755-1315/1101/2/022045

Rani, P., Arora, V., & Sharma, N. (2022). An overview of integration of PV system into electric grid. *IOP Conference Series Materials Science and Engineering*, *1228*(1), 012017.

Reiche, D. (2010). Renewable energy policies in the Gulf countries: A case study of the carbon neutral "Masdar City" in Abu Dhabi. *Energy Policy*, *38*(1), 378–382.

Riaz, M., Hanif, A., Hussain, S., Ali, M., & Zafar, A. (2021). An optimization-based strategy for solving optimal power flow problems in a power system integrated with stochastic solar and wind power energy. *Applied Sciences*, *11*(15), 6883. https://doi.org/10.3390/app11156883

Salam, M. A., & Khan, S. A. (2017). Transition towards sustainable energy production – A review of the progress for solar energy in Saudi Arabia. *Energy Exploration &Amp; Exploitation*, *36*(1), 3–27. https://doi.org/10.1177/0144598717737442

Saprykina, N. (2018, June). Forecasting technology as a method of modeling and building smart city concept. *In IOP conference series: Materials science and engineering* (Vol. 365, No. 2, p. 022068). IOP Publishing.

Sitna, H. H., Assaf, A., & Asyarif, L. L. (2021). The level of achievement for assessment of smart city in smart technology: A literature review. In *E3S Web of Conferences* (Vol. 328, p. 04013). EDP Sciences.

Shaaban, K., & Adalbi, M. A. (2021, June). Smart city transportation system in developing countries: The case of Lusail City, Qatar. In T. Ahram, W. Karwowski, & D. Schmorrow (Eds.), *International conference on applied human factors and ergonomics* (pp. 445–452). Springer International Publishing.

Steensma, G., Román, R., Marshall, C., Bermejo, J., Iyer, K., Al-Hajraf, S., & Al-Qattan, A. (2019, July). Shagaya renewable energy park project. In *AIP conference proceedings* (Vol. 2126, No. 1). AIP Publishing.

Vargas-Hernández, J., & Zdunek-Wielgołaska, J. (2020). Urban green infrastructure as a tool for controlling the resilience of urban sprawl. *Environment Development and Sustainability*, *23*(2), 1335–1354. https://doi.org/10.1007/s10668-020-00623-2

Wu, P., Song, Y., Wang, J., Wang, X., Zhao, X., & He, Q. (2017). Regional variations of credits obtained by LEED 2009 certified green buildings – A country level analysis. *Sustainability*, *10*(2), 20. https://doi.org/10.3390/su10010020

Chapter 15

Innovating Sustainable Urban Development: Evaluating the Social Acceptance of Residential Solar Integration for the Prosperity of Abu Dhabi

Hamed M. Hussain[a] *and Khalil Rahi*[b]

[a]*Petroleum and Chemical Engineering, United Arab Emirates University, Al Ain City, United Arab Emirates*
[b]*Abu Dhabi University, Dubai, United Arab Emirates*

Abstract

This study examines the social acceptance and potential for residential solar integration in Abu Dhabi to support sustainable urban innovation and prosperity. A survey of 216 Abu Dhabi residents analyzed four scenarios for incentivizing household solar panel adoption. Results demonstrate high interest and preference for governmental financial support mechanisms to spur solar integration. This research contributes an assessment framework and citizen engagement insights that could inform smart city development strategies focused on renewable energy transformation across the Gulf region. As urban centers like Abu Dhabi strive to enact their visions for economic diversification and environmental sustainability, understanding citizen perspectives on innovations like distributed solar will be critical to align technology advancement with societal needs and values. The methodology and findings of this Abu Dhabi case study provide a model to catalyze stakeholder buy-in for new energy solutions, with potential applications for other Gulf cities aiming to transition their energy systems and build integrative renewables infrastructure to power next-generation smart development. This solar integration acceptance

The Emerald Handbook of Smart Cities in the Gulf Region:
Innovation, Development, Transformation, and Prosperity for Vision 2040, 251–287

doi:10.1108/978-1-83608-292-720241015

research lays the groundwork for continued scholarly discourse and policymaker collaboration around smart cities in the Gulf that realize multidimensional progress for society, economy, and environment.

Keywords: Social acceptance; solar integration; sustainable urban innovation; smart city; Gulf region

1. Introduction

The fast-paced urbanization of Arabian Gulf countries, such as the UAE, has ushered in a new era of energy demand, placing substantial pressure on conventional energy resources. Smart cities in this region are increasingly leaning on renewable energy sources, with solar power emerging as a particularly viable option due to the area's abundant sunlight.

The integration of Artificial Intelligence (AI) into these solar power systems has unlocked significant advantages. AI algorithms optimize the energy harvest by predicting solar radiation, automating the angle of solar panels to the sun's trajectory, and ensuring regular maintenance through predictive analytics (Abdullah & Jeanty, 2011). This digital intelligence not only boosts energy yield but also ensures a consistent power supply, reducing the reliance on grid electricity and minimizing interruptions. Recent studies highlighted the efficacy of AI algorithms in forecasting solar radiation in the UAE using various meteorological inputs, providing a blueprint for enhancing solar power harnessing (Abu Dhabi Distribution Company (ADDC), 2019). Furthermore, it elucidated the transformational potential of AI in enhancing the sustainability quotient of smart cities in the Gulf region, with particular emphasis on solar power optimization (Abdullah & Jeanty, 2011).

As Arabian Gulf nations prioritize sustainability in their vision statements and national strategies, the convergence of AI and solar technology stands as an exemplar of innovation meeting necessity, with the potential to redefine the energy landscape of the region. Abu Dhabi is well known for its scorching sun and hot climate, which is naturally rich in solar energy, making it an interesting environment for green energy if all suitable conditions are available. The same applies to other emirates in the UAE. For instance, the city of Dubai generates 44% of its electricity from renewable energy, which is almost half of its consumption, meaning that the gas consumption that used to take place in powering the rest of the 66% of city power has become obsolete, in hopes of saving power and saving the environment (Abuzaid et al., 2022).

However, the current situation in Abu Dhabi city proves that there is a high dependency on natural gas combustion to generate power, which can reach up to 80% of the power generated in the capital city (Abuzaid et al., 2022). Therefore, based on this data, opportunities must be seized for the betterment of the renewable energy field, and particularly solar power projects, as they can majorly reduce the consumption of fuel when generating electricity through photovoltaic solar panels. Solar power is among one of the most widely used renewable energy resources (RES) to generate power in an eco-friendly process. Furthermore, the behavioral nature of local homeowners must be taken into consideration as it is analyzed in this research study. Hence, it will be put into an analysis after using various methods to evaluate the development of solar power resources.

It is important to mention that the distribution of solar panels should be certified by the water and electricity authorities, which is done by Abu Dhabi Distribution Company (ADDC), to eliminate possible future risks of quality and durability in the photovoltaic panels' cells (Abuzaid et al., 2022). The current standardized amount of electricity usage as per ADDC is 400 kWh daily (based on the local tariff), which is considered to be eco-friendly usage, and any consumption above 400 kWh is considered with a higher tariff to ensure the importance of using electricity efficiently. Moreover, a small number of solar panels can generate roughly 10%–20% of this electricity, resulting in a reduction of gas-backed electricity consumption to 80%–90% (ADDC, 2020).

The main basis for this study is the hypothesis that we can generate electricity through photovoltaic solar panels on private premises. The main assessment will be made on several possible scenarios and formulas to generate solar power in residential houses, which will make this project feasible in the city of Abu Dhabi, as it is the case for many solar power projects worldwide (Afework et al., 2021). Thus, the research targets numerous possibilities to enhance the use of solar power consumption in the city of Abu Dhabi, which can be achieved by different scenarios, such as allowing private solar panel distribution to villas and townhouses throughout Abu Dhabi.

The study will focus on the utilization of sustainable energy resources in the city, along with introducing homeowners to the technology of photovoltaic solar panels. The statistical status in Abu Dhabi proves that there is a high dependency on natural gas combustion to generate power, which can reach up to 80% of the power generated in the capital city. Hence, increased utilization of solar power in the city will allow for the savior of energy and the environment by reducing fossil fuel combustion. The major drawbacks of excessive fossil fuel use are further depicted in the examples used during the study. Nonetheless, such project cannot be validated without understanding all the prominent aspects in the renewable energies field (Agterbosch et al., 2009). Additionally, the discussion of renewable energy and its applicability in the city of Abu Dhabi is thoroughly investigated.

The objective is to clarify the most possible and applicable scenarios where solar energy can be utilized accordingly in the city of Abu Dhabi. Also, the research sheds light on electricity generation and energy consumption risks that might be challenging, such as the possible degradation modes that can take place on the solar panel cells when being directly affected by constant natural incidents such as humidity (Al Dhafra Solar Project, Abu Dhabi, 2020). Then solar power is discussed as a type of renewable energy, along with the current social acceptance and preference for solar panel installation and implantation by homeowners. The offered formula in the survey resembles the potential that Abu Dhabi city has to follow the path of other cities which are adapting solar power projects, as it is the case in many countries worldwide, that aim to adapt micro-grid systems demonstrated in hybrid models for homeowners (Alba Energy, 2018). Similar projects are adapted in India and in Sweden in the form of self-sufficient systems too, which can function dependently on the solar power generated by solar panels only and stored in batteries. Consequently, the current tariff details in the UAE (specifically through ADDC) are explained in relation to the survey questions and statistical data that have been proven along with the references to solidify and support the main research study objectives evidently.

In brief, the smart cities concept entails the integration of advanced technologies and innovative strategies to optimize urban operations and enhance the quality of life for residents. In the Gulf Region, where rapid urbanization and environmental concerns pose significant challenges, smart city initiatives are crucial for sustainable development. The summarized research is intricately linked to the smart cities paradigm in the Gulf Region. By focusing on the implementation of solar panels in residential buildings in Abu Dhabi City, the research addresses key aspects of smart cities such as energy efficiency, sustainability, and environmental resilience. Moreover, solar energy utilization not only reduces dependence on fossil fuels but also mitigates carbon emissions, aligning with the overarching goals of smart cities in promoting sustainable and eco-friendly urban environments. Furthermore, by tailoring solutions to the specific context of Abu Dhabi City, this research contributes directly to the advancement of smart city initiatives in the Gulf Region, fostering innovation and progress toward a more sustainable future.

2. Literature Review

2.1. Electricity Generation Challenges

2.1.1. Main Challenges

Electrical power is a practical necessity for the economic success of society. The globe confronts severe energy issues at the start of the 21st century. The notion of sustainable development has emerged for a livable future in which human needs are addressed while nature is kept in balance. In the beginning, electricity generation methods, such as direct current systems, were first known in the late 19th century. Consequently, smaller electricity generation methods were marketed for international enterprises for industrial and mining usage, both in metropolitan areas and in distant underdeveloped areas. The major attempts to electrify the planet were led by Thomas Edison and Werner von Siemens. Although this invention energized industry and only a few districts, it failed to cover all the inhabitants.

Hence, it was evident that in order to convert an idea to provide enough electricity that covers entire towns into reality, the electricity would have to be generated in one location (such as a huge river with significant hydropower potential) and transmitted to the city (Albattah & Attoye, 2021). Originally, when rural families lacked access to electricity, it was generated at far-off hydroelectric plants or by the combustion of fossil fuels in populated areas, supplying it to local structures and then use the leftover heat to create steam to heat those same structures. As coal plants grew over time, pollution pressure forced them to be located away from population areas. Many kilometers of transmission wires transmitted power with a 10%–15% loss. (Alkahtani et al., 2023). The heat was evacuated since it is impractical to convey waste heat across large distances. The wasted heat was a trade-off for cleaner air in cities because there was no practical technology for clean, local generation. The power industry eventually built a sizable system and put new generators on remote sites far from its clients. Every plant was individually created and constructed on-site, resulting in cost savings. Building big plants costs less per unit of generation than building smaller ones (Albattah & Attoye, 2021).

From 1910 to 1960, these conditions prevailed, and everyone in the energy industry and the federal government began to think that distant, controlled generation was perfect and that it would generate electricity for the least amount of money compared to alternative options (Alkahtani et al., 2023).

2.1.2. Financial Struggles

The main source of power that puts electricity into existence is gas, as it is considered the most used energy source to generate electricity worldwide. Hence, gas is as important as water and food because without gas generating electricity, we would not be able to perform our daily life activities such as turning the light in the room or charging our mobile phones or using the stove, so basically gas has been reshaped in theory from being a fossil fuel type that is created after the decomposition of plants and animals, into a fundamental basis in our civilization and a necessity to all human beings one earth (Alnaser & Alnaser, 2019).

Everything around us operates through an electrical device or partially electrical device, and that leads us to investigate the current state of electricity generation and if there are any risks of losing it. Electricity is generated by burning natural gas after being extracted from underground in offshore and onshore locations (Alwadei et al., 2022). The struggle that needs to be addressed is the linear increase in gas prices over the last decade worldwide (Andrei & Sammarco, 2017), consequently raising concerns about the possible risks that may come into existence in the near future if gas prices were to increase further without setting a limit (Alnaser & Alnaser, 2019). Gas prices control the permitted consumption and electricity tariffs in all countries, increased prices might not affect organizations, but it will affect individuals, as some might face the challenge of keeping up with the increasing utilities bill that with time will be challenging to low- and middle-class households and eventually to high class as well if it were to increase nonstop. Another aspect that affects gas prices badly is political events and wars, which can lead to a disruption state in the supply and demand ratio of gas, forming many predictable bad scenarios for gas prices and usage afterwards, subsequently altering the chosen default options to generate electricity (Alnaser & Alnaser, 2019). In simpler terms, if gas is expensive, it is more likely for people to use it less, and the scale can narrowly affect individuals and widely affect countries.

2.1.3. Climate Changes

Fossil fuels currently meet most of the world's energy needs. But the great weight of the scientific data points to the unrestrained usage of fossil resources that is creating global climate change with potentially disastrous consequences. Until the 20th century, everybody in the electricity sector and the authorities assumed that distant, pivotal electricity generation was ideal and that it delivered energy at the cheapest prices compared to other options (Albattah & Attoye, 2021). They are susceptible to supply chain disruptions due to their high level of integration. This supply paradigm is becoming less appealing, especially when it comes to power. In addition to their fragility, centralized energy supply systems are becoming less popular because of a variety of bothersome issues, like the exhaustion of fossil fuels and the impact they have on climate change, as well as the insecurity of the infrastructure for energy transportation, investors' desire to reduce risks, and the use of more compact, modular generation and transmission systems (Albattah & Attoye, 2021). Therefore, there has been an obvious increase in fuel prices internationally due to the realistic fears of fuel depletion in the future. In addition, the constant increase in the population worldwide overloads the demand ratio over the supply for energy generation as well as the occurrence of unnatural global events such as global warming, which is majorly caused by large releases of carbon emissions along with other air pollutants that interact with the atmosphere (Bakos, 2009).

2.1.4. Need for Sustainability

Electricity is mainly generated by the combustion of natural gases, which can generate electric power. Nonetheless, as it is a global fact, gas combustion has resulted in many environmental crises, such as global warming and an increased amount of carbon emissions (Bakos, 2009). Here comes the eco-friendly initiative of green energy, which is meant to have power from natural elements rather than one-time-use sources like gas and petrol (Bazmi & Zahedi, 2011). Sustainability has been applied in many sectors in our daily lives as it has many forms and functions. Sustainable energy can sometimes refer to green energy. An example of sustainable energy can be solar power and windmills. However, these are defined as a capacity to withstand in a continuous manner throughout many sectors of life (Borchers et al., 2007). Hence, making sustainable energy different from renewable energy, such as crop planting, where renewable energy is always renewed, sustainability is always sourced from natural elements such as the sun (Casten & Downes, 2005). By transforming the forces of nature into usable forms of energy, renewable energy sources offer marketable energy. The power of the sun, its direct and indirect effects on the globe (such as solar radiation, wind, falling water, and a diversity of plants, such as biomass), gravitational forces (such as tides), and the heat from the earth's core are all factors that are used to power these devices (geothermal). These resources offer a huge amount of energy potential, but they are frequently scattered, not always accessible, many of them intermittent, and there is a lot of geographical variability. At the present time, great development is being achieved in terms of boosting collecting and conversion efficiency, cutting operation expenses, and enhancing dependability. Over the last two decades, there has been a worldwide effort to create and conduct research in the area of renewable energy sources and systems utilizing modeling, optimization, and simulation technologies (Albattah & Attoye, 2021).

2.1.5. Self-sufficient Houses

A model-based study was carried out for solar panel installations in Swedish households using real-world household energy consumption data from 2,104 single-family dwellings (Chalvatzis & Hooper, 2009). Furthermore, to investigate the extent to which a battery could help to increase photovoltaic electricity self-consumption, which is the amount of photovoltaic generated electricity that is consumed in-house, the study aimed at self-sufficiency, which is the fraction of electricity used by the household that is not bought from the grid. The study aimed to optimize self-sufficiency, and as a result, photovoltaic power self-consumption, for 122 alternative combinations of solar panel capacities and battery capacities for the 2,104 households. A battery with a relative battery capacity of 2.5–4.0 kWh (defined as the battery energy storage capacity in kWh divided by the expected annual electricity output of the solar panels in MW/h) can increase the PV electricity self-consumption of Swedish households by a practical maximum of 18–48 percentage points. As an example, a family home with a yearly power use of 20 MW/h and a photovoltaic system of 7 kW would need batteries of a large capacity, perhaps in the range of 15–24 kWh, in order to accommodate this difference in relative battery capacity, that is actual usable capacity as shown in Table 1 (Chalvatzis & Hooper, 2009). It is fair to expect that the highest capacity of batteries that may be put into Swedish households to increase self-consumption and self-sufficiency will fall somewhere in the range described above, hence a relative battery capacity of 2.5–4 kWh.

2.2. Social Acceptance

2.2.1. Ecological Behavior

In democratic communities, the problem is that rational individual behavior may not be consistent with the common good and limit the effective use of public

Table 1. Self-Sufficient Households Study (Hussain et al., 2022).

Solar Panels (7 kW power)	Max. Impact
Self-consumption with battery	Up to 50%
Self-sufficiency with battery	Up to 30%
Yearly average consumption	20 MWh
Average Battery Capacity	15–24 kWh

resources. Expert opinions as well as public sentiment and perspectives are considered as inputs to the planning and decision-making process. (Chang et al., 2003; Climate Council, 2022). Hardin notably pointed this out in his landmark book, *The Prisoner's Dilemma*, a game-theoretic model, is used in Tragedy of the Commons (1968), and it illustrates how the equilibrium solution, which is preferred by individual players, varies from the cooperative solution, which is optimal for society (De Benedetti et al., 2018). However, if emotions or opinions, for example, perspective, did not influence conduct, that can be irrelevant. As a result, an intriguing issue emerges those merits more investigation: Is conduct related to attitude?

Paravantis (Del Río & Burguillo, 2009) provides a concise summary of the linked literature. The term "attitude" refers to a person's evaluative opinion of a certain notion or item (Diakoulaki & Karangelis, 2007). The beliefs, values, and actions that a person has about environmental concerns shape their mental attitude (Divya et al., 2023). It is difficult to define environmental attitudes precisely, making it challenge to be utilized to forecast ecologically sociable behavior (Elrahmani et al., 2021). According to the concepts of deliberate action and behavior, behavioral intention comes before explicit behavior (Eneh, 2011; Evans et al., 2009). Environmental knowledge, environmental values, and intended ecological conduct are all considered to be components of the complex construct known as environmental attitude. For predicting ecological behavior, the theory of planned behavior emphasizes aspects of conduct that are beyond willed control (Eveloy & Gebreegziabher, 2019). There is a high association between environmental attitude and desire to engage in ecological behavior.

According to the idea of planned behavior, there is a relationship between environmental values and ecological behavior, which is most likely mediated by a third component called ecological behavior intention (Gambone, 2019). To summarize, people progress along a continuum of environmental knowledge competency, with phases including awareness, concern, comprehension, and action (Del Río & Burguillo, 2009; Grilli, 2017). Yes, behaviors do, in the end, follow attitudinal views. Consequently, it is established that it is critical to understand electricity users' views since their attitudes serve as the foundation for their subsequent behavior (Gwandu & Creasey, 1995). Three particular criteria appear to underpin public behavior: public knowledge, public views, and anxiety, as an example: worry which worsens with unawareness (Chang et al., 2003; Herath & Mittal, 2022; Hussain et al., 2017). These should be looked at further. Let us turn the focus to the public's perceptions regarding RES. To be sure, local people, regional governments, financial institutions, nongovernmental organizations (NGOs), and localized communication systems appear to be participating in RES initiatives (Hussain et al., 2022).

Despite their disparate opinions and competing interests, these players must find ways to collaborate. Cohesion, the removal of personal interests, and having accurate information and being represented, as in the case of including all parties in the decision-making procedure, are all requirements for collaboration (Hussain et al., 2017; 2022). Participants from the private and governmental sectors are encouraged to develop commercially viable and socially acceptable sustainable technology. According to Assefa and Frostell, social impacts have an impact on the entire society as well as on each individual member. They are followed by public acceptance, which is a critical element of sustainability. The NIMBY (Not in My Back Yard) phenomenon, on the other hand, refers to opposition to a project that is unpopular with the neighborhood (Hussain et al., 2017; International Trade Administration, 2020; James, 2015), and it can lead to social strife and economic losses (Climate Council, 2022).

The opposing phenomena is known as PIMBY (Please in My Back Yard), and it is thought to happen when a plan is perceived as useful (e.g., as a source of money) and favorably appreciated by similar societies (Climate Council, 2022; Kaaya et al., 2019). One critical factor that should be discussed is the current social acceptance for green energy utilities in Abu Dhabi. In previous years, there have been events and initiatives in the city of Abu Dhabi held by private and public establishments to introduce green energy to companies and governments as a futuristic substitute for the energy that is generated from fossil fuels (ADDC, 2020). In return, solar acceptance has been fairly accepted by people in Abu Dhabi through public events such World Future Energy Summit (WFES) which is held on a yearly basis in the city, it focuses on new technologies and methods to generate energy that is efficient and eco-friendly. Such events have many participating companies and visitors as well as potential investors interested in green energy (ADDC, 2020).

2.2.2. Interest Versus Awareness

According to one study that is investigating thoroughly the awareness level of clean energy, such as solar energy, using photovoltaic panels, demonstrates the high awareness among residents of the UAE to further utilize clean energy means. However, despite the high awareness highlighted in the study, further promotion is required to increase the interest in solar energy (Kazerani & Tehrani, 2020). While awareness is thought to be high, social interest is considered quite low among residents, which must be motivated to achieve a sustainable city. In sequence with the 2050 vision of the UAE, the study sheds light on the aim to reduce fossil fuel consumption and to successfully achieve 50% clean energy utilization by 2050 (Kazerani & Tehrani, 2020). Findings of the study imply that motivating interest can play a critical factor for residents to increase their utilization of photovoltaic panels.

As per the investigation, additional utilization of solar energy can lead to a 73% decline in the current reliance on electricity grid consumption. Nevertheless, further elaboration is made on the lack of awareness among developers and electrical service providers, which subsequently influences decision makers negatively in choosing clean energy sources. The study also indicates that promoting clean energy means to the public may encourage the willingness of residents to buy and install photovoltaic solar panels. The study concentrates on the level of acceptance in building integrated photovoltaics, because it is believed to have the potential to affect the design stage in new buildings for investors and owners. In summary, the study discovered that the acceptance of solar power integration with buildings is directly associated with public awareness of green energy methods and public interest as well, indicating that the relationship addressed is to be proportional in terms of influence on UAE residents (Kazerani & Tehrani, 2020). Green energy is more likely to fascinate locals when public awareness is constantly promoted among the public to achieve renewable energy sources for new buildings.

2.2.3. Current Electricity Tariff

Founded in 2005, ADDC is a subsidiary of TAQA Group, which was established in 2005. It was mainly known as Abu Dhabi National Energy Company (ADDC, 2020). ADDC is regularly known to be semi-governmental entity that is entitled to distribute water and electricity services in three categorized locations; Abu Dhabi Island (which is the capital city), Eastern Abu Dhabi (Baniyas, Shahamah, and Mussafah), and Western Abu Dhabi (Liwa and Silla which is also known as Western region of the UAE). The current solar energy initiative that is proposed sheds light on the pre-approval of solar panel installations that can be done at houses. Similarly, a new initiative has been launched by ADDC that plans to educate tenants and owners to use less water and electricity by reducing excessive and unrequired consumption (ADDC, 2020). The initiative is called "Tarsheed," which is an Arabic word that means to use less and efficiently. It aims to reduce the total energy consumption in the city of Abu Dhabi to sustain the natural resources available and utilize them in a sense that makes generated energy sufficient for everyone living in the city. To further examine the electricity consumption in Abu Dhabi city, local tariffs were intensively explored to emphasize the allowed and actual consumption rates, reflecting either a green or a red band on the bills as per ADDC.

As it is stated on the website,

> Our residential tariffs are designed to encourage sustainable consumption. They are divided into two different rates that affect your total bill. The green band is charged at a lower rate, up to a set daily allowance based on a calculated "ideal average" consumption. If you use more than this limit, you will be charged at a higher red band rate. So, if you stay in the green, you'll pay less for each unit. (ADDC, 2020)

The accepted average daily consumption is referred to as "green brand," which refers to sustainable consumption. The consumption of electricity for houses is up to 400 kWh. Green band consumption refers to the government-backed electricity tariff since it is considered eco-friendly usage, making any consumption above that to be calculated as red band consumption, which will result in an extra charged tariff that is a little bit higher. The green consumption tariff is 6.7 fils for 1 kWh and is usable for up to 400 kWh for houses in local tariffs. The red band consumption tariff is 7.5 fils for every extra 1 kWh. That is counted to be the red rate that is charged to usage crossing the green usage band, as shown in Table 2.

Table 2. Green and Red Consumption Tariff (Hussain et al., 2022).

Consumption Type	**Apartments**	**Villas**	**Tariff (1 Dhs/100fils)**
Green Consumption	30 kWh/day	400 kWh/day	6.7 fils/kWh
Red Consumption	Above 30 kWh/day	Above 400 kWh/day	7.5 fils/kWh

2.2.4. Opportunities and Ambitions

One of the advantages the UAE has is being the only country in the world with the most sunlight exposure. This fact drives new and current opportunities to be further researched and developed into tangible methods to generate solar energy. Hence, reducing fossil fuel consumption through generating alternative clean energy that produces electricity more efficiently (Kempe, 2006). According to a recent study that reviews the clean energy projects in the country, solar energy is believed to be the most abundant form of clean energy in the UAE. Relatively, photovoltaic technologies are noted as a growing source of power that is achieving the current local requirements for energy consumption. In further demonstration, the UAE government is constantly proliferating clean energy generation projects in the country to familiarize locals and to achieve a sustainable energy system in the future (Kempe, 2006).

In return, the ambitions also expand to achieve a clean environment with reduced carbon prints and other pollutants that negatively affect the ecological system. Among increasing concerns about energy generation risks worldwide and a predicted energy catastrophe, which might become inevitable to face if the current clean energy opportunities were not taken into further consideration, it is very important to indicate the serious steps that are essential to transition into clean energy generation methods. According to the study, which discusses UAE solar energy projects, Abu Dhabi covers 84% of the total land area in the country, ultimately enabling the capital to utilize the high sunlight exposure to convert it into electrical power to be consumed by residents in the city (Kempe, 2006). Moreover, the review study stresses the need to widen research on solar energy in universities and institutions as a necessity to evolve the current solar energy technology and applications in the country. Photovoltaic panels can comparably reduce the current costs of generating electricity due to the direct process of solar panels to capture sunlight and the ability they have to transform sunlight particles into electricity with low costs because they only require the solar material set up. Which, in sequence, requires further research work to seize the maximum amount of renewable energy possible.

2.2.5. Current Global Implications

Across the globe, the amalgamation of AI and solar energy is fast transforming the landscape of renewable energy harnessing, leading to optimized efficiency, reduced costs, and heightened grid stability. AI, with its predictive and analytical prowess, has enhanced various aspects of solar power utilization. One of the chief applications is in predictive maintenance, where AI identifies potential issues in solar installations, thus ensuring consistent performance. According to a study, AI-driven predictive analytics can increase solar power plant efficiency by up to 10% by pre-emptively identifying areas needing maintenance and optimizing energy output (Kirpichnikova & Chirov, 2022). Moreover, AI algorithms, by leveraging historical and real-time meteorological data, can forecast solar radiation and energy production with heightened accuracy.

This aids grid operators in better managing and distributing energy, especially during peak demand periods. A study employed deep learning models, revealing that they outperform traditional forecasting methods by reducing prediction errors by as much as 20% (Knier, 2008). Additionally, AI optimizes solar panel orientations by constantly adjusting panel angles in real-time in response to the sun's position and prevailing weather conditions. This ensures maximum energy harvest throughout the day. The implementation of such smart tracking systems has been shown to enhance energy capture by up to 15% (Krishnamoorthy et al., 2020). The above applications, propelled by AI's capabilities, underpin a transition toward smarter, efficient, and more resilient solar power infrastructures, echoing the global pivot to sustainable and intelligent energy solutions.

2.3. Solar Energy Applications

2.3.1. Modern Day Applications

Current-day solar panels have become very popular worldwide since they generate electrical power from sunlight, which is converted to a sort of energy, scientifically called solar energy (Lazarou et al., 2007). Solar panels were first noted by a French scientist in the 19th century. He was able to observe the light particles, known as photons, in sunlight that affect a specific type of surface differently, such as silicon. This discovery allowed further investigations to take place into this matter.

As years followed, it was also developed further by Albert Einstein, who was able to identify the photoelectric effect, as he named it. The effect takes place differently depending on the surface hosting the light particles of sunlight. The discovery of the effect observed the possibility of absorbing the photons and capturing the electric charges on a semi-conductor surface, resulting in an electric current. Hence the name, the photoelectric effect (Lazarou et al., 2007).

One aspect to emphasize is that this effect has been utilized in mini semiconductor materials where it allows the capturing of light particles to create an electric current and voltage that can pass to a connected battery to save the captured electrons and create a voltage. These semiconductors are called photovoltaic cells, which make up solar panels (Lenzen, 2010). Solar panels consist of small photovoltaic cells connected to each other, creating solar panels that generate solar energy from voltages derived from the photons of sunlight (Lazarou et al., 2007).

More importantly, when combining these solar panels together, we create a solar array that includes multiple solar panels. On a larger scale, there are real applications of solar energy, which have been used in many countries in different sectors. For instance, it has been used privately in the United States of America by farm owners in the south, where they were given the chance to invest in their farming lands (LG, 2021). So instead of planting and cropping, they created "solar farms" that included hundreds of solar panels being laid down next to one another, generating solar power that goes directly to the utility grid and large batteries to be sold to the government. In Europe, solar panels have been regularly bought by homeowners to be implanted on the rooftops of their houses to generate solar power that is sold to the governmental electricity grid (Li et al., 2009). In this case, solar panel owners generate and sell electricity and make a profit out of it. However, they pay for their electricity bills as normal.

People nowadays are having more confidence in investing in sustainability initiatives as more global electronics companies have been manufacturing and selling solar panels and investing in green energy. For instance, the famous electronics company LG has been selling solar panels at different options to its customers in the United Kingdom. In addition to LG, a competitive number of companies have been enrolled in the same field of green energy as LG. This allowed for more research and development in this area, as well as it allowed concerned scientists to further explore and investigate the efficiency and liability of the panel (Mirasgedis et al., 2000).

In the United Arab Emirates, local initiatives have taken place in some sectors to save the environment, for example, the applications of photovoltaic panels in the UAE can be seen in many agricultural projects in the city of Abu Dhabi. As an example, the agricultural system that takes place on Hafeet Mountain relies solely on solar panel arrays due to the bright sunlight that hits the mountain without any blockage, which is supposed to generate enough electricity to power the water pump connected to the tanks above placed on the 1,200-meter-high mountain. Its' main purpose is to absorb sunlight and generate electricity that will pass through a distribution board to regulate the electric current flowing to power the water tank pump, which is connected to multiple pipes, creating the water network that is used in this unique agriculture project.

One of Abu Dhabi's sustainability visions is to host the largest solar farm in the world. The upcoming Al Dhafra solar project is estimated to power roughly 160,000 houses across the city of Abu Dhabi. It will be the first independent power producer to generate 2 Gigawatts of power from photovoltaic cells. The land is estimated to reach a length of 35 kilometers, making it the largest solar power generation in the whole world. Similarly, a solar project is currently under process in Masdar City, the sustainability-powered city inside of Abu Dhabi. The project is supposed to generate solar power and store it in large batteries for future usage.

2.3.2. Resources Competition

In the current situation, some of the few ways to generate renewable energy would be through either nuclear energy or other RES. Nuclear energy, contrary to fossil fuels, does not generate electricity through combustion processes but rather through fission processes that produce heat, which is simultaneously used to create steam that will be later used in a turbine generator to finally produce electricity (Naam, 2013). Fission refers to the chemical process of splitting atoms of uranium. The required uranium for this process is a highly unique type and is also non-renewable (Alwadei et al., 2022).

However, despite its convenience, nuclear energy does produce radioactive waste, and although it will not pollute the air or produce greenhouse gases, it is extremely harmful to both human health and the environment, hence the need to keep it in safe long-term storage. The mentioned challenges shed light on the importance of researching new ways to generate electricity, perhaps with less cost and fewer risks to health and the environment, such as renewable energy sources (Naam, 2013). There are various types that have been utilized in the past decades. As per the United Nation article on renewable energy, there are six major types of natural energy sources: solar power, windmill energy, hydropower, geothermal energy, ocean energy, and bioenergy. This article points out as well that solar energy is the only one existing in large amounts in comparison to the other renewable energy sources, making it an abundant source of renewable energy (Naam, 2013).

2.3.3. Abu Dhabi Solar Energy Plan

The ADDC is constantly working on implementing solar energy, especially for those individuals who are concerned about sustainable energy and are willing to install solar panels on the rooftop of their homes. The procedure that is officially registered on their website states that the process will be like applying for a new electricity meter, where the premise owner is supposed to contact a certified contractor, and then the contractor will carry on the regular requests with ADDC until the service is powered. In normal electricity meter installation cases, the first step for the responsible owner will be to contact an ADDC certified contractor to work collaboratively along with ADDC engineers to activate the service at home. The procedure will include pre-inspection visits to ensure that there is sufficient and convenient space prior to the installation; then the contractor will agree with a certified consultant to design the installation, which will be submitted to ADDC for an Internal Approval; and after the solar panel meter is installed, the final request to be applied for by the contractor will be to Activate and Supply Energy. Once the supply starts running, electricity will be metered regularly, and readings will be taken monthly to issue the electricity bills.

2.3.4. Humidity Impacts on Solar Panels

It is necessary to consider two possibilities while studying the impact of humidity. The first scenario involves the impact of water vapor particles on the amount of sunlight irradiance, while the second scenario involves humidity infiltration into the enclosure of the solar cell. There are three possible outcomes when light strikes water droplets. It might reflect, diffract, or be refracted. The direct solar radiation component's reception level is reduced by these factors. Irradiance is affected nonlinearly by humidity, and irradiance produces small linear fluctuations in voltage and huge nonlinear variations in current. Hence, the efficiency decreases when knowing that dampness reduces current but has no impact on voltage, the power output (Ng et al., 2023). The reasons for the nonlinear variations in irradiance with relative humidity are the non-uniform distribution and the vast range of water vapor particle sizes in the environment. Smaller water vapor particles provide greater scattered angles.

Greater water vapor particles in the atmosphere also cause more diffraction. It is undeniably true that tropical nations like Malaysia will see a depressingly greater decline in irradiance levels due to their much higher relative humidity levels. Relative humidity is negatively impacted by wind speed, which has an impact on received irradiance (Ng et al., 2023). Moisture ingression will be examined in the second strategy for understanding how humidity affects solar cell performance. Long-term exposure to humidity will result in some performance loss for photovoltaic cells (Nyholm et al., 2016). In general, depending on the kind of PV technology used, there are two module failure modes. Crystalline silicon cells may fail at the cell interconnections or develop fractured cells, and failure at the scribe lines is the main factor degrading cell thin film modules. Therefore, for crystalline silicon cells and thin films, respectively, embrittlement of the encapsulant material and corrosive moisture are the key causes of the aforesaid degradations. These degradation processes may be accelerated by hot and humid temperatures (Oikonomou et al., 2009).

Passivated PV cell surfaces cause current deterioration, which results in performance reduction, but have little to no impact on voltage (Nyholm et al., 2016). Voltage may sometimes spike dramatically, which is unexpectedly weird. The fact that water droplets are confined within the cell and function as magnifying lenses to concentrate the sun is a plausible explanation for these phenomena. In a similar vein, water vapor particles serve as barriers that slow the pace at which electrons are produced by the cell by preventing incoming sunlight from exiting (Nyholm et al., 2016). The use of edge sealants and the use of low ionic conductivity materials for the encapsulant are two preventative measures that may be implemented to lessen the infiltration of moisture into cell chambers (Onen, 2021).

2.3.5. Environmental Impact of AI Integration

The integration of AI into solar energy systems in Arabian Gulf countries, particularly the UAE, heralds' substantial environmental benefits, fostering a transition from carbon-intensive energy sources to greener alternatives. As the region grapples with the ramifications of climate change, exacerbated by rising temperatures and desertification, renewable energy adoption offers a sustainable reprieve. In the UAE, where solar irradiance can reach values as high as 2,200 kWh/m^2 annually, optimizing solar energy harnessing through AI is a strategic imperative (Peng et al., 2023).

AI-driven solar systems ensure optimal panel orientation, predictive maintenance, and accurate energy output forecasting. These optimizations translate to higher energy yields with the same infrastructure, reducing the necessity for expansive solar farms, thereby preserving natural landscapes. Furthermore, by enhancing energy efficiency, the dependency on fossil fuels – primarily natural gas in the region – is curtailed, leading to reduced greenhouse gas emissions.

Moreover, the deployment of AI-enhanced solar installations in the UAE has the potential to mitigate its CO_2 emissions by 30% by the year 2030, as part of its vision (Rahi & Bourgault, 2022). This not only aligns with the UAE's commitment to reduce its carbon footprint but also establishes a template for neighboring countries in the Gulf to emulate. In the broader context, the synergy of AI and solar technology in the region underscores a proactive approach to environmental stewardship, marrying technological innovation with sustainable practices to combat the escalating threats posed by climate change.

3. Materials and Methods

3.1. Research Objectives

This exploratory investigation research aims to enhance the dependency on renewable energy consumption in the city of Abu Dhabi by exploring the social acceptance of solar panels within villas and townhouse owners in the city of Abu Dhabi.

Therefore, the main objectives of the research project are to:

1. Research the most up-to-date technologies related to solar power and renewable energy.
2. Research the solar power generation scenarios applied around the world for homeowners to install solar panels and generate electricity.
3. Conduct a quantitative analysis to examine the social acceptance of solar panel installation and renewable energy utilization in the city of Abu Dhabi by presenting four prominent scenarios to be elected by the target audience in Abu Dhabi city.

The main research question for this research project is:

To what extent is renewable energy socially accepted among villa and townhouse owners in Abu Dhabi, and what is their preferred scenario for installing solar panels?

3.2. Innovative Scenarios

Accordingly, after comprehensive research on the literature review of the application of solar power in the United Arab Emirates, there are several scenarios that can be taken into consideration when generating solar energy. The formula and scenario will be based on specific criteria such as the solar panel owners, implanting place, power required, and main purpose for the photovoltaic solar panels. Demonstrated below are the different types of scenarios to be implemented to get the most out of solar energy generation.

In scenario no. 01, the solar panels will generate electricity to power a residential house only. The responsible person for buying and maintaining the panels will be the homeowners, which is the currently proposed option by ADDC as per their

Solar Energy Plan. Hence, homeowners buy and install solar panels for personal use at their own expense if this will help them save money on their electricity bills.

In scenario no. 02, homeowners buy and install solar panels for personal use, at their own expense, because this will help save the environment. This allows homeowners to give priority to eco-friendly initiatives that take place regularly in the city of Abu Dhabi. It further elaborates on the importance of taking into consideration the environmental status of the previous energy generation sources.

In scenario no. 03, the solar panels installed will generate power, delivering it directly to power the utility grid, which is the situation in many houses in Germany, as the government supports the use and installments of solar panels. Owners buy and install solar panels at their own expense, sell the produced power to the authorities, and pay the balance of their electricity usage on a yearly basis (this scenario is less expensive than the previous scenarios because no batteries are needed to store the electricity produced).

In scenario no. 04, homeowners buy and implant solar panels for personal use if the cost is supported by the government (e.g., discount, easy and affordable installment plans, etc.). Subsequently, the scenarios addressed and surveyed intensively overview the four different possibilities to implant and utilize solar panels to develop energy generation in houses in the city of Abu Dhabi, whether it will be powering the utility grid or houses alone.

3.3. Homeowners Survey

Because there is little information available about the societal acceptance of solar energy in the UAE, this study was created as an exploratory investigation. More information about this subject might significantly increase the use of solar energy. The goal of this research is threefold. Firstly, this investigation investigates popular attitudes about the use of solar energy. Secondly, this study investigates popular perceptions of the challenges of utilizing solar power. Thirdly, the study examines civic perceptions of boosting solar power utilization in Abu Dhabi city. The inquiry in this chapter makes use of the main and minor data in cooperation with the replies from the ground were used as main data by using a survey approach.

The information was gathered by delivering survey questionnaires to Abu Dhabi residents and homeowners. The poll was taken between September 2022 and October 2022. The respondents' rate of comprehension of solar power was disregarded when surveys were sent. Because survey answers may come from a variety of backgrounds and have diverse amounts of experience on this topic, the language used in the questionnaires was plain and intelligible in both English and Arabic, with only a few technical terminologies. Respondents with a rudimentary awareness of renewable energy, on the other hand, were favored. Two criteria for participation in the exercise were considered in the survey's sample design. The first criteria were that respondents live in the wider Abu Dhabi area.

The second criterion was that participants be at least 18 years old and property owners. Some generic questions were asked of respondents during the poll, such as their interest in environmental concerns, knowledge of global warming, familiarity with renewable energy technology, interest in solar energy, and so forth. As a result, only those who responded to the preceding interest with relatable data were chosen to answer the survey. A survey has been developed to further demonstrate the social awareness of solar power in Abu Dhabi. The questionnaire involved essential questions such as gender, age, and marital status of the surveyed person. It also inspects further the current electricity consumption for villa owners and whether it is labeled under ADDC green band consumption or red band consumption, depending on the type of premise they are using electricity in, the devices used at their homes, and the air conditioning system currently being used. The survey highlights the general knowledge as well as whether the respondent is aware of solar energy at all and if they are willing to support it by considering the use of solar panels soon.

Another crucial question that was added was the energy depletion incident that is taking place due to gas consumption and how inevitable it is for us to use a substitutional source of energy such as solar energy. Gas emissions have also been added deliberately into the questionnaire to further understand the viewpoints and range of knowledge of the respondents in general.

The main audience that was targeted were homeowners, mostly post-graduate students and faculty, along with married and single men and women, to deeply identify the social acceptance status currently available in the city of Abu Dhabi among interested intellectuals. Using the data collected from the survey, a detailed result analysis has been extracted that is neatly presented with charts, which are, by default, demonstrating the information collected while also shedding light on the concluding projections of this survey's data, based on homeowners' opinions in the city of Abu Dhabi.

3.4. Study Limitations

The research study focuses particularly on investigating the social acceptance of solar power utilization for homeowners. The study is supported by a survey that was sent to 216 homeowners, which offered them 4 prominent solar power installation scenarios among other environmental aspects. The mentioned scenarios are applicable worldwide and are examined in the literature review. Therefore, the elected scenario is the key to stimulating the adaptation of solar power systems in the houses of residents in the city of Abu Dhabi. Nonetheless, the role of research and development in the sustainability field always needs to be further observed and discussed to meet the higher substantial goals in the United Arab Emirates.

3.5. Challenges and Issues

The fusion of AI and solar power holds immense promise for revolutionizing the renewable energy sector, especially in regions like the Arabian Gulf, characterized by abundant sunlight. However, as a solar power systems expert, it is crucial to understand and underscore the inherent challenges that the Gulf region, and notably the UAE, faces in implementing these AI-integrated systems. One primary concern is the environmental conditions specific to the region. The UAE, like much of the Gulf, experiences frequent sand and dust storms. A study points out that accumulated dust particles on solar panels can attenuate incoming solar radiation leading to significant efficiency losses (Rahi et al., 2021).

While AI can schedule predictive maintenance, the high frequency of such events necessitates a more holistic approach to system design and maintenance. Furthermore, the temperature is another critical factor. Solar panels, typically optimized for temperate conditions, witness reduced efficiencies in the high ambient temperatures of the Arabian desert. For every degree Celsius rise above 25 °C, the efficiency of silicon solar panels drops by approximately 0.5%.

Furthermore, recent observations highlighted the pronounced impact of elevated temperatures on solar panels deployed with peak summer temperatures causing a decrease in efficiency (Ramachandran et al., 2022). Technologically, the integration of AI with solar systems mandates advanced data analytics infrastructure and expertise. A gap exists between the rapid advancements in AI-driven solutions and the availability of skilled professionals to deploy and manage them. Consequently, it was emphasized that fostering local expertise and bridging the talent gap is imperative for the region's renewable energy aspirations (Salimi et al., 2022). Incorporating AI also heightens the importance of cybersecurity. With solar grids becoming smarter and more interconnected, their vulnerability to cyber-attacks increases, making grid security a paramount concern.

4. Results and Discussion

4.1. General Characteristics of the Respondents

The survey that was developed in this research project had been conducted between September 2022 and October 2022 with the method of information gathering and surveyhero.com as data collection platform. 186 responses were received out of 219 surveys that were further analyzed using JASP 0.16.0.0, and Microsoft Excel. The utilization of solar energy has been surveyed accordingly based on this background. The respondents' characteristics, awareness level in solar energy, ability to buy and implant solar panels, expectations on the governmental support to utilize solar energy in Abu Dhabi has been discussed and investigated in this section.

4.2. The Interest and Importance of Solar Energy

Consequently, the respondents involved concerned local individuals in the development of solar energy in Abu Dhabi, which involved respondents in the range of 18–59 years old. 2% of the respondents comprised of individuals between 50 and 59, 5% between 40 and 49, whereas 21% were between 30 and 39, and 72% were aged between 18 and 29 years old, as shown in Fig. 1.

In terms of gender, the respondents comprised 43% females and 57% males out of 186 responses, that are in the province of Abu Dhabi. As for areas, only 2% of the respondents are in the Western region of Abu Dhabi, whereas 8% are

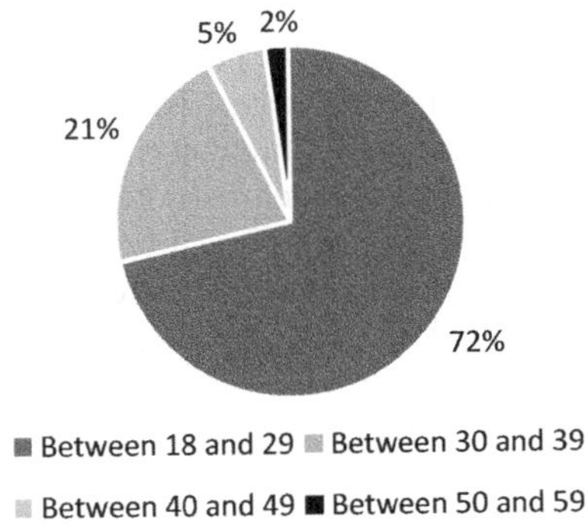

Fig. 1. Respondents Age Pie Chart (Hussain et al., 2022).

in Al Ain, 43% are in the Eastern region, and the rest of the respondents are in the main city of Abu Dhabi comprising 47% making them largest group in the survey, as shown in Figs. 2 and 3.

The level of education of the respondents was also considered through the survey to assess their awareness on solar energy accordingly. The general level of education is as follows: 1% of the respondents have a degree lower than a high school diploma, 2% have a PhD or DBA, 9% have been into colleges but hold no degree, 10% holding a high school diploma, 22% holding a master's degree, and 56% are holding a bachelor's degree, as shown in Fig. 4.

The financial status of the respondents was also included in the survey to demonstrate the financial ability of the interested respondents to buy and implant solar energy panels in their homes. The average household income per month of the respondents comprised 4% receiving a monthly salary of 60,000 to 89,999 AED, 6% receiving a monthly salary above 90,000 AED, 14% receiving a monthly salary of 40,000 to 59,999 AED, 32% receiving 20,000 to 39,999 AED as a monthly salary, with 44% receiving a monthly salary of 0 to 19,999 AED representing the majority of the respondents, as shown in Fig. 5.

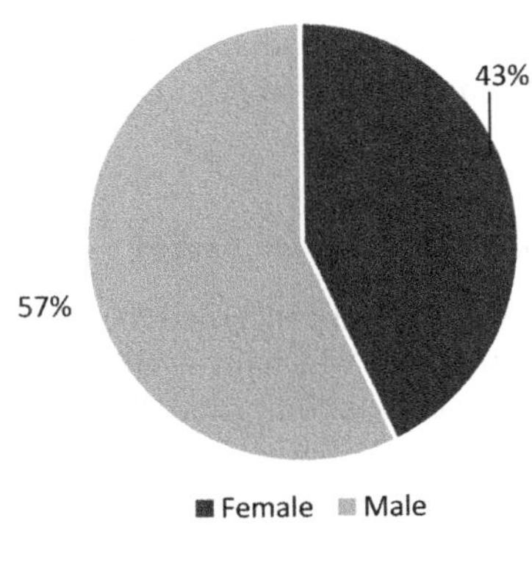

Fig. 2. Respondents Gender Pie Chart (Hussain et al., 2022).

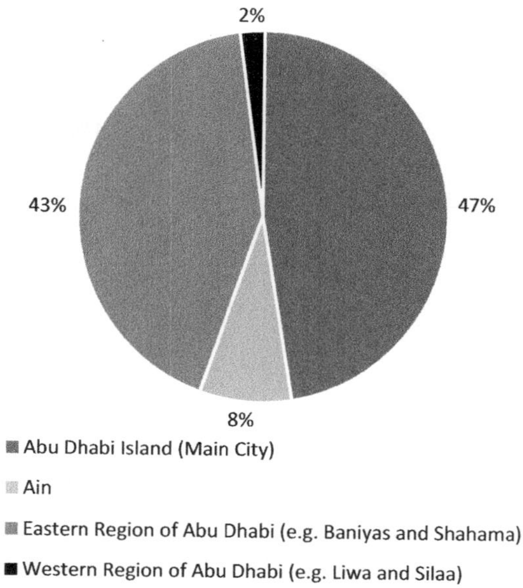

Fig. 3. Respondents Area Within Abu Dhabi Pie Chart (Hussain et al., 2022).

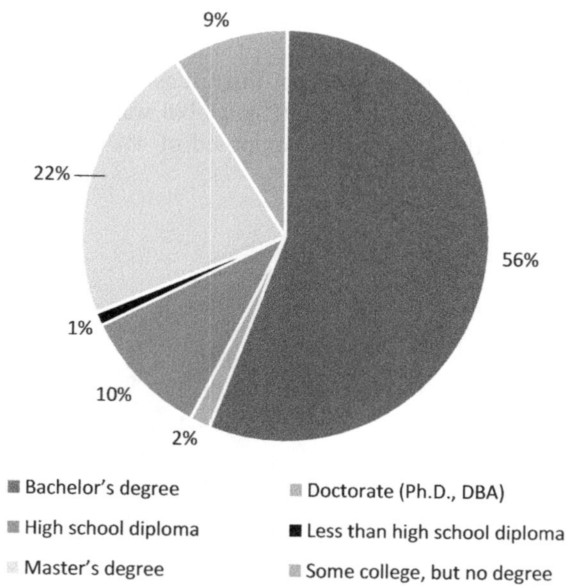

Fig. 4. Respondents Level of Education Pie Chart (Hussain et al., 2022).

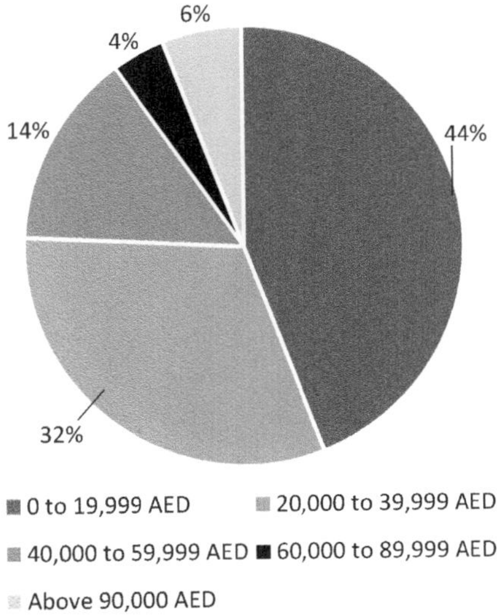

Fig. 5. Respondents Average Monthly Income Pie Chart (Hussain et al., 2022).

The residential tariffs in Abu Dhabi consists of two types as discussed previously, with variations in terms of electricity consumption tariffs between locals and expats. In this survey, 60% have local tariff and 40% are having Expat tariff, as shown in Fig. 6.

The survey examined the general level of interest as well in the following subjects: environmental issues such as pollution, global warming, and overpopulation. Through which the respondents composed of 4% only being not at all

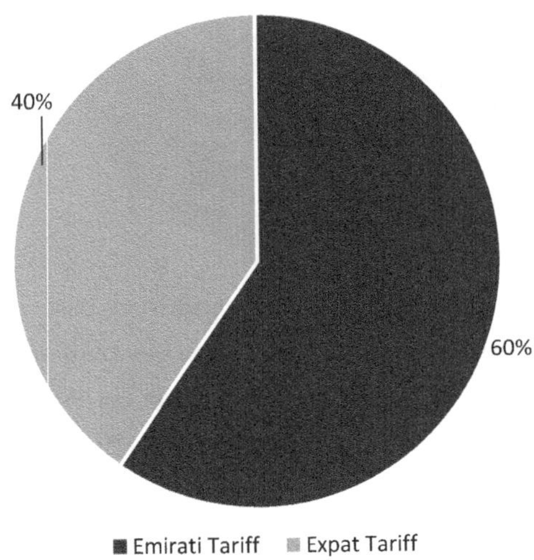

Fig. 6. Respondents Residential Tariff Pie Chart (Hussain et al., 2022).

interested, 12% being not at all interested, 47% being moderately interested, and 41% of the respondents being extremely interested, as shown below in Fig. 7.

The rate of awareness of solar power as demonstrated in the pie chart below shows the majority of the respondents being extremely interested as the responses consist of only 9% being not at all interested in solar energy, 41% being moderately interested, and 50% being extremely interested in solar energy displaying the majority of the respondents in the city of Abu Dhabi, as shown in Fig. 8.

The survey highlights the opinion of the public on solar energy utilization schemes in Abu Dhabi. Many of the answers to the survey demonstrate an interest in solar power development for Abu Dhabi City and Al Ain. Remarkably, the respondents majorly believe that the government is the most important organization to promote solar energy in Abu Dhabi City, followed by research centers,

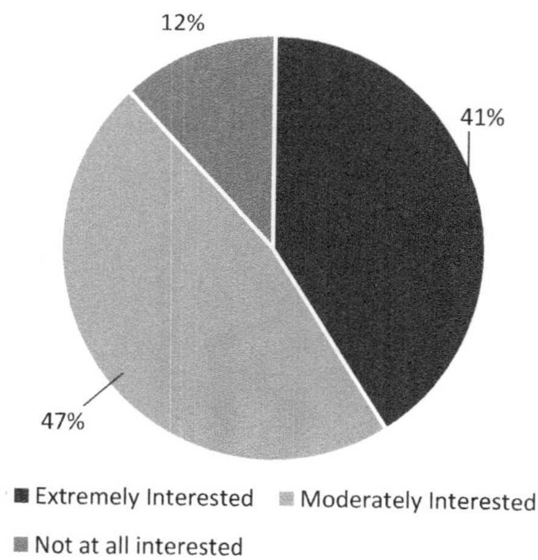

Fig. 7. Respondents Level of Interest in Environmental Issues Pie Chart (Hussain et al., 2022).

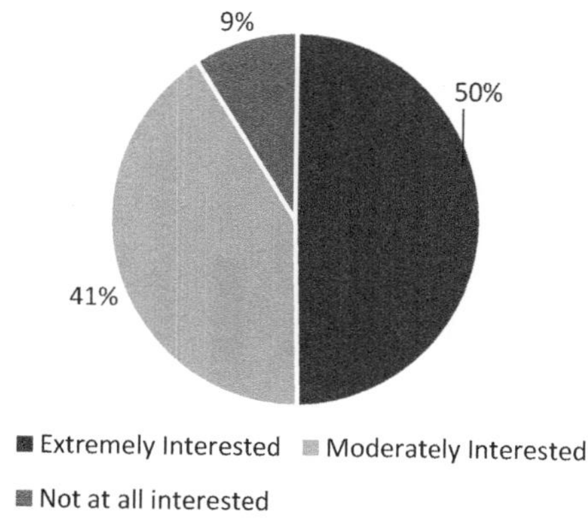

Fig. 8. Respondents Rate of Interest Pie Chart (Hussain et al., 2022).

higher education institutions, and lastly, private organizations that have been proven, as per the surveyed respondents, to be the least important organization type to promote solar energy.

Additionally, the majority of the respondents suggest that knowing the right information on solar energy materials is the most important factor for them to decide on buying and installing solar energy materials, whereas other respondents suggest that knowing the initial costs and return on investment from installing the solar energy materials are the most important factors to make their decision. However, only a minority of the respondents suggest that knowing the impact of solar energy materials on the environment is a prominent factor for them to decide (Figs. 9 and 10).

The survey also questioned the respondents on the level of importance of the factors that can enhance the formulation of new solar energy policies in the city of Abu Dhabi. As per the majority of the respondents, one of the main factors in terms of importance is to increase public knowledge and awareness of solar

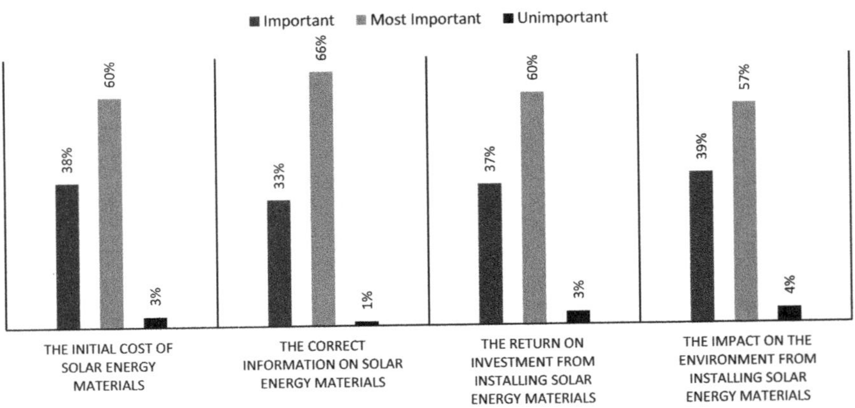

Fig. 9. Respondents Level of Importance to Decide on Purchasing and Installing Solar Energy Materials Graph (Hussain et al., 2022).

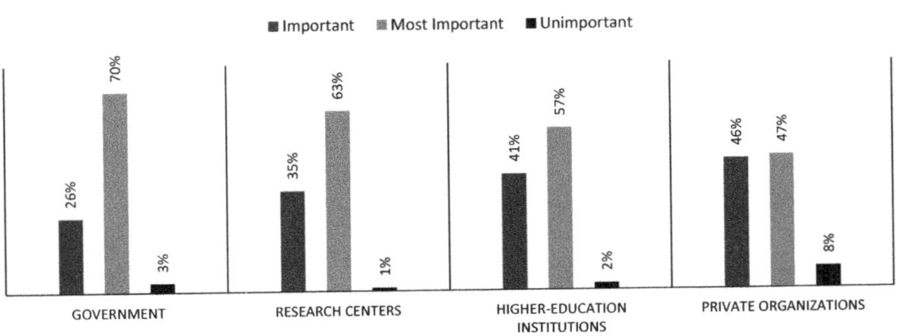

Fig. 10. Respondents Level of Importance of the Following Organizations in Promoting Solar Energy Graph (Hussain et al., 2022).

energy, which will demand new policies to facilitate the expected and realistic scenarios to practically utilize solar energy.

Likewise, the second major factor is to provide incentives for solar energy usage that will motivate the public accordingly, which will subsequently give birth to new policies that further regulate and develop the solar energy sector, while some respondents claim that establishing collaborations with developed countries to learn from their experience in solar energy is an important factor for developing new solar energy policies in Abu Dhabi. Notably, the respondent's most favorable scenario is to buy and implant solar panels for personal use if the cost is supported by the government through green energy initiatives that focus on solar energy in the city.

However, the second favorable scenario is for homeowners to buy and install solar panels at their own expense if it will help them lower their monthly electricity bills and save money consequently. The third scenario is for homeowners to buy and install solar panels, at their own expense, and then sell the produced solar energy to the government utility grid and pay the balance of their electricity usage on a yearly basis without having the need to consume the generated solar energy nor save it in batteries. Lastly, the least favored scenario among the four scenarios is for homeowners to buy and install solar panels at their own expense if this practice will help save the environment as well as save natural resources that are at risk of depletion.

Figs. 11 and 12 give a visual representation of the factors to develop new solar energy policies and the respondents' level of interest in buying and implanting solar panels to generate electricity.

4.3. Statistical Data and Majors Findings

The main objective of the survey was to understand the level of solar awareness and interest for homeowners in Abu Dhabi city. And more importantly, to analyze

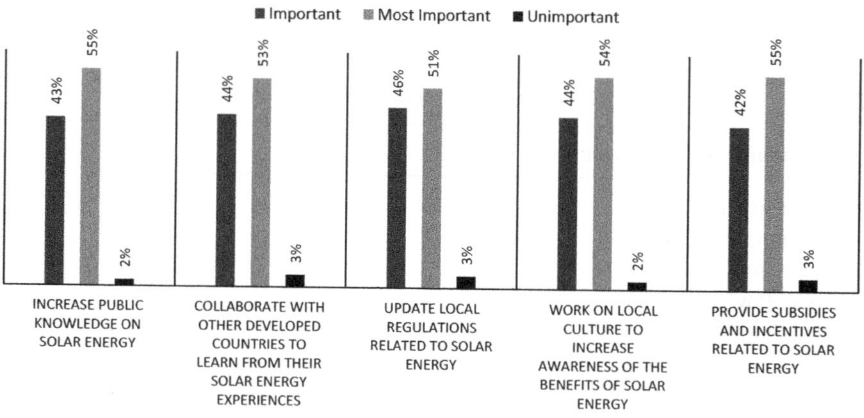

Fig. 11. Respondents Level of Importance of the Factors to Develop New Solar Energy Policies Graph (Hussain et al., 2022).

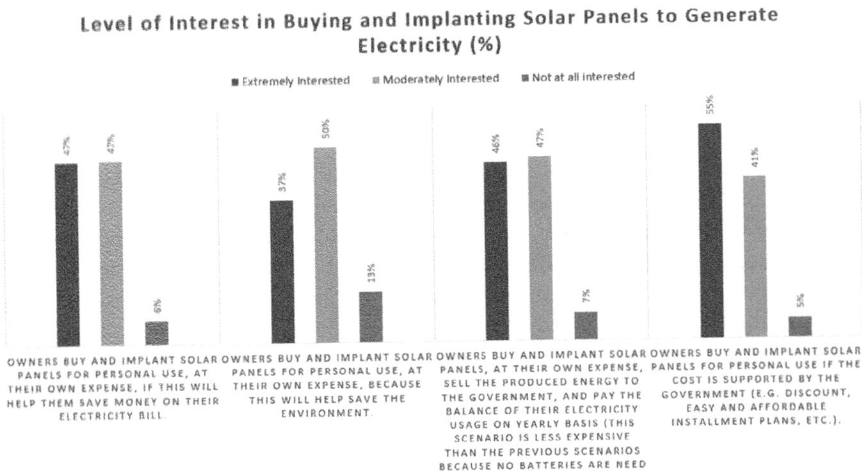

Fig. 12. Respondents Level of Interest in Buying and Implanting Solar Panels to Generate Electricity Graph (Hussain et al., 2022).

the most appealing scenario that will stimulate homeowners to install solar power systems in their private homes. The survey focused on several aspects, as demonstrated in Table 3, to ensure balanced results that support the preferred scenario. For instance, social status questions were provided in the survey as well, which is a survey practice that affirms the participation results to be coming from individuals of different backgrounds, genders, and financial statuses. The diversity of the surveyed individuals allowed the research paper to have a robust opinion about the most optimal scenario, established by dissimilar individuals instead of a similar group of people. Nonetheless, only interested individuals in the research study were selected for the survey, hence the survey included a limited number of 216 individuals only. Moreover, the survey results do not aim to represent all the homeowners in Abu Dhabi city, but rather display a unified point of view that has the potential to be applied in the near future by homeowners.

Table 3. Aspects Covered in the survey (Hussain et al., 2022).

Major Survey Aspects
• Social Background
• Financial Status
• Environmental Issues
• Solar Power Systems
• Investing in Solar Power
• Investing in Awareness
• Spreading Accurate Information
• Governmental Contribution
• More R&D projects

Different financial and social status information was included in the survey. The income, for example, was important to highlight, as the solar panels differ in prices depending on the electricity capacity as it was addressed previously. Thus, individuals with low income are less likely to invest in solar panels. In addition, respondents' gender was also required to affirm balanced survey results between male and female homeowners in Abu Dhabi city. The age range, as well, was required in the survey to further analyze the social acceptance levels of solar applications between the respondents to be diversified among different age groups.

Aside from that, current electricity tariffs are provided to present an idea of the electricity generation costs and optimal usage, which are impacting homeowners nowadays. This study provides an overview of the current solar power applications in a specific region and how it can be further developed to be utilized by homeowners in Abu Dhabi city. In addition, it analyses the social acceptance of homeowners for solar applications. The knowledge in this research paper is proven by official statistical data on the optimal electricity consumption for the residents of the city, and the survey results as well, which are discussed in detail, to pave the way for future research to be initiated in accordance with the current research paper that can allow for more utilization of solar panel systems in the private homes of Abu Dhabi city.

According to the survey results, the majority of the respondents showed a respectable acceptance level for solar power. After identifying the general characteristics of the respondents, the survey displayed a set of questions regarding solar power awareness. Furthermore, the most appealing scenario that sounded applicable and suitable for homeowners was for the water and electricity authorities to provide a paid service for the installation of solar panels at private houses. Moreover, respondents found it to be very convenient for interested people in solar power if the authority would provide a financial support program for the same group of people that, ultimately, will inspire more homeowners to invest in solar system installation at their private premises. Following the first scenario, respondents were relatively interested in scenarios that will allow them to cut electricity tariff costs, as well as being able to sell the produced solar power. Lastly preferred was the scenario to install solar panels as a contribution to environmental initiatives, as shown in Table 4.

Table 4. The Most and Least Accepted Scenarios (Hussain et al., 2022).

Presented Scenarios	**Interest Levels**	**Combined Percentages**
Government Support Scenario	Extremely Interested	96%
Cutting Cost Scenario	Moderately Interested	94%
Selling Solar Power Scenario	Moderately Interested	93%
Environmental Initiative Scenario	Not at All Interested	87%

4.4. Recommendations and Future Research

The trajectory of AI-integrated solar power systems in the Arabian Gulf, especially in countries like the UAE, suggests a transformative shift in renewable energy paradigms. The desert regions of the Gulf, with their abundant year-round sunlight, can harness an average solar energy of approximately 5–6 kWh/m^2 per day, positioning them uniquely to be global leaders in solar energy production (Stigka et al., 2014). By integrating AI into solar systems, we can unlock several groundbreaking opportunities. First, adaptive solar tracking using AI can significantly enhance the efficiency of solar panels. Predictive models, trained on historical data, can adjust panel orientations in real-time, maximizing energy capture.

Recent studies have shown that smart tracking can potentially increase energy yield by 15–25% compared to fixed installations (Sugimura et al., 1990). Furthermore, AI's predictive analytics capabilities can lead to optimized energy storage and distribution strategies. With the incorporation of advanced algorithms, energy storage systems can predict peak demand times, ensuring stored energy is dispatched when needed most. This could be particularly transformative for grid operations, minimizing blackouts and optimizing grid supply. AI-integrated systems also offer a promising avenue for maintenance. Utilizing machine learning for fault detection can predict equipment failures and maintenance needs, reducing downtime and increasing the overall lifespan of the installation. Recent research has indicated that AI-driven predictive maintenance can reduce operational costs by as much as 20% (Kirpichnikova & Chirov, 2022).

However, it's worth noting that for these advancements to bear fruit, a collaborative approach involving policy support, industry participation, and academic research is essential. The UAE, with its vision, emphasizes sustainable development and technological integration, stands as a beacon for the Gulf, and the future indeed looks bright. Despite Abu Dhabi's vast solar energy potential, just a few open locations have been utilized to harvest solar electricity. Institutes' and organizations' non-participation was ascribed to a shortage of motivation. Furthermore, the difficulty was regarded as a crucial obstacle for UAE authorities to overcome unless communal engagement was fostered (Tan et al., 2010).

The government must explore removing consumer hurdles. Other consumer hurdles indicated by respondents include a demand for more information to raise awareness and a relative lack of interest that needs to be stimulated to encourage stakeholders to invest in photovoltaic systems. Solar energy awareness projects may be done through media channels such as publications, television, and social media. Nonetheless, efficient financial assistance is critical as well. The electricity authority in Abu Dhabi can help by enacting subsidies and price measures, as consumption stimulation boosts demand and has the potential to broaden the market. Additionally, price strategy may involve bid rigging, which minimizes approval time when vendors meet predetermined prices and speeds up contract signing and project start-ups.

Moreover, partnerships between governmental and commercial firms might be used, as well, to emphasize and support the establishment of a research and development infrastructure for solar energy studies in Abu Dhabi city (Tan et al., 2010). The perception of consumers is highly dependent on the current aspects of the developing environmental and financial challenges, which are adhering to the electricity generation processes that are being adopted in the city of Abu Dhabi. Evidently, the future vision of the UAE to enhance renewable energy consumption allows for more work to be invested in developing new solar energy technologies, to be utilized to achieve the goals the government has set to lessen carbon discharges and upsurge the production of sustainable energy, to meet the needs of consumers for electricity. Similarly, raising awareness and sharing insights on solar energy can allow developers and electricity component providers to broaden their provided services as well as to encourage the integration of solar energy systems in newly built buildings.

4.5. Implication to the Gulf Region and the Smart Cities Agenda

The results of this study have important implications for major stakeholders in Abu Dhabi that help shape the built environment and energy landscape. Urban planners and architects can leverage the willingness for community solar needs to incorporate distributed infrastructure in neighborhood plans. Grid operators can model integration scenarios to optimize interconnected renewable operations. Policymakers can enact residential solar subsidies and develop regulatory frameworks, while creating skills training programs to engage key influencing stakeholders such as governmental entities, academic institutions, and other for-profit and nonprofit organizations. In fact, through collaborative action across these domains, findings-driven approaches can progress a solar-powered smart city vision.

Additionally, this research holds particular relevance for renewable smart city development across the Arabian Gulf. The Gulf region has significant potential for solar energy but currently heavily relies on fossil fuels. Therefore, by implementing distributed generation – producing energy closer to where it's consumed – and actively involving citizens, the Gulf region can simultaneously advance sustainability goals and address economic priorities.

Furthermore, strategic adoption incentives emerge as a principal opportunity – these could be modeled between Gulf Cooperation Council (GCC) member states to share best practices. Standardized frameworks can ease cross-country solar integration, benefiting interconnected grid operations. Regional utility authorities can jointly pioneer grid management systems to orchestrate renewable fluctuations.

Specific initiatives recommended based on this study include:

1. Establishing a joint task force between municipal and national agencies to design solar adoption incentives for citizens. This top-down policy push can actualize the willingness shown by residents.
2. Creating sandbox environments for testing grid integration models, enabled through partnerships between authorities, academia, and industry. Lessons can inform wider integration standards.
3. Launching community-based educational hubs to offer hands-on solar training for students and wider publics. This can build societal capabilities to support the energy transition.

Furthermore, the research advocates the following strategic policy directions to actualize Abu Dhabi's goals:

1. Mainstreaming renewables funding through long-range governmental planning budgets
2. Streamlining regulation on distributed generation testing and validation
3. Incentivizing community-led solar pilot projects through competitive research and entrepreneurship grants.

The unique insights generated in this study provide a framework to unlock societal potential for transformative, yet accepted energy innovations; an imperative for realizing future-ready smart environments through participatory means. The output enables evidence-based decision-making for those guiding context-specific sustainable development.

In fact, academic entities like the Arab Forum for Environment and Development (AFED), alongside architectural bodies such as the Gulf Organization for Research and Development, hold unique convening power to align smart city stakeholders around solar roadmaps resonant with public sentiment. Progress relies on context-attuned sociotechnical co-creation.

5. Conclusion

The integration of AI with solar power systems in the smart cities of the Arabian Gulf, including the UAE, signifies a paradigm shift in addressing urban energy demands while mitigating environmental impacts. Smart cities, inherently data-driven, stand to benefit immensely from AI's predictive and analytical prowess. The expansive solar irradiance of the region, which averages around 5–6 kWh/m^2 per day, forms an ideal backdrop for such systems (Stigka et al., 2014).

The fusion of AI with solar installations has been shown to enhance energy capture efficiency, allowing cities to derive more energy from the same solar footprint (Sugimura et al., 1990). This not only bolsters the region's energy security but also underscores its commitment to sustainable urban development. Furthermore, AI's role in predictive maintenance, grid optimization, and adaptive energy storage ensures that the power infrastructure is both resilient and adaptable, critical attributes for the rapidly growing cities of the Gulf (Tarsheed, 2020).

As the Arabian Gulf continues its journey toward a sustainable and digitized future, the symbiotic relationship between AI and solar energy emerges as a linchpin, holding the promise of cleaner, smarter, and more efficient urban landscapes. The study restates the crucial importance of exploring different scenarios for homeowners in the city of Abu Dhabi to possess and utilize solar panels, since the awareness of sustainable energy utilization is constantly increasing in the country. However, most of the electricity generated in the city of Abu Dhabi is dependent on natural gas combustion, which has several drawbacks such as durability and pollution.

The study sheds light on the current electricity generation methods and risks. It further investigates the different possible scenarios to generate solar energy independently at homes, along with displaying the durability risks as well as addressing the social acceptance level of local residents to install solar panels in their houses. In addition, the study reveals public expectations and governmental policies on the implantation of solar panels for the government of Abu Dhabi. Furthermore, the survey results and discussion confirm the high public interest and awareness of solar energy. Based on the surveyed responses from the residents of Abu Dhabi City, the most preferred scenario that is deemed suitable by homeowners is the financial support and guidance from the authoritative parties of water and electricity in the city. Nonetheless, major findings included several points, such as:

1. The majority of the respondents are interested in general environmental issues such as pollution and global warming.
2. 90% of the respondents in Abu Dhabi are interested in solar energy.
3. The majority of respondents are interested in buying and implanting solar panels if the cost is supported by the government through installment plans, easy payment plans, and discounts.
4. The majority of the respondents believe that increasing public knowledge of solar energy is important to develop new solar energy policies in Abu Dhabi.
5. The majority of the respondents suggest that knowing the correct information on solar energy materials is important to decide on buying and installing solar energy panels, while others suggest knowing the initial cost.
6. The majority of the respondents claim that governmental organizations are important to promote solar energy, followed by research centers, higher education institutions, and lastly, private organizations.
7. The current solar energy policies worldwide indicate that there is room for enhancement in the solar energy sector.
8. The importance of solar energy research in research and development centers to further analyze and develop new ways to utilize solar energy.

So, in practice, the findings of this study supply actionable evidence to advance solar infrastructure build-outs across Abu Dhabi's urban environment. Residents positivity and acceptance to adopt solar power enable targeted policy and urban development initiatives. Recommended solar subsidy programs, grid management optimizations and community demonstration projects can now be designed reflecting ground-truth insights.

For industry, the results point to opportunities in next-generation solar products and partnership platforms tailored to community integration. From AI-enabled solar tracking mechanisms to peer-to-peer electricity trading networks, the citizen openness invites smart city infrastructure that interweaves digital and power systems cohesively across neighborhoods. Governance frameworks can also benefit, with study implications on streamlining regulation and administrative burdens limiting current adoption.

In terms of future directions, this solar acceptance analysis would significantly gain from integration with geospatial modeling on district-level distributed generation capacity. Mapping theoretical solar potential against the social readiness highlights locales primed for pilot projects. Additionally, comparative research across other Gulf cities could illuminate contextual nuances in perceptions – jointly progressing regional renewables agendas. At the heart of smart city realization lies the complex yet harmonious unification of technology, society, and environment – this work crucially bridges these dimensions.

6. Human Participant

This study describes research on humans. Therefore, an approval from the Institutional Review Board of Abu Dhabi University was provided on the 19th of September 2022. The approval code is COE-22-09-00030.

7. Conflicts of Interest

The authors declare no conflict of interest.

8. Teaching Orientation

8.1. Suggested Teaching Assignments

To build on the solar adoption research, the following student assignments offer practical ways to apply and contextualize the key learnings around renewable innovations in smart cities:

1. Research Project: Survey residents in your city regarding perspectives on adopting solar panels or other renewable energy innovations. Compare attitudes against government sustainability targets to analyze gaps and opportunities.
2. Policy Report: Investigate solar incentives and programs in a Gulf or Middle East country. Assess the landscape, and draft recommendations for advancing the adoption of solar panels.

8.2. Recommended Readings

The paper's discourse on AI-solar integration and social acceptance provides helpful background. The following available readings offer further perspectives across smart cities, solar adoption, and Gulf renewable landscapes:

1. Batty, M. (2018). Artificial intelligence and smart cities. *Environment and Planning B: Urban Analytics and City Science*, *45*(1), 3–6.
2. Parks, D. (2020). Promises and techno-politics: Renewable energy and Malmö's vision of a climate-smart city. *Science as Culture*, *29*(3), 388–409.
3. Elrahmani, A., Hannun, J., Eljack, F., & Kazi, M. K. (2021). Status of renewable energy in the GCC region and future opportunities. *Current Opinion in Chemical Engineering*, *31*, 100664.

4. Alnaqbi, S. A., & Alami, A. H. (2023). Sustainability and Renewable Energy in the UAE: A Case Study of Sharjah. *Energies*, *16*(20), 7034.
5. Mezher, T., Dawelbait, G., & Abbas, Z. (2012). Renewable energy policy options for Abu Dhabi: Drivers and barriers. *Energy Policy*, *42*, 315–328.

8.3. Case Study

The following case scenario allows application of the solar adoption considerations from this Abu Dhabi study to a hypothetical Gulf city environment:

The municipality of Al Ain city has set targets to fulfill 10% of electricity demand via distributed solar by 2030. You are the Head of the Renewable Energy Department – develop a high-level strategy to incentivize and streamline residential solar adoption aligned with the goals. Outline policy mechanisms, partnerships, and demonstration projects with consideration for citizen willingness and grid capacity factors.

8.4. Titles for Research Essays

Additional smart city topics resonating with dimensions in this project that could inform rich research inquiries include:

1. Assessing Public Perspectives on Electric Vehicles to Inform Smart Transportation Infrastructure Development
2. Harnessing AI for Optimized Water Management in Desert Climate Smart Cities
3. Driving the Renewable Energy Transition in the Gulf Cooperation Council through Regional Policy Harmonization.

8.5. Recommended Projects URL

Ongoing smart city initiatives in Abu Dhabi providing inspirational models:

1. Masdar City: https://masdarcity.ae/
2. MBR Solar Park – Phase 3: https://masdar.ae/en/renewables/our-projects/mohammed-bin-rashid-al-maktoum-solar-park-phase-3

References

Abdullah, S., & Jeanty, P. W. (2011). Willingness to pay for renewable energy: Evidence from a contingent valuation survey in Kenya. *Renewable and Sustainable Energy Reviews*, *15*(06), 2974–2983.

Abu Dhabi Distribution Company (ADDC). (2019, March 3). *ADDC*. (Online). https://www.addc.ae/en-US/home/Pages/AboutUs.aspx

Abu Dhabi Distribution Company (ADDC). (2020). *Abu Dhabi Distribution Company* (Online). https://www.addc.ae/en-US/residential/Pages/RatesAndTariffs2020.aspx

Abuzaid, H., Awad, M., & Shamayleh, A. (2022). Impact of dust accumulation on photovoltaic panels: A review paper. *International Journal of Sustainable Engineering, 15*(1), 264–285.

Abuzaid, H., Moeilak, L. A., & Alzaatreh, A. (2022). Customers' perception of residential photovoltaic solar projects in the UAE: A structural equation modeling approach. *Energy Strategy Reviews, 39*(2), 100778.

Afework, B., Boechler, E., Hanania, J., Stenhouse, K., Yyelland, B., & Donev, J. (2021). *Photovoltaic cell* (p. 1). Energy Education.

Agterbosch, S., Meertens, R. M., & Vermeulen, W. J. V. (2009). The relative importance of social and institutional conditions in the planning of wind power projects. *Renewable and Sustainable Energy Reviews, 13*(02), 393–405.

Al Dhafra Solar Project, Abu Dhabi. (2020, August 17). *Power technology*.

Alba Energy. (2018, May 2). Solar farms are popping up all across Texas. (Online). Retrieved February 28, 2024, from https://albaenergy.com/solar-farms-are-popping-up-all-across-texas/

Albattah, M., & Attoye, D. E. (2021). A quantitative investigation on awareness of renewable energy building technology in the United Arab Emirates. *Sustainability, 13*(12), 1–20.

Alkahtani, H., Aldhyani, T. H. H., & Alsubari, S. N. (2023). Application of artificial intelligence model solar radiation prediction for renewable energy systems. *Sustainability MDPI, 8*(6973), 15.

Alnaser, W. E., & Alnaser, N. W. (2019). The impact of the rise of using solar energy in GCC countries. *Renewable Energy and Environmental Sustainability, 4*(1), 7. DOI: 10.1051/rees/2019004.

Alwadei, S., Farahat, A., Ahmed, M., & Kambezidis, H. D. (2022). Prediction of solar irradiance over the Arabian Peninsula: Satellite data, radiative transfer model, and machine learning integration approach. *Applied Sciences, 12*(2), 717.

Andrei, M., & Sammarco, G. (2017). Gas to wire with carbon capture & storage: A sustainable way for on-site power generation by produced gas. In *Abu Dhabi International Petroleum Exhibition & Conference*, Abu Dhabi.

Bakos, G. C. (2009). Distributed power generation: A case study of small scale PV power plant in Greece. *Applied Energy, 86*(09), 1757–1766.

Bazmi, A. A., & Zahedi, G. (2011). Sustainable energy systems: Role of optimization modeling techniques in power generation and supply – A review. *Renewable and Sustainable Energy Reviews, 15*(8), 3480–3500.

Borchers, A. M., Duke, J. M., & Parsons, G. R. (2007). Does willingness to pay for green energy differ by source?*Energy Policy, 35*(06), 3327–3334.

Casten, T. R., & Downes, B. (2005). Critical thinking about energy: The case for decentralized generation of electricity. *Skeptical Inquirer, 29*(01), 25–33.

Chalvatzis, K. J., & Hooper, E. (2009). Energy security vs. climate change: Theoretical framework development and experience in selected EU electricity markets. *Renewable and Sustainable Energy Reviews, 13*(09), 2703–2709.

Chang, J., Leung, D. Y. C., Wu, C. Z., & Yuan, Z. H. (2003). A review on the energy production, consumption, and prospect of renewable energy in China. *Renewable and Sustainable Energy Reviews, 7*(05), 453–468.

Climate Council. (2022). Why are gas and electricity prices going up. (Online). https://www.climatecouncil.org.au/resources/explained-gas-electricity-prices-going-up/

De Benedetti, M., Leonardi, F., Messina, F., Santoro, C., & Vasilakos, A. (2018). Anomaly detection and predictive maintenance for photovoltaic systems. *Neurocomputing, 310*, 59–68.

Del Río, P., & Burguillo, M. (2009). An empirical analysis of the impact of renewable energy deployment on local sustainability. *Renewable and Sustainable Energy Reviews*, *13*(6–7), 1314–1325.

Diakoulaki, D., & Karangelis, F. (2007). Multi-criteria decision analysis and cost-benefit analysis of alternative scenarios for the power generation sector in Greece. *Renewable and Sustainable Energy Reviews*, *11*(04), 716–727.

Divya, S., Panda, S., Hajra, S., Jeyaraj, R., Paul, A., Park, S. H., Kim, H. J., & Oh, T. H. (2023). Smart data processing for energy harvesting systems using artificial intelligence. *Nano Energy*, *106,* Article 108084.

Elrahmani, A., Hannun, J., Eljack, F., & Kazi, M.-K. (2021). Status of renewable energy in the GCC region and future opportunities. *Current Opinion in Chemical Engineering*, *31*, 100664.

Eneh, O. C. (2011). A review on petroleum: Source, uses, processing, products and the environment. *Journal of Applied Sciences*, *11*, 2084–2091.

Evans, A., Strezov, V., & Evans, T. J. (2009). Assessment of sustainability indicators for renewable energy technologies. *Renewable and Sustainable Energy Reviews*, *13*(05), 1082–1088.

Eveloy, V., & Gebreegziabher, T. (2019). Excess electricity and power-to-gas storage potential in the future renewable-based power generation sector in the United Arab Emirates. *Energy*, *166*, 426–450.

Gambone, S. (2019). *LG Solar Panels vs. SunPower Solar Panels*. Paradise Energy Solutions.

Grilli, G. (2017). Renewable energy and willingness to pay: Evidences from a meta-analysis. *Economics and Policy of Energy and the Environment*.

Gwandu, B. A. L., & Creasey, D. J. (1995). Humidity: A factor in the appropriate positioning of a Photovoltaic power station. *Renewable Energy*, *6*(3), 313–6.

Herath, H. M., & Mittal, M. (2022). Adoption of artificial intelligence in smart cities: A comprehensive review. *International Journal of Information Management Data Insights*, *2*(1), 100076.

Hussain, A., Arif, S. M., & Aslam, M. (2017). Emerging renewable and sustainable energy technologies: State of the art. *Renewable and Sustainable Energy Reviews*, *71*, 12–28.

Hussain, H. M., Rahi, K., Al Tarawneh, M., & Preece, C. (2022). Developing applicable scenarios to install and utilize solar panels in the Houses of Abu Dhabi City. *Sustainability*, *14*(22), 15361.

International Trade Administration. (2020, September 12). United Arab Emirates – Renewable Energy. (Online). https://www.trade.gov/country-commercial-guides/united-arab-emirates-renewable-energy

James, P. (2015). *Urban sustainability in theory and practice: Circles of sustainability* (pp. 19–21). Routledge.

Kaaya, I., Koehl, M., Mehilli, A. P., de Cardona Mariano, S., & Weiss, K. A. (2019). Modeling outdoor service lifetime prediction of PV modules: Effects of combined climatic stressors on PV module power degradation. *IEEE Journal of Photovoltaics*, *9*(4), 1105–1112.

Kazerani, M., & Tehrani, K. (2020). Grid of hybrid AC/DC microgrids: A new paradigm for smart city of tomorrow. *IEEE*, *15*.

Kempe, M. D. (2006). Modeling of rates of moisture ingress into photovoltaic modules. *Solar Energy Materials and Solar Cells*, *90*(16), 2720–2738.

Kirpichnikova, I., & Chirov, D. (2022). Investigation of the influence of elevated ambient temperatures on the operation of photovoltaic modules. In *IEEE*, Sochi.

Knier, G. (2008, August 06). Chapter 11: Onboard systems. In D. Doody (Ed.), *Basics of spaceflight* (pp. 1–4). NASA.

Krishnamoorthy, M., Vimal Raj Periyanayagam, A. D., Santhan Kumar, Ch., Praveen Kumar, B., Srinivasan, S., & Kathiravan, P. (2020, July). Optimal sizing selection, and techno-economic analysis of battery storage for PV/BG-based hybrid rural electrification system. *IETE Journal of Research*, *68*(6), 4061–4076.

Lazarou, S., Pyrgioti, E., & Agoris, D. (2007). The latest Greek statute laws and its consequences to the Greek renewable energy source market. *Energy Policy*, *35*(08), 4009–4017.

Lenzen, M. (2010, March). Current state of development of electricity-generating technologies: A literature review. *Energies*, *3*(3), 462–591.

LG. (2021). LG solar module limited warranty. (Online). Retrieved September 15, 2021, from https://www.lg.com/us/business/download/resources/BT00002151/BT00002151_2573.pdf

Li, H., Jenkins-Smith, H. C., Silva, C. L., Berrens, R. P., & Herron, K. G. (2009). Public support for reducing US reliance on fossil fuels: Investigating household willingness-to-pay for energy research and development. *Ecological Economics*, *68*(03), 731–742.

Mirasgedis, S., Diakoulaki, D., Papagiannakis, L., & Zervos, A. (2000). Impact of social costing on the competitiveness of renewable energies: The case of Crete. *Energy Policy*, *28*(01), 65–73.

Naam, R. (2013). *The infinite resource: The power of ideas on a finite planet.* University Press of New England.

Ng, A. W., Wu, A., & Wut, E. T. M. (2023). *AI optimized solar tracking system for green and intelligent building development in an urban environment* (pp. 1935–1951). Springer.

Nyholm, E., Goop, J., Odenberger, M., & Johnsson, F. (2016). Solar photovoltaic-battery systems in Swedish households – Selfconsumption and self-sufficiency. *Applied Energy*, *183*, 148–159.

Oikonomou, E. K., Kilias, V., Goumas, A., Rigopoulos, A., Karakatsani, E., Damasiotis, M., Papastefanakis, D., & Marini, N. (2009). Renewable energy sources (RES) projects and their barriers on a regional scale: The case study of wind parks in the Dodecanese islands, Greece. *Energy Policy*, *37*(11), 4874–4883.

Onen, A. (2021). *Role of artificial intelligence in smart grids*. Springer.

Peng, T., Li, Y., Song, Z., Fu, Y., Nazir, M. S., & Zhang, C. (2023). Hybrid intelligent deep learning model for solar radiation forecasting using optimal variational mode decomposition and evolutionary deep belief network – Online sequential extreme learning machine. *Journal of Building Engineering*, *76*, 107227.

Rahi, K., & Bourgault, M. (2022). Validation of a new project resilience scale in the IT Sector. *Project Management Journal*, *53*(6). https://doi.org/10.1177/87569728221114321

Rahi, K., Bourgault, M., & Preece, C. (2021). Risk and vulnerability management, project agility and resilience: A comparative analysis. *IJISPM - International Journal of Information Systems and Project Management*, *9*(4), 5–21.

Ramachandran, T., Mourad, A.-H. I., & Hamed, F. (2022). A review on solar energy utilization and projects: Development in and around the UAE. *Energies*, *15*(10), 1–27.

Salimi, M., Hosseinpour, M., & Borhani, T. N. (2022). Analysis of solar energy development strategies for a successful energy transition in the UAE. *Processes*, *10*(7), 1338.

Stigka, E. K., Paravantis, J. A., & Mihalakakou, G. K. (2014). Social acceptance of renewable energy sources: A review of contingent valuation applications. *Renewable and Sustainable Energy Reviews*, *32*, 100–106.

Sugimura, R. S., Wen, L. C., Mon, G. R., & Ross, R. S. (1990). Test techniques for voltage/humidity induced degradation of thin-film photovoltaic modules. *Solar Cells*, *28*(2), 103–114.

Tan, C. M., Chen, B. K. E., & Toh, K. P. (2010). Humidity study of a-Si PV cell. *Microelectronics Reliability*, *50*(9-11), 1871–1874.

Tarsheed. (2020). Tarsheed AD. (Online). Retrieved October 14, 2021, from https://www.tarsheedad.com/en-us/Pages/electricityconservation.aspx

United Nations. (2022). *What is renewable energy?*

US Energy Information Administration. (2022). *Energy explained.* (Online). https://www.eia.gov/energyexplained/electricity/electricity-in-the-us.php

Varun, Prakash, R., & Bhat, I. K. (2009). Energy, economics, and environmental impacts of renewable energy systems. *Renewable and Sustainable Energy Reviews*, *13*(09), 2716–2721.

Verona, R. (2022). *Economic analysis of rooftop solar PV panels.* Social Sciences DataLab.

Wang, Q., Chen, Xi, Jha, A. N., & Rogers, H. (2014). Natural gas from shale formation – The evolution, evidences and challenges of shale gas revolution in United States. *Renewable and Sustainable Energy Reviews*, *30*, 1–28.

Wiser, R. H. (2007). Using contingent valuation to explore willingness to pay for renewable energy: A comparison of collective and voluntary payment vehicles. *Ecological Economics*, *62*(3–4), 419–432.

World Future Energy Summit (WFES). (2021, April 05). (Online). https://www.worldfutureenergysummit.com/en-gb/exhibit.html

Zarnikau, J. (2003). Consumer demand for 'green power' and energy efficiency. *Energy Policy*, *31*(15), 1661–1672.

Zografakis, N., Sifaki, E., Pagalou, M., Nikitaki, G., Psarakis, V., & Tsagarakis, K. P. (2010). Assessment of public acceptance and willingness to pay for renewable energy sources in Crete. *Renewable and Sustainable Energy Reviews*, *14*(03), 1088–1095.

Chapter 16

Strategic Planning for the Development of Smart Cities in Oman

Hussin A. M. Yahia[a], *Azzan Majid Al-Shukaili*[a], *Ram Kishore Manchiryal*[a], *Taleb Eissa*[b] *and Ali Ahmed Mohammed*[c]

[a]*Department of Civil and Mechanical Engineering, Middle East Collage, Knowledge Oasis Muscat, Al Rusayl, Sultanate of Oman*
[b]*Department of Civil Engineering, Florida Institute of Technology Melbourne, Florida, USA*
[c]*Department of Civil Engineering, Lassonde School of Engineering, York University, Toronto, Canada*

Abstract

The global urban population is expanding rapidly, leading to higher energy consumption, air pollution, and traffic jams in cities that require more efficient solutions. Smart cities are anticipated to address the efficiency of urban infrastructure. In the Sultanate of Oman, the government launched the idea of establishing smart cities to promote large-scale planned urban expansion, especially in the capital, Muscat, as well as to promote accelerated development and growth. As the Sultanate embarks on this journey to establish smart cities "such as the Sultan Haitham City Project" to drive development in the country. The objective of this study is to ascertain the principles and facets of sustainable development and its contribution to the formulation of urban development policies. Additionally, this study seeks to evaluate the notion of smart cities along with their distinctive features and constituents. As the Sultanate of Oman embarks on this journey to create engines and alternatives to develop the economy and push the wheel of development in the country according to Oman Vision 2040, this chapter outlines the way to make this dream a reality. The chapter provides strategies for planned urban expansion, especially in the capital, Muscat,

The Emerald Handbook of Smart Cities in the Gulf Region: Innovation, Development, Transformation, and Prosperity for Vision 2040, 289–304

doi:10.1108/978-1-83608-292-720241016

and guidelines for the development of smart cities. It explains in detail the difficulties faced by some smart cities in the world.

Keywords: Smart cities; Muscat; urbanization; infrastructure; development

Introduction

At the prevailing time, over 50% of the arena's population is dwelling in urban regions, which is around three billion, according to the United Nations Population Fund cited, and this number is predicted to grow to 70% by 2050 (Messerli et al., 2019). This implies that the major social issues affecting society will also grow (Olén & Lehsten, 2022). Currently, the main issues facing large cities are population growth, pollution, traffic, infrastructure constraints, and new energy and energy requirements. That is why cities around the world are looking for ways to move forward and improve their lives by adopting intelligent and innovative technologies that solve these issues and not only enhance people's lives but also help them become more sustainable spaces, more energy efficient, and less polluting (Abdoullaev Azamat, 2011). This led to the emergence of the term "smart cities" in response to the need to adopt modern urban development that leverages the growth of digital technologies.

As technology and information have advanced rapidly since the start of the 21st century, they have permeated many areas of life. This has led to a new society built on advanced technology, where people are using modern methods of communication and information instead of traditional methods. Developing smart cities is a way to keep up with the current social trends (Dameri & Cocchia, 2013). The concept of smart cities varies depending on the applications in the city and can be applied to both existing and historic buildings that cannot be removed due to their design. Another approach is to build and equip a smart city with all modern communication methods, which is in line with the current needs of modern cities. Planning plays a crucial role in developing future solutions that cater to the requirements of intelligent cities in the forthcoming years (Sukarno & Putri, 2022).

Various innovations have been implemented for smart cities that rely heavily on technology and the information network, including digital, electronic, virtual, and intelligent cities (Qonita & Giyarsih, 2023). The main goal of smart cities is to provide efficient services to individuals. Electronic services stand out due to their exceptional efficiency, eliminating spatial limitations and saving valuable time and effort. By leveraging the principles, features, and implementation of smart cities, it establishes the scientific groundwork and fosters the development of new smart cities. Moreover, it will transition existing cities into smart and sustainable urban areas (Al Sulaimani & Yahia, 2021).

Cities have a significant impact on the environmental, social, and economic aspects of our lives. With the progress in technology and science, human activities have spread across different sectors. The remarkable growth in information technology has transitioned the world from the Industrial Age to the era of information technology. This advancement has facilitated the rapid progress of information and communication technology (ICT), which is intertwined with all urban activities. Terms and concepts such as e-commerce, e-government and e-learning, have become prevalent in societal life, contributing to the emergence of the Information Society. With the onset of the COVID-19 pandemic, societies worldwide have increasingly relied on ICTs to sustain their diverse activities. Presently, no city is without wired and wireless communication networks, offering services and support to individuals across various domains (Kutty et al., 2022). The swift evolution of information technology has prompted certain cities to embrace technological solutions, leading to the development of innovative models in urban planning and building sustainability, culminating in the concept of smart cities where infrastructure and technology harmoniously coexist (Estevez et al., 2021). However, a smart city's potential remains untapped without the active participation of individuals who can leverage the innovative services tailored for them.

Smart Sustainable Cities

Today, the concept of smart cities is widely discussed as it aims to create a digital environment that is both environmentally friendly and stimulating for learning and creativity. It strives to establish a sustainable setting that promotes well-being and happiness. However, there is no universally agreed-upon definition of smart cities among global organizations, making it challenging to measure their success. In academia, smart city definitions vary, reflecting the diverse trends and social movements that have contributed to their emergence. According to Droege (2018), the concept of smart cities was initially associated with virtual cities that represented simulated versions of real cities. It encompasses digital, wired, wireless, informational, cognitive, and electronic cities. Smart cities utilize a wide range of technologies to enhance the quality of life for individuals residing in urban areas. As explained by Komninos (2009), a smart city refers to a regional system that combines knowledge-based activities and institutions to foster education, creativity, and digital spaces that encourage interaction and problem-solving within the city.

The distinguishing characteristic of a smart city is its high level of creativity, as creativity is considered a fundamental aspect of intelligence. By leveraging advanced innovations from modern technologies and communications, traditional institutions and administrations in smart cities become more efficient in serving their residents and businesses (Ferrer et al., 2017). Smart cities are also predictive in nature, utilizing monitoring devices with data analysis capabilities to anticipate future trends and activities (Halegoua, 2020). In smart cities, citizens enjoy a prosperous economy and a high quality of life (Zhuhadar et al., 2017). These cities prioritize sustainability criteria to ensure a promising future for generations to come.

Despite the various interpretations of smart cities, most emphasize the importance of communication infrastructure, information technology, human intelligence, sustainability, and quality of life. Merely having a digital representation of a city through ICTs is not sufficient to label it as a smart city. To encompass all these essential aspects, smart cities can be defined as urban areas that are supported by networks and digital technologies, providing interactive electronic services across various sectors (Backhouse & al Hadhrami, 2022). These cities can tackle challenges by enhancing the intelligence of people, organizations, and technologies, while also promoting social and environmental sustainability. They also rely on a knowledge-based economy to enhance competitiveness and serve as a development model for cities and regions. The key to achieving sustainable economic growth and improving quality of life involves utilizing ICT infrastructure, as well as investing in the intelligence of individuals, organizations, and technologies to drive innovation and problem-solving capabilities (Locurcio et al., 2023).

Smart Cities Levels

To create a series of multi-level common components, it is essential to thoroughly examine various theories and research on smart cities and adhere to the fundamental principles of the commonly used framework (Halegoua, 2020). Smart cities are structured by integrating three levels, which encompass collective and human artificial intelligence:

- Individuals,
- Institutions, and
- Technology.

The initial level pertains to individuals: it serves as the foundational tier of the smart city, comprising the productive groups within the city engaged in manufacturing and services. This level encompasses individuals who determine the organization of work and the city's development mechanism, namely the creators' category. At this level, it is imperative to facilitate the exchange of information among individuals. Sharing information enhances the efficiency of a smart city. This level is closely linked to city residents who are intelligent and innovative. Cities strive to attract creative and skilled individuals to foster economic and social growth and advancement (Lyu et al., 2022).

The second level focuses on collaborative institutions and the collective intelligence of the city's inhabitants. It encompasses institutional mechanisms that govern the flow of knowledge and collaboration in education and innovation, incorporating entities that support research and development, creativity, training facilities, technology exchange, and intellectual property sharing. These institutions and organizations generate knowledge for individuals and establish organizational regulations that aid in cultivating a workforce equipped with research and innovation skills (Samad & Azar, 2019).

The third level comprises communication and information infrastructure, along with digital tools that establish a virtual environment reliant on interactive technologies and multimedia tools, offering services across various domains such as technology transfer and management. Its significance lies in overseeing novel initiatives and delivering services that enhance the city's functionality (Pan et al., 2022).

Dimensions of a Smart City

The dimensions of a smart city pertain to the progress and expansion of urban areas, encompassing aspects like transportation, economy, natural resources, quality of life, and governance. These dimensions serve as a framework for enhancing various sectors within the city. An intelligent city is characterized by six dimensions: smart economy, smart life, smart environment, smart people, smart mobility, and smart governance.

Urbanization is rapidly increasing worldwide. Even though cities cover a small portion of the Earth's surface, over half of the global population resides in urban areas. The United Nations predicts that by 2050, urban dwellers will make up 64.8% of the population, rising to 88.4% in high-income nations. With urban population growth on the rise, the importance of efficient urban management is more crucial than ever (Burki, 2021). This issue is particularly pertinent for the Gulf Cooperation Council (GCC) countries, which are highly urbanized, with over 90% of their population expected to live in cities by 2050 (Al-Yousfi et al., 2019).

Research Methodology

The primary aim of gathering and evaluating information is to understand the factual details relevant to the research. Nayak and Singh (2021) highlighted the significance of data collection in research, which encompassed both quantitative and qualitative methods. The study outlines the procedures involved in collecting and analyzing data. Quantitative data was gathered through a questionnaire distributed to engineers, while qualitative data were obtained through interviews with experts knowledgeable about smart cities and sustainability. The sample questionnaire was analyzed using SPSS, and the interviews with experts were analyzed using the constant comparative method.

This research emphasizes the challenges faced by Muscat City as a case study and offers specific strategies to effectively guide its future growth and prevent common issues. It aims to develop ideas and proposals that will contribute to future development in this direction. The objective of this study is to strategically plan for the advancement of smart cities in line with Oman Vision 2040. The key goals are as follows:

1. Recognize the principles and elements of sustainable development and how they influence urban development policies in Oman.
2. Assess the idea of smart cities, examining their features and elements through a review of international examples of creating and evolving smart cities.
3. Explore strategic methods for advancing smart, sustainable cities and communities in Oman.

Results and Discussion

206 questionnaires were distributed to engineers and academics specializing in smart cities in selected areas of Muscat, such as Madinat Al Irfan, Sultan Qaboos

University, and Petroleum Development Oman. The questionnaire questions focused on the following topics:

- General information.
- Knowledge of smart cities.
- How satisfied are you with smart cities?
- Smart cities challenges.
- Oman strategy toward the smart city.

The purpose of the questionnaire is to assess people's knowledge of smart cities, their characteristics, and their relationship to sustainability, in addition to participants' satisfaction with smart cities and infrastructure readiness in the Sultanate, as well as the most important challenges facing the establishment of smart cities in Oman. The participants' responses are shown in Table 2.

The total number of respondents was 206, with 135 males answering at 65%. Seventy-one females answered at 34.5%. The age range of respondents ranged from 31 to 40 years old, with a 50.5% response rate (Table 1). The age group of 18–30 years accounted for 42.2% of respondents, and the age group of

Table 1. The Participants' Responses.

Variables	**Category**	**Results**	
		F	**%**
Gender	Male	135	65.5
	Female	71	34.5
Nationality	Omani	190	92.2
	Non-Omani	16	7.8
Age Group	18–30 years old	87	42.2
	31–40 years old	104	50.5
	41–50 years old	14	6.8
	>=51 years old	1	0.5
Education	High School Diploma	8	3.9
	University Diploma	24	11.7
	Bachelor	145	70.4
	Master's degree or above	29	14.1
Occupation	Government employees	67	32.5
	Businessmen	16	7.8
	Employees in the private sector	83	40.3
	Student	16	7.8
	Others	24	11.7

Table 1. (*Continued*)

Variables	Category	Results	
		F	%
How long is your experience	No experience	34	16.5
	1–5 years	76	36.9
	6–12 years	69	33.5
	More than 12 years	27	13.1

Table 2. Descriptive Statistics of Knowledge of Smart Cities.

Factors	Strongly Agree	Agree	Neutral	Disagree	Strongly Disagree	Mean	Std. Deviation
Smart cities are urban areas that support networks and digital technologies.	40.3	45.1	9.7	4.4	0.5	1.8	0.825
There is a relationship between sustainability and smart cities.	34.5	52.9	10.7	1.9	0.00	1.80	.701
There is a relationship between smart cities and ICT.	35.9	42.2	16.5	4.9	0.5	1.92	.871
There is easy integration of society into the world of information technology.	28.6	46.6	17.5	7.3	0.00	2.03	.869
Smart cities mainly aim to promote science and knowledge.	31.6	50.5	10.7	7.3	0.00	1.94	.844
Smart cities provide an intelligent transportation system.	40.8	48.5	6.8	3.9	0.00	1.74	.752

41–50 years accounted for 6.8%. The age group of 51 + answered at 0.5%. The highest percentage of respondents had a bachelor's degree, ranging from 70% to 80%. The lowest percentage had a high school diploma (3.9%). The second part of the survey focused on respondents' understanding of smart cities and how they relate to sustainability. The respondents' responses are presented in Table 2.

Most participants confirmed that there is a relationship between sustainability and smart cities. Smart cities also provide a smart, pollution-free transportation system that keeps pace with the requirements of the times. Smart cities are a type of urban development that meets the new needs of institutions, companies, and citizens in the Sultanate of Oman, attracts investors and capital, and helps in the economic development of the country. Most participants also confirmed, as shown in Table 3, that the infrastructure in Oman is ready for transformation into smart cities.

Table 3. Satisfaction with Smart Cities.

Factors	**Strongly Agree**	**Agree**	**Neutral**	**Disagree**	**Strongly Disagree**	**Mean**	**Std. Deviation**
Do you agree with the establishment of smart cities in Oman?	48.5	36.4	10.2	4.9	0.00	1.71	0.838
Infrastructure is ready in Oman to convert to smart cities	13.1	35	20.9	24.3	6.8	**2.77**	**1.158**
ICT is a prerequisite for the success of smart cities and the effectiveness of their services.	28.6	52.9	13.6	4.9	0.00	**1.95**	**0.785**
A feature of a smart city is that it offers an appropriate level of culture.	18.9	57.3	18.4	5.3	0.00	2.10	0.761
Smart cities are linked to sustainability.	26.7	51.5	16.0	4.9	1.0	2.02	0.844
Smart cities are a kind of urban development that meets new needs of enterprises, companies and citizens.	28.2	59.7	7.8	4.4	0.0	1.88	0.723

Table 3. (*Continued*)

Factors	Strongly Agree	Agree	Neutral	Disagree	Strongly Disagree	Mean	Std. Deviation
Smart cities attract investors and raise the economic level.	35.9	50.0	9.7	4.4	0.00	1.83	0.777
Developing the government work system through technologies and providing government services through electronic channels.	31.6	55.8	9.7	2.4	0.5	1.84	0.729
The government plays a major role in adopting smart cities.	35.9	48.1	12.1	1.9	1.9	1.86	0.847

Smart Cities Challenges

The participants stressed that smart cities need a wide range of technical technologies capable of working together in light of the presence of a strong Internet that keeps pace with new generations. Many of the challenges facing the establishment of smart cities in the Sultanate of Oman were clarified, as shown in Table 4.

The challenges facing many countries in creating smart cities that keep pace with urban development were also clarified.

1. **Economic Challenges**: Developing a smart city requires the utilization of ICTs. Efficient energy usage, pollution reduction, decreased transportation expenses, boosted GDP and economic development, enhanced tourism, solar-powered waste compactors, energy consumption monitoring through smart meters and monitors, sustainable waste management, as well as seeking support from both national and international entities; all necessitate a substantial budget for execution (Behzadfar et al., 2017).
2. **Technology Challenges**: Smart cities require a diverse array of technologies that can collaborate to form an integrated system and develop solutions. Various types of technological devices are utilized for sensing, capturing, storing, and utilizing data from a variety of sources, both fixed and mobile (ISO-IEC standard, smart cities, 2015).
3. **Theoretical Challenges**: Nations lack a standardized understanding of what constitutes a smart city, including its definition and architectural design. Consequently, countries are unable to collaborate or establish a common

Table 4. Challenges Smart Cities.

Factors	Strongly Agree	Agree	Neutral	Disagree	Strongly Disagree	Mean	Std. Deviation
Network speed	50.0	31.1	10.2	6.8	1.9	1.80	1.006
Satellites Internet	32.5	47.1	12.5	5.3	2.4	1.98	0.942
Privacy	35.0	47.1	15.0	2.4	0.5	1.86	0.791
Difficulties use	18.0	41.7	27.7	10.7	1.9	2.37	0.963
Implementation cost	27.7	49.5	15.0	6.8	1.0	2.04	0.888
Operation and maintenance cost	27.7	51.9	14.1	4.9	1.5	2.00	0.864
Private sector participation	32.5	46.1	16.0	3.9	1.5	1.96	0.880
Maintaining skilled human resources	31.1	48.5	14.1	5.3	1.0	1.97	0.869
Need better understanding of how to get started	34.5	43.2	17.0	4.9	0.5	1.94	0.867
Need more internal capacity	37.9	45.1	11.7	2.9	2.4	1.87	0.904
Need more supportive policies	34.5	47.6	15.0	2.9	0.00	1.86	0.772
Budget limitations	36.4	47.6	9.7	6.3	0.00	1.86	0.835
Need more supporting infrastructure	36.9	49.0	9.2	4.4	0.5	1.83	0.808
Too reliant on legacy systems	27.2	49.5	18.4	3.9	1.0	2.02	0.838
Need more long-term vision or plan	30.1	50.5	15.0	1.9	2.4	1.96	0.866
Need more project management capabilities	35.0	49.0	11.2	4.4	0.5	1.86	0.815
Need to gain community support	30.1	50.0	17.0	1.9	1.0	1.94	0.797

smart city framework. Therefore, further theoretical research is essential to address the challenges associated with smart cities (Cui et al., 2018).

4. **Infrastructure**: The advancement of technology is rapidly progressing, and numerous developed nations possess the expertise in urban planning to create smart cities. However, inadequate ICT infrastructures and deficiencies in crucial skills pose a challenge in implementing these smart city initiatives (Giffinger et al., 2007).
5. **People's Challenges**: According to Behzadfar et al. (2017), individuals may not embrace every positive and negative feature of a smart city, or they may find virtual life, atmospheric waves, and low network security issues unsatisfactory.
6. **Cultural Challenges**: Utilizing technology can aid in preserving the unique heritage and overall cultural identity of a community. Neglecting the traditional culture of a region may inadvertently lead to the emergence of new cultural norms facilitated by technology.

Oman Strategy in the Smart City

A smart city is a location created to utilize smart devices in order to access real-time data on various aspects such as economy, environment, transportation, and education. The main goal is to enhance people's health by providing smart care, promoting well-being, and enabling smart living (Safinia et al., 2017). The intelligent devices are focused on analyzing, predicting, understanding, and enhancing the following areas:

- City logistics and mobility
- Smart living
- Environment and energy
- Citizen e-participation and engagement
- Big Data analytical solutions
- ICT infrastructure and IoT
- Smart buildings
- Smart education and economy
- eMarketing strategies.

It is essential to thoroughly analyze the effective strategies carried out in various cities around the globe in order to learn from their achievements. Nevertheless, it is unwise to simply copy these approaches without consideration, as every city has its own unique background, location, and cultural beliefs.

- **Sultan Haitham City**: constitutes a futuristic urban icon, a treasure for the future, and a sustainable gift from Sultan Haitham bin Tarik to future generations and all segments of society. Sultan Haitham City is the result of Oman's comprehensive transformation approach, aiming to achieve sustainable urban development and benefit thriving communities. It is a prime example of the country's vision for the future.

This city serves as a blueprint for building sustainable cities, embodying the aspirations of Oman's youth. It embraces modern living while prioritizing sustainability. Spanning an impressive 14,800,000 square meters, Sultan Haitham City has been meticulously planned to incorporate green spaces covering 2,900,000 square meters. It is designed to accommodate 100,000 residents, offering a variety of housing options including separate villas, semi-attached villas, townhouses, and flats.

The city is divided into 19 integrated residential neighborhoods as shown in Fig. 1, each equipped with a range of facilities and services to cater to the needs of its residents. It boasts 23 mosques, including a grand mosque, as well as 11 health facilities. These health facilities include two centers dedicated to serving 20,000 patients and six centers with a capacity to accommodate 10,000 patients. Additionally, the city will feature a referral hospital with 1,200 beds, a private hospital, and a center specifically designed to support individuals with disabilities and elderly citizens. Education is also a priority, with 39 government and private schools covering all levels of education. Sultan Haitham City is a testament to Oman's commitment to sustainable urban development and its vision for the future. It sets a new standard for cities in the region and beyond (Al Sulaimani & Yahia, 2021).

- **The city of Nizwa**: serves as a prime example of the country's urban landscape. Nizwa is widely recognized for its rich historical significance, notably highlighted by its iconic castle and bustling central market, catering to both local residents and visitors from nearby cities. Additionally, the city is home to a collection of ancient clay houses. However, when these historic dwellings were abandoned by their occupants, various challenges emerged, including unauthorized occupation, which caused concern among residents and disrupted the overall stability of the area. In response to this situation, the community initiated an impressive endeavor to repurpose these houses into heritage inns, effectively revitalizing the city's economy and creating employment

Fig. 1. Sultan Haitham City. https://wfmmedia.com/oman-sultan-haitham-city-futuristic-urban-marvel/

Fig. 2. The Sustainable City Yiti. https://esgmena.com/2024/01/18/the-sustainable-city-yiti-shares-strategies-to-become-net-zero-emissions-community-by-2040/

opportunities for the local population. By embracing its "heritage character," Nizwa has successfully positioned itself as a sustainable city, aligning with the Vision 2040's objective of transforming Oman's urban centers into sustainable hubs, each with its distinct identity shaped by its geographical location, cultural heritage, historical significance, and, most importantly, its residents (Laila AL Hadhrami).

- **The Sustainable City Yiti**: The Yiti Integrated Tourism Development serves as a representation of urbanization as shown in Fig. 2 and sustainability in the Sultanate of Oman. Situated in the capital city of Muscat by the tranquil waters of the Arabian Sea, this forthcoming sustainable destination is characterized by its abundant environmental variety, distinctive sustainable activities, and promising prospects. Through the utilization of renewable energy, water conservation, waste management, sustainable food production, eco-friendly transportation, and improved air quality, The Sustainable City - Yiti strives to achieve net zero emissions by 2040 in alignment with Oman's sustainability objectives. With a focus on its residents, the city's strong sense of community remains steadfast as we collaborate with our inhabitants to spearhead the transition toward a more promising future.

Conclusions

The study provided a clear explanation of smart cities, outlining their dimensions, prerequisites, and uses. It also delved into the challenges hindering the implementation of smart city technologies, proposing a roadmap for transitioning traditional cities into smart ones based on scientific criteria derived from theoretical research. The research thoroughly examined the issue at hand through thorough analysis and inference, culminating in a methodology for developing smart cities that incorporates a local case study as an essential component of smart infrastructure. Consequently, the study established its goals, addressed the key questions, and arrived at a series of findings and suggestions.

The primary aim of this research was to investigate strategies for developing smart, sustainable cities and communities in Oman. One of the key recommendations outlined in the approach for establishing smart cities in Oman is to align the plan with the Oman Vision 2040 strategy, requiring all entities to collaborate toward achieving the objective. The following points, derived from the findings, contribute to the strategy:

- Enhancing the communication infrastructure and constructing an advanced, reliable network to deliver continuous and efficient services with appropriate redundancy. This includes regular assessments to keep pace with the rapid advancements in communication and information technologies.
- Creating a legislative, legal, and investment environment that attracts local, regional, and international investments supporting smart technologies.
- Developing and upskilling the workforce through professional training, preparing skilled workers, and equipping university graduates with the latest technologies.
- Ensuring that smart city applications cater to the needs of the entire community, especially the most vulnerable social groups, and are user-friendly.
- Emphasizing the importance of collaboration and coordination among different stakeholders, fostering partnerships between the public and private sectors.
- Implementing a phased time plan to allow for greater flexibility in execution.
- Promoting electronic applications to facilitate societal integration into digital interactions, ensuring easy access to essential services and guidance on their utilization.

Authors' Contribution

This chapter is the result of the joint work of the authors. "Abstract" and "Introduction" by Azzan Majid Al-Shukaili. Hussin Yahia wrote "Methodological Approach" and Analysis Data. Dr Ram wrote "Conclusions." Taleb and Ali coordinated and supervised this chapter.

Acknowledgements

This study was supported by the Department of Civil and Mechanical Engineering in Middle East College, Oman.

References

Abdoullaev Azamat. (2011). Keynote: "A Smart World: A Development Model for Intelligent Cities". In *The 11th IEEE international conference on computer and information technology (CIT-2011)*.

Al Sulaimani, S. H. K., & Yahia, H. (2021). Evaluating the impact of change orders on construction projects in Oman. *Journal of Student Research, 10*(3). https://doi.org/10.47611/jsr.v10i3.1296.

Al-Yousfi, W. M., Yahia, H. A., & Kishore, R. (2019). Instigating smart city infrastructure in Oman: Considering social benefits and economic values. *Journal of Student Research*. https://doi.org/10.47611/jsr.vi.10

Backhouse, J., & al Hadhrami, L. (2022). Developing smart city ambassadors in Oman. In P. Fitsilis (Ed.), *Building on smart cities skills and competences: Human factors affecting smart cities development* (pp. 201–216). Springer International Publishing.

Behzadfar, M., Ghalehnoee, M., Dadkhah, M., & Haghighi, N. M. (2017). International challenges of smart cities. *Armanshahr Architecture & Urban Development, 10*(20), 79–90.

Burki, T. (2021). European Commission classifies obesity as a chronic disease. *The Lancet Diabetes & Endocrinology, 9*(7), 418.

Cui, L., Xie, G., Qu, Y., Gao, L., & Yang, Y. (2018). Security and privacy in smart cities: Challenges and opportunities. *IEEE Access, 6*, 46134–46145.

Dameri, R. P., & Cocchia, A. (2013). Smart city and digital city: Twenty years of terminology evolution. In *ITAIS Italian conference of information systems* (p. 4).

Droege, P. (Ed.). (2018). *Urban energy transition: Renewable strategies for cities and regions*. Elsevier.

Estevez, E., Cenci, K., Fillottrani, P., & Janowski, T. (2021). Review of international standards and policy guidelines for smart sustainable cities. In *Smart cities and smart governance: Towards the 22nd century sustainable city* (pp. 69–99).

Ferrer, J. N., Taranic, I., Veum, K., Van dem Oosterkamp, P., & Wilson, C. (2017). *The making of a smart city: Policy recommendations*. Center for European Policy Studies (CEPS) & Energy Research Centre (ECN). EU Smart Cities Information System. https://smart-cities-marketplace.ec.europa.eu/sites/default/files/2021-04/the_making_of_a_smart_city_-_policy_recommendations.pdf

Giffinger, R., Fertner, C., Kramar, H., Kalasek, R., Pichler-Milanovic, N., & Meijers, E. J. (2007). Smart cities. *Ranking of European medium-sized cities. Final Report*.

Halegoua, G. (2020). *Smart cities*. MIT press. https://2u.pw/ilQjAqAI

ISO-IEC JTC 1 N11712 CESI. (2014). *contribution on possible work on Smart Cities*. ISO/IEC JTC 1

Komninos, N. (2009). Intelligent cities: Towards interactive and global innovation environments. *International Journal of Innovation and Regional Development, 1*(4), 337–355.

Kutty, A. A., Kucukvar, M., Abdella, G. M., Bulak, M. E., & Onat, N. C. (2022). Sustainability performance of European smart cities: A novel DEA approach with double frontiers. *Sustainable Cities and Society, 81*, 103777. https://doi.org/10.1016/j.scs.2022.103777

Locurcio, M., Tajani, F., & Anelli, D. (2023). Sustainable urban planning models for new smart cities and effective management of land take dynamics. *Land, 12*, 621. https://doi.org/10.3390/land12030621

Lyu, J., Hamzah, H., & Tedong, P. A. (2022). A framework for the analysis of urban innovation in smart cities: Literature review findings. *Planning Malaysia, 20*.

Messerli, P., Murniningtyas, E., Eloundou-Enyegue, P., Foli, E. G., Furman, E., Glassman, A., & Hernández Licona, G. (2019). *Global sustainable development report 2019: the future is now–science for achieving sustainable development*. New York.

Nayak, J. K., & Singh, P. (2021). *Fundamentals of research methodology problems and prospects*. SSDN Publishers & Distributors.

Olén, N. B., & Lehsten, V. (2022). High-resolution global population projections dataset developed with CMIP6 RCP and SSP scenarios for year 2010–2100. *Data Brief, 40*, 107804. https://doi.org/10.1016/j.dib.2022.107804

Pan, H., Kwak, Y., & Deal, B. (2022). Participatory development of planning support systems to improve empowerment and localization. *Journal of Urban Technology*, *29*(2), 33–54. https://doi.org/10.1080/10630732.2022.2031431

Qonita, M., & Giyarsih, S. R. (2023). Smart city assessment using the Boyd Cohen smart city wheel in Salatiga, Indonesia. *GeoJournal*, *88*(1), 479–492.

Safinia, S., Al-Hinai, Z., Yahia, H. A., & Abushammala, M. F. (2017). Sustainable construction in sultanate of Oman: Factors effecting materials utilization. *Procedia Engineering*, *196*, 980–987.

Samad, W. A., & Azar, E. (2019). *Smart cities in the Gulf*. Springer.

Sukarno, M., & Putri, S. A. G. (2022, December). Smart environment planning for smart city based on regional medium-term development plan Surabaya City 2021-2026. In *IOP conference series: Earth and environmental science* (Vol. 1105, No. 1, p. 012023). IOP Publishing.

The current leading academic non-vendor text is perhaps Townsend, supran X. http://www.cersi.it/itais2013/pdf/119.pdf

Zhuhadar, L., Thrasher, E., Marklin, S., & de Pablos, P. O. (2017). The next wave of innovation – Review of smart cities intelligent operation systems. *Computers in Human Behavior*, *66*, 273–281.

Chapter 17

Toward Kuwait Seaport's Sustainability 2040: The Implication of Digital Twins Technology on Terminal Containers' Transformation

Zeinab Gamal[a,b], *Ahmed Aboualam*[c] *and*
Munther Abbas Alkandari[d]

[a]*Higher Institute of Computer Science and Business Administration, Egypt*
[b]*Onaizah Colleges, Saudi Arabia*
[c]*Canadian International College, Cape Breton University, Egypt*
[d]*Higher Institute of Telecommunication and Navigation, Kuwait*

Abstract

Seaports seek to find innovative technological solutions to deal with the ever-increasing growth of transportation and increasing the intensity of competition through the use of emerging technology such as digital twin technology to improve the quality of their logistics operations. Despite the success of digital twins in many industries, there is still a lack of their application in the field of seaports where ports play a central role as part of global transportation chains. Seaports sustainability comprises three main aspects: the social aspect that encompasses more job opportunities, the economic aspect that enhances foreign trade, and the environmental aspect that refers to the process of managing and operating ports in a way that saves the environment. This chapter discusses how to apply digital twins' technology on the imported Twenty equivalent foot unit (TEUs) taking into consideration the population growth, and the capacity of the storage area of the container terminals in an attempt to explore the impact on Kuwait's seaports sustainability. The study provides a framework for capacity management in an attempt to initiate the next generation of smart

The Emerald Handbook of Smart Cities in the Gulf Region:
Innovation, Development, Transformation, and Prosperity for Vision 2040, 305–326

doi:10.1108/978-1-83608-292-720241017

seaports cities and consequently impact society, economy, and well-being in Kuwait and Gulf region. The results of the study showed that there is a strong correlation between population growth and imported TEUs growth during the essential stage of the study. The correlation factor was 0.97, and this correlation will support the prediction until Kuwait vision 2040.

Keywords: Kuwait 2040; digital twins' technology; container terminal; TEUs; population growth rate; seaports sustainability; smart cities

1. Introduction

Emerging technologies such as digital twins have become essential for rapid development and employment in the revolution (Industry 5.0), especially with the increasing reliance on Internet of Things technologies, which helps in utilizing limited resources to enable safe transportation of materials (Sasikumar et al., 2023). Digital twins are an emerging technology surrounded by many promises and possibilities to reshape the future of industries and society in general; it is also a system of systems that go far beyond traditional computer-based simulations and analysis. This system replicates all the elements, processes, dynamics, and firmware of a physical system in a digital counterpart. The two systems (physical and digital) exist side by side, sharing all inputs and processes using real-time data communications, information transfer, monitoring, analysis, evaluation, and forecasting of the supply chain (Mihai et al., 2022). Digital twins technology enables companies to quickly improve business processes performance (Negri et al., 2020).

The world's leading seaports have realized the potential of digital technologies in providing high-quality seaport logistics services, and some seaports have begun to invest in their implementation (Agatić & Kolanović, 2020). Ports are also the center of more local economies, as they are a gateway to global trade, support local networks, and are thus a catalyst for the economic development of entire regions and countries (Klar et al., 2023a; Platias & Spyrou, 2023). Ports are considered an essential link in the logistical services chain, and many economic, social, and environmental factors require ports to apply digital technology in order to achieve sustainability. Therefore, ports must transform into smart ports by applying new technologies of the revolution (Industry 4.0) to ensure their sustainability and achieve a competitive advantage for seaports (Belmoukari et al., 2023). The result will be an initiation of more smart cities in Kuwait and the Gulf region. Port sustainability refers to approaches and methods that a port adopts to meet present and future requirements of those who use it while maintaining and keeping human and natural assets. (Hossain et al., 2021). There are three main pillars of port sustainability: environmental, economic, and social sustainability (Purvis et al., 2019). Traditional port impact research papers have two chief shortages. First, they exhibit port transactions as constant objects that concentrate only on port effect, ignoring the evolving impact of port's future development in processes (e.g., automation), ambiguousness (e.g., climate change impacts), and the impact of port on city. Secondly, they overrate the advantages of ports and their social, economic, and environmental impact, while underrate their unfavorable effect (Musso et al., 2011; Halpern et al., 2012; Van den Houten, 2017; Couling & Hein, 2020, as cited in Ogara et al., 2023). Smart city is one that adopts the smart technology in all business operation and management including infrastructure, building, transportation, and communication. Port sustainability is one of the most important topics nowadays. As it has been discussed widely in the literature, it encompasses economic, social, and environmental aspects. In this study, the authors examine the impact of digital twins' technology dimensions on seaports sustainability in Kuwait and consequently, put the cornerstone of smart cities establishment in Kuwait and Gulf region. There is a triangle relationship between three variables, population is the main reason behind trades activities, and there is a derived demand for transport, growth in population produce growth in trade, and for sure growth in transportation. The annual statistics which are issued by the public authority for civil information in Kuwait shows that there is an annual growth in population. Also, the annual statistics of Kuwait Port Authority shows that there is an annual growth in imported TEUs in Shuwaikh and Shuaiba Ports. In this study, the authors are concentrating on container terminals in Kuwait.

2. Significance of the Research

1. The study provides insights into the implementation of digital twins' technology which is one of the main aspects of smart cities particularly smart ports.
2. From a practical point of view, the study explains how sustainability achieved as a result of digital twins' implementation in Kuwait seaports and consequently port city begins to become smarter which has reflected on other cities in Gulf region to transform into smart cities.
3. The study adds to the smart cities' agenda in Gulf region as the digital twins' technology is one of the core technologies that assist in implementing more sustainable and smart cities.

3. Theoretical Background

3.1. Digital Twins' Technology

Mihai et al. (2022) defined digital twins as a digital transformation technology that relies on data, accurately visualizes the physical system, helps in monitoring, and improving simulation and forecasting processes. Therefore, it provides functional services and meets application requirements with high efficiency. Digital twins refer to the digital virtualization of a physical product; it expands the concept of storage of real-time data and proactive simulation (Thürer et al., 2022).

Today, technological progress such as the Internet of Things, artificial intelligence, and cloud computing has led to the development of digital twin technology and its use in many industries by simulating an almost complete copy of the physical world. Digital twins are changing the way businesses operate, and the technology is expected to expand to include more applications, use cases and its applications in different industries (Attaran & Celik, 2023).

Ding et al.'s (2023) study indicated that digital twins in seaports and container terminals could facilitate high-fidelity simulations by designing a decision support system based on digital twin and big data technologies to provide real-time monitoring and integrated decision support.

It provides optimal operation plans and a standard for early warnings of ship delays through various simulations for resource allocation at the planning level and real-time operational decision-making through real-time monitoring and analysis of the efficiency of shipping operations (Ding et al., 2023). Also, Elmay et al. (2023) added that the digital twins provide updates on the status of the shipment and the progress of the trip, by tracking container movements, and attaching documents for government approvals and certificates, issuing customs clearance certificates, remote inspection of pending and approved permits. In Fig. 1, Elmay et al. (2023) have suggested an outline of a representative shipping process during container shipping by providing open communications and data flow with complete confidence and security between parties involved "such as design manager, exporter, and the shipping line agent" to take the right decisions.

This study depended on the dimensions of digital twins identified by Tao et al. (2018), which are "physical factors, virtual factors, and connected data" because they are consistent with the objectives of the study. Physical model: represents a real component that can be observed in digital twins, and may be a practical activity, a physical system, or product. Virtual model reflects the physical system with respect to behavior, geometry, rules, and properties. Connected data is a two-way communication between all specific factors to enable cooperation between them.

3.2. Port Sustainability

Due to their involvement in international production and distribution networks and their yearly trading volume of approximately 10.3 billion metric tons, ports

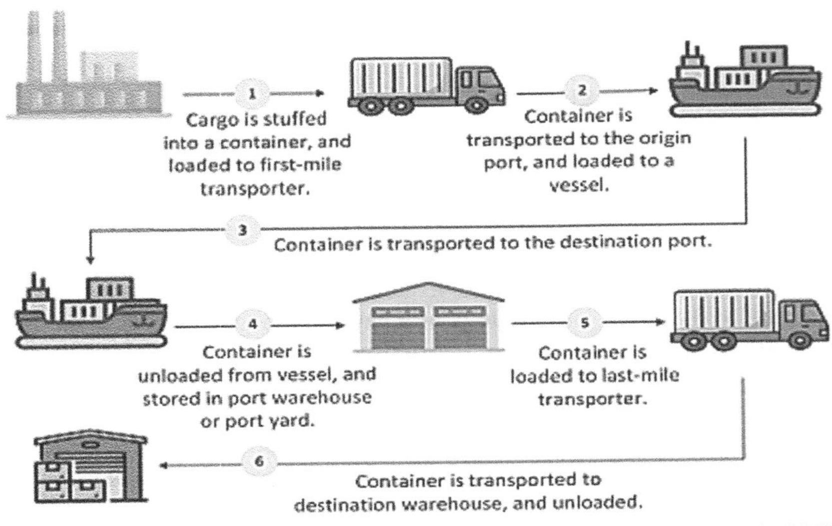

Fig. 1. Outline of a Representative Shipping Process. *Source*: Elmay et al. (2023).

are significant players in the world economy (UNCTAD, 2017). Thirteen of the top 20 cities ranked by Human Capital Indicators as port cities, and 36 of the top 50 competitive cities globally in 2012 were port cities (Girard, 2013 as cited in Lozano et al., 2018). 11 billion tons of seaborne trade were carried by 98,140 ships in 2018; this represents 80% and 60–70% of global trade volume and value, respectively (UNCTAD, 2019 as cited in Alamoush et al., 2021). Currently, trade by sea accounts for over 80% of global trade (Zheng et al., 2020). Seas offer the vital ecosystem service of transportation channels, and port cities acquire necessities (Costanza, 1999; Haase, 2015; Wang et al., 2021 as cited in Ogara et al., 2023).

Taking into consideration the previous figures, the need to handle very large ships, competition from new ports, environmental problems like ship pollution of the air, land, and water, and transportation bottlenecks, preventing people, goods, and raw materials from moving freely between the land and the sea are just a few of the many difficulties that the port industry faces worldwide and these difficulties also include business models and competitive advantage of ports which are under significant pressure from these problems (Carpenter et al., 2018). Ports are confronted with the issue of devising innovative approaches to use their resources, such as waterfront zones, in an economically, environmentally, and socially meaningful manner (Daamen & Vries, 2013 as cited in Carpenter et al., 2018) which is one of the most important challenges in nowadays for terminal containers in ports.

According to Stein and Acciaro (2020), ports must adopt a balance between social, environmental, commercial, and economic growth (Carpenter et al., 2018). In general, ports with strong environmental policies have higher economic efficiency (Castellano et al., 2020 as cited in Alamoush et al., 2021). While reducing negative effects remains the main goal of port environmental management (Walker, 2016), some ports implement environmental management programs to comply with regulations, while others see sustainability as a crucial component of their overall business plan (Hua et al., 2020 as cited in Hossain et al., 2021). Port sustainability is significant for both commercial and social aspects.

The three primary categories of economic, social, and environmental indicators are included in more recent thematic-based frameworks for port sustainability (Xiao & Lam, 2017). These methods are in line with internationally accepted sustainability reporting instruments, such as Warhurst (2002), Global Reporting Initiative (GRI) (2011), and CSD United Nations Commission of Sustainability Development (2007 as cited in Ogara et al., 2023).

By combining social, economic, and environmental advantages to achieve sustainable development, oceans, and coasts are regarded as development regions that can produce wealth and support human well-being (Okafor-Yarwood et al., 2020 as cited in Ogara et al., 2023). There are many seaports already taking steps in attaining sustainability in their operations, especially in terminal containers.

According to the Brundtland Report from 1987, sustainability is defined as the idea that considers how the present generation can use resources to meet their needs without jeopardizing the ability of future generations to meet their requirements. (Ogara et al., 2023). According to Denktas-Sakar and Karatas-Cetin (2012) and Oh et al. (2018), port sustainability is defined as business strategies and actions that safeguard and sustain human and natural activities while meeting port's current and future demands of stakeholders. However, ports must acknowledge that their current acts have an impact on future generations' lives as well as the surroundings in which they work and live (Alamoush et al., 2021). Therefore, ports can only function sustainably when decisions are made with the long-term health of the economy in mind, demonstrating a deep commitment to environmental stewardship, and incorporating community goals into company operations (I2S2 2013 as cited in Alamoush et al., 2021).

The American Association of Port Authorities (AAPA) defines port sustainability as plans and initiatives that preserve and enhance natural and human resources while meeting the needs of port stakeholders both now and in the future (AAPA, 2007 as cited in Hossain et al., 2021).

So-called port sustainability includes the ideas of reducing port environmental externalities, guiding economic growth, and attending to societal demands (Cheon, 2017; Cheon et al., 2017; Laxe et al., 2017 as cited in Alamoush et al., 2021). It highlights social, economic, and environmental facets as a trinity of intended effects that need a clear plan for promoting and maintaining their sustainability (Ogara et al., 2023). Economic, social, and environmental dimensions are the three main components of terminal containers and port sustainability.

In view of various studies, the first gap that can be gathered is that sustainability activities and measures are handled mostly within the environmental dimensions (e.g., Darbra et al., 2009; Lirn et al., 2013; Lam & Notteboom, 2014; Acciaro et al., 2014; Davarzani et al., 2016 as cited in Alamoush et al., 2021). Although few studies examined the three sustainability dimensions, such as (Laxe et al., 2017; Lim et al., 2019; Oh et al., 2018; Roh et al., 2016; Shiau & Chuang, 2013; Sislian et al., 2016), the economic and social aspects of ports are not thoroughly covered in the literature (Alamoush et al., 2021). Second, whereas other pertinent sectors (such as land transport and shipping) are not frequently addressed, port sustainability within internal operations is the focus of numerous research (Roh et al. 2016; Lim et al., 2019; Castellano et al., 2020; Hossain et al., 2021 as cited in Alamoush et al., 2021).

However, Couling and Hein (2020) provided that most port development studies focus on the commercial and logistical aspects of port cities and frequently overlook the crucial roles that oceans play in the formation of urban and regional agglomerations (Ogara et al., 2023).

There is an absence of studies that focus on terminal containers sustainability; therefore, further research is required to address environmental, social, and economic aspects of terminal containers in seaports.

Alamoush et al. (2021) stipulated that the three dimensions of port sustainability were established by a recent study, while interaction of the three dimensions was shown by Lim et al. (2019 as cited in Alamoush et al., 2021), showing that they all interact collectively or in pairs. Nevertheless, because sustainability characteristics are interrelated, they cannot be studied in isolation (Alamoush et al., 2021).

Environmental issues may have an impact on social issues, and ports' financial support may enhance environmental features (Shiau & Chuang, 2013). Prior studies heavily emphasized the environmental aspects of sustainability (65%), integrating the social (3%), and economic (14%) aspects only somewhat, while just 17% of research included technical reports and the three dimensions of sustainability (Alamoush et al., 2021). The wider urban redevelopment of port cities and surrounding urban areas, such as Rotterdam, Barcelona, Liverpool, Tokyo, and Hamburg, should consider economic, environmental, and social factors (Carpenter et al., 2018). This will lead to a variety of new, creative, and innovative developments, both in the ports and in the wider cities of which they are a part (Girard, 2013 as cited in Carpenter et al., 2018). The same applies to terminal containers of ports in the Arab region.

In literature, only one research that has shown port sustainability by addressing the environmental dimension, which includes the social and economic aspects, or, alternatively, by addressing the three dimensions and taking steps to improve the social aspects (employees, labor, and communities), mitigate and eliminate port environmental externalities (preserving the integrity of the environment), and simultaneously work to strengthen port economic benefits (Alamoush et al., 2021).

The growth of ports and port cities around the world offers significant social and economic potential, but there are obstacles in their way, for example, one ton of port throughput is equivalent to $100 worth of economic value added in terms of the ports' economic production (Ogara et al., 2023). However, a one-million-ton increase in port throughput is linked to a short-term boost of 300 jobs and an average of 220–1,500 jobs per million tons of port cargo (Merk & Dang, 2013 as cited in Ogara et al., 2023).

Engaging local stakeholders in port operations, planning and growth is an important part of port sustainability programs (Hendricks, 2017). Port operations have an impact on host communities in both positive and negative ways, therefore it's critical to work to keep a good rapport with them (Adams et al., 2009). Port stakeholder involvement is an important practice within ports identified as "sustainable" (Lam et al., 2013; Puig et al., 2017). According to Dinwoodie et al. (2012), port authorities can respond to stakeholder complaints and/or customer and competition pressure by maintaining open and continuous communication. Stakeholders include government, employees, customers, shipping companies, port administrators and shareholders. Port 2050 will include a complete set of indicators that are meant to be tracked and reported on a yearly basis about the effects on animals, habitat, noise levels, air pollutants, and water quality (Hossain et al., 2021).

3.2.1. Digital Twins' Technology and Sustainability

According to Fur et al. (2023), the sustainability performance of international seaports was examined over a period of seven years, based on the financial, environmental, and social dimensions of sustainability. Possible reasons for inequality in seaports sustainability performance for all three dimensions of sustainability are discussed. It was concluded that the use of digital twins' technology has an influential role in achieving sustainability of seaports.

Clemente et al. (2023) confirmed the important role of smart seaports in pioneering the application of modern technology such as the Internet of Things and digital twins to enhance the efficiency and reliability of seaports by digitizing port core systems. To provide green energy in seaports and improve management and handling, this promotes a noteworthy reduction in the environmental footprint of seaport activities.

Significantly, Singh et al. (2023) confirmed that the recently discovered modern and innovative technology has a positive and strong impact on the sustainability of maritime shipping in terms of three pillars of sustainability. The study recommended decision-makers to facilitate the implementation of procedures to achieve sustainable practices in the transport of maritime goods. Zhou et al. (2021) showed the ability of digital twins' technology to provide the accuracy required to realistically predict seaport performance to face any future disturbances. This study relied on the policy of allocating the computing budget to real-scale applications through digital twin applications to improve the efficiency of seaport performance.

Additionally, Moshood et al. (2021) investigated the impact of digital twins' technology on the visibility of the supply chain. The results showed that digital twins' technology helps firms to improve and develop predictive metrics, physical assets, and forecasts their companies' logistics services. This study also provided some solutions to overcome obstacles and challenges of applying digital twins' technology in the logistic services industry.

Furthermore, Klar et al. (2023b) showed that the digital twins' technology can analyze comprehensive data to make smart decisions and enhance governance and cooperation among stakeholders to help achieve sustainability for seaports.

3.2.2. Terminal Seaport and Digital Technology

In fact, seaports that handle containers continue to be the most important seaport type in terms of supporting global trade as sea transportation is the most environmentally friendly mode of transportation, but when adopting new technology, such as multipurpose containers, they forced to replace their antiquated equipment with more advanced and adaptable models and introducing some extremely novel ideas, like smart ports and green seaports, and change, or rather modernize, their company ownership and financing structures (Mohamed & Salah Eldin, 2020). In addition to promoting paperless trade and decreasing human interactions, digitalization, the Internet of Things (IoT), and big data platforms also enhance data analytics for better decision-making and sustainability performance monitoring (Alamoush et al., 2021). Digital twins' technology may provide advanced technological and sustainable solutions to forecast demand for terminal containers. In order to implement a smart seaport concept, seaport administration would need to take advantage of contemporary technological advancements and marvels in almost all operations and activities conducted within the ports (Mohamed & Salah Eldin, 2020). In addition, the goal of smart seaports is to remove technological obstacles and the associated traffic jams, allowing freighted products to be safely delivered to even the most remote markets in the shortest amount of time and money (Beltrán, 2015 as cited in Mohamed & Salah Eldin, 2020). One of the most advanced technologies that can facilitate the implementation of smart seaports, particularly for terminal containers, is the digital twins' technology.

4. Methodology

Due to containers' dominance in maritime transport, this study concentrates on container terminal in Kuwait (Shuwaikh & Shuaiba container terminal), therefore, we are going to explore the correlation factor between population and number of imported TEUs in the essential stage (from 2010 until 2019). The study excludes the coronavirus period and its effect on trade and transportation (from 2020 until 2022).

According to the nature of the correlation factors, we can calculate the number of imported TEUs in the prediction stage (from 2020 until 2040 the years of the vision of the new Kuwait).

Digital twins can be applied on the portion of each terminal from TEUs in the prediction stage, in comparison with the capacity of the storage area for each terminal, with respect to container dwell time in each terminal.

1. Analyzing the growth rate of the population and handling cargoes during the essential stage:

   Table 1 shows the population in Kuwait from 2010 to 2019 according to the statistics shown on the official site of the public authority for civil information (PACI, 2023).

   From the previous table, we find out that the average growth rate of the population in the essential stage is 3.25, and from our point of view, the growth rate is high in comparison with most other countries.
2. Analyzing the growth rate of Imported TEUs in container terminals during the essential stage.

   Table 2 shows the Imported TEUs in container terminals from 2010 to 2019 according to the annual statistical reports of the Kuwait Ports Authority (KPA, 2010–2019).

   From the previous table, we find out that the average growth rate of imported TEUs in the essential stage is 3.42, and from our point of view, it is close to the growth rate of the population considering the correlation factor between the two variables (population) and (handled cargoes) during the essential stage by using the previous data.

Table 1. The Population in Kuwait from 2010 to 2019.

Year	Kuwaiti	Non-Kuwaiti	Total
2010	1,148,363	2,433,691	3,582,054
2011	1,183,185	2,514,107	3,697,292
2012	1,212,436	2,611,292	3,823,728
2013	1,242,499	2,722,645	3,965,144
2014	1,275,857	2,816,136	4,091,993
2015	1,307,605	2,931,401	4,239,006
2016	1,337,693	3,073,431	4,411,124
2017	1,370,013	3,130,463	4,500,476
2018	1,403,113	3,218,525	4,621,638
2019	1,432,045	3,344,362	4,776,407

Source: The official site of the public authority for civil information (PACI, 2023).

Table 2. The Number of TEUs in Container Terminals from 2010 to 2019.

Year	**Imported TEUs**
2010	402,206
2011	403,692
2012	410,695
2013	417,379
2014	452,938
2015	496,416
2016	497,508
2017	522,365
2018	528,059
2019	542,097

Source: KPA (2010–2019).

3. Correlation factor between population and imported TEUs during the essential stage:
   According to data in Tables 1 and 2, the correlation factor shown in the flowing diagram:

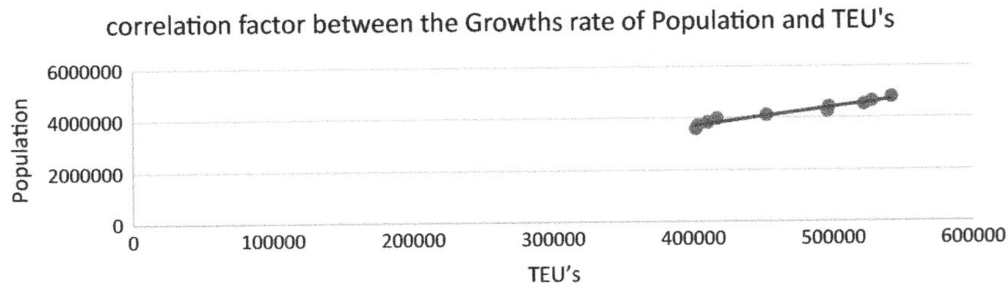

Source: Authors construction based on data in Tables 1 and 2.

The correlation factor is 0.97, which means that there is a strong correlation between the population and TEUs growth during the essential stage of the study; there is increase in both variables. This strong correlation factor will support our calculation in the prediction stage, but before the prediction stage, the average per capita should be calculated.

4. The average per capita from imported TEUs during the essential stage:
   To calculate the annual amount of imported TEUs during the prediction stage, average per capita must be calculated during the essential stage, to apply it in the prediction stage, as shown in Table 3.
   From the previous table, the average per capita during the essential stage is 0.112 TEU.
5. Population in Kuwait during the prediction stage:
   By applying the average growth rate in the essential stage: 3.25, Table 4 shows the population of Kuwait during the prediction stage.

Table 3. The Average Per Capita the Essential Stage from 2010 Until 2019.

Year	Imported TEUs	Population	Per Capita
2010	402,206	3,582,054	0.112284
2011	403,692	3,697,292	0.109186
2012	410,695	3,823,728	0.107407
2013	417,379	3,965,144	0.105262
2014	452,938	4,091,993	0.110689
2015	496,416	4,239,006	0.117107
2016	497,508	4,411,124	0.112785
2017	522,365	4,500,476	0.116069
2018	528,059	4,621,638	0.114258
2019	542,097	4,776,407	0.113494

Source: Authors construction.

Table 4. Population of Kuwait During the Prediction Stage.

Year	Predicted Population	Year	Predicted Population
2020	4,931,640	2031	7,011,033
2021	5,091,918	2032	7,238,892
2022	5,257,405	2033	7,474,156
2023	5,428,271	2034	7,717,066
2024	5,604,690	2035	7,967,871
2025	5,786,842	2036	8,226,827
2026	5,974,914	2037	8,494,199
2027	6,169,099	2038	8,770,260
2028	6,369,595	2039	9,055,293
2029	6,576,607	2040	9,349,590
2030	6,790,347		

Source: Authors construction based on population of Kuwait during the prediction stage.

As shown in the table, Kuwait population in year 2040 is 9,349,590.

6. Number of imported TEUs in the prediction stage:

   By multiplying the predicted population for each year by the average per capita in the essential stage, we can obtain the number of imported TEUs in the prediction stage as shown in Table 5.

   As shown in the table, the number of imported TEUs in year 2040 is 1,047,154 in both container terminals (Shuwaikh and Shuaiba), the prediction stage must figure out the portion for each terminal from TEUs in the next step.

Table 5. Number of TEUs in the Prediction Stage.

Year	Predicted Pop. × 0.112	Predicted Imported TEUs	Year	Predicted Pop. × 0.112	Predicted Imported TEUs
2020	4,931,640 × 0.112	552,344	2031	7,011,033 × 0.112	785,236
2021	5,091,918 × 0.112	570,295	2032	7,238,892 × 0.112	810,756
2022	5,257,405 × 0.112	588,829	2033	7,474,156 × 0.112	837,105
2023	5,428,271 × 0.112	607,966	2034	7,717,066 × 0.112	864,311
2024	5,604,690 × 0.112	627,725	2035	7,967,871 × 0.112	892,402
2025	5,786,842 × 0.112	648,126	2036	8,226,827 × 0.112	921,405
2026	5,974,914 × 0.112	669,190	2037	8,494,199 × 0.112	951,350
2027	6,169,099 × 0.112	690,939	2038	8,770,260 × 0.112	982,269
2028	6,369,595 × 0.112	713,395	2039	9,055,293 × 0.112	1,014,193
2029	6,576,607 × 0.112	736,580	2040	9,349,590 × 0.112	1,047,154
2030	6,790,347 × 0.112	760,519			

Source: Authors construction.

Table 6. The Portion of Each Terminal in the Essential Stage.

Year	Imported TEUs	Shuwaikh Terminal	Suuwaikh Portion	Shuaibah terminal	Shuaibah Portion
2010	402,206	363,017	90.26	39,189	9.74
2011	403,692	362,466	89.79	41,226	10.21
2012	410,695	360,700	87.83	49,995	12.17
2013	417,379	345,229	82.71	72,150	17.29
2014	452,938	367,591	81.16	85,347	18.84
2015	496,416	367,247	73.98	129,169	26.02
2016	497,508	334,548	67.24	162,960	32.76
2017	522,365	274,757	52.60	247,608	47.40
2018	528,059	321,375	60.86	206,684	39.14
2019	542,097	332,565	61.35	209,532	38.65

Source: Authors construction.

7. The portion for each terminal from TEUs in prediction stage:

   To find out the portion for each terminal in the prediction stage, we need to calculate the portion for each terminal in the essential stage, then applying the average of this portion on the number of TEUs which found in the last step, as shown in Table 6.

From the previous table, the average portion in the essential stage for Shuwaikh container terminal is 74.78%, while the portion of Shuaiba container terminal is 25.22%. By implementing the average portion for each terminal on the number of imported TEUs in prediction stage, we will find the portion of each terminal during the prediction stage in Table 7.

8. The number of TEUs in the container terminal according to dwell time:

   According to the data that taken during the interview with the supervisor of operation department in Kuwait Port Authority captain Yousof Al Dammak, the Dwell time for each imported container to leave the container terminals after discharged from the ship is eight days, and by dividing each year which consist of 365 on 8 days, we will find that we have 45.63 units in each year, so if we divide the number of imported TEUs in each year during the prediction stage, we will find the number of TEUs in all the eight days. Table 8 is showing the number of imported TEUs in each terminal according to dwell during each year:

   From the previous table, number of imported TEUs in Shuwaikh container terminal according to the dwell time during 2040 is 17,161, at the same time, number of imported TEUs in Shuaiba container terminal according to the dwell time during 2040 is 5,788.

9. Applying digital twins on the container terminals according to the capacity of storage area:

   During the interview with the supervisor of operation department in Kuwait port authority captain Yousof Al Dammak, we figure out that the

Table 7. The Portion of Each Terminal in the Prediction Stage.

Year	Predicted Number of Import TEUs	Shuwaikh Portion	Shuaiba Portion	Year	Predicted Number of Import TEUs	Shuwaikh Portion	Shuaiba Portion
2020	552,344	413,043	139,301	2031	785,236	587,199	198,036
2021	570,295	426,466	143,828	2032	810,756	606,283	204,473
2022	588,829	440,327	148,503	2033	837,105	625,987	211,118
2023	607,966	454,637	153,329	2034	864,311	646,332	217,979
2024	627,725	469,413	158,312	2035	892,402	667,338	225,064
2025	648,126	484,669	163,457	2036	921,405	689,026	232,378
2026	669,190	500,421	168,770	2037	951,350	711,420	239,931
2027	690,939	516,684	174,255	2038	982,269	734,541	247,728
2028	713,395	533,477	179,918	2039	1,014,193	758,413	255,779
2029	736,580	550,815	185,765	2040	1,047,154	783,062	264,092
2030	760,519	568,716	191,803				

Source: Authors construction.

Table 8. The Number of TEUs in the Container Terminal According to Dwell Time.

Year	Portion of Shuwaikh Container Terminal	Number of TEUs in Each Unit According to Dwell Time (45.63)	Year	Portion of Shuaiba Container Terminal	Number of TEUs in Each Unit According to Dwell Time (45.63)
2020	413,043	9,052	2020	139,301	3,053
2021	426,466	9,346	2021	143,828	3,152
2022	440,327	9,650	2022	148,503	3,255
2023	454,637	9,964	2023	153,329	3,360
2024	469,413	10,287	2024	158,312	3,469
2025	484,669	10,622	2025	163,457	3,582
2026	500,421	10,967	2026	168,770	3,699
2027	516,684	11,323	2027	174,255	3,819
2028	533,477	11,691	2028	179,918	3,943
2029	550,815	12,071	2029	185,765	4,071
2030	568,716	12,464	2030	191,803	4,203
2031	587,199	12,869	2031	198,036	4,340
2032	606,283	13,287	2032	204,473	4,481
2033	625,987	13,719	2033	211,118	4,627
2034	646,332	14,165	2034	217,979	4,777
2035	667,338	14,625	2035	225,064	4,932
2036	689,026	15,100	2036	232,378	5,093
2037	711,420	15,591	2037	239,931	5,258
2038	734,541	16,098	2038	247,728	5,429
2039	758,413	16,621	2039	255,779	5,606
2040	783,062	17,161	2040	264,092	5,788

Source: Authors construction.

capacity of the storage area of Shuwaikh container terminal is 23,000 TEUs, and the capacity of the storage area of Shuaiba container terminal is 10,089 TEUs, Therefore, the storage area for each terminal can receive the number of TEUs in the prediction stage until 2040.

Digital twins can depend on the capacity of the storage area and the number of TEUs in the prediction stage to give alert before reaching the total capacity by sufficient time so that it can give the opportunity for planning and implementing the development plan to increase the capacity of the storage area.

5. Discussion

The implications of the study contribute to smart cities body of knowledge, agenda in Kuwait vision 2040 and Gulf region in terms of economic, social, environmental, informational, and technological sustainability:

The study elaborates the impact of the digital twins' technology on the imported TEUs taking into consideration the population growth, and the capacity of the storage area of container terminals and how it impacts Kuwait's seaports sustainability. Ports that adopt digital twins (physical, virtual parts, and connected data) sooner will be at an advantage due to improving the environmental performance of ports to achieve sustainability by achieving the integrated concept of the smart port and overcoming barriers to prepare and enter earlier for products.

The application of digital twins' technology in container terminals and its effectiveness to figure out the time of developing the storage area of container terminals to meet the need of increasing of population from trade. It is crucial for ports, particularly container terminals, to adopt and implement compatible technology such as digital twins not only to secure sustainable and efficient operations, but also, to compete in the growing sphere of maritime transportation, taking into consideration Kuwait Port Authority annual statistical report.

6. Findings

When cities become smarter, it influences different stakeholders due to the radical change in the way cities manage and operate their activities. Each group of stakeholders may positively impact as follows:

- The average growth rate of population in Kuwait during the essential stage is 3.25. and the average growth rate of imported TEUs in Kuwait during the essential stage is 3.42. as a consequence, there is a strong correlation between population and TEUs growth during the essential stage which is 0.97 and the average per capita during the essential stage is 0.112 TEU. This can impact smart port managers and the ways of making decisions regarding future storage capacity.
- Kuwait population in year 2040 is expected to be 9,349,590 and the number of imported TEUs in year 2040 will be 1,047,154 in both container terminals (Shuwaikh and Shuaiba). This will lead to Shuwaikh container terminal handling 74.78% from Imported TEUs, while Shuaiba container terminal will handle 25.22%. This can impact policy makers and government officers' decisions on how to facilitate the transformation process to smarter ports.
- In year 2040, the number of imported TEUs in Shuwaikh container terminal will be 783,062, while in Shuaiba container terminal will be 264,092. According to average dwell time for each container, in year 2040, there will be 17,161 TEUs in Shuwaikh container terminal in every and each of the eight days in average, while there will be 5,788 TEUs in Shuaiba container terminal. This can impact urban planners and architects to redesign ports in more smart and sustainable structures for effective operations.

- As a result of the expected population growth in Kuwait, seaports terminal containers can be operated in a more sustainable way when the appropriate technology adopted. Thus, digital twins' technology is one of the powerful tools to achieve port sustainability and consequently port employees and travelers can be positively impacted.
- A significant relationship exists between digital twins and port sustainability as a result of using digital twins' model to forecast the increase in numbers for both populations and TEUs. Therefore, there is a need for more investigation regarding the correlation between digital twins' technology and terminal containers. Such an investigation can be done by students in maritime studies and faculty members who are interested in this area.
- There is a significant relationship between the increase in imported TEUs and the population. The prediction will be supported by the correlation until Kuwait Vision 2040, with a correlation factor of 0.97. These findings will help decision makers in Kuwait to establish a plan that will enable the container terminal to provide an efficient logistics service by supplying physical assets in response to the rise in imported TEUs in light of population growth and the need for strategic development plans that aim to realize Kuwait's 2040 vision.

7. Recommendations

- It is recommended for policy makers to recognize the importance smart cities and its sustainable impact not only on Kuwait but also on Gulf area.
- Policy makers are required to understand and promote the impact of smart cities on ports operations and infrastructure to meet the competition resulted from the growing number of other counterpart's smart ports.
- It is recommended that policy makers precisely study and choose the appropriate model of smart cities that can be implemented to match the increase in maritime transport and trade in Gulf region and that my attained by adopting a convenient technology such as digital twins.

8. Conclusions

Smart city development is outstanding concept to manage sustainability within the cities. According to smart city condition, digital twin integrate both virtual and real-world elements of the smart city. Digital twins' technologies such as IoT technologies, sensor and artificial intelligence are applicable tools to enhance sustainable cities.

The rapid development of digital twin technologies has helped create a new model for building smart cities. Currently, digital twins' technologies can lead to the transformation of the current urban development paradigm toward smart cities. Using the digital twin technology saves time and money and improves seaports performance through real-time analysis and monitoring for transporting goods in seaports. Digital twins were originally intended to improve manufacturing processes by simulations that have high delicate models of individual components. It is a digital representation of a physical process, system, place, person or device. Smart cities with digital twins, improve community disaster management. However, with an increase in large building information and big data created from IoT sensors in a smart city, it is possible to generate digital twin smart cities.

This study contributes to the body of knowledge of smart cities a framework on how to apply digital twins' technology on seaports cities in Kuwait as an approach to enhance its sustainability and consequently build smart seaports cities in Kuwait and other GCC countries.

The study contributes to practice by an innovative new technological solution to cope with the increasing growth of transport, while at the same time improving its sustainability. Through inputs from real-time sensors, that provide environmental and operational data, a user can recognize patterns that drove to inefficiencies in the past, then achieve a comprehensive and realistic vision of the current operating conditions in the seaports. Simulation is achieved through scenarios to predict the future situation to reach port sustainability.

The study contributes to the applied industry as the result of the potential increase in population impacts the number of TUEs, thus, smart port design, implementation, and governance will be affected. Therefore, decision makers in ports are required to change the way they manage ports and container terminals to meet the evolving demand of both populations and imported containers. This can be achieved when properly choosing the appropriate technology which facilitates smooth operations and at the same time avail economic, social, and environmental sustainability. This result is supported by Fur et al. (2023), and the study of Clemente et al. (2023).

Seaport sustainability has been approved in the study in terms of demand forecasting, meeting any deficiencies, efficiency of logistic and container terminal operations, as a result of the application of digital twins. Such results were confirmed by Singh et al. (2023), Alamoush et al. (2021), Zhou et al. (2021), and Moshood et al. (2021).

There is a strong relationship between sustainability and technology such as digital twins as the capacity of container terminal can be predicted when applying digital twins technology considering the pattern of population increase and consequently, sustainability occurs. That result was supported by Alamoush et al. (2021), Klar et al. (2023), and Singh et al. (2023).

9. Limitations and Directions for Future Research

This chapter investigated the roles of digital twins (physical, virtual parts, and connected data) on seaport sustainability. The study has some limitations that point to future research directions:

- The study is limited to the imported TEUs in container terminals from 2010 to 2019 and restricted to Shuwaikh and Shuaiba ports only. Further research is required to include other ports in Kuwait to assess the impact of digital twins' technology on its operations.
- Also, the study is limited to the effect of digital twins' technology on Kuwait's seaports sustainability taking into consideration the population growth, and

the capacity of the storage area of the container terminals. There is a need for more research to include other technologies such as artificial intelligence and Internet of Things.

- One limitation is the lack of hypotheses and empirical testing. Therefore, it is recommended that future research can involve empirical, qualitative, and quantitative data investigation of various hypotheses and variables to better represent the studied population in Kuwait.
- The study is limited to the sustainability of container terminals. It is recommended that future studies focus on other seaport terminals rather than containers such as car terminals, dry bulk, and liquid bulk terminals.
- The study is limited to economic sustainability; therefore, more studies are needed to explore the impact of digital twins' technology on other dimensions of sustainability such as social and environmental sustainability.

Few studies discussed the sustainability of container terminals in Arab regions, therefore, there is a need for more studies that focus on terminal container's environmental, social, and economic dimensions. That complies with Ogara et al.'s (2023) study and Alamoush et al.'s (2021) study. There are huge opportunities for Arab regions ports in general and in GCC in particular to competitively expand business opportunities including logistic services, shipping, storage, and accommodation of more containers when using more port's sustainability approach. This complies with Girard (2013) as cited in Lozano et al. (2018).

It is important to include the three dimensions of sustainability: economic, social, and environmental as some seaport's stakeholders focus on the economic side of sustainability and ignore other dimensions. Consequently, future research is required to install adequate technology that considers all sustainability dimensions. This complies with Lozano et al.'s (2018) study and the studies by Shiau and Chuang (2013), Sislian et al. (2016), Roh et al. (2016), Laxe et al. (2017), Oh et al. (2018), Lim et al. (2019).

When taking decisions concerning seaport technology and sustainability, it is essential to involve all relevant stakeholders therefore future research is important in this area as their satisfaction is important to guarantee the success of container terminal sustainable operations. That is supported by the studies of Hendricks (2017), Lam et al. (2013), and Puig et al. (2017).

References

AAPA. (2007). *Sustainable port development: A practitioner's perspective*. http://aapa.files.cmsplus.com/SeminarPresentations/08FINANCE_Degens_Sebastian.pdf

Acciaro, M., Vanelslander, T., Sys, C., Ferrari, C., Roumboutsos, A., Giuliano, G., Lam, J., & Kapros, S. (2014). Environmental sustainability in seaports: A framework for successful innovation. *Maritime Policy & Management*, *41*(5), 480–500.

Adams, M., Quinonez, P., Pallis, A. A., & Wakeman, T. (2009). *Environmental issues in port competitiveness*. Dalhousie University, Halifax. Atlantic Gateway Initiative Centre for International Trade and Transportation Working Paper 7.

Agatić, A., & Kolanović, I. (2020). Improving the seaport service quality by implementing digital technologies. *Pomorstvo, 34*(1), 93–101.

Alamoush, A. S., Ballini, F., & Ölçer, A. I. (2021). Revisiting port sustainability as a foundation for the implementation of the United Nations Sustainable Development Goals (UN SDGs). *Journal of Shipping and Trade, 6*(1), 1–40.

Annual Statistical Report The Public Authority for Civil Information. (2023, November 24). www.paci.gov.kw

Attaran, M., & Celik, B. G. (2023). Digital twin: Benefits, use cases, challenges, and opportunities. *Decision Analytics Journal*, 100165.

Belmoukari, B., Audy, J. F., & Forget, P. (2023). Smart port: A systematic literature review. *European Transport Research Review, 15*(1), 1–12.

Beltrán, J. (2015). Working on "Smart Port". Concept. In *Atlantic stakeholders platform conference*.

Carpenter, A., Lozano, R., Sammalisto, K., & Astner, L. (2018). Securing a port's future through circular economy: Experiences from the Port of Gävle in contributing to sustainability. *Marine Pollution Bulletin, 128*, 539–547.

Castellano, R., Ferretti, M., Musella, G., & Risitano, M. (2020). Evaluating the economic and environmental efficiency of ports: Evidence from Italy. *Journal of Cleaner Production, 271*, 122560.

Cheon, S. (2017). The economic–social performance relationships of ports: Roles of stakeholders and organizational tension. *Sustainable Development, 25*(1), 50–62.

Cheon, S., Maltz, A., & Dooley, K. (2017). The link between economic and environmental performance of the top 10 US ports. *Maritime Policy & Management, 44*(2), 227–247.

Clemente, D., Cabral, T., Rosa-Santos, P., & Taveira-Pinto, F. (2023). Blue seaports: The smart, sustainable and electrified ports of the future. *Smart Cities, 6*(3), 1560–1588.

Costanza, R. (1999). The ecological, economic, and social importance of the oceans. *Ecological Economics, 31*(2), 199–213.

Couling, N., & Hein, C. (2020). *The urbanization of the sea: From concepts and analysis to design*.

CSD United Nations Commission of Sustainability Development. (2007). *Indicators of sustainable development: Guidelines and methodologies*. United Nations.

Daamen, T. A., & Vries, I. (2013). Governing the European port-city interface: Institutional impacts on spatial projects between city and port. *Journal of Transport Geography, 27*, 4–13.

Darbra, R. M., Pittam, N., Royston, K. A., Darbra, J. P., & Journee, H. (2009). Survey on environmental monitoring requirements of European ports. *Journal of Environmental Management, 90*(3), 1396–1403.

Davarzani, H., Fahimnia, B., Bell, M., & Sarkis, J. (2016). Greening ports and maritime logistics: A review. *Transportation Research Part D: Transport and Environment, 48*, 473–487.

Denktas-Sakar, G., & Karatas-Cetin, C. (2012). Port sustainability and stakeholder management in supply chains: A framework on resource dependence theory. *The Asian Journal of Shipping and Logistics, 28*(3), 301–319.

Ding, Y., Zhang, Z., Chen, K., Ding, H., Voss, S., Heilig, L., Chen, Y., & Chen, X. (2023). Real-time monitoring and optimal resource allocation for automated container terminals: A digital twin application at the Yangshan Port. *Journal of Advanced Transportation, 2023*.

Dinwoodie, J., Tuck, S., Knowles, H., Benhin, J., & Sansom, M. (2012). Sustainable development of maritime operations in ports. *Business Strategy and the Environment, 21*(2), 111–126.

Elmay, F. K., Madine, M., Salah, K., & Jayaraman, R. (2023, March). NFTs for trusted traceability and management of digital twins for shipping containers. In *2023 IEEE international conference on pervasive computing and communications workshops and other affiliated events (PerCom Workshops)* (pp. 433–438). IEEE.

Fur, S., Heithoff, M., Michael, J., Netz, L., Pfeiffer, J., Rumpe, B., & Wortmann, A. (2023). Sustainable digital twin engineering for the internet of production. In *Digital twin driven intelligent systems and emerging metaverse* (pp. 101–121). Springer Nature Singapore, Singapore.

Girard, L. F. (2013). Toward a smart sustainable development of port cities/areas: The role of the "Historical Urban Landscape" approach. *Sustainability*, *5*, 4329–4348

Global Reporting Initiative (GRI). (2011). *GRI 3.1 sustainability reporting guidelines*. Global Reporting Initiative (GRI), Amsterdam, The Netherlands.

Haase, D. (2015). Reflections about blue ecosystem services in cities. *Sustainability of Water Quality and Ecology*, *5*, 77–83.

Halpern, B. S., Longo, C., Hardy, D., McLeod, K. L., Samhouri, J. F., Katona, S. K., Kleisner, K., Lester, S. E., O'Leary, J., Ranelletti, M., Rosenberg, A. A., Scarborough, C., Selig, E. R., Best, B. R., Brumbaugh, D. R., Chapin, F. S., Crowder, L. B., Daly, K. L., Doney, S. C., … Zeller, D. (2012). An index to assess the health and benefits of the global ocean. *Nature*, *488*(7413), 615–620.

Hendricks, K. R. (2017). *How people green the port: Sustainability in Canadian ports*. Simon Fraser University.

Hossain, T., Adams, M., & Walker, T. R. (2021). Role of sustainability in global seaports. *Ocean & Coastal Management*, *202*, 105435.

Hua, C., Chen, J., Wan, Z., Xu, L., Bai, Y., Zheng, T., & Fei, Y. (2020). Evaluation and governance of green development practice of port: A sea port case of China. *Journal of Cleaner Production*, *249*, 119434.

I2S2. (2013). Environmental initiatives at seaports worldwide: A snapshot of best practices. *International Institute for Sustainable Seaports (I2S2)*, VA, USA.

Klar, R., Fredriksson, A., & Angelakis, V. (2023a). Digital twins for ports: Derived from smart city and supply chain twinning experience. *IEEE Access. arXiv preprint arXiv*:2301.10224.

Klar, R., Fredriksson, A., & Angelakis, V. (2023b, March). Assessing the maturity of digital twinning solutions for ports. In *2023 IEEE International conference on pervasive computing and communications workshops and other affiliated events (PerCom Workshops)* (pp. 552–557). IEEE.

Kuwait Port Authority. (2010–2019). *Annual statistical reports*. Kuwait Port Authority.

Lam, J. S. L., Ng, A. K. Y., & Fu, X. (2013). Stakeholder management for establishing sustainable regional port governance. *Research in Transportation Business and Management*, *8*, 30–38.

Lam, J. S. L., & Notteboom, T. (2014). The greening of ports: A comparison of port management tools used by leading ports in Asia and Europe. *Transport Reviews*, *34*(2), 169–189.

Laxe, F. G., Bermúdez, F. M., Palmero, F. M., & Novo-Corti, I. (2017). Assessment of port sustainability through synthetic indexes. Application to the Spanish case. *Marine Pollution Bulletin*, *119*(1), 220–225.

Lim, S., Pettit, S., Abouarghoub, W., & Beresford, A. (2019). Port sustainability and performance: A systematic literature review. *Transportation Research Part D: Transport and Environment*, *72*, 47–64.

Lirn, T. C., Jim Wu, Y. C., & Chen, Y. J. (2013). Green performance criteria for sustainable ports in Asia. *International Journal of Physical Distribution & Logistics Management*, *43*(5/6), 427–451.

Lozano, R., Suzuki, M., Carpenter, A., & Tyunina, O. (2017). An analysis of the contribution of Japanese business terms to corporate sustainability: Learnings from the 'looking-glass' of the east. *Sustainability*, *9*(12), 1–17. https://doi.org/10.3390/su9020188

Merk, O., & Dang, T.-T. (2013). *The effectiveness of port-city policies: A comparative approach*.

Mihai, S., Yaqoob, M., Hung, D. V., Davis, W., Towakel, P., Raza, M., & Nguyen, H. X. (2022). Digital twins: A survey on enabling technologies, challenges, trends and future prospects. *IEEE Communications Surveys & Tutorials*.

Mohamed, S., & Salah Eldin, M. (2020). Evaluating the Sustainable Green Seaports (SGP) in Egypt: Case study of Alexandria and Eldekhila Seaports. *Journal of Alexandria University for Administrative Sciences, 57*(1).

Moshood, T. D., Nawanir, G., Sorooshian, S., & Okfalisa, O. (2021). Digital twins driven supply chain visibility within logistics: A new paradigm for future logistics. *Applied System Innovation, 4*(2), 29.

Musso, E., Benacchio, M., Ferrari, C., & Haralambides, H. E. (2011). On the economic impact of ports: Local vs. national costs and benefits. In *Proceedings of the 9th world conference on transport research, Seul (Corea del Sur)*.

Negri, E., Berardi, S., Fumagalli, L., & Macchi, M. (2020). MES-integrated digital twin frameworks. *Journal of Manufacturing Systems, 56*, 58–71.

Ogara, D. A., Morishita, J., Davies, P. J., Mbui, M., Gamoyo, M., Njoroge, N., Isaac Chann, I. O., Igesa, B. S., & Ochieng, S. O. (2023). An indicator-based approach to assess sustainability of port-cities and marine management in the Global South. *Frontiers in Marine Science, 10*, 1052128.

Oh, H., Lee, S. W., & Seo, Y. J. (2018). The evaluation of seaport sustainability: The case of South Korea. *Ocean & Coastal Management, 161*, 50–56.

Okafor-Yarwood, I., Kadagi, N. I., Miranda, N. A., Uku, J., Elegbede, I. O., & Adewumi, I. J. (2020). The blue economy–cultural livelihood–ecosystem conservation triangle: The African experience. *Frontiers in Marine Science, 7*, 586.

Platias, C., & Spyrou, D. (2023). EU-funded energy-related projects for sustainable ports: Evidence from the Port of Piraeus. *Sustainability, 15*(5), 4363.

Puig, M., Michail, A., Wooldridge, C., & Darbra, R. M. (2017). Benchmark dynamics in the environmental performance of ports. *Marine Pollution Bulletin, 121*(1–2), 111–119.

Purvis, B., Mao, Y., Robinson, D. (2019). Three pillars of sustainability: In search of conceptual origins. *Sustainability Science, 14*, 681–695

Roh, S., Thai, V. V., & Wong, Y. D. (2016). Towards sustainable ASEAN port development: Challenges and opportunities for Vietnamese ports. *The Asian Journal of Shipping and Logistics, 32*(2), 107–118.

Sasikumar, A., Vairavasundaram, S., Kotecha, K., Indragandhi, V., Ravi, L., Selvachandran, G., & Abraham, A. (2023). Blockchain-based trust mechanism for digital twin empowered Industrial Internet of Things. *Future Generation Computer Systems, 141*, 16–27.

Shiau, T. A., & Chuang, C. C. (2013). Social construction of port sustainability indicators: A case study of Keelung Port. *Maritime Policy & Management, 42*(1), 26–42. https://doi.org/10.1080/03088839.2013.863436

Singh, S., Dwivedi, A., & Pratap, S. (2023). Sustainable maritime freight transportation: Current status and future directions. *Sustainability, 15(8)*, 6996.

Sislian, L., Jaegler, A., & Cariou, P. (2016). A literature review on port sustainability and ocean's carrier network problem. *Research in Transportation Business & Management, 19*, 19–26.

Statista. (2023, October 23). www.statista.com.

Stein, M., & Acciaro, M. (2020). Value creation through corporate sustainability in the port sector: A structured literature analysis. *Sustainability, 12*(14), 5504.

Tao, F., Cheng, J., Qi, Q., Zhang, M., Zhang, H., & Sui, F. (2018). Digital twin-driven product design, manufacturing and service with big data. *The International Journal of Advanced Manufacturing Technology, 94*, 3563–3576.

The Public Authority for Civil Information (PACI). (2023, November 24). www.paci.gov.kw.

Thürer, M., Li, S. S., & Qu, T. (2022). Digital twin architecture for production logistics: The critical role of programmable logic controllers (PLCs). *Procedia Computer Science, 200*, 710–717.

UNCTAD. (2017). The review of maritime transport. *UNCTAD/RMT/2017*. http://unctad.org/en/PublicationsLibrary/rmt2017_en.pdf

UNCTAD. (2019). Review of maritime transport 2019. In *United Nations conference on trade and development*, Geneva, Switzerland.

Van den Houten, J. (2017). *A system dynamics exploration of port-city development: The case of Tema, Ghana*.

Walker, T. R. (2016). Green marine: An environmental program to establish sustainability in marine transportation. *Marine Pollution Bulletin, 105*(1), 199–207.

Wang, C. N., Nguyen, N. A. T., Fu, H. P., Hsu, H. P., & Dang, T. T. (2021). Efficiency assessment of seaport terminal operators using DEA Malmquist and epsilon-based measure models. *Axioms, 10*(2), 48.

Warhurst, A. (2002). Sustainability indicators and sustainability performance management. *Mining, Minerals and Sustainable Development [MMSD] Project Report, 43*, 129.

Xiao, Z., & Lam, J. S. L. (2017). A systems framework for the sustainable development of a Port City: A case study of Singapore's policies. *Research in Transportation Business & Management, 22*, 255–262.

Zheng, Y., Zhao, J., & Shao, G. (2020). Port city sustainability: A review of its research trends. *Sustainability, 12*(20), 8355.

Zhou, C., Xu, J., Miller-Hooks, E., Zhou, W., Chen, C. H., Lee, L. H., & Li, H. (2021). Analytics with digital-twinning: A decision support system for maintaining a resilient port. *Decision Support Systems, 143*, 113496.

Chapter 18

Shifting Toward Digitalization and Smart Cities: Evidence from Kuwait

Rashed Alhaimer

Faculty of Business, Arab Open University, Kuwait

Abstract

In an era where technology rapidly reshapes society, Kuwait stands at the forefront of digitalization and the realization of smart cities. This chapter examines the impact of this technological revolution across diverse sectors within Kuwait. From integrating digital infrastructure in urban development to the role of social media in political, health, and corporate realms, it provides a comprehensive exploration of digitalization's influence. The chapter begins by highlighting Kuwait's $10 billion investment in 2024, marking a significant step toward becoming a smart city. It investigates the evolving role of social media in political campaigns, demonstrating how digital platforms shape political narratives and voter behavior. In public health, the chapter underscores the Kuwaiti government's strategic use of social media to disseminate vital health information, particularly during and after the COVID-19 pandemic, influencing public behavior and community engagement. Furthermore, the discussion focuses on Kuwait's Vision 2035, examining how governmental social media campaigns mold public perception and awareness. The chapter also explores the private sector, showing how companies leverage social media for innovation, performance enhancement, and talent acquisition. Additionally, it scrutinizes the influence of social media on political candidates' brand personalities and voter behavior, providing insights into digital political communication dynamics. Finally, the chapter addresses the surge in online shopping during emergencies, discussing its implications for e-commerce, consumer behavior, and the challenges it presents in terms of infrastructure, cybersecurity, and the digital divide.

The Emerald Handbook of Smart Cities in the Gulf Region:
Innovation, Development, Transformation, and Prosperity for Vision 2040, 327–343

doi:10.1108/978-1-83608-292-720241018

Keywords: Kuwait; digitization; digital transformation; virtual campaigns; public health; political campaigns

1. The Introduction of Digitalization and Smart Cities in Kuwait

Smart cities could be best defined as those cities that utilize various types of information and communication technology in order to improve the kind of infrastructure that has been installed within the said city. In addition to that many scholars and researchers have argued that various cities around the world have now acclaimed what is known as digital infrastructure and therefore can be called smart cities (Elian & Kisswani, 2023). However, there are many cities that are yet to achieve the status of being a smart city but are working their way toward proper digitalization and being termed as a smart city. The primary example is Kuwait where the government has invested 10 billion US dollars toward the process of digitalization in the year 2024 (LinkedIn. 2023). To define digitalization, it could be implied that it is a culmination of various means through which countries would compete with one another. In the case of Kuwait, national Eid cards have been introduced as the first step toward the concept of digitalization within the city. With the introduction of this card, the people within would enjoy the status of being digitally verified without much hassle. This entire process includes the likes of attaining digital signatures as well as other types of electronic configurations to smooth out the process. Kuwait understands that digitalization is the next step to compete on the world stage against other countries that have developed into smart cities we'll also work toward increasing the speed of work that they might get done within a day and become an advanced nation. In order to facilitate the process of digitalization the country has formed a partnership with companies like Google to use their cloud services as a way to commit toward the process of digitalization. This has been effective for the country to develop its awareness regarding the entire process of being digitally aware amongst various employees in different kinds of sectors while also drastically increasing the number of digital activities within the nation. Digitalization not only makes the country more forward thinking but also ensures that they are developed when it comes to technology in order to survive against various types of competition on the world stage. That is why this research is primarily focused when gaining an understanding regarding the very concept of what digitalization is and how it affects the nation of Kuwait in detail while also diving into various factors that might directly affect the process of digitalization in the nation.

2. The Growing Trend of Using Social Media Platforms and Applications to Promote Political Campaigns

It comes as no surprise that social media platforms have developed greatly in the last few years and have become the norm to be used in various kinds of ways around the world in both personal and professional sectors. One of the ways in which social media platforms are used is political campaigns through which various political parties celebrate their candidates when they're standing in an election. This is especially true when it comes to the nation of Kuwait where the campaigning landscape in the country has witnessed some major changes and transformations with the implementation of various kinds of social media applications and platforms (Bukari et al., 2023). With that being said, in today's world social media has become one of the most prominent platforms in Kuwait when it comes to promoting various kinds of election campaigns as it enables various political candidates to react and properly interact with various kinds of brand personalities. These candidates are able to interact in various kinds of ways which has only been made possible through the means of social media. These means would not have been possible if the political candidates stuck to traditional means of campaigning. A fine instance of this would be to consider the 2012 parliamentary election where the nation witnessed a massive shift from the means of traditional campaigning toward social media campaigning for the purpose of elections in the nation of Kuwait (Mesbah, 2022). Ever since that historical moment, various social media platforms such as Twitter Instagram and Snapchat have been pivotal and have played crucial roles when it comes to shaping various kinds of political events in the country while also massively influencing how the voters behave during an election and their behaviors in general.

The political candidates in the nation of Kuwait utilize various social media platforms in strategic manners which supports them to form a connection with the electorate as well as enables them to form chains of interaction with various specific attributes that the citizens of the country might resonate with a fine example of this would be how a candidate could utilize various social media platforms such as Instagram and Twitter to portray themselves and their behaviors as being honest, trustworthy and down to earth which appeals to the citizens of the nations or the voters who relate to these values. Additionally, using this strategic measure the voters would be able to value the candidate as being filled with integrity and authenticity which improves the vote bank for that political party. In addition to that, here the theory of brand personality model could utilized when returning various kinds of strategic positioning with the potential voters as it influences the words as well as identifying with the various kinds of projections made by the political candidates of the involved political parties. It comes as no surprise that Kuwait has a unique political landscape where the religious and tribal aspects are very important due to the fact that it moves the voters toward a particular vote bank (Freer, 2024). In this regard, social media campaigning enables political candidates to use these religious and tribal aspects in the portrayal of their personalities such that the voters of the citizens can easily be manipulated through the matters concerned with various religions and tribal aspects within the country (Reisach, 2021). For instance, if a political candidate hails from a special tribe or a minority religious background, they can attract and bring voters of the same religious or tribal background to their side which supports their political status. On the split side however, many researchers and scholars have discussed that the use of social media platforms for any kind of political matter could result in an extremely challenging position for the candidate due to the fact that the authenticity and how real the person is can be majorly scrutinized when portrayed through social media platforms (Alhaimer, 2023b). This happens due to the fact that any kind of difference between the candidate as a real person and how they're choosing to portray themselves through social media platforms or on an online platform could lead to severe distrust from the side of the voters or the people and result in disengagement from the side of the people toward the political party and the candidate involved (Marozzo & Bessi, 2017). However, even with great risks like this social media platforms are majorly utilized by political candidates in Kuwait for the purpose of gaining the trust and attention of the wood bank. Applications such as Instagram and Twitter are used all the time by political candidates where they would engage in real-time communication with the vote bank to understand their concerns as well as to the confidence of the people in gaining an advantage (Stier et al., 2020). Instagram and Twitter have become the main social media platforms used by political parties and candidates. Research in this field shows that these two platforms buttress the voters as well as the candidates to resonate and relate to one another and to feel like there's an involvement of emotions from both parties (Sahly et al., 2019). On the contrary, however, research has also highlighted that these social media platforms used for political purposes may only be used to alter and manipulate the feelings and emotions of the voters and the people to gain their confidence and to improve the vote bank for the political party involved. However, it has been proved many times that after gaining confidence and getting the votes the political candidates and the party start showing real emotions and faces which could be vastly different from their online identity created to gain the confidence of the people or the voters in the country (Hendricks & Schill, 2017).

Here, it is also important to note that one of the most important social media applications for platforms for the people of Kuwait is Instagram as people enjoy getting to see and know about the various trends on the platform.

According to public data a detailed overview of Instagram users in Kuwait as of June 2022, totaling 2,618,400 users. The age group distribution highlights that the majority of users are between 25 and 34 years old, comprising 37.1% of the user base, with males making up 23.4% and females 13.7%. The 18–24 age group follows, representing 21.2% of users, where 12.2% are males and 9% are females. The 35–44 age group accounts for 22.3% of users, with 13.7% males and 8.6% females. For the 13–17 age group, users constitute 5.1%, with females slightly higher at 2.8% compared to males at 2.3%. The 45–54 age group makes up 9.2% of users, with a higher percentage of females (5.7%) compared to males (3.5%). The smallest user groups are the 55–64 age range at 3.6% and those aged 65 and above at 1.4%, with females dominating both categories at 2.2% and 0.9%, respectively, compared to males at 1.4% and 0.5%. This demographic breakdown showcases the predominant engagement of young adults on Instagram in Kuwait.

According to other data, 62% of men and 38% of women are all engaged in Instagram which is precisely why every potential political candidate is majorly focused toward this social media platform in order to make their social media political campaigns successful. Political candidates are known to strategically use the visual nature of this social media application in order to bring the attention of the voters toward their political parties and political campaigns. In addition to that political candidates also use Instagram to bring attention toward various kinds of behind this in activities and stances for community events which enables making the voters belief that they're part of the political party and would greatly relate to it. Furthermore, the utilization of personal stories of these political candidates is also updated and posted on a regular basis to humanize themselves such that the voters or the people of Kuwait can relate to them and offer them and their political parties their vote of confidence (Petrova et al., 2021). However, research in this field has shown that numerous political stances provided or portrayed by political candidates over social media platforms such as Instagram could be severely misleading to the public and the voters if the actions of these political candidates and the digital image that they have created over social media do not match once the elections are over (Oparaugo, 2021).

Fig. 1 illustrates the demographic distribution of Instagram users in Kuwait as of January 2021, with a total of 2,329,100 users. The data reveal that men constitute a larger portion of the user base at 58.2%, while women make up 41.8%. The most significant age group is 25–34 years, accounting for 47.2% of users, with 27.9% being men and 19.3% women. The 18–24 age group represents 16.7% of the total users, with males at 9.6% and females at 7.1%. The 35–44 age group makes up 22.8%, where 13.3% are men and 9.4% are women. The youngest group, aged 13–17, comprises 3.4% of users, with a slight male majority at 1.7% compared

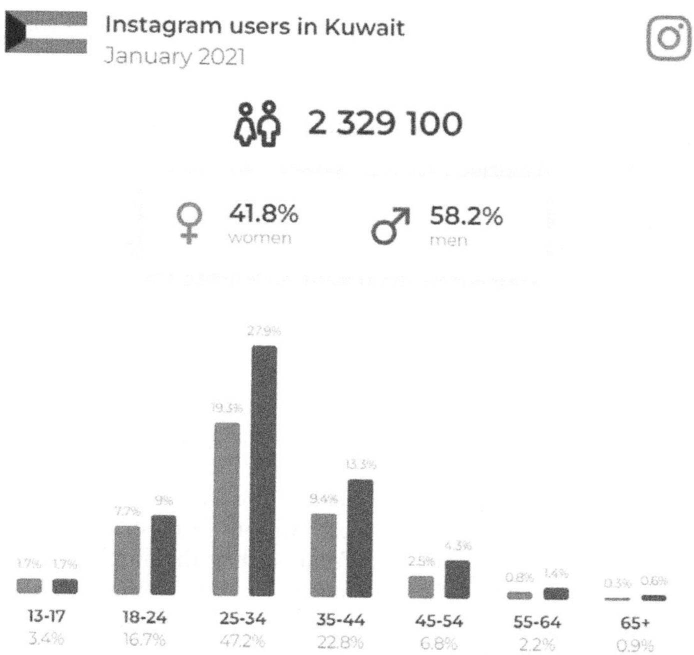

Fig. 1. Social Media Advertising in Kuwait in Recent and Upcoming Years.
Source: Digital Marketing Community.

to females at 1.7%. The 45–54 age group represents 6.8%, with men making up 4.3% and women 2.5%. Finally, the smallest user groups are the 55–64 age range at 2.2% and those aged 65 and above at 0.9%, with men leading in both categories at 1.4% and 0.5%, respectively, compared to women at 0.8% and 0.4%. This demographic breakdown highlights the dominance of young adults, particularly males, on Instagram in Kuwait.

In addition to Instagram, there are other social media platforms that are used for political purposes in the nation of Kuwait such as Facebook Twitter and Snapchat a great number of people enjoy spending their time browsing through their social media applications. Research has shown that over 3 million people in the nation of Kuwait use Facebook while around 2 million people have been recorded to use Twitter (Alhaimer, 2022). It is also important to note that over 200 million US dollars are spent by political parties toward social media political campaigning in order to gain the confidence of the voters in their vote bank (Statista. 2023). It is important to understand that today's generation is very much into social media applications than traditional means of political campaigns which is precisely why governments and political parties all around the world use social media platforms to gain the confidence of people and make an impact on their vote bank. Most of the money spent on social media political campaigns is used toward mobile advertising due to the fact that most people use their mobile phones to access these social media platforms. Thus, it could be deduced that political parties and political candidates in the nation of Kuwait are Privy toward using their social media platforms to reach the vote banks and to gain the confidence of the audience. This is precisely what makes social media applications and platforms such a strong tool for political campaigns.

3. Social Media as Platform for Spread Governmental Public Health Messages

The spread of coronavirus in 2020 made the people in Kuwait extremely great and concerned about their lives and their health resulting in a nationwide health care concern. The government in the nation of Kuwait was responsible for letting the people know how they would prevent the spread of this virus through various public health messages and how they could care for their loved ones during the spread of coronavirus. It is important to note here that COVID-19 was not the only prominent threat that the nation was facing as even after the passing of the virus the government in the nation was determined to push its citizens toward following the revised health guidelines which were provided by the World Health Organization in order for them to stay healthy and lead a fit lifestyle (Alhaimer, 2022). In order to spread this health concern various forms of public health messages were delivered through additional forms such as the situation where people were made aware of the news of the spread of the virus. However, it is important to note that these traditional forms of delivering public messages by the government like television were not enough for the citizens to know and understand the adversities that came alongside the spread of coronavirus (Anwar et al., 2020). That is precisely why the Kuwaiti government decided to seek the use of various social media platforms given how important it is as a tool to form chains of communication with people all over the world making it very easy and convenient to spread public health messages. similar to other nations in the world Kuwait also came to the realization that social media platforms and applications are very important when it comes to spreading public health messages related to coronavirus during and after the pandemic to ensure that the citizens were safe and they were aware what their roles are during these dire times (Alhaimer, 2022). With the use of social media platforms various important pieces of information related to the ongoing pandemic, we spread in order to keep engaging with the citizens and to aid them with accurate and real-time data that could severely influence their behavior during the outbreak. The government in Kuwait has reconciled various social media platforms to spread knowledge to the public regarding the pandemic and the consequences that came alongside it. The use of various social media platforms and digital platforms such as Instagram, Facebook, and Twitter were used rarely to share important data related to the coronavirus outbreak with the people of the nation. It is important to note that people need to be made aware of the preventive measures that could minimize the spread of the virus as well as enable them to take care of themselves if they were affected by it. Therefore, the use of social media platforms or digital platforms made it extremely easy and convenient to share this valuable information with the people which in turn might caused an increasing in the survival ratio in Kuwait (Mohammed, 2023). In addition to that the Ministry of Health in Kuwait used Twitter to spread the message about what the pandemic had done to the world and to the Kuwaiti people such that these reasons were prepared to face it and survive the ongoing pandemic (Kuwait Times. 2023). Some of the ways through which the government within the nation has spread messages have been discussed in depth below.

3.1. The Use of Twitter to Spread Updates Regarding COVID-19

As mentioned above, the Ministry of Health in Kuwait made use of Twitter to spread daily information and knowledge about the updates regarding the coronavirus to the citizens of the nation. Upon investigation, it was revealed that these particular tweets had enabled users to read them to properly understand and analyze the pandemic and its nature in order to take various actions and decisions to survive the spread of the virus (Alomari et al., 2021). These messages increase understanding of the importance of following the guidelines presented by the World Health Organization as it would directly impact their survival. Here it is important to note that the official Twitter account of the Kuwaiti government enabled them to reach people on a much larger scale than they would have managed through traditional means of television and mail. Well, also this instilled a feeling of urgency about the importance of health and wellness for the people in Kuwait. in addition to that the Kuwaiti government's decision to employ various social media channels in order to educate the people on what the vaccination methods would be for the coronavirus enable them to gain a better understanding of the virus as well as the preventive measures. These official social media platforms across various applications contain multiple sources of information such as videos and graphical illustrations which contain interactive content which results in a much better chain of communication which is efficient for individuals to comprehend the importance of vaccination and preventive measures (AlAwadhi et al., 2021). Through the use of social media content, the people of Kuwait grew particularly understanding of why they should be wearing masks and practicing safe social distancing before they were vaccinated and the need for proper sanitation.

3.2. Spreading Awareness of the Coronavirus Vaccinations Through Instagram Campaigns

Alongside Twitter and Facebook, the Kuwaiti Ministry of Health also used Instagram to launch campaigns in order to raise awareness regarding vaccinations and this was considered to be an important strategy in order to promote health and wellness during the time of the pandemic. The government ran various health campaigns on Instagram to empower the citizens of Kuwait to understand how the vaccines worked (Rodrigues et al., 2023). Social media platforms contained various engaging visuals and audio posts from various healthcare professionals around the world which educated them about the safety and importance of the vaccines created solely for preventing the spread of coronavirus we're also discussing the efficiency of these vaccines in order to contain the spread of disease. The Kuwaiti Ministry of Health used the visual appeal that Instagram provides in order to spread this invaluable information to the citizens of the nation motivating them toward taking the important steps to prevent the spread of the coronavirus and to care for themselves. Additionally, it was also noticed that a sense of solidarity had been formed amongst the people of Kuwait due to the influence of the Kuwaiti government through various social media platforms which resulted in a better community that backed each other during these times. The health messages spread by the government of Kuwait were practical in promoting a sense of compassion amongst the people in addition to improving a sense of empathy and adherence to the recognized safety measures by the World Health Organization. Altogether this enabled the government of Kuwait to bring the people together into a sense of collective consciousness which was a useful issue in tackling the virus and its effects.

3.3. The Use of Facebook for the Engagement of Different Communities

People in the nation of Kuwait are more involved across different social media platforms than they are in traditional communication means which was effective for healthcare organizations to educate the people about the coronavirus and its effects. Research has shown that the people of Kuwait use Facebook as the most popular social media application after Instagram (Alenezi & Brinthaupt, 2022). The type of engaging content that is posted on this social media platform enables people to interact with each other from all over the world which makes it a very popular social media tool that could be utilized by the government to spread the message that they want. Facebook was crucially utilized by healthcare organizations in Kuwait in order to create various initiatives to bring the community together through means of community engagement. The feature of Facebook Live was massively utilized by healthcare organizations in the nation to have live interactive sessions with the audience (Alenezi & Brinthaupt, 2022). This enables understanding the nature of the pandemic and the results that came with it. These live sessions conducted by healthcare experts were a support to the audience enabling them to protect themselves from the effects of the coronavirus pandemic as well as the aftercare measures if someone were affected by it. In addition to that numerous Facebook live sessions were also arranged for Q&A purposes where the people could clarify their doubts and gain a better understanding about how they might react to the pandemic. Additionally, town hall meetings were also organized to improve direct communication and interaction with the citizens of Kuwait. These efforts resulted in the people understanding how the government was functioning and how they were trying to make the lives of the people better by spreading information that was trustworthy and reliable.

3.4. Preventing the Spread of Misinformation on Twitter

Healthcare organizations and health authorities in Kuwait actively fought against the spread of misinformation on Twitter regarding the coronavirus pandemic by providing numerous clarifications and credible sources against misinformation spread by people that could have resulted in a sense of panic during the spread of the virus (Bajouk & Ferré-Pavia, 2023). This effort put by the health authorities massively increased the wellbeing of users to understand which information was false while also supporting them realize how committed the government was toward truthfulness and transparency while promoting safety for all of its citizens (Shin et al., 2018). Therefore, the government of Kuwait used these social media platforms very efficiently and effectively for the spread of various kinds of health messages to influence and shift the behavior of the people during the times of the pandemic. Spreading useful information one time to the public facilitated in bringing the communities together and engaging them while also addressing numerous concerns regarding the pandemic which empowered the public to create a safe and healthy society (Chen et al., 2023). These examples of the Kuwaiti government portray the multifaceted strategy adopted by the government when it comes to the usage of social media platforms in promoting and shaping the chains of communication of public health for any given situation, be it a current or a future scenario (Ghenai & Mejova, 2018).

4. Using Various Governmental Social Media Campaigns to Change the Attitude of the Citizens Toward Kuwait's Vision 2035

The vision of 2035 out of Kuwait has been idealized in order to transform the nation into a country that is based on knowledge and primarily focuses on the field of diversification (see Fig. 2). This particular vision provides view of how

Fig. 2. Some Aspects of Kuwait's Vision 2035. *Source*: Public Documents, The World Bank, 2023.

dependent Kuwait is on its oil revenue alongside their environmental and social goals. Kuwait Vision 2035 primarily focuses on growing the private sector as well as improving the services for the public sector (Arif & Aldosary, 2023). Vision 2035 wants to develop Kuwait as a new nation that could foster innovation and improve the quality of life for all people. To spread awareness regarding Vision 2035, the Kuwaiti government has taken to Twitter and created an official account for this purpose (Alhaimer, 2023a). This account is used to share tweets about the progress and achievements of Vision 2035 to keep the people updated regarding the work of the government as well as the progress that it has gained so far. Researchers have claimed that tweeting about Vision 2035 would be useful for people come to terms with the benefits that this vision just to achieve. In addition to that, Vision 2035 on Twitter is highlighting numerous forms of data in order to spread awareness regarding the vision and to effective for persons to understand what it is all about.

As mentioned before, Instagram is a social media platform which is dependent on visually oriented data and posts which enables the Kuwaiti government to interact with the people of the nation through various forms of storytelling. In the case of Instagram visual storytelling enables the government to empower population to understand the truthfulness and transparency of the information that is presented to them. Visual storytelling includes the use of video infographics and carousel posts that break down and make it easier to understand the complicated aspect of Vision 2035 which makes it greatly accessible for people ranging from a diverse background (Sunarso & Mustafa, 2023). The finance of this would be how info graphic posts on Instagram could be used to present important data and figures about Kuwait's Vision 2035 in a very simple yet understandable manner which is useful for people to grasp the vision better. In addition to that these kinds of visually attractive posts are utilized to form proper chains of communication with people regarding the success stories of various people and groups who have profited from the utilization of the social media apps by the government. The sharing of this kind of data with the people assists in humanizing the impact of Vision 2035 which makes it more relatable for all the citizens in the country.

The government of Kuwait posts numerous discussions about Vision 2035 through the use of Facebook. The citizens of Kuwait are encouraged to raise questions and voice their concerns through frequent life Q&A sessions alongside other discussion mediums such as town hall discussions and through the use of interesting articles. This kind of two-way communication has enabled the government to tackle various kinds of issues alongside acquiring information and knowledge about how the people perceive the vision's execution as well as how well it has been executed. The Ministry of Planning in Kuwait hosts numerous Facebook live events where people are encouraged to come and discuss Vision 2035 by asking any question that they might have related to the vision regarding the success of the country and the progress that everyone is supposed to enjoy. Research in this field has proved time and time again that the involvement of citizens when it comes to the betterment of a country or a region enables the facilitation of this process by increasing the confidence of the citizens toward the government (Amosun et al., 2022). Therefore, the use of Facebook Live to incorporate people into the decision-making and to understand how they feel about Vision 2035 is a strategic manner that has supported the government to understand the minds of the citizens and to know what requirements they have. During these live sessions, the audience has the ability to interact with all kinds of government officials and community leaders alongside topic specialists who would address the questions and concerns that they have regarding Vision 2035. This kind of direct engagement results in the promotion of transparency and responsiveness amongst the citizens and the government (Hlpf.un.org. 2023). In addition to that the planning ministry in Kuwait ensures that they organize frequent live sessions over Facebook regarding Vision 2035 to address any queries that the people might have. Furthermore, the government organizes FAQs, blog posts and videos that explain various concepts regarding the vision and various other aspects such that the citizens have a better understanding (Ottesen et al., 2023). These spots encourage participation from the citizens and enable them to ask questions and queries through feedback and survey mechanisms.

The government also runs interactive surveys over Twitter and Instagram in order to gather feedback from the citizens on various aspects of Vision 2035. By doing this the government is incorporating individuals and their opinions when it comes to making choices (Park et al., 2022). Doing this supports government gain knowledge and essential data for the vision as well as promotes that the goal of the government fulfills the goal for the people who their various feedbacks and viewpoints should be taken into account. However, it is also important to note that respective of great efforts by the government and the planning ministry, misinformation and misconceptions regarding 35 have shown up on their social media platforms as noticed by several researchers (Suárez-Lledó & Álvarez-Gálvez, 2021). The government works aggressively toward eliminating any of these challenges recently responding with correct information and providing fact checks for every misconception that there could be over social media platforms.

To conclude the poetic government uses social media campaigns to influence and change the opinions of the people toward Vision 2035 which demonstrates the strategy and extensive the government has been. It is also crucial to note that utilizing various kinds of avenues promotes communication while also enabling the government to solve certain issues, the government is able to create a strong narrative that would appeal to the citizens of the nation (Nash, 2021). This example primarily portrays how social media is a powerful instrument used by governments of the world roll touring public attitudes and behavior while also gaining the confidence of the people when it comes to national developmental goals.

5. The Utilization of Social Media Platforms to Promote the Innovation and Performance of Companies in the Private Sector

Social media is a very powerful tool that aids businesses to improve their performance as well as innovation. The private sector in Kuwait is no different as it uses social media platforms such as Instagram and Twitter to improve their performances as well as how they interact with the audience and reach them. One of the most important online platforms used by organizations in the private sector is LinkedIn where people from various fields showcase their expertise and their network of connections (Alhaimer, 2021b). Various insights and regular updates regarding private companies in Kuwait are shared through these websites or platforms with the customers such that there is a level of transparency between the two parties. In addition to that private organizations in Kuwait also use social media platforms such as Instagram and Twitter to not only acquire customers but also understand where they are standing through feedback forms and surveys. This enables the brand to not only grow but also understand how the public reacts to its services. However, there are several limitations to this as well. The image of a customer or a private company portrayed over a social media platform or online media such as Instagram, Twitter, or LinkedIn may alter significantly when it comes to real-life scenarios. What it means is that the online image created by customers or private companies might be different to how they are in real life. This might create multiple discrepancies and issues when it comes to hiring or providing service. In order to counter these limitations companies and customers need to provide proper information as the basic first step crucial for any kind of online communication. Furthermore, a proper screening procedure might determine the best companies and services alongside the best customers enabling businesses in the private sector as well as the customers to gain what they desire without having to worry about false portrayal of services on these social media platforms (Khan et al., 2020).

6. The Influence of Social Media on the Brand Personality of Political Candidates and Behaviors of Voters in Kuwait

It comes as no surprise that social media has a great impact on the brand personalities of political candidates in the nation of Kuwait which greatly impacts the voter and their behavior (Alhaimer, 2023b). It has been noticed that over platforms like Facebook and Instagram, political candidates usually portray different traits that might boost participation and connectedness with the people or the voters (Boulianne & Larsson, 2023). These candidates use various kinds of channels to spread their political agendas and to influence public behavior toward their favor. In doing so, the voters on the other hand are exposed to previous kinds of information about political parties as well as political candidates that might influence their thinking their behavior as well as encourage them to actively take part in the procedure. The use of social media platforms for political purposes enables real-time feedback and participation from the voters of the people which would directly impact the decision-making process in political scenarios. However, there are many limitations to the use of social media platforms by political candidates when it comes to engaging in political campaigns and other political purposes. Researchers in this field have noticed a trend where political candidates have not been truthful in the portrayal of their image or their political agendas and how they wish to work for the people that they're gaining confidence from (Jenkins et al., 2016). Most of the time it has been seen that political candidates would often flip from what they have portrayed over social media platforms in real life after the voting has been concluded. This false image portrait online through political campaigns on social media platforms has resulted in people losing their confidence in political parties and candidates which has further resulted in the people refusing to take social media political campaigns seriously (Lane et al., 2019). This gap between what's being portrayed online and how political candidates really are in real-life scenarios is a major challenge of using social media platforms for political propaganda. Social media has a great impact on the brand personality of political candidates, however portraying lies and deceit results in a major challenge when it comes to this particular aspect (Kim et al., 2016).

7. The Utilization of E-commerce Websites and Online Shopping in Kuwait During Emergencies

With emergencies such as pandemics or natural disasters, the situation pushes people to quarantine themselves for their own safety and to buy everything online through e-commerce websites in order to minimize contact with other people as well as stay safe during times like the coronavirus pandemic. During the COVID-19. What is form of trading as well as E-commerce platforms have thrived in Kuwait which has also highlighted the importance of these platforms during times like this (Alhaimer, 2021a). However, there are numerous limitations when it comes to e-commerce and online shopping in Kuwait regardless of the dependence that the people of Kuwait are starting to develop in online shopping. the first limitation that the people in Kuwait face when it comes to online shopping during times of emergencies is the unavailability of proper transportation and delivery systems which results in the delay of delivery as well as the prominence of digital literacy lacks act as a huge barriers that rupture smooth flow of goods and services from one place to another (Abdallah et al., 2024). In addition to that another limitation is the lack of a proper regulatory agency that might monitor and control online transactions in order to prevent any kind of fraudulent activity when it comes to e-commerce and online shopping in Kuwait. This has resulted in multiple counts of fraud as well as increased risks toward data privacy and cyber security risks. In addition to that Kuwait is slowly digitalizing which means that the e-commerce websites and the sellers on this website are not yet up to the mark and could not deliver proper services most of the time (Alhaimer, 2021b). Finally, the absence of a proper regulatory and authoritative body when it comes to online shopping and e-commerce results in sellers and distributors charging immensely more than they would in physical shopping scenarios, which makes it very difficult for communities hailing from different backgrounds to purchase products online.

8. Conclusion

This chapter serves as an important learning framework for getting an understanding of various dynamics that are present within a digitalization context, the adoption of the powerful social media tool when it comes to political campaigns, governmental communication which employs public health alongside Kuwait's vision 2035 and the process of its implementation through social media platforms. In addition to that, it also allows a deeper understanding of the private sector and renovations in this sector including the likes of serving the public through the use of social media platforms as well as allowing people to shop online during times of emergencies in the nation of Kuwait. The research and evaluation portray there are advantages as well as disadvantages present within these borders which proves that it is very important to have proactive intervention as well as certain levels of collaboration between various stakeholders to ensure that there is a state of positive potential that enables organizations to avoid any kind of losses. It is also important to note that there are prominent gaps present in the current literature when it comes to the understanding of the long-term social impacts of digitalizing the nation. Additionally, other aspects also involve the implementation of digitalization, improving operational efficiency through the use of social media campaigns, and influencing and altering the perception of the people in the country toward the authorities or the government visions while accepting universally accepted ethical conduct when operating online during the time of an emergency. Therefore, there must be policies and principles developed in order to breach these prevalent gaps as informed by policymaking alongside resilient development within the country as per the requirement of the evolving and digital Kuwait. Finally, delving deep into cross-thematic entanglements between various kinds of themes as well as how they are implied toward the oppressed communities might shine a light on various ways that might promote sustainable and moralistic development followed by certain guidelines presented by the government and world organizations.

References

Abdallah, W., Al-Fakeh, B., & McNally, M. (2024). Assessing customer loyalty with the online and in-store shopping convenience of private pharmacies in Kuwait. In *Assessing customer loyalty with the online and in-store shopping convenience of private pharmacies in Kuwait* (pp. 521–532). https://doi.org/10.1007/978-3-031-36895-0_42

AlAwadhi, E., Zein, D., Mallallah, F., Bin Haider, N., & Hossain, A. (2021). Monitoring COVID-19 vaccine acceptance in Kuwait during the pandemic: Results from a national serial study. *Risk Management and Healthcare Policy*, 1413–1429. https://doi.org/10.2147/RMHP.S300602

Alenezi, W., & Brinthaupt, T. M. (2022). The use of social media as a tool for learning: Perspectives of students in the Faculty of Education at Kuwait University. *Contemporary Educational Technology, 14*(1), ep340. https://doi.org/10.30935/cedtech/11476

Alhaimer, R. (2021a). Fluctuating attitudes and behaviors of customers toward online shopping in times of emergency: The case of Kuwait during the COVID-19 pandemic. *Journal of Internet Commerce, 21*(1), 26–50. https://doi.org/10.1080/15332861.2021.1882758

Alhaimer, R. S. (2021b). The role of social media in the innovation and performance of Kuwaiti enterprises in the food sector. *Journal of Theoretical and Applied Electronic Commerce Research, 16*(6), 1960–1972. https://doi.org/10.3390/jtaer16060110

Alhaimer, R. (2022). The health belief model: Evaluating governmental public health messages on social media aimed at preventing a COVID-19 epidemic in Kuwait. *Cogent Business & Management, 9*(1), 2031682. https://doi.org/10.1080/23311975.2022.2031682

Alhaimer, R. (2023a). Impact of governmental social media campaigns on citizens' attitudes towards Kuwait's vision 2035. *Global Knowledge, Memory and Communication*. https://doi.org/10.1108/GKMC-06-2023-0209

Alhaimer, R. S. (2023b). Unveiling the digital persona image: The influence of social media on political candidates' brand personality and voter behaviour in Kuwait. *Humanities and Social Sciences Communications, 10*(1), 1–12. https://doi.org/10.1057/s41599-023-02420-4

Alomari, E., Katib, I., Albeshri, A., & Mehmood, R. (2021). COVID-19: Detecting government pandemic measures and public concerns from Twitter Arabic data using distributed machine learning. *International Journal of Environmental Research and Public Health, 18*(1), 282. https://doi.org/10.3390/ijerph18010282

Amosun, T. S., Chu, J., Rufai, O. H., Muhideen, S., Shahani, R., & Gonlepa, M. K. (2022). Does e-government help shape citizens' engagement during the COVID-19 crisis? A study of mediational effects of how citizens perceive the government. *Online Information Review, 46*(5), 846–866. https://doi.org/10.1108/OIR-10-2020-0478

Anwar, A., Malik, M., Raees, V., & Anwar, A. (2020). Role of mass media and public health communications in the COVID-19 pandemic. *Cureus, 12*(9). https://doi.org/10.7759/cureus.10453

Arif, M., & Aldosary, A. S. (2023). Urban spatial strategies of the Gulf Cooperation Council: A comparative analysis and lessons learned. *Sustainability, 15*(18), 13344. https://doi.org/10.3390/su151813344

Bajouk, H., & Ferré-Pavia, C. (2023). Government social media engagement and health perceptions during the COVID-19 crisis: A cross-sectional study. *International Journal of Strategic Communication, 17*(5), 457–474. https://doi.org/10.1080/1553118X.2023.2231417

Boulianne, S., & Larsson, A. O. (2023). Engagement with candidate posts on Twitter, Instagram, and Facebook during the 2019 election. *New Media & Society, 25*(1), 119–140. https://doi.org/10.1177/14614448211009504

Bukari, Z., Hamid, A. B., Md. Som, H., Hossain Uzir, Md. U., Bashiru, S., Agyepong, L., Kafui Tsagli, S., & Al Halbusi, H. (2023). Determinants of voters behaviour and voting intention: The mediating role of Social Media. *Cogent Business & Management, 10*(2). https://doi.org/10.1080/23311975.2022.2140492

Chen, S., Xiao, L., & Kumar, A. (2023). Spread of misinformation on social media: What contributes to it and how to combat it. *Computers in Human Behavior, 141*, 107643. https://doi.org/10.1016/j.chb.2022.107643

Elian, M. I., & Kisswani, K. M. (2023). Smart Cities: GCC and Kuwait experience. In *Smart cities: Social and environmental challenges and opportunities for local authorities* (pp. 339–358). Springer International Publishing. https://doi.org/10.1007/978-3-031-35664-3_1

Freer, C. (2024). *The resilience of parliamentary politics in Kuwait: Parliament, rentierism, and society*. Oxford University Press

Ghenai, A., & Mejova, Y. (2018). Fake cures. *Proceedings of the ACM on Human-computer Interaction, 2*(CSCW), 1–20. https://doi.org/10.1145/3274327

Hendricks, J. A., & Schill, D. (2017). The social media election of 2016. In *Springer eBooks* (pp. 121–150). https://doi.org/10.1007/978-3-319-52599-0_5

Hlpf.un.org. (2023). *Kuwait and its city planning*. https://hlpf.un.org/sites/default/files/vnrs/2023/VNR%202023%20Kuwait%20Report.pdf

Jenkins, J. C., Słomczyński, K. M., & Dubrow, J. K. (2016). Political behavior and big data. *International Journal of Sociology, 46*(1), 1–7. https://doi.org/10.1080/00207659.2016.1130409

Khan, M. B., Iqbal, S., & Hameed, I. (2020). Economic development vision of the Gulf Cooperation Council. *Research, Innovation and Entrepreneurship in Saudi Arabia: Vision 2030*, 17.

Kim, T., Atkin, D., & Lin, C. A. (2016). The influence of social networking sites on political behavior: Modeling political involvement via online and offline activity. *Journal of Broadcasting & Electronic Media, 60*(1), 23–39. https://doi.org/10.1080/08838151.2015.1127242

Kuwait Times. (2023). *kuwaiti-scientist-says-award-gives-her-greater-sense-of-responsibility*. https://kuwaittimes.com/kuwaiti-scientist-says-award-gives-her-greater-sense-of-responsibility

Lane, D. S., Lee, S. S., Liang, F., Kim, D. H., Shen, L., Weeks, B. E., & Kwak, N. (2019). Social media expression and the political self. *Journal of Communication, 69*(1), 49–72. https://doi.org/10.1093/joc/jqy064

LinkedIn. (2023). *Digital transformation in Kuwait*. https://www.linkedin.com/pulse/digital-transformation-kuwait-mohammad-j-sear-3f/

Marozzo, F., & Bessi, A. (2017). Analyzing polarization of social media users and news sites during political campaigns. *Social Network Analysis and Mining, 8*(1). https://doi.org/10.1007/s13278-017-0479-5

Mesbah, H. (2022). Tweeted attitudes towards women parliamentary candidates in Kuwait: A social dominance perspective. *Journal of Digital Social Research, 4*(1), 98–127. https://doi.org/10.33621/jdsr.v4i1.114

Mohammed, Y. (2023). How effective is digital marketing for government organisations in times of crisis: the case of Kuwait's fire force during COVID-19. *International Journal of Emergency Management, 18*(1), 1–22. https://doi.org/10.1504/IJEM.2023.129407

Nash, K. (2021). *Interactive documentary: Theory and debate*. Routledge.

Oparaugo, B. (2021). *Media and politics: Political communication in the digital age*. Available at SSRN 3780554. https://dx.doi.org/10.2139/ssrn.3780554

Ottesen, A., Thom, D., Bhagat, R., & Mourdaa, R. (2023). Learning from the future of Kuwait: Scenarios as a learning tool to build consensus for actions needed to realize vision 2035. *Sustainability, 15*(9), 7054. https://doi.org/10.3390/su15097054

Park, Y. J., Chung, J. E., & Kim, J. N. (2022). Social media, misinformation, and cultivation of informational mistrust: Cultivating COVID-19 mistrust. *Journalism*, *23*(12), 2571–2590. https://doi.org/10.1177/14648849221085050

Petrova, M., Sen, A., & Yıldırım, P. (2021). Social media and political contributions: The impact of new technology on political competition. *Management Science*, *67*(5), 2997–3021. https://doi.org/10.1287/mnsc.2020.3740

Reisach, U. (2021). The responsibility of social media in times of societal and political manipulation. *European Journal of Operational Research*, *291*(3), 906–917. https://doi.org/10.1016/j.ejor.2020.09.020

Rodrigues, F., Ziade, N., Jatuworapruk, K., Caballero-Uribe, C. V., Khursheed, T., & Gupta, L. (2023). The impact of social media on vaccination: A narrative review. *Journal of Korean Medical Science*, *38*(40). https://doi.org/10.3346%2Fjkms.2023.38.e326

Sahly, A., Shao, C., & Kwon, K. H. (2019). Social media for political campaigns: An examination of Trump's and Clinton's frame building and its effect on audience engagement. *Social Media + Society*, *5*(2), 205630511985514. https://doi.org/10.1177/2056305119855141

Shin, J., Jian, L., Driscoll, K., & Bar, F. (2018). The diffusion of misinformation on social media: Temporal pattern, message, and source. *Computers in Human Behavior*, *83*, 278–287. https://doi.org/10.1016/j.chb.2018.02.008

Statista. (2023). *Digital advertising and social media advertising*. https://www.statista.com/outlook/dmo/digital-advertising/social-media-advertising/kuwait#ad-spending

Stier, S., Bleier, A., Lietz, H., & Strohmaier, M. (2020). Election campaigning on social media: Politicians, audiences, and the mediation of political communication on Facebook and Twitter. In *Routledge eBooks* (pp. 50–74). https://doi.org/10.4324/9780429202483-4

Suárez-Lledó, V., & Álvarez-Gálvez, J. (2021). Prevalence of health misinformation on social media: Systematic review. *Journal of Medical Internet Research*, *23*(1), e17187. https://doi.org/10.2196/17187

Sunarso, B., & Mustafa, F. (2023). Analysing the role of visual content in increasing attraction and conversion in MSME Digital Marketing. *Journal of Contemporary Administration and Management (ADMAN)*, *1*(3), 193–200. https://doi.org/10.61100/adman.v1i3.79

Chapter 19

United Arab Emirates: Integration of Smart Technology – An Innovative Solution for Climate Change Mitigation

Prisilla Jayanthi Gandam[a], Xi Chen[b], Muralikrishna Iyyanki[c], Utku Kose[d,e] and Valentina Emilia Balas[f]

[a]*CMR Institute of Technology, Hyderabad, Telangana, India*
[b]*University of California, Santa Cruz, USA*
[c]*Dr Raja Ramanna DRDO Distinguished Fellow, India Academic Programs, India*
[d]*Suleyman Demirel University, Isparta, Turkey*
[e]*University of North Dakota, USA*
[f]*Aurel Vlaicu University of Arad, Romania*

Abstract

Smart cities are where dreams are made true for the future. Abu Dhabi, UAE has been given the title of smartest city in the Middle East region in the "Smart City Index 2021." UAE is known for its rich natural resources that established much business connectivity and developed the country economically and socially. It built an innovative infrastructure with equipment for healthcare and connected people through smartphones avoiding patient travel. This enhances the patient's life expectancy and mortality rate. UAE's net zero emission by 2050 will reduce the carbon footprint from its industries. UAE, on the other hand, is building sustainable, innovative, smart, and energy-efficient cities. It is the leading country in the world with digital transformation in the Arab world. In this study, the scope of Gulf countries moving toward smart cities are analyzed with air pollution. The slope of regression for PM_{10} from linear regression was Khadija (0.9442), CI 0.9237 to 0.9647 and Khalifa City (0.9745), CI 0.9591 to 0.9900. In 2022, the CO_2 per capita emissions of UAE (25.8t) are higher over the

The Emerald Handbook of Smart Cities in the Gulf Region:
Innovation, Development, Transformation, and Prosperity for Vision 2040, 345–362

doi:10.1108/978-1-83608-292-720241021

world (4.7t). However, PM_{10} and AQI seem to be pretty good in a few cities, enhancing the living style, and climate change mitigation.

Keywords: Smart city; air quality; healthcare; climate change; UAE

1. Introduction

Cities everywhere are under stress, and in many of them and many, growth has led to growing pains, including traffic, and pollution to name a few. However, the world's preeminent laboratories are cities for analyzing solutions. The data and technology are applied for better decision-making. The city infrastructures are more efficient with digital systems, and cities are becoming smarter with sensor installation. Moreover, the usage of technology and data delivers a better quality of life. Cities could improve human health, and reduce the emission of carbon, and speed up the emergency response time more well-organized and available. The people are more connected through digital apps and platforms to their community [https://www.trinitymobility.com/segment/smart-city]. Instant information helps everyone make better decisions. But there's a wave of innovation worldwide. As cities get smarter, they're becoming comfortable and responsive. Smart cities enable to improvement of the environment, transport, public commitment, and effective decision-making with data. Smart city enhances service levels, citizen well-being, sustainability, and economic development with communication technology. Smart cities are in the process of reducing their carbon footprint by implementing renewable energies.

Cities are becoming smarter every year by year. Many governments around the globe are racing to infuse technology into just about every aspect of their city's operation. Smart cities include public transit, IT connectivity, water and power resources, hygiene and solid waste management, efficient city mobility, e-governance, and citizen participation. Let's consider how a smart city works. For instance, consider Singapore, the standardized city-state with extensive efforts to collect data regularly. The government is initiating to deploy the equipment that can capture people smoking in prohibited zones. In 2014, Singapore tossed its smart nation program to add cameras on the roadways. Thereby, the government can monitor crowd density, sanitation of public areas, and the registered vehicle movement. The data collected is fed into Virtual Singapore (an online platform) that gives the government access to understand the functioning of the city in real-time. This helps the government to predict how the public can react to a shopping mall explosion or the spread of infectious diseases.

Similarly in Barcelona, a research firm assessed that installing smart systems in the city would save billions of dollars per year in energy costs. (1) Smart street lights. The public light dims with no activity and brightens up while sensors detect motion. (2) Parking sensors. Instead of locating the parking spot by driving cars in circles, the drivers can get real-time information on an app that identifies free parking spots. Sensors on the street use lighting and metal detectors to know if a parking spot is occupied. Finally, the city engaged a few garbage sensors placed in such a way, the containers will have a vacuum network through a pipe that sucks up trash below ground. Hence, the waste is collected automatically, which not only lowers noise pollution from garbage trucks but keeps bad pollution away. However, to save money, one needs to first spend it.

2. Fourth Industrial Revolution – Healthcare – Smart City

Industry Revolution 4.0 has increasingly shaped the social and economic growth of the country (Ćwiklicki et al., 2020). IR 4.0 has connected the cities, and healthcare to becoming sustainable entities. The development of AI, IoTs (Castro & Faria, 2020; Kotzias et al., 2023), AVs, AR, big data, neuro-technologies, biotechnology, and quantum have challenged positive and negative externalities (Gonçalves et al., 2021; Popov et al., 2022). According to Park (2016), IR 4.0 involves the transformation of the entire globe including companies, industries, and society. IR 4.0 has integrated telemedicine and homes; by monitoring, recording, and centralization of patient data (Fujita et al., 2021; Kotzias et al., 2023; Lee & Lim, 2017). Healthcare happens to be a technology-driven industry, with innovations improving the patient care quality with digital and paperless. Medical 4.0 is transforming health care from mobile computing applications to cloud computing. Haleem et al. (2022) Medical 4.0 promises a patient-centered paradigm.

The smart cities promote healthcare 4.0; including innovative health, online health, mHealth, mobile health, eHealth, and telemedicine (Ahsan & Siddique, 2022). Abu Dhabi, UAE rich in natural resources is a young state that reformed its healthcare centers. It focused on the redesigning, financing, and affordable provision of healthcare for better quality of patient care and increasing the patient's living age (Koornneef et al., 2012). Healthcare 4.0 serves faster for remote, rural, and unreachable areas with better quality services (Jayanthi et al., 2020). The terms Medical 4.0, health industry 4.0, and healthcare 4.0 are interchangeable and signify the same. The smart proposal for reducing carbon in healthcare is to redesign Health 4.0 that improve outcomes and reduce costs. The pentacle model (Fig. 1) is designed (Jayanthi et al., 2022) to achieve better outcomes with experts in clinical experience giving improved patient experience with less costs and zero carbon emission.

3. UAE – Smart City Enhances Business Partnerships

In a smart city, data-sharing technologies are implemented to improve energy efficiency, minimize GHG emissions, and improve the citizen life quality with IoTs and ICTs. The United Nations Economic Commission for Europe (UNECE) and the International Telecommunication Union (ITU); have jointly stated smart sustainable cities to be innovative city that uses ICTs to improve life quality, urban operation, and services efficiently. Also, it ensures the needs of present and future generations considering economic, social, and environmental aspects [https://earth.org/]. Smart cities will impact the citizen lives, and their business and partnership. It adopts an approach for innovative governance; provides a centralized

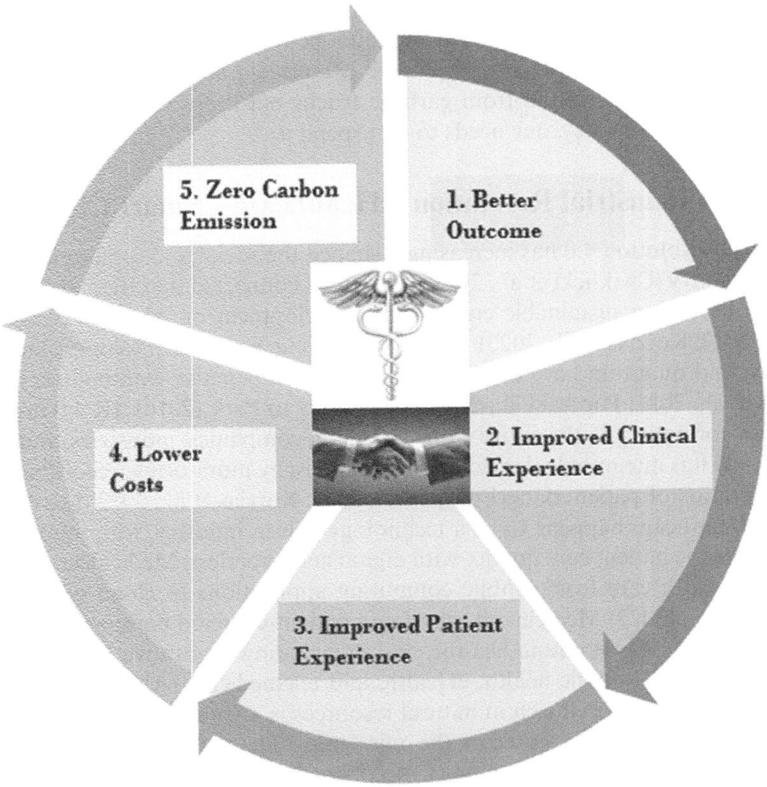

Fig. 1. The Pentacle Model for Healthcare 4.0 (Jayanthi et al., 2022).

single platform solution, and the information flow to maximize resource use and waste reduction. According to Khatib et al. (2023), it helps to establish digital leadership and collaboration between the stakeholders. UAE's leadership addresses the interconnected technologies that enhance the style of urban living including transportation, energy, healthcare, and security. UAE's smart infrastructure initiative includes intelligent lighting systems and smart grid networks to name a few. The innovation enhances energy efficiency and conserves the environment. UAE promotes solar street lights by the adoption of renewable energy sources [https://www.go-globe.com/the-future-of-smart-cities-in-the-uae/]. UAE's resource-based industries (oil and gas) have made a remarkable economic development within a short span of period (1973 to 1982). This had a huge structural growth in transforming agriculture, industrialization, demographic transition, urbanization, production, employment, and foreign trade (Shihab, 2001).

UAE ranked fourth globally in the Global Entrepreneurship Index (GEI) 2020 (Elmassah et al., 2022). The university-industry collaboration for enhancing knowledge-based industries (Ryan & Daly, 2019) and knowledge diffusion is promoted by UAE (Al-kfairy et al., 2022) and 95% of the businesses are from the

SME sector (Anand et al., 2020). Science and Technology Innovation is referred to as a new driver for the economy (Hameed et al., 2016). UAE had an interest in non-oil resources and development projects for economic growth (Alnassai, 2023) moreover Abu Dhabi's wealth fund is the Abu Dhabi Investment Authority (ADIA) (Yaw, 2010).

4. UAE – CRUDE MORTALITY RATE – HIGHER LIFE EXPECTANCY

Crude mortality rate indicates the number of deaths occurring during the year, per 1,000 populations estimated at the midyear. According to the WHO mortality database, total deaths recorded were 10,357; deaths were caused by intentional injuries (2.17) and unintentional injuries (2.08). It clearly shows intentional deaths were more common in the UAE compared to unintentional deaths (Zaki, 2023). The CDR of worldwide is 8/1000 whereas in the UAE, it is between 1 and 2/1000 (Blair & Sharif, 2012). Since 1971, the UAE has foreseen, healthcare has benefitted from rapid economic growth and improved the quality of patient healthcare (Koornneef et al., 2017). However, the total CDR recorded in UAE (10357) for 2020, with a crude rate of (1.12) according to the UAE statistical annual report 2020 in Table 1.

The trend of CDR declines for UAE (1.9), Arab World (5.8), and Qatar (1.3) for the years 1960–2021 is shown in Fig. 2(a–c), respectively. Whereas, the trend in India (Fig. 2d) declines from the year 1960 to 2017 (6.59) and rises in 2021 (9.4) [https://data.worldbank.org/]. It clearly shows that India being a developing country would not reduce CDR. However, the Gulf countries are capable of setting themselves as smart cities and reducing CDR to a minimum. The trend of under-five CDR (UAE), declines from the year 1990 to 2021, shown in Fig. 3, taken from

Table 1. Births and Deaths (UAE) for the Year 2020 [UAE Statistical Annual Report 2020].

	Citizen				**Non-Citizen**				**Grand-Total**			
	M	*F*	**Un known**	**Total**	*M*	*F*	**Un known**	**Total**	*M*	*F*	**Un known**	**Total**
Reg. Live Births	16,341	15,263	0	**31,604**	33,899	32,069	0	**65,968**	50,240	47,332	0	**97,572**
Reg. Still Births	62	45	4	**111**	188	158	8	**354**	250	203	12	**465**
Tot. Reg. Deaths	1,374	1,020	0	**2,394**	6,307	1,655	1	**7,963**	7,681	2,675	1	**10,357**

*Reg – Registered

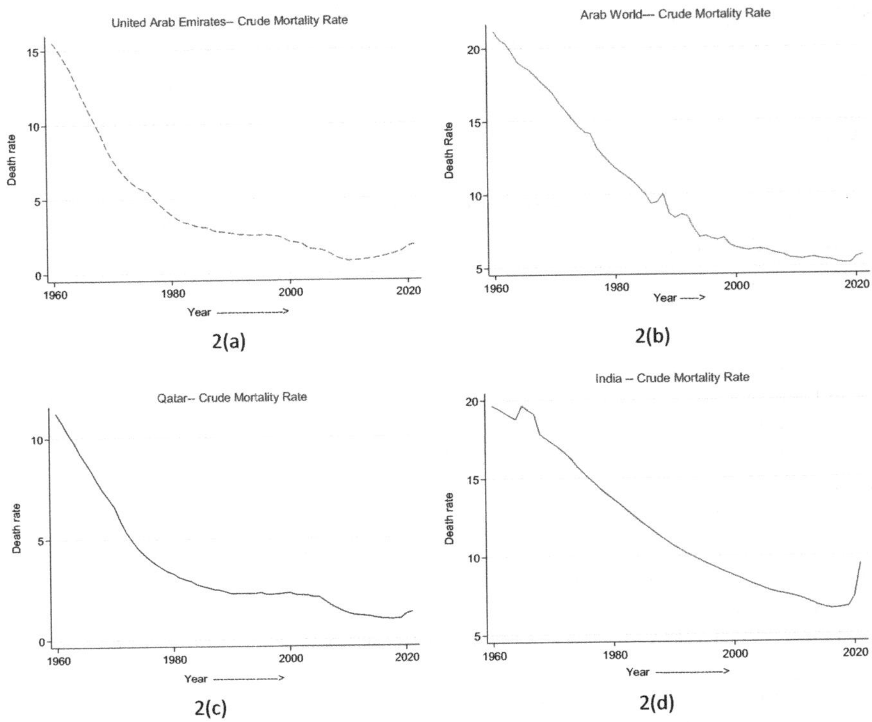

Fig. 2. CDR of UAE, Arab World, Qatar, and India [*Source*: https://data.worldbank.org/].

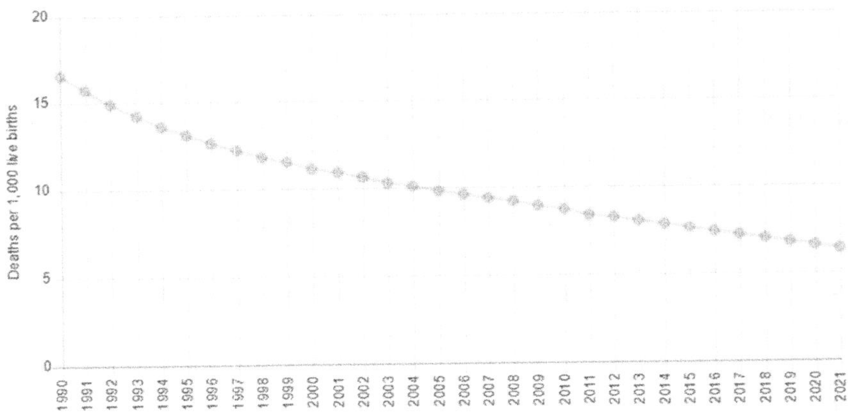

Fig. 3. Trends in Under-Five Mortality Rate in the United Arab Emirates [*Source*: https://data.unicef.org/country/are].

https://data.unicef.org/country/are/. Similarly, Fig. 4 indicates the chart of birth rate (UAE) from the year 2011 to 2021 decreases in the year 2019, and one can see the increase from 2020 to 2021. The chart of CDR of Abu Dhabi from the year 1985 to 2016 is shown in Fig. 5, with less number of death rates in 2016.

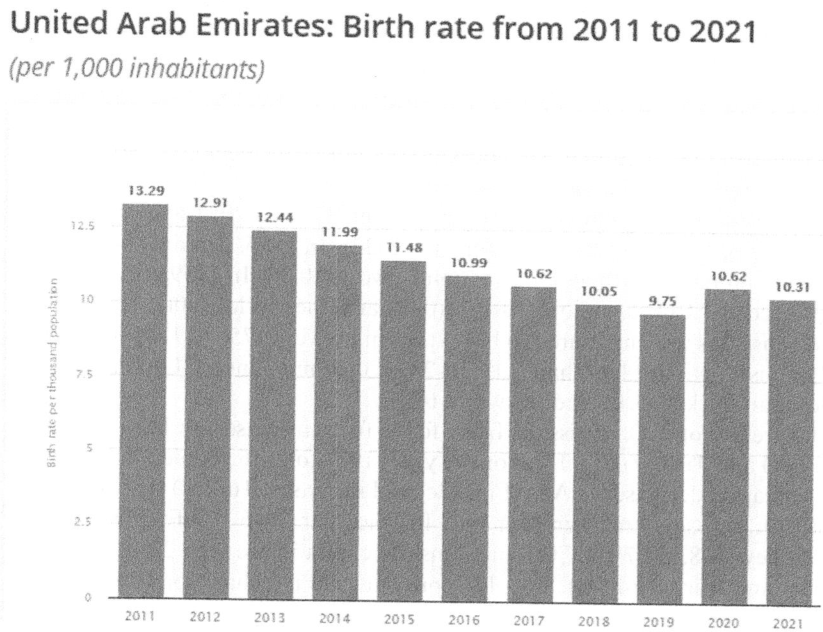

Fig. 4. The Graph of UAE – CBR [*Source*: https://www.statista.com/statistics/977328/crude-birth-rate-in-the-united-arab-emirates/].

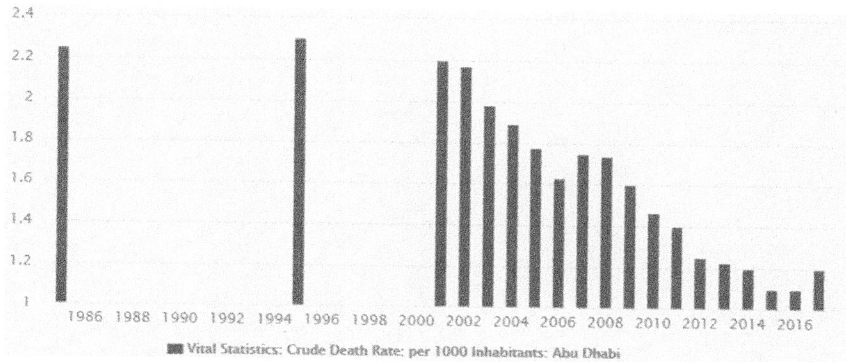

Fig. 5. CDR of Abu Dhabi from 1985 to 2016 in the Chart (*Source*: www.ceicdata.com).

5. ABU DHABI AND TURKEY – AIR QUALITY STUDY

The world map representing CO_2 emissions per capita from fossil fuel and industry is shown in Figs. 6–8. The data source is taken from our world in data. Fossil CO_2 includes emissions from coal, oil, gas, cement, steel, and other industrial process. However, fossil emissions do not include land use change, deforestation, or vegetation. In the year 1870 (Fig. 6), such data are unknown in a few countries. However, the emissions of CO_2 are represented by the variation of the color from 0t (low) to 20t (high). In the year 1945 (Fig. 7), most of the countries reached the peak of 20t represented in a deep dark color, and in the year 2022 (Fig. 8), the increase in the carbon dioxide is observed.

The carbon dioxide emission per capita from 1750 to 2022 is represented in Fig. 9 (1873), Fig. 10 (1966), and Fig. 11 (2022). In 1865, Turkey (<0.1t) commenced to emit CO_2 with global emission of world (0.3t). In 1959, UAE (< 0.1t) just started the CO_2 emission, Turkey (0.6t) and the world (3.0t). Up to 1842, world emissions were measured to be 0.1t, from the year 1750 to 1841, the overall world emissions were less than 0.1t. In 2022, the emissions of UAE (25.8t) are higher than Turkey (5.1t) and the world (4.7t).

The trends of CO_2 emission from fossil fuels increase for both the UAE (Fig. 12a) and Turkey (Fig. 12b) for the years 1971 to 2016. While the CO_2 emission per capita decreases (UAE) (Fig. 12c) and increases (Turkey) (Fig. 12d). The fossil CO_2 emissions measured in tons for the year 2016, UAE (218,788,684), and Turkey (368,122,740). CO_2 emissions per capita in the year 2016 were UAE (24.33t) and Turkey (4.54t). Fig. 13 represents the pie chart for CO_2 emission

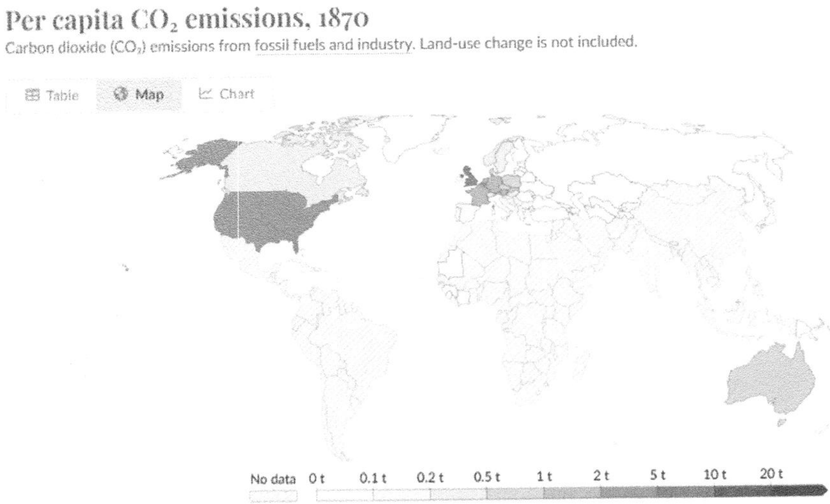

Fig. 6. The World Map in the Year 1870 Representing CO_2 Emissions [https://ourworldindata.org/co2-and-greenhouse-gas-emissions].

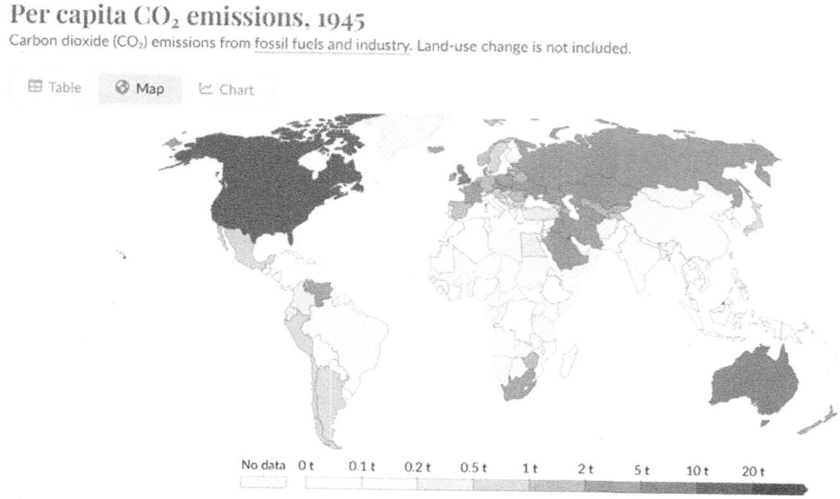

Fig. 7. The World Map in the Year 1945 Representing CO_2 Emissions [https://ourworldindata.org/co2-and-greenhouse-gas-emissions].

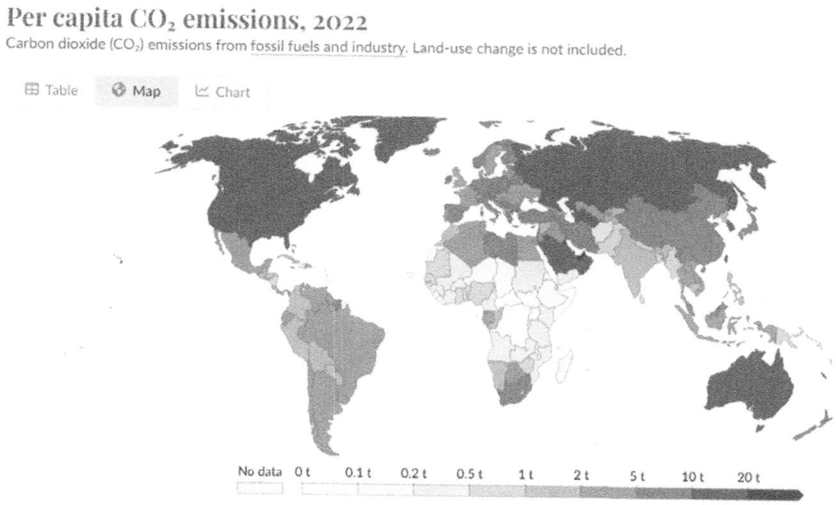

Fig. 8. The World Map in the Year 2022 Representing CO_2 Emissions [https://ourworldindata.org/co2-and-greenhouse-gas-emissions].

from the sources viz. power industry, industrial combustion, non-combustion, transport, and buildings for UAE (Fig. 13a) and Turkey (Fig. 13b).

The pollutant particulate matter (PM_{10}) and air quality index (AQI) data was collected from the World Air Quality Index for two cities in Abu Dhabi namely, Khalifa City and Khadija Primary School. The trends for PM_{10} and AQI of Khalifa city are shown in Fig. 14a and Khadija (Fig. 14b) (Fig. 14c and 14d)

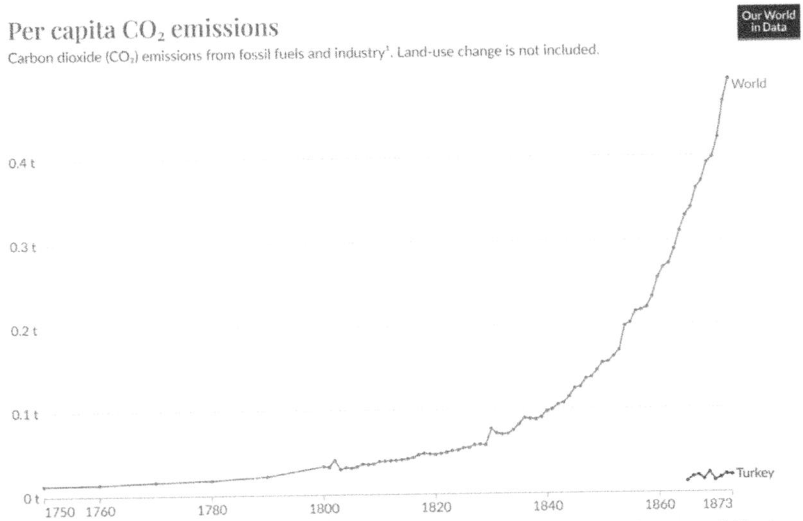

Fig. 9. Global CO_2 Emission Per Capita for 1873 [https://ourworldindata.org/co2-and-greenhouse-gas-emissions].

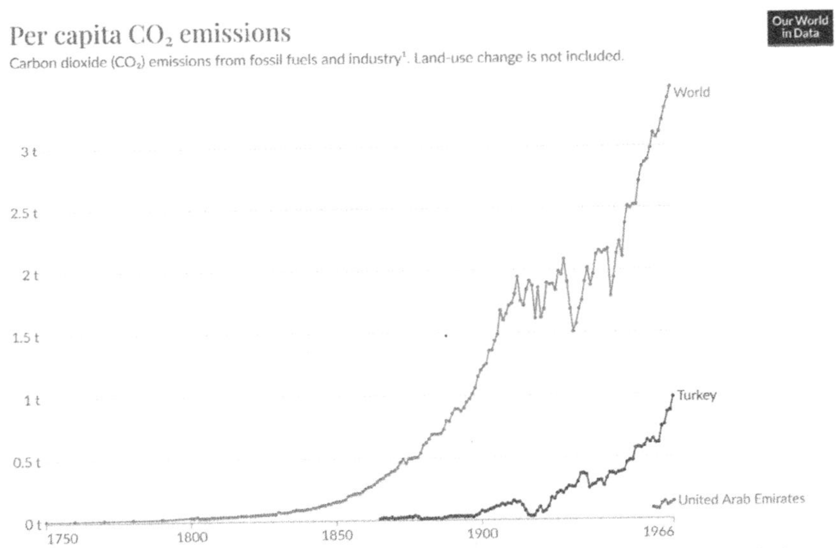

Fig. 10. Global CO_2 Emission Per Capita for 1966 [https://ourworldindata.org/co2-and-greenhouse-gas-emissions].

from the period of 2018 to 2024. The linear regression models were developed, and it was found that the *R*-squared value for Khadija (46185.06) and Khalifa city (63854.22) and the model was significantly fit in Table 2. The model (Table 3) was developed with AQI (dependent) and independent variables as PM_{10}, O_3,

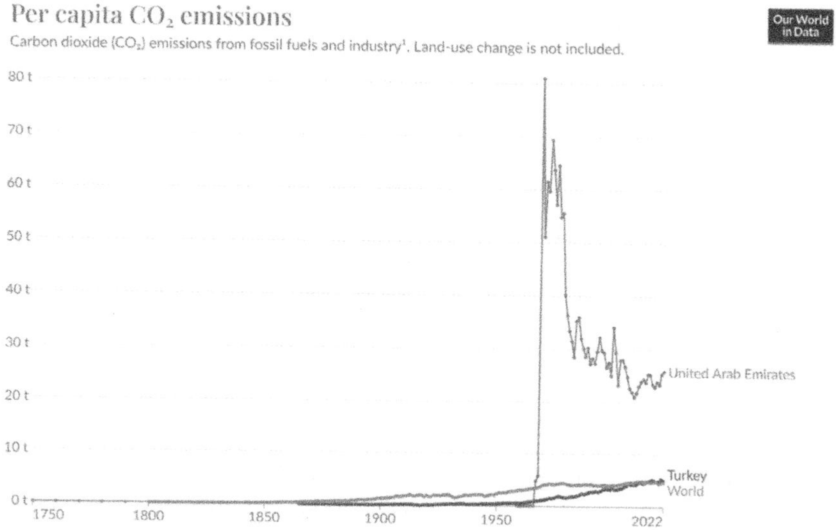

Fig. 11. Global CO_2 Emission Per Capita for 2022 [https://ourworldindata.org/co2-and-greenhouse-gas-emissions].

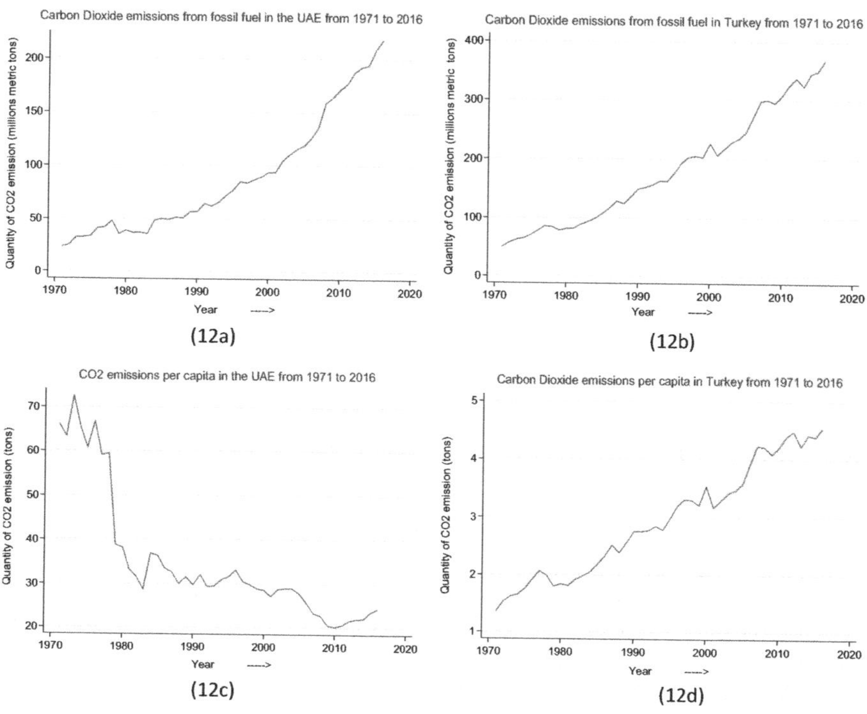

Fig. 12. CO_2 Emission in UAE and Turkey for the Years 1971 to 2016.

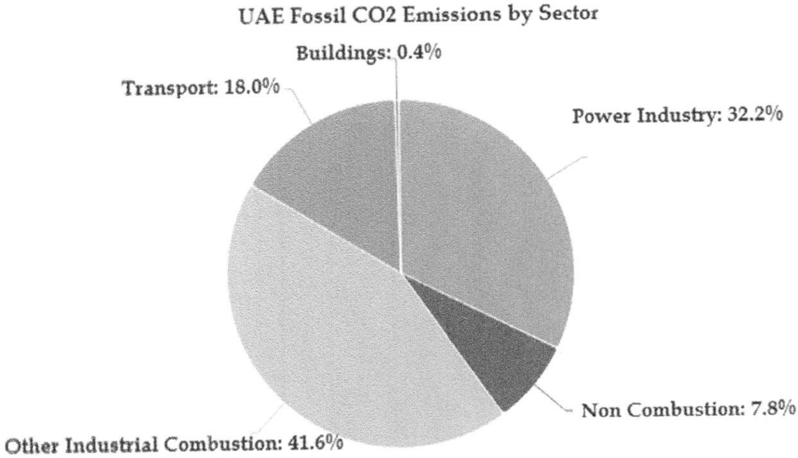

Fig. 13a. Pie Chart Representing the Percentage of Emissions in UAE from Various Sources (1970 –2010) [https://www.worldometers.info/co2-emissions].

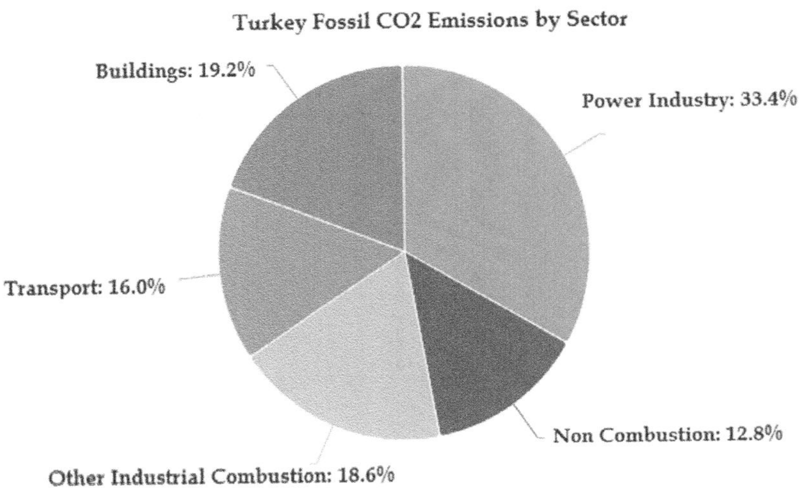

Fig. 13b. Pie Chart Representing the Percentage of Emissions in UAE from Various Sources (1970 –2010) [https://www.worldometers.info/co2-emissions].

NO_2, and SO_2. The PM_{10} maximum (195) and minimum (32) were found in Khadija city mentioned in Table 4. The higher AQI is severe, and bad quality air is observed, which can make people prone to various respiratory diseases. Based on the data, one can conclude air quality is much better in smart cities like Khadija. Even in Khalifa city, the quality of air seems to be healthy, except for the period April 2022 to August 2022 where the range seems to be 115 to 174. Inhaling severe PM is very dangerous for humans.

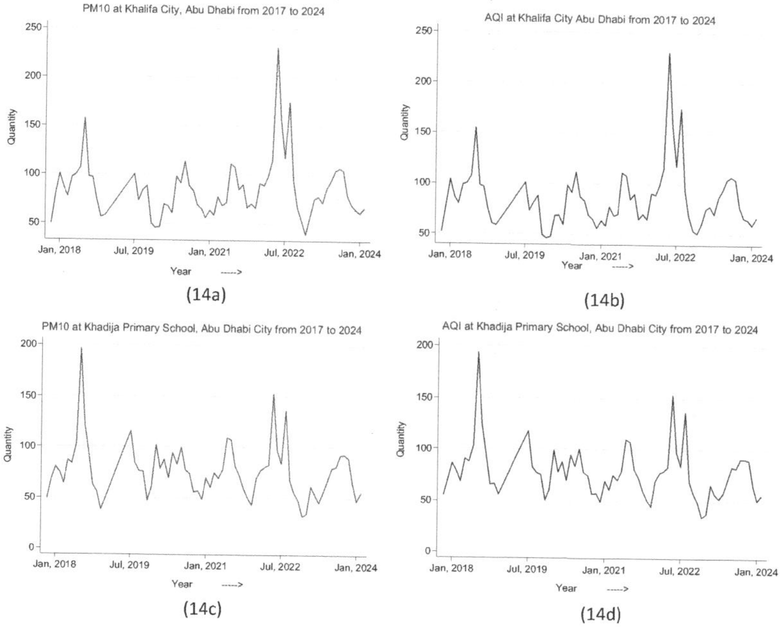

Fig. 14. The Trends of AQI and PM_{10} for Two Cities of Abu Dhabi.

6. UAE – Smart City And its Impact on Climate Change Mitigation

The majority of cities are liable for the world's economic activities, energy consumption, and GHGs. Moreover, the cities are working to reduce energy use and renewable energy. The cities depend on the use of transport efficiently. A smart city helps to reduce GHG emissions in the transport sector by minimizing electricity and heat production. However, the transport sector produces global GHG emissions (14%), whereas the contribution of electricity and heat production is 25% [https://www.frontier-economics.com/]. Smart energy systems focus on climate change mitigation, to reduce emissions, while the adaptation of smart water systems to avoid future floods and droughts (Obringer & Nateghi, 2021).

The smart city of Abu Dhabi has involved many digital government services and free Wi-Fi connectivity in public transport (K'Akumu, & Alhamoudi, 2023). Masdar City, a smart and sustainable city, has initiated technology-driven goals associated with an eco-friendly environment, self-generated energy availability through solar, energy efficiency and zero-carbon in transportation, physical infrastructure, and water management (Sankaran & Chopra, 2020), carbon-neutral, and zero-waste (Randeree & Ahmed, 2019). Masdar City provides the community with the highest standard of quality of life with the minimum carbon footprint (Alameri & Wagle, 2011) and the implementation and commercialization of smart-clean technology (Cugurullo & Ponzini, 2018). Moreover, the city

Table 2. Models for Two Cities in Abu Dhabi.

Model	SS	MS	*R*-sq	Adj *R*-sq	RMSE	Prob > *F*
Khadija	46185.06	11546.26	0.99	0.99	2.16	0.00
Khalifa City	63854.22	15963.55	0.99	0.99	1.67	0.00

Table 3. Linear Models for Abu Dhabi Cities.

LR-Model	AQI	Coef.	Std. Err.	*t*	*P*>|*t*|	95% CI
Khadija	PM_{10}	0.9442	0.010	91.86	0.000	0.9237–0.9647
	O_3	0.1954	0.0315	6.20	0.000	0.1324–0.2583
	NO_2	0.0781	0.0774	1.01	0.317	−0.0765–0.2329
	SO_2	−0.1065	0.1299	−0.82	0.415	−0.3659–0.1529
Khalifa City	PM_{10}	0.9745	0.0077	126.03	0.000	0.9591–0.9900
	O_3	0.0953	0.0213	4.47	0.000	0.0527–0.1378
	NO_2	−0.0173	0.0601	−0.29	0.774	−0.1374–0.1028
	SO_2	−0.0582	0.1230	−0.47	0.638	−0.3039–0.1875

Table 4. The Maximum and Minimum Values of PM_{10} and AQI (*Source*: World Air Quality Index).

City		Min	Max
Khalifa City	PM_{10}	39 (Dec, 2022)	230 (May, 2022)
	AQI	46 (Dec, 2019)	230 (May, 2022)
Khadija Primary School	PM_{10}	32 (Dec, 2022)	195 (July, 2018)
	AQI	34 (Dec, 2022)	193 (July, 2018)

manages the natural resources in a more sustainable manner (Alzaabi et al., 2019; Manandhar et al., 2020) and is designed based on ecofriendly water, air, and soil (Madakam & Ramaswamy, 2016). Ringel (2021) states that smart homes use technology and policy for better energy efficiency, reductions in energy consumption, and effects of GHG and climate change (Rio et al., 2021; Tok et al., 2014).

Smart technologies contribute to climate change mitigation (Obringer & Nateghi, 2021; Sharifi & Khavarian-Garmsir, 2023) and the integration of climate strategies (Fernández & Peek, 2020). Cities decide the future of climate change (Tomlinson, 2021). Smart city application reduces harmful risks (Huang-Lachmann, 2019), enhance life quality, reduce carbon dioxide emissions (Jawad, 2019), and minimize air pollution (Abdullah, 2023). Artificial Intelligence (AI), edge computing, and IoT are used effectively in monitoring energy consumption, detecting the emissions of

CO_2 and nitrogen, measuring pH levels and sulfur oxides, and monitoring transport systems accurately in smart cities (Razmjoo et al., 2022). Smart cities provide cleaner energy sources and more efficient energy services (Zadeh & Garay-Rondero, 2023).

7. Conclusion

The countries are working their best to make their cities smarter. The smart city has several technologies to integrate and are interconnected to run safer and more smoothly. The city with business connectivity promotes employment and healthcare and helps economic growth. Hence, people migrate from one end to another improving social connectivity. The smart cities monitor air pollution, use renewable energy, and hence reduce the GHG emission. The smarter the city, the better the quality of life, and also, improves climate by minimizing carbon footprint. Finally, the developed linear regression model produced RMSE for Khadija (2.16) and Khalifa City (1.67) in Abu Dhabi.

References

Abdullah, A. (2023). The concept of smart cities: A sustainability aspect for future urban development based on different cities. *Frontiers in Environmental Science*, *11*. https://doi.org/10.3389/fenvs.2023.1241593

Alameri, M., & Wagle, G. (2011). Abu Dhabi efforts in facing global warming challenges through urban planning. WIT Transactions on Ecology and the Environment. *Sustainable Development and Planning*, *150*, 29–38. https://doi.org/10.2495/SDP110031

Al-kfairy, M., Majdalawieh, M., & Alrabaee, S. (2022). Conceptualising the role of the UAE Innovation Strategy in university-industry knowledge diffusion process. *Proceedings of the 23rd European Conference on Knowledge Management*, *23*(2).

Alnassai, J. M. I. A. (2023). A study on the barriers to entrepreneurship in the UAE. *Journal of Risk and Financial Management*, *16*, 146. https://doi.org/10.3390/jrfm16030146

Alzaabi, M. A., Rizk, Z. S., & Mezher, T. (2019). Linking smart cities concept to energy-water-food nexus: The Case of Masdar City in Abu Dhabi, UAE: Current state, opportunities, and challenges. In *Smart cities in the Gulf*. https://doi.org/10.1007/978-981-13-2011-8_6

Ahsan, M. M., & Siddique, Z. (2022). Industry 4.0 in healthcare: A systematic review. *International Journal of Information Management Data Insights*, *2*(1). 100079. ISSN 2667-0968. https://doi.org/10.1016/j.jjimei.2022.100079

Anand, T., Banerjee, A., & Riaz, S. (2020). Factors leading to sustained growth of SMES in the U.A.E: A Concept Paper. *Journal of Advanced Research in Economics and Administrative Sciences*, *1*(2). https://doi.org/https://doi.org/10.47631/jareas.v1i2.85

Blair, I., & Sharif, A. A. (2012). Population structure and the burden of disease in the United Arab Emirates. *Journal of Epidemiology and Global Health*, *2*(2), 61–71. https://doi.org/10.1016/j.jegh.2012.04.002

Castro, E. M. J., & Faria, A. N. M. (2020). Impact of the Fourth Industrial Revolution on the health sector: A qualitative study. *Healthcare Informatics Research*, *26*(4), 328–334. https://doi.org/10.4258/hir.2020.26.4.328

Cugurullo, F., & Ponzini, D. (2018). The transnational smart city as urban eco-modernisation: The case of Masdar City in Abu Dhabi. In *Inside smart cities: Place, politics and urban innovation*. https://doi.org/10.4324/9781351166201

Ćwiklicki, M., Klich, J., & Chen, J. (2020). The adaptiveness of the healthcare system to the fourth industrial revolution: A preliminary analysis. *Futures*, *122*, 102602, ISSN 0016-3287. https://doi.org/10.1016/j.futures.2020.102602

Elmassah, S., James, R., & Bacheer, S. M. (2022). Ethnic entrepreneurial success factors: Evidence from the United Arab Emirates. *Heliyon*, *8*(6). https://doi.org/10.1016/j.heliyon.2022.e09639

Fernández, G. C., & Peek, D. (2020). Smart and Sustainable? Positioning adaptation to climate change in the European Smart City. *Smart Cities*, *3*, 511–526. https://doi.org/10.3390/smartcities3020027

Fujita, T., Yamamoto, S., Onodera, R., Yuzurihara, M., Sato, K., Suzue, Y., Okai, H., Soble, J., & Miyata, H. (2021). *Healthcare and the Fourth Industrial Revolution: Realizing the 'greatest happiness of the greatest diversity'* [Briefing Paper. World Economic Forum].

Gonçalves, G. D. L., Filho, L. W., Neiva, S. D. S., Deggau, A. B., Veras, M. de O., Ceci, F., Lima, M. A. de., & Guerra, J. B. S. O. de A. (2021). The impacts of the Fourth Industrial Revolution on smart and sustainable cities. *Sustainability*, *13*, 7165. https://doi.org/10.3390/su13137165

Haleem, A., Javaid, M., Singh, R. P., & Suman, R. (2022). Medical 4.0 technologies for healthcare: Features, capabilities, and applications. *Internet of Things and Cyber-Physical Systems*, *2*, 12–30. ISSN 2667-3452. https://doi.org/10.1016/j.iotcps.2022.04.001

Hameed, I., Khan, M.B., Shahab, A., Hameed, I., & Qadeer, F. (2016). Science, technology and innovation through entrepreneurship education in the United Arab Emirates (UAE). *Sustainability*, *8*, 1280. https://doi.org/10.3390/su8121280

Huang-Lachmann, J. T. (2019). Systematic review of smart cities and climate change adaptation. *Sustainability Accounting, Management and Policy Journal*, *10*(4), 745–772. https://doi.org/10.1108/SAMPJ-03-2018-0052

Jawad, W. K. (2019). Case study of Dubai as a Smart city. *International Journal of Computer Applications*, *178*(40), 35–37. https://doi.org/10.5120/ijca2019919291

Jayanthi, P., Iyyanki, M. K., Mothkuri, A., & Vadakattu, P. (2020). Fourth Industrial Revolution: An impact on health care industry. *AISC*, *965*, 58–69. https://doi.org/10.1007/978-3-030-20454-9_6

Jayanthi, P., Rai, B. K., & Muralikrishna, I. (2022). The potential of quantum computing in healthcare. In *Technology road mapping for quantum computing and engineering* (pp. 81–101). IGI Global. https://doi.org/10.4018/978-1-7998-9183-3.ch006

K'Akumu, O. A., & Alhamoudi, A. M. (2023). The smart city as policy concept in the Arab World: A critical research into the strategic visions of urban development in Abu Dhabi. *Journal of Asian and African Studies*, *0*(0). https://doi.org/10.1177/00219096231192316

Khatib, M. E., Maazmi, F. A, Alhosani, A. I., & Aljaberi, K. A. (2023). Smart cities in UAE: Governance, innovation and quality. *International Journal of Business Analytics and Security*, *3*(2), 40–53. https://doi.org/10.54489/ijbas.v3i2.254

Koornneef, E. J., Robben, P. B. M., Al Seiari, M. B., & Al Siksek, Z. (2012). Health system reform in the Emirate of Abu Dhabi, United Arab Emirates. *Health Policy*, *108*(2–3). 115–121. https://doi.org/10.1016/j.healthpol.2012.08.026

Koornneef, E., Robben, P., & Blair, I. (2017). Progress and outcomes of health systems reform in the United Arab Emirates: A systematic review. *BMC Health Services Research*, *17*, 672. https://doi.org/10.1186/s12913-017-2597-1

Kotzias, K., Bukhsh, F. A., Arachchige, J.J., Daneva, M., & Abhishta, A. (2023). Industry 4.0 and healthcare: Context, applications, benefits and challenges. *IET Software,17*, 195–248. doi: 10.1049/sfw2.12074

Kotzias, K., Bukhsh, F. A., Arachchige, J. J., et al. (2023). Industry 4.0 and healthcare: Context, applications, benefits and challenges. *IET Software, 17*,195–248. https://doi.org/10.1049/sfw2.12074

Lee, J., & Lim, J. (2017). The prospect of the Fourth Industrial Revolution and home healthcare in super-aged society. *Annals of Geriatric Medicine and Research, 21*(3), 95–100. https://doi.org/10.4235/agmr.2017.21.3.95

Madakam, S., & Ramaswamy, R. (2016). Sustainable smart city: Masdar (UAE) (A City: Ecologically Balanced). *Indian Journal of Science and Technology, 9*(6). https://doi.org/10.17485/ijst/2016/v9i6/87669

Manandhar, P., Bande, L., Tsoupos, A., Marpu, P.R., & Armstrong, P. (2020). A study of local climate zones in Abu Dhabi with urban weather stations and numerical simulations. *Sustainability, 12*, 156. https://doi.org/10.3390/su12010156

Obringer, R., & Nateghi, R. (2021). What makes a city 'smart' in the Anthropocene? A critical review of smart cities under climate change. *Sustainable Cities and Society, 75*, 103278. https://doi.org/10.1016/j.scs.2021.103278

Park, H. (2016). Are we ready for the Fourth Industrial Revolution? *IMIA Yearbook of Medical Informatics* http://dx.doi.org/10.15265/IY-2016-052

Popov, V. V., Kudryavtseva, E. V., Katiyar, N. K., Shishkin, A., Stepanov. S. I., & Goel, S. (2022). Industry 4.0 and digitalisation in healthcare. *Materials, 15*, 2140. https://doi.org/10.3390/ma15062140

Randeree, K., & Ahmed, N. (2019). The social imperative in sustainable urban development: The case of Masdar City in the United Arab Emirates. *Smart and Sustainable Built Environment, 8*(2), 138–149. https://doi.org/10.1108/SASBE-11-2017-0064

Razmjoo, A., Gandomi, A. H., Pazhoohesh, M., Mirjalili, S., & Rezaei, M. (2022). The key role of clean energy and technology in smart cities development. *Energy Strategy Reviews, 44*, 100943. https://doi.org/10.1016/j.esr.2022.100943

Ringel, M. (2021). Smart city design differences: Insights from decision-makers in Germany and the Middle East/North-Africa Region. *Sustainability, 13*, 2143. https://doi.org/10.3390/su13042143

Rio, D. D. F. D., Sovacool, B. K., & Griffiths, S. (2021). Culture, energy and climate sustainability, and smart home technologies: A mixed methods comparison of four countries. *Energy and Climate Change, 2*, 100035. https://doi.org/10.1016/j.egycc.2021.100035

Ryan, J. C., & Daly, T. M. (2019). Barriers to innovation and knowledge generation: The challenges of conducting business and social research in an emerging country context. *Journal of Innovation & Knowledge, 4*(1), 47–54. https://doi.org/10.1016/j.jik.2017.10.004.

Sankaran, V., & Chopra, A. (2020). Creating global sustainable smart cities (A Case Study of Masdar City). *Journal of Physics, 1706*, 012141.

Sharifi, A., & Khavarian-Garmsir, A. R. (2023). Chapter 5 – Smart city solutions and climate change mitigation: An overview. In *Urban climate adaptation and mitigation* (pp. 93–116). ISBN 9780323855525. https://doi.org/10.1016/B978-0-323-85552-5.00017-8

Shihab, M. (2001). *Economic development in the UAE. United Arab Emirates: A new perspective* (pp. 249–259). Trident Press Ltd.

Tok, E., Mohammad, F. A., & Merekhi, M. A. (2014). Crafting smart cities in the gulf region: A comparison of Masdar and Lusail. *European Scientific Journal, 2*. https://doi.org/10.4018/978-1-4666-8205-4.ch020

Tomlinson, R. (2021). Smart cities and climate change: The evolution and curation of knowledge. *Urban Sustainability Laboratory Working Paper Series*. Wilson Center.

UAE Statistical Annual Report. (2020). *Statistics and Research Center Ministry of Health and Community*.

Yaw, N. (2010). *The United Arab Emirates: Some lessons in economic development* [WIDER Working Paper No. 2010/11]. The United Nations University World Institute for Development Economics Research (UNU-WIDER). ISBN 978-92-9230-246-7.

Zadeh, I. S. B., & Garay-Rondero, C. L. (2023). Enhancing urban sustainability: Unravelling carbon footprint reduction in smart cities through modern supply-chain measures. *Smart Cities*, *6*, 3225–3250. https://doi.org/10.3390/smartcities6060143

Zaki, H. A., Hamdi, A. H., Elgassim, M., et al. (2023). Unmasking the underlying causes: A cross-sectional analysis of mortality from intentional and unintentional injuries in the United Arab Emirates. *Cureus*, *15*(10), e46567. https://doi.org/10.7759/cureus.46567

Chapter 20

The Social, Economic, and Sustainability Agenda of Smart Cities in the Kingdom of Bahrain

Omer Hag Hamid

Arab Open University, Bahrain

Abstract

This chapter presents a comprehensive framework for advancing smart cities in Bahrain, focusing on innovation, development, transformation, and prosperity. The framework integrates technological advancements with sustainable practices to enhance societal well-being, economic growth, and overall development. It aligns with Bahrain's 2030 Vision, aiming to create Next Generation Smart Cities. The chapter emphasizes enhancing social impact, healthcare access, and promoting a healthy lifestyle, driving economic growth through digital infrastructure, and promoting sustainability through renewable energy and smart urban planning. It provides a roadmap for integrating smart technologies into urban planning, infrastructure development, and service delivery, improving residents' quality of life and promoting economic growth. The framework highlights the importance of sustainability, emphasizing environmentally friendly practices and energy-efficient solutions. Bahrain's smart city initiatives are multifaceted, encompassing social, economic, and sustainability agendas. Socially, the focus is on promoting cohesion and community engagement through smart parks, community centers, and digital platforms that facilitate communication and collaboration. Economically, Bahrain aims to drive growth and innovation by investing in digital infrastructure and deploying smart technologies across sectors like finance and tourism. Additionally, the kingdom is developing a skilled workforce in emerging technologies to enhance competitiveness. Sustainability is a key pillar, with Bahrain investing in

The Emerald Handbook of Smart Cities in the Gulf Region:
Innovation, Development, Transformation, and Prosperity for Vision 2040, 363–377

doi:10.1108/978-1-83608-292-720241023

renewable energy sources and smart grid technologies to reduce reliance on fossil fuels and improve energy efficiency. These initiatives collectively aim to enhance the quality of life, drive economic prosperity, and promote environmental sustainability in Bahrain's smart cities.

Keywords: Smart cities; Bahrain; social agenda; economic agenda; sustainability; well-being; quality of life; social inclusion

Introduction

The Kingdom of Bahrain, an island in the Arabian Gulf, holds a rich tapestry of history, culture, and modernity. Situated in a strategic location between the east and the west, Bahrain has been a crossroads of civilizations for millennia, shaping its unique identity and character. Bahrain's history dates back to ancient times, with archaeological evidence suggesting human habitation as far back as the Dilmun civilization around 3,000 BC. The island has been ruled by various powers throughout history, including the Assyrians, Babylonians, Persians, and Portuguese, each leaving their mark on Bahrain's cultural landscape (Meero et al., 2020). The discovery of oil in Bahrain in the early 20th century transformed the kingdom's economy and society. Bahrain was one of the first Gulf countries to discover oil, leading to rapid modernization and urban development. The oil boom brought newfound wealth to Bahrain, leading to the development of infrastructure, healthcare, and education systems. Despite its small size, Bahrain has played a significant role in regional politics and economics. The kingdom was a founding member of the Gulf Cooperation Council (GCC) in 1981, a political and economic alliance of Gulf countries aimed at promoting cooperation and integration among its members (Alsabbagh & Alnaser, 2023). The kingdom's culture is reflected in its traditional music, dance, and cuisine, as well as its architecture and art. Bahrain is also known for its tolerance and religious diversity, with a sizable expatriate population living and working in the kingdom. The country is home to various religious sites, including the Al-Fateh Mosque, one of the largest mosques in the world, and the Bahrain World Trade Centre, a symbol of modernity and sustainability. Today, Bahrain is a modern and cosmopolitan kingdom, known for its welcoming hospitality and forward-thinking outlook. The kingdom has made significant strides in diversifying its economy away from oil, focusing on sectors such as finance, tourism, and technology (Wirayuda et al., 2022).

In recent years, Bahrain has undergone a significant transformation, emerging as a regional hub for finance and technology. This transformation has been driven by the kingdom's strategic initiatives and commitment to fostering innovation and sustainability. One of the key factors contributing to Bahrain's emergence as a regional hub is its business-friendly environment. The kingdom offers a conducive ecosystem for businesses, with streamlined regulations, tax incentives, and a supportive government that encourages entrepreneurship and investment. This has attracted a diverse range of businesses, from start-ups to multinational corporations, looking to establish a presence in the region (Meero et al., 2020). Additionally, Bahrain's strategic location in the Gulf region has played a crucial role in its development as a hub for finance and technology. The kingdom serves as a gateway to the lucrative Gulf market, providing businesses with access to a large consumer base and opportunities for growth and expansion. Bahrain's commitment to innovation has also been a driving force behind its transformation. The kingdom has invested heavily in developing its technology infrastructure, including high-speed internet connectivity, digital payment systems, and data centers. This has enabled businesses to leverage technology to improve efficiency, drive growth, and enhance their competitiveness in the global market. Furthermore, Bahrain has prioritized sustainability in its development strategy, recognizing the importance of environmental conservation and resource management (Elgabiri et al., 2020). The kingdom has implemented various initiatives to promote sustainability, such as investing in renewable energy sources, promoting energy efficiency, and implementing sustainable urban planning practices. These efforts have not only contributed to environmental conservation but have also helped to enhance the quality of life for Bahraini citizens. Bahrain's emergence as a regional hub for finance and technology is a testament to its commitment to innovation, sustainability, and economic development. The kingdom's strategic initiatives and business-friendly environment have positioned it as a leader in the region, paving the way for a prosperous and promising future.

Smart Cities in Bahrain

The Kingdom of Bahrain's continuous advancement in information and communication technologies (ICTs) is significantly contributing to enhancing sustainability, economic viability, and citizen well-being. This progress is propelling Bahrain toward the development of smart cities, which are urban areas leveraging advanced digital and engineering technologies to improve residents' economic and social conditions. The Telecommunication Regulatory Authority (TRA) has spearheaded several smart city initiatives, such as smart traffic lights. These lights integrate traditional traffic controls with sensors and artificial intelligence to manage vehicle and pedestrian traffic effectively. Furthermore, the Electricity and Water Authority (EWA) is promoting the use of solar energy among the public, offering affordable loans for installing full solar energy systems. The University of Bahrain has also played a role in advancing smart city initiatives by hosting the 3rd Smart Cities Symposium in 2020 (Wirayuda et al., 2022). This symposium aimed to raise awareness about the future prospects of smart cities and provide a platform for the exchange of ideas among stakeholders. Bahrain's smart cities movement aligns with the UN HABITAT's New Urban Agenda, aiming to make urban areas more efficient, livable, and sustainable with the active participation of public administration, citizens, and businesses. This movement serves as an accelerator for achieving the Sustainable Development Goals (SDGs), particularly SDG 11, which focuses on sustainable cities and communities. The National Real Estate Plan 2021–2024 in Bahrain encourages the development of smart and environmentally friendly real estate projects. The ongoing dialogue among public and private sectors, technology providers, and NGOs in Bahrain aims to implement the Smart Real Estate concept in real estate projects. The Bahrain Smart Cities Summit organizes online panel discussions to address these issues (Alsabbagh & Alnaser, 2023).

Smart traffic lights, intelligent traffic lights that combine traditional controls with sensors and artificial intelligence, are used throughout Bahrain to intelligently route vehicle and pedestrian traffic. Bahrain is also utilizing smarter cooling infrastructure to address the high demand for cooling in its hot climate. Areas like Diyar Al Muharraq are using direct cooling systems, resulting in over a 30% reduction in electricity consumption. This system provides cold water to buildings, reducing the use of conventional air conditioning systems and lowering electricity demand (AlZayani et al., 2023). In the banking sector, new bank accounts can be opened easily online via mobile apps without visiting a physical branch. Personal identity and information can be captured and verified virtually, in compliance with the kingdom's financial regulations and policies, with all private information protected by law. Bahrain has also been hosting events focusing on smart cities, such as the Bahrain Smart Cities Summit. This summit, dedicated to smart and sustainable solutions for cities of today and the future, serves as a platform for cross-sector collaborations in Bahrain, fostering innovation and transformation toward a smarter kingdom (Katbi et al., 2020). Additionally, the 3rd Smart Cities Symposium held in 2020 aimed to raise awareness about smart cities' prospects, exchange ideas internationally, and emphasize universities' role in promoting smart cities through consultancy, building smarter ideas, or creating innovative solutions.

The Kingdom of Bahrain has set its sights on transforming into a smart city, using technology and innovation to uplift the quality of life for its residents while fostering sustainable economic growth. This chapter delves into Bahrain's social, economic, and sustainability agenda within the smart city framework, shedding light on the pivotal initiatives and strategies driving these objectives forward. In pursuit of its smart city vision, Bahrain has been strategically deploying cutting-edge technologies and innovative solutions. One of the key focuses has been on enhancing social well-being and quality of life. The kingdom has been leveraging advancements in healthcare technology, such as telemedicine and digital health records, to improve healthcare access and services. Initiatives like smart parks, community centers, and cultural events have also been introduced to promote social cohesion and active citizen engagement, enriching the social fabric of Bahraini society.

On the economic front, Bahrain's smart city initiatives are geared toward stimulating growth and fostering innovation. The kingdom has been investing significantly in digital infrastructure, including high-speed internet and data centers, to attract businesses and entrepreneurs (Umar, 2021). Moreover, smart technologies are being deployed across various sectors, such as finance, tourism, and logistics, to drive efficiency and productivity. Bahrain's focus on nurturing a skilled workforce in emerging technologies further underscores its commitment to economic advancement through innovation. Sustainability lies at the core of Bahrain's smart city agenda, with the kingdom actively pursuing environmentally friendly practices. Bahrain has been investing in renewable energy sources like solar and wind power to reduce its reliance on fossil fuels (Freije et al., 2019). The implementation of smart grid technologies is another step toward enhancing energy efficiency and curbing carbon emissions. Additionally, sustainable urban planning and transportation solutions are being explored to alleviate congestion and pollution, ensuring a greener and more sustainable future for Bahrain's cities. Bahrain's journey toward becoming a smart city is driven by a vision to enhance the well-being of its citizens, drive economic growth, and promote environmental sustainability. Through a strategic blend of technology, innovation, and sustainable practices, Bahrain is not only shaping a brighter future for its residents but also setting a compelling example for smart city development across the globe.

Social Agenda of Smart Cities

The social agenda of smart cities is diverse, aiming to improve quality of life, foster social inclusion, and provide equitable opportunities for all citizens. These initiatives focus not only on technological advancements but also on delivering tangible benefits that address social issues and enhance overall well-being. One crucial aspect is promoting social inclusion and well-being, particularly among vulnerable populations like older adults. Research emphasizes the importance of creating "age-friendly" smart cities that cater to the needs of older individuals, highlighting technology's role in enhancing social connections and quality of life (Li & Woolrych, 2021) Additionally, the concept of smart healthcare within smart cities underscores prioritizing health and well-being as key social objectives, showing a holistic approach to urban development (Gagliardi & Albergo, 2023). Citizen participation is another crucial element. Ensuring that citizens can engage with and benefit from smart city initiatives is essential for promoting social justice and equality. Studies highlight the need to address institutional barriers and promote citizen engagement to foster a more inclusive and participatory urban environment (Pansera et al., 2022). Furthermore, integrating lean thinking and social well-being indices in smart city frameworks emphasizes prioritizing citizens' well-being and social needs. By focusing on social well-being indicators, cities can ensure that their initiatives align with residents' values and priorities (Ismail et al., 2022).

In urban planning, smart cities are seen as sociotechnical innovations that aim to enhance quality of life, city operations' efficiency, and competitiveness while meeting present and future generations' diverse needs. This holistic approach considers economic, social, environmental, and cultural aspects to create sustainable and inclusive urban environments (Toli & Murtagh, 2020). Smart cities have the potential to significantly impact social inclusion by leveraging technology and innovative approaches. Research indicates that social inclusion indicators play a crucial role in building citizen-centric smart cities (Malek et al., 2021). However, there are challenges in achieving social and cultural inclusiveness in technology-centric smart cities, which can lead to disadvantages (Huang et al., 2021). Studies emphasize that higher levels of social awareness and participation can promote the growth of smart cities and contribute to social inclusion (Szarek-Iwaniuk

& Senetra, 2020). In the context of social well-being, particularly for older individuals, smart city interventions can enhance social inclusion by prioritizing social participation and engagement (Li & Woolrych, 2021). Despite the focus on environmental goals in smart city projects through ICT implementation, there is a need to address social equity implications for comprehensive sustainability (Sengupta & Sengupta, 2022). Smart cities are influenced not only by technological factors but also by social factors, highlighting the importance of considering social aspects in their development (Morozova & Yatsechko, 2022). Gender mainstreaming policies in smart city development are essential for addressing social inclusion challenges, as some frameworks may overlook social inclusion aspects (Wahyudi & Elanda, 2023). The essence of smart cities lies in optimizing resources, energy efficiency, and overall sustainability, which can contribute to social and economic development (Suvarna et al., 2020). Achieving social well-being and economic performance in smart cities requires a balance between environmental sustainability and social sustainability (Shee et al., 2021). Incorporating considerations of social inequality and citizen participation from the project's initiation to assessment can lead to more equitable and inclusive smart cities, enhancing residents' quality of life (Chen et al., 2022). However, a predominant focus on digital urban renewal in smart city development can deepen social polarization and gentrification, emphasizing the need for inclusive approaches (Praharaj, 2021). Understanding institutional logics and engaging in public participation are crucial for the social construction of smart cities in specific contexts (Pansera et al., 2022).

Bahrain Social Agenda

Bahrain's smart city initiatives are rooted in the vision of Vision 2040, which seeks to transform the kingdom into a hub of innovation, creativity, and prosperity. Central to this vision is the goal of enhancing the well-being and quality of life of Bahraini citizens, ensuring that they have access to high-quality healthcare services, promote a healthy lifestyle, and enhance social inclusion. (Alnaser & Alnaser, 2019). One of the key initiatives in Bahrain's smart city agenda is the deployment of smart healthcare technologies. Telemedicine, for example, allows patients to consult with healthcare providers remotely, reducing the need for in-person visits and improving access to care, especially for those in remote areas. Wearable health devices enable individuals to monitor their health in real-time, providing valuable data to healthcare providers for more personalized care. Digital health records ensure that patient information is easily accessible and secure, further improving healthcare delivery and accessibility (Alqahtani & Alareeni, 2020). In addition to healthcare, Bahrain's smart city initiatives also focus on promoting a healthy lifestyle among its citizens. Smart parks equipped with fitness trackers and outdoor exercise equipment encourage physical activity, while smart buildings and homes are designed to promote health and well-being. These initiatives not only improve the overall health of Bahraini citizens but also contribute to a more vibrant and active community (Alqahtani & Alareeni, 2020). Furthermore, Bahrain is implementing initiatives to enhance social inclusion and promote active citizenship. Smart community centers serve as hubs for social activities and cultural events, bringing people together and fostering a sense of belonging. Digital platforms and mobile applications are being developed to facilitate communication and collaboration among community members, further strengthening social cohesion (Katbi et al., 2020). Bahrain's smart city initiatives are transforming the kingdom into a more inclusive, innovative, and sustainable society. By leveraging technology and innovation, Bahrain is enhancing the well-being and quality of life of its citizens, driving economic growth, and ensuring a prosperous future for generations to come.

The recognition received by Diyar Al Muharraq at the Bahrain Smart Cities Summit 2024 serves as a notable example of initiatives promoting social inclusion within the Kingdom of Bahrain. Diyar Al Muharraq's commitment to community engagement and inclusivity aligns with broader national efforts to enhance social cohesion and create a more inclusive society. One key aspect of Diyar Al Muharraq's approach to social inclusion is its focus on catering to different segments of the local community, including citizens and residents, with various living standards. This approach reflects a commitment to ensuring that all members of the community feel welcome and included, regardless of their background or circumstances. Additionally, Diyar Al Muharraq's efforts to embed the authentic Bahraini identity across its projects and preserve Arabic and Islamic traditions contribute to creating a sense of belonging and cultural inclusion among residents. By inaugurating mosques and mixed-use halls that serve as community gathering spaces, Diyar Al Muharraq fosters a tight-knit community where individuals can connect with their heritage and each other. Furthermore, Diyar Al Muharraq's provision of parks, pedestrian walkways, courts, and gyms in its residential projects demonstrates a commitment to promoting a healthy lifestyle and building community ties. These amenities not only enhance residents' quality of life but also encourage social interaction and engagement, fostering a sense of community among residents. Diyar Al Muharraq's initiatives highlight the importance of social inclusion in smart city development. By prioritizing community engagement, cultural preservation, and the provision of inclusive amenities, Diyar Al Muharraq contributes to creating a more inclusive and cohesive society in Bahrain.

Economic Agenda of Smart Cities

Smart cities are designed to enhance the quality of life for residents while optimizing resources, energy efficiency, and overall sustainability, thereby fostering economic, social, and environmental development (Suvarna et al., 2020). These cities aim to leverage ICTs to improve urban operations, services, and competitiveness, ensuring they meet the needs of present and future generations across various aspects, including economic, social, environmental, and cultural dimensions (Toli & Murtagh, 2020). The economic benefits of smart city projects are often highlighted in public discourse, emphasizing job creation, revenue generation, and cost reductions (Sengupta & Sengupta, 2022). While smart cities offer potential economic development and urban operational improvements, there is a need for more research on their sustainability, particularly in terms of long-term impacts and outcomes (Chen et al., 2022). The construction of smart city projects is expected to bring significant benefits, not only to investors but also in terms of enhancing citizen happiness, addressing urban challenges, promoting economic growth, and advancing related industries (Yuan, 2022). Additionally, smart city initiatives prioritize economic sustainability, although there is a call for more attention to environmental sustainability alongside economic considerations (Vainio & Sankala, 2022). Furthermore, the development of smart cities is seen as a means to drive social and economic development, contributing to overall sustainable economic and urban growth (Park, 2022). The smart city concept encompasses a vision that supports economic growth, social progress, and urban development through the implementation of smart projects (Romanelli, 2021). Overall, smart cities aim to balance smartness and sustainability, with a focus on economic competitiveness, innovation leadership, and the contribution of ICT to GDP as driving factors for their development (Fernández & Peek, 2020).

Bahrain Economic Agenda

Smart cities in Bahrain are part of a global urban agenda aimed at enhancing the quality of life for citizens and promoting economic development through the integration of technology and sustainable practices (Agbali et al., 2019). The smart city concept is increasingly relevant for policymakers as a strategy to improve the quality of life and achieve sustainability goals. By leveraging technology and digitalization within the city's infrastructure, smart cities can create economic and social opportunities while enhancing overall development (Kumar, 2019). The economic agenda of smart cities in Bahrain involves utilizing human and technological advancements to increase economic opportunities and social well-being. These cities promise to generate economic value by seamlessly connecting urban services and infrastructure through digital technologies (Evans et al., 2019). Additionally, smart city projects are designed to support evidence-based policy decision-making, which can lead to economic and environmental sustainability and an improved quality of life (Mills et al., 2021). Furthermore, the smart city agenda in Bahrain is embedded within a broader political regime, emphasizing the need to consider the political context in which smart city initiatives are developed (Akbari, 2022). The development of smart cities in Bahrain aligns with the global urban agenda, focusing on managing resources efficiently and providing better living conditions for all citizens (Hoogen et al., 2020). In conclusion, the economic agenda of smart cities in Bahrain is intertwined with social well-being, sustainability goals, and technological advancements, all aimed at enhancing the overall quality of life for residents and promoting economic development within the city.

On the economic front, Bahrain's smart city initiatives are geared toward driving growth and innovation. The kingdom is investing in digital infrastructure, such as high-speed internet and data centers, to attract businesses and entrepreneurs (Al-Ansari et al., 2022). By providing a robust digital infrastructure, Bahrain aims to create a conducive environment for businesses to thrive, thereby boosting economic growth. Furthermore, Bahrain is leveraging smart technologies to enhance productivity and efficiency in key industries such as finance, tourism, and logistics. For example, in the finance sector, Bahrain is implementing fintech solutions to streamline processes and improve services. Similarly, in the tourism sector, smart technologies are being used to enhance the visitor experience and attract more tourists to the kingdom (Nebrida & Gomba, 2023). Moreover, Bahrain is focusing on developing a skilled workforce in emerging technologies to ensure its competitiveness in the global economy. The kingdom is investing in education and training programs to equip its citizens with the necessary skills to succeed in the digital age. By developing a skilled workforce, Bahrain aims to attract more businesses and investors, further driving economic growth and innovation (Bunagan & Sison, 2019). Bahrain's smart city initiatives are comprehensive and forward-thinking, aiming to enhance the well-being and quality of life of its citizens while driving economic growth and innovation. By leveraging technology and digital infrastructure, Bahrain is positioning itself as a smart city of the future, poised for sustainable development and prosperity.

Sustainability Agenda of Smart Cities

The sustainability agenda of smart cities is a crucial aspect of urban development, focusing on achieving environmental, social, and economic sustainability through technological innovation and strategic planning. Smart cities aim to integrate sustainability principles into their design and operations to ensure long-term viability and resilience (Gao et al., 2023). The sustainability component of smart cities becomes increasingly prominent at the city level, emphasizing the importance of incorporating sustainable practices into urban planning and management (Treude, 2021). In the context of Finnish smart city initiatives, there is a clear emphasis on economic sustainability, although environmental sustainability is an area that requires further attention and research (Vainio & Sankala, 2022). While smart cities have the potential to contribute to environmental sustainability, there are challenges related to policy coherence and reducing CO_2 emissions that need to be addressed to fully realize this potential (Bibri, 2019). Smart city projects often focus on environmental sustainability aspects such as emergencies, air quality, climate change, and waste management, highlighting the importance of addressing environmental concerns in urban development (Agbali et al., 2019). By aligning smart and sustainable city initiatives with the United Nations SDGs, cities can work toward achieving a more sustainable and inclusive urban environment (Kutty et al., 2020). Education plays a crucial role in enhancing the sustainability of smart cities, with a focus on developing relevant curriculums and programs to support the smart city approach (Issa et al., 2019). The adoption of smart city agendas at various levels, including the UN and EU, underscores the global trend toward sustainable urban development (Akbari, 2022). Smart city initiatives leverage technology and data to enhance sustainability, efficiency, and living standards in urban areas, emphasizing the importance of integrating sustainability principles into urban planning and development (Bakhtiar, 2023).

By promoting data-driven decision-making, smart cities can reduce operational costs, identify risks, and enhance overall sustainability (Lamola, 2023).

Bahrain Sustainability Agenda

The sustainability agenda of smart cities in Bahrain is a multifaceted concept that encompasses environmental, economic, and social dimensions. Smart sustainable cities in Bahrain aim to contribute to sustainability by integrating smart technologies and practices that address environmental challenges, promote economic growth, and enhance social well-being (Janik et al., 2019). These cities prioritize sustainable communities, neighborhoods, and cities by considering criteria that support sustainable development (Umdu & Alakavuk, 2022). Stakeholders in smart city initiatives focus on environmental sustainability aspects such as emergencies, air quality, climate change, and waste management to meet citizens' needs and ensure sustainable outcomes. The development of smart sustainable cities involves the instrumentation, datafication, and computerization of urban systems to drive sustainability and innovation (Bibri et al., 2023).

Sustainability is a critical pillar of Bahrain's smart city agenda, reflecting the kingdom's commitment to environmental stewardship and long-term resource management. Bahrain recognizes the importance of sustainable practices in ensuring the well-being of its citizens and the health of its environment. As such, the kingdom is investing in various initiatives aimed at promoting sustainability across different sectors, including energy, urban planning, and transportation (Nebrida & Gomba, 2023). One of the key areas of focus for Bahrain's sustainability agenda is the transition to renewable energy sources. The kingdom is investing in solar and wind power projects to reduce its reliance on fossil fuels and decrease its carbon footprint. These renewable energy sources not only help reduce greenhouse gas emissions but also contribute to energy security and independence (Al-Alawi et al., 2023). Smart grid technologies are being deployed to improve energy efficiency and optimize energy distribution. These technologies enable better management of energy resources, reduce energy wastage, and lower overall energy costs (Al-Ansari et al., 2022). By adopting smart grid solutions, Bahrain aims to create a more sustainable and resilient energy infrastructure for the future. In addition to renewable energy, Bahrain is also focusing on sustainable urban planning and development. The kingdom is implementing smart city solutions to create more efficient and livable urban spaces. This includes the use of green building practices, such as energy-efficient designs and materials, to reduce energy consumption and environmental impact (Abbas et al., 2020). Furthermore, Bahrain is investing in sustainable transportation solutions to reduce congestion and pollution in its cities. The kingdom is promoting the use of public transportation, cycling, and walking as viable alternatives to private car usage. Smart transportation systems, such as intelligent traffic management systems and electric vehicle charging infrastructure, are being deployed to enhance mobility and reduce the environmental impact of transportation. Bahrain's sustainability efforts are not only focused on environmental conservation but also on promoting social and economic sustainability (Dhaen, 2019).

The kingdom recognizes that sustainable development requires a holistic approach that considers the needs of all stakeholders, including citizens, businesses, and the environment. Bahrain's sustainability agenda is a key component of its smart city initiatives, reflecting the kingdom's commitment to creating a more sustainable and resilient future. By investing in renewable energy, smart grid technologies, sustainable urban planning, and transportation solutions, Bahrain is laying the foundation for a more sustainable and prosperous future for its citizens.

Conclusion

In Bahrain, smart city initiatives are not just about technological advancement but also about enhancing the quality of life, promoting economic growth, and ensuring environmental sustainability. These initiatives are part of a broader vision to transform Bahrain into a hub of innovation, creativity, and prosperity. The kingdom's commitment to smart city development is evident in its strategic investments and policies aimed at leveraging technology to address urban challenges and improve the well-being of its citizens.

One key aspect of Bahrain's smart city agenda is the focus on social well-being and inclusivity. The kingdom recognizes the importance of ensuring that all segments of society, including vulnerable populations like older adults, are included and benefit from smart city initiatives. Initiatives such as smart healthcare and age-friendly smart cities are designed to cater to the specific needs of these populations, enhancing their quality of life and promoting social inclusion. Smart healthcare, for example, enables remote consultations and monitoring, reducing the need for in-person visits and improving access to healthcare services, especially for those in remote areas. Wearable health devices further empower individuals to take control of their health, providing valuable data to healthcare providers for more personalized care. These initiatives not only improve healthcare delivery but also promote a healthier lifestyle among citizens. Furthermore, Bahrain's smart city initiatives extend to promoting social cohesion and community engagement. Smart parks, community centers, and cultural events serve as hubs for social activities, bringing people together and fostering a sense of belonging. Digital platforms and mobile applications facilitate communication and collaboration among community members, strengthening social bonds and creating a more cohesive society.

On the economic front, Bahrain's smart city initiatives are geared toward driving growth and fostering innovation. The kingdom's investments in digital infrastructure, such as high-speed internet and data centers, aim to attract businesses and entrepreneurs, creating a conducive environment for economic development. Smart technologies are being deployed across various sectors, including finance, tourism, and logistics, to drive efficiency and productivity, further stimulating economic growth. Additionally, Bahrain is focusing on developing a skilled workforce in emerging technologies to ensure its competitiveness in the global economy. The kingdom's investments in education and training programs aim to equip its citizens with the necessary skills to succeed in the digital age, attracting more businesses and investors and driving economic growth and innovation.

Sustainability lies at the core of Bahrain's smart city agenda, with the kingdom actively pursuing environmentally friendly practices. Bahrain's investments in renewable energy sources like solar and wind power aim to reduce its reliance on fossil fuels and decrease its carbon footprint. Smart grid technologies are being deployed to enhance energy efficiency and optimize energy distribution, further contributing to environmental sustainability. Bahrain's smart city initiatives are comprehensive and forward-thinking, aiming to enhance the well-being and quality of life of its citizens while driving economic growth and innovation. By leveraging technology and innovation, Bahrain is positioning itself as a smart city of the future, poised for sustainable development and prosperity. Through strategic investments and policies, Bahrain is laying the foundation for a more inclusive, innovative, and sustainable society for generations to come.

References

Abbas, S. Y., Kirwan, K., & Lu, D. (2020). Exploring enablers and barriers to Municipal Solid Waste (MSW) management technologies adoption in the Kingdom of Bahrain. *Journal of Environmental Protection*. https://doi.org/10.4236/jep.2020.115022

Agbali, M., Trillo, C., Ibrahim, I. A., Arayici, Y., & Fernando, T. (2019). Are smart innovation ecosystems really seeking to meet citizens' needs? Insights from the stakeholders' vision on smart city strategy implementation. *Smart Cities*. https://doi.org/10.3390/smartcities2020019

Akbari, A. (2022). Authoritarian smart city: A research agenda. *Surveillance & Society*. https://doi.org/10.24908/ss.v20i4.15964

Al-Alawi, A. I., Messaadia, M., Mehrotra, A. A., Sanosi, S. K., Elias, H., & Althawadi, A. H. (2023). Digital transformation adoption in human resources management during COVID-19. *Arab Gulf Journal of Scientific Research*. https://doi.org/10.1108/agjsr-05-2022-0069

Al-Ansari, A., Mohsen, H. A., & Coutts, C. E. (2022). Critical factors in promoting sustainable innovation in Bahrain. *Journal of Sustainable Business and Economics*. https://doi.org/10.30564/jsbe.v5i2.6

Alnaser, N. W., & Alnaser, W. E. (2019). Analyzing the impact of Bapco 5 MW Solar PV grid-connected project on Bahrain's outlook for energy-mix production. *International Journal of Modern Nonlinear Theory and Application*. https://doi.org/10.4236/ijmnta.2019.83006

Alqahtani, H., & Alareeni, B. (2020). Evaluation of sustainable buildings construction in the Kingdom of Bahrain. *Journal of Sustainable Construction Materials and Technologies*. https://doi.org/10.29187/jscmt.2020.49

Alsabbagh, M., & Alnaser, W. E. (2023). Assessment of climate change mitigation readiness in the Kingdom of Bahrain. *International Journal of Climate Change Strategies and Management*. https://doi.org/10.1108/ijccsm-08-2021-0096

AlZayani, F., Mohammed, A., & Shoaib, H. M. (2023). The impact of smart technologies on SMEs' sustainability: The mediation effect of sustainability strategy. *Competitiveness Review*. https://doi.org/10.1108/cr-09-2022-0136

Bakhtiar, I. S. (2023). Comparative review of smart city from an urban planning perspective in Johor Bahru City and Petaling Jaya City. *IOP Conference Series Earth and Environmental Science*. https://doi.org/10.1088/1755-1315/1274/1/012016

Bibri, S. E. (2019). The anatomy of the data-driven smart sustainable city: Instrumentation, datafication, computerization and related applications. *Journal of Big Data*. https://doi.org/10.1186/s40537-019-0221-4

Bibri, S. E., Alahi, A., Sharifi, A., & Krogstie, J. (2023). Environmentally sustainable smart cities and their converging AI, IoT, and Big Data technologies and solutions: An integrated approach to an extensive literature review. *Energy Informatics*. https://doi.org/10.1186/s42162-023-00259-2

Bunagan, V., & Sison, S. M. (2019). Entrepreneurial culture and orientations of SME owners/managers in the Kingdom of Bahrain. *Journal of Small Business and Entrepreneurship Development*. https://doi.org/10.15640/jsbed.v7n1a5

Chen, T., Gil-Garcia, J. R., & Gasco-Hernandez, M. (2022). Understanding social sustainability for smart cities: The importance of inclusion, equity, and citizen participation as both inputs and long-term outcomes. *Journal of Smart Cities and Society*. https://doi.org/10.3233/scs-210123

Dhaen, F. S. A. (2019). The advantages of using 5G in live radio and TV Broadcasting at the Ministry of Information Affairs in Bahrain. *EPH – International Journal of Science and Engineering*. https://doi.org/10.53555/eijse.v5i2.91

Elgabiri, M., Palmer, D., Buflasa, H. A., & Thomson, M. (2020). Offshore wind energy potential for Bahrain via multi-criteria evaluation. *Wind Engineering*. https://doi.org/10.1177/0309524x20925399

Evans, J., Karvonen, A., Luque-Ayala, A., Martin, C., McCormick, K., Raven, R., & Palgan, Y. V. (2019). Smart and sustainable cities? Pipedreams, practicalities and possibilities. *Local Environment*. https://doi.org/10.1080/13549839.2019.1624701

Fernández, C. G., & Peek, D. (2020). Smart and sustainable? Positioning adaptation to climate change in the European Smart City. *Smart Cities*. https://doi.org/10.3390/smartcities3020027

Freije, A., Naser, H. A., & Abdulla, K. H. (2019). Attitudes and opinions towards public littering in the Kingdom of Bahrain. *Arab Journal of Basic and Applied Sciences*. https://doi.org/10.1080/25765299.2019.1628688

Gagliardi, A. R., & Albergo, F. (2023). The rise of smart healthcare in smart cities: A Bibliometric literature review and avenue for a research agenda. *ITM Web of Conferences*. https://doi.org/10.1051/itmconf/20235103002

Gao, C., Wang, F., Hu, X., & Martinez, J. (2023). Research on sustainable design of smart cities based on the Internet of Things and ecosystems. *Sustainability*. https://doi.org/10.3390/su15086546

Hoogen, A. v. d., Scholtz, B., & Calitz, A. P. (2020). Using theories to design a value alignment model for smart city initiatives. https://doi.org/10.1007/978-3-030-44999-5_5

Huang, K., Luo, W., Zhang, W., & Li, J. (2021). Characteristics and problems of smart city development in China. *Smart Cities*. https://doi.org/10.3390/smartcities4040074

Ismail, N., Kalach, M., & Kleib, Y. (2022). Towards a citizen centered smart city: Integrating lean thinking and social wellbeing. https://doi.org/10.24928/2022/0149

Issa, H.-E. A., Shukaili, A. A., & Fattah, F. A. M. A. (2019). Enhancing education for smart cities: Evidence from Omani higher education institutions. *Journal of Engineering and Applied Sciences*. https://doi.org/10.36478/jeasci.2020.851.854

Janik, A., Ryszko, A., & Szafraniec, M. (2019). Smart and sustainable cities: In search of comprehensive theoretical framework. *Scientific Papers of Silesian University of Technology Organization and Management Series*. https://doi.org/10.29119/1641-3466.2019.140.10

Katbi, A., AlAmmari, J., & Al-Soufi, A. (2020). The demand side of open government data: A case study of Kingdom of Bahrain. *International Journal of Managing Information Technology*. https://doi.org/10.5121/ijmit.2020.12203

Kumar, B. (2019). The (In) Security O smart cities: Vulnerabilities, risks, mitigation and prevention. *International Journal of Engineering and Advanced Technology*. https://doi.org/10.35940/ijeat.e1097.0785s319

Kutty, A. A., Abdella, G. M., Küçükvar, M., Onat, N. C., & Bulu, M. (2020). A system thinking approach for harmonizing smart and sustainable city initiatives with United Nations sustainable development goals. *Sustainable Development*. https://doi.org/10.1002/sd.2088

Lamola, K. (2023). Building sustainable smart cities and the green building agenda. *Resourceedings*. https://doi.org/10.21625/resourceedings.v3i1.952

Li, M., & Woolrych, R. (2021). Experiences of older people and social inclusion in relation to smart "age-friendly" cities: A case study of Chongqing, China. *Frontiers in Public Health*. https://doi.org/10.3389/fpubh.2021.779913

Malek, J. A., Lim, S. B., & Yiğitcanlar, T. (2021). Social inclusion indicators for building citizen-centric smart cities: A systematic literature review. *Sustainability*. https://doi.org/10.3390/su13010376

Meero, A., Rahiman, H. U., & Rahman, A. A. A. (2020). The Prospects of Bahrain's entrepreneurial ecosystem: An exploratory approach. *Problems and Perspectives in Management*. https://doi.org/10.21511/ppm.18(4).2020.32

Mills, D. E., Pudney, S., Pevcin, P., & Dvorak, J. (2021). Evidence-based public policy decision-making in smart cities: Does extant theory support achievement of city sustainability objectives? *Sustainability*. https://doi.org/10.3390/su14010003

Morozova, I. A., & Yatsechko, S. S. (2022). The risks of smart cities and the perspectives of their management based on corporate social responsibility in the interests of sustainable development. *Risks*. https://doi.org/10.3390/risks10020034

Nebrida, J. A., & Gomba, F. E. (2023). Sustainable construction strategies for building construction projects in the Kingdom of Bahrain: A model. *Sustainable Engineering and Innovation*. ISSN 2712-0562. https://doi.org/10.37868/sei.v5i1.id193

Pansera, M., Marsh, A., Owen, R., López, J. A. F., & De Alba Ulloa, J. L. (2022). Exploring citizen participation in smart city development in Mexico City: An institutional logics approach. *Organization Studies*. https://doi.org/10.1177/01708406221094194

Park, S.-C. (2022). Global megatrend of the fourth industrial revolution in the digital economy: How to realize it with smart cities as a practical measure? *International Organisations Research Journal*. https://doi.org/10.17323/1996-7845-2022-02-06

Praharaj, S. (2021). Area-based urban renewal approach for smart cities development in India: Challenges of inclusion and sustainability. *Urban Planning*. https://doi.org/10.17645/up.v6i4.4484

Romanelli, M. (2021). Driving smart cities through smart projects. https://doi.org/10.4108/eai.4-12-2020.2303264

Sengupta, U., & Sengupta, U. (2022). SDG-11 and smart cities: Contradictions and Overlaps Between Social and Environmental Justice Research Agendas. *Frontiers in Sociology*. https://doi.org/10.3389/fsoc.2022.995603

Shee, H., Miah, S. J., & Vass, T. d. (2021). Impact of Smart Logistics on smart city sustainable performance: An empirical investigation. *The International Journal of Logistics Management*. https://doi.org/10.1108/ijlm-07-2020-0282

Suvarna, M., Büth, L., Hejny, J., Mennenga, M., Li, J., Ng, Y. T., Herrmann, C., & Wang, X. (2020). Smart manufacturing for smart cities – Overview, insights, and future directions. *Advanced Intelligent Systems*. https://doi.org/10.1002/aisy.202000043

Szarek-Iwaniuk, P., & Senetra, A. (2020). Access to ICT in Poland and the co-creation of urban space in the process of modern social participation in a smart city – A case study. *Sustainability*. https://doi.org/10.3390/su12052136

Treude, M. (2021). Sustainable smart city – Opening a black box. *Sustainability*. https://doi.org/10.3390/su13020769

Toli, A. M., & Murtagh, N. (2020). The concept of sustainability in smart city definitions. *Frontiers in Built Environment*. https://doi.org/10.3389/fbuil.2020.00077

Umar, T. (2021). Challenges of BIM implementation in GCC construction industry. *Engineering Construction & Architectural Management*. https://doi.org/10.1108/ecam-11-2019-0608

Umdu, D. C., & Alakavuk, E. (2022). Sustainable communities, neighborhoods, cities and their criteria. *The European Journal of Research and Development*. https://doi.org/10.56038/ejrnd.v2i2.66

Vainio, T., & Sankala, I. (2022). Exploring the balance between smartness and sustainability in Finnish smart city initiatives during the 2010s. *Current Urban Studies*. https://doi.org/10.4236/cus.2022.103024

Wahyudi, R., & Elanda, Y. (2023). Challenges and strategies for gender mainstreaming policy in smart city development in Indonesia. *Gender Equality International Journal of Child and Gender Studies*. https://doi.org/10.22373/equality.v9i1.17192

Wirayuda, A. A. B., Al-Mahrezi, A., & Chan, M. F. (2022). Factors impacting life expectancy in Bahrain: Evidence from 1971 to 2020 data. *International Journal of Social Determinants of Health and Health Services*. https://doi.org/10.1177/00207314221129052

Yuan, L. (2022). Research on construction of the smart city. https://doi.org/10.2991/assehr.k.220105.228

Chapter 21

Transforming Ecological Footprint Through Circular Economy Innovation Toward Sustainable Well-Being in the Development of (Islamic) Smart Cities in GCC Countries

Khairunnisa Musari

Kiai Haji Achmad Siddiq State Islamic University, Indonesia

Abstract

The concept of smart city has been adopted by various cities across the world. As urbanization is enlarging at a flying pace, the number of cities that engage the smart city concept is to rise, including in Gulf Cooperation Council (GCC) countries. This chapter discusses issues related to the development of (Islamic) smart cities in the GCC countries to create sustainable well-being through ecological and environmental resilience. This issue starts from the conflict between economic and human development on the one hand and ecological and environmental resilience on the other hand as one of the fundamental challenges of sustainable development. The higher the Human Development Index (HDI),, the higher the ecological footprint. Likewise, in the GCC region, all member countries are classified as Very High Human Development; however, all of these countries have a high ecological footprint. Therefore, through a literature study, this chapter focuses on investigating the development of (Islamic) smart cities in the GCC countries in reducing ecological footprints through a circular economy innovation with the support of technology and a digitally-enabled community in creating ecological and environmental resilience. This chapter provides valuable insight into the implementation of the (Islamic) smart cities in the GCC region in taking a global strategic role in green

The Emerald Handbook of Smart Cities in the Gulf Region:
Innovation, Development, Transformation, and Prosperity for Vision 2040, 379–403

doi:10.1108/978-1-83608-292-720241026

transition through the circular economy to transform their region toward sustainable well-being and become a symbol of sustainable smart global Islamic civilization.

Keywords: Circular economy; ecological footprint; smart city; climate change; digital transformation; transformation; development; green transition; sustainable; well-being

Introduction

The urbanization process is estimated to keep up for decades. More people will be living in urban areas. In 1950, according to the United Nations (UN) – Department of Economic and Social Affairs, Population Division (2019), about 29.53% of the population in the world was urban. In 2018, over half of the world's population, about 55.3%, had lived in urban areas. The proportion is set to reach 60.47% in 2030 and 68.37% in 2050. A prime consequence of urbanization is a leap up in the size of land area and population of urban areas. Asia has become the region with the highest urban population in the world.

As urbanization is enlarging at a flying pace globally, the number of cities that engage in the concept of smart city is to rise. Asia is also recognized as the global leader in designing smart cities. Smart cities in this region are at the forefront of deploying technology and digital transformation (Musari & Naaz, 2023; Panda Media Center, 2017; Power, 2017). The leaders of the cities in Asia are among the world's most forward-thinking regarding smart cities (Prakash, 2021). Apart from being a dynamic economic and technological growth region, most Asian cities have utilized digital information to improve operational efficiency (Damian & Phan, 2022; Doagoo, 2022; Joo & Tan, 2020).

Several countries in the Asian region are the countries of the Gulf Region. UN – Department of Economic and Social Affairs, Population Division (2019) mapped the Gulf Cooperation Council (GCC) countries, namely Bahrain, Kuwait, Oman, Qatar, Saudi Arabia, and the United Arab Emirates (UAE), as part of Western Asia. Since the first 2000s (Petrucci & AlSheikh, 2022), the issue of smart cities in GCC countries have received attention given the high urbanization and economic growth in this region (Asmyatullin et al., 2020; Saxena & Al-Tamimi, 2018). Table 1 shows the annual percentage of population at mid-year residing in urban areas during 1950–2050.

Related to the existence of (Islamic) smart cities in the GCC region, this chapter discusses the development of these cities in green transition by creating ecological and environmental resilience. This issue arises from anomalies between economic and human development that are not in harmony with ecological and environmental resilience as one of the fundamental challenges of sustainable development (Musari & Khan, 2023). The conflict between HDI and the ecological footprint is reflected in the higher the HDI which is followed by the higher the ecological footprint (Khan, 2022, 2023). The higher

Table 1. The Annual Percentage of Population at Mid-Year Residing in Urban Areas (1950–2050).

Subregion, Country	1950	1960	1970	1980	1990	2000	2010	2020	2030	2040	2050
WORLD	29.6	33.8	36.6	39.3	43.0	46.7	51.7	56.2	60.4	64.5	68.4
Western Asia	28.9	36.4	44.8	52.1	61.1	63.8	68.2	72.3	75.4	78.5	81.4
Bahrain	64.4	82.3	83.8	86.1	88.1	88.4	88.6	89.5	90.8	92.1	93.2
Kuwait	61.5	74.9	85.7	94.8	98.0	99.0	100.0	100.0	100.0	100.0	100.0
Oman	8.6	16.4	29.7	47.6	66.1	71.6	75.2	86.3	91.7	93.9	94.9
Qatar	80.5	85.3	88.4	89.4	92.8	96.3	98.5	99.2	99.5	99.6	99.7
Saudi Arabia	21.3	31.3	48.7	65.9	76.6	79.8	82.1	84.3	86.5	88.5	90.4
UAE	54.5	73.5	79.8	80.7	79.1	80.2	84.1	87.0	89.2	90.9	92.4

Source: UN – Department of Economic and Social Affairs, Population Division (2019), modified.

the income, the higher the footprint, and the higher the relative share of energy consumption (or carbon footprint) in that footprint (Wackernagel & Beyers, 2019). No doubt, most of the pollution is coming from the developed countries, which are most seriously implementing circular economy (Sancar, 2023). Historically, the United States (US) and European Union have contributed the most to emissions (Tekdoğan, 2023).

Likewise, in the GCC region, all GCC countries have been ranked as having very high human development countries, but also have a high ecological footprint. By considering the empirical study that urbanization increases the ecological footprint, while human capital reduces it (Ahmed, Asghar, et al., 2020; Ahmed, Zafar, et al., 2020), therefore, by taking the (Islamic) smart cities in the GCC region that are included in the top 100 smart cities in 2023, this chapter through literature study will focus on investigating the development of (Islamic) smart cities in the GCC countries in reducing ecological footprints through a circular economy innovation with the support of technology and a digitally-enabled community to transform into sustainable well-being. Data were gathered from different sources including the publications and official websites of related state ministries, the United Nations (UN), smart city specialized platforms, and several international reports.

(Islamic) Smart Cities in the GCC Countries

In Asia, a study by Musari and Naaz (2023) found the existence of an Islamic smart city. The study concluded that Islamic smart cities also pay attention to fostering inclusive growth opportunities and combating climate change through programs such as the smart economy, smart business, and smart environment initiatives. In Muslim-majority countries, particularly countries whose government has indeed declared the country as an Islamic state, the concept of Smart City and Islamic Smart City will merge in the future. Even if the labeling "Islamic" is not used, the "Smart City" concept that will be developed will carry Islamic values and cultures.

All GCC member countries have smart cities. However, not all of them are in the top 100 smart cities in the world. Referring to the Institute for Management Development (IMD) Smart City Index Report 2023, the smart cities in the GCC region that are included in the top 100 in 2023 are: Abu Dhabi and Dubai in the UAE, then Riyadh, Mecca, Jeddah, and Madinah in Saudi Arabia, as well as Doha in Qatar and Muscat in Oman. Table 2 shows the smart cities in the GCC region that are ranked in the top 100 smart cities in 2023. All smart cities have City HDI and State HDI of more than 0.8, which means they have a very high level of human development cities and countries.

The GCC countries are a symbol of Islamic civilization. GCC Islamic civilizational heritage underpins behavioral norms such as *khalifa* (trusteeship), *meezan* (natural laws and cosmic balance), *la-tabzeer* (zero-extravagances), *la-israf* (zero-waste), *wasatiyya* (moderation), and embedded framework like *maqasid al Shari'ah* (objectives of Islamic law) (Musari & Khan, 2023). Therefore, even though smart cities in this region are not proclaimed as "Islamic Smart City,"

Table 2. Smart Cities of the GCC Countries on the Top 100 Smart Cities in 2023.

Country	Country HDI 2021	City	City HDI 2021	Rank 2019	Rank 2020	Rank 2021	Rank 2023	Rating 2023
UAE	0.911	Abu Dhabi	0.911	16	14	12	13	BB
UAE	0.911	Dubai	0.911	13	19	14	17	BB
Saudi Arabia	0.875	Riyadh	0.9	55	44	39	30	BB
Saudi Arabia	0.875	Mecca	0.871	–	–	–	52	B
Saudi Arabia	0.875	Jeddah	0.871	–	–	–	56	B
Qatar	0.855	Doha	0.855	–	–	–	59	BB
Saudi Arabia	0.875	Madinah	0.871	–	–	79	85	CCC
Oman	0.816	Muscat	0.816	–	–	–	96	B

Source: IMD Smart City Index Report 2023, modified.

these cities actually are "(Islamic) Smart City," namely smart cities located in Islamic countries where Islamic values and cultures are coloring these smart cities. The religious nuances of this faith become thick and embed the smart city while still providing space and comfort for adherents of other religions (Musari & Naaz, 2023).

Clearly, the majority of the population in the GCC countries is Muslim, therefore it can be understood that Islamic values and culture become the basis of their societies' responses to many things, including facing the existence of smart city. Then, considering the study by Mutambik et al. (2023) which concluded that the eventual accomplishment of smart city developments in GCC countries does not only need transparency and wide participation. There is a general recognition that the achievement of the society potential is not just a matter of technologies and infrastructures but is basically relying on large-scale social change and development, also requires political change.

Ecological Footprints in GCC Countries

All GCC countries have integrated environmental concerns and economic diversification into their respective national visions. However, there is a conflict between economic and human development on the one side as well as ecological and environmental resilience on the other side (Musari & Khan, 2023). Indeed, the economic development and environmental sustainability nexus has long been a fiercely debated issue (ElMassah & Hassanein, 2023). One manifestation of this issue is the higher HDI is apparently followed by an increasingly higher ecological footprint (Khan, 2022, 2023) and GCC countries have been facing a high ecological footprint per capita (Ansari et al., 2020).

Furthermore, an empirical study indicated that human capital in the GCC countries does not help in preventing the degradation of the environment in this region (Amer et al., 2022). It could be because of very low general awareness of environmental issues and low civil-society capacity (Hertog & Luciani, 2009). Referring to the 2012 Annual Report of the Arab Forum for Environment and Development (AFED), the Arab region engaged in a steady ecosystem deficit since 1979 so the level of consumption for life-supporting goods and services, in 2012, was more than twice what local ecosystems can provide. This has been embraced by a doubling in the regional ecological footprint and a decrease of nearly four times in freshwater availability. (AFED, 2012)

Table 3 shows the HDI, national footprint, biocapacity, and ecological footprint in the GCC Region. The data show that the ecological for the GCC countries is a deficit with an increasing trend between 2019 and 2022. Likewise in ecological footprint number of earths in the GCC countries has increased trend. Referring to Musari and Khan (2023), the ecological footprint underscores the urgency for the GCC countries to address the ecological and natural capital crisis and the transitioning of their economies to more restorative, sustainable, regenerative, and intergenerationally equitable economic models.

Ecological footprint can be defined as a comprehensive measure for assessing environmental sustainability and human activity effect (Rees & Wackernagel, 1996); an accounting system that documents the ecological performance; a highly aggregated measure of people's use of nature (Wackernagel & Beyers, 2019); a measure of how much of the Earth's biologically productive land and water area is needed to support a population's consumption and waste production (Musari & Khan, 2023). Previously, the terms used "regional capsule" (Rees, 1986) and "suitable carrying capacity." The term "ecological footprint" began to emerge when Rees criticized economic models of resource use by arguing that more attention must be paid to the amount of land needed by urban areas to survive (Collins & Flynn, 2015).

Referring to the Global Footprint Network, ecological footprint is a resource counting-up instrument that measures the amount of the regenerative capacity of the earth needed by a population, individual, or activity. It is a measure of the territory of productive water and land biologically required to generate all the consumed resources, facilitate the infrastructure for urban areas, and sponge up the produced waste. Without further specification, ecological footprint generally refers to the ecological footprint of consumption. On the supply side, biocapacity is the capacity of ecosystems to generate biological materials for people and to sponge up the waste material produced by people, under the scheme of management and extraction technologies. From year to year, biocapacity may change to suit the management, climate, and quantity reflected on the inputs to the economic activities of people.

Furthermore, the difference between the ecological footprint and biocapacity of a country or region can be a deficit or reserve. Ecological reserve occurs when biocapacity exceeds the footprint of a population in an area. In contrast, an ecological deficit occurs when the footprint of a population exceeds the biocapacity of the area available to that population. If a national or regional ecological

Table 3. HDI, Ecological Footprint, Biocapacity, and Ecological Deficit/Reserve in GCC Countries.

Country	HDI 2021	Total Ecological Footprint (Global Hectares per Person)				Total Biocapacity (Global Hectares per Person)		Ecological (Deficit) or Reserve		Ecological Footprint Number of Earths Required (2022)	
		Production		Consumption							
		2019	2022	2019	2022	2019	2022	2019	2022	2019	2022
Bahrain	0.88	8.0	8.2	8.1	8.2	0.5	0.6	(7.5)	(7.6)	5.2	5.4
Kuwait	0.83	7.6	8.7	7.7	8.6	0.8	0.8	(6.9)	(7.7)	4.9	5.7
Oman	0.82	6.6	7.3	6.4	7.3	1.6	1.8	(4.8)	(5.5)	4.1	4.8
Qatar	0.86	11.8	13.4	12.0	13.1	1.0	1.0	(11.0)	(12.1)	7.7	8.7
Saudi Arabia	0.88	5.6	5.5	5.8	5.7	0.7	0.7	(5.0)	(5.0)	3.7	3.8
UAE	0.91	7.0	7.1	8.9	8.7	0.6	0.5	(8.3)	(8.2)	5.7	5.8
Average	0.86	7.77	8.37	8.15	8.60	0.87	0.90	(7.25)	(7.68)	5.22	5.70

Source: Human Development Report 2021/2022; National Footprint and Biocapacity Accounts, 2023 Edition; modified.

deficit occurs, this means that the region is dumping waste into global commons or liquidating regional ecological assets, or importing biocapacity through trade.

GCC countries contribute to 57% of all available biocapacity of marine fisheries and climate change. Pollution, habitat destruction, and overfishing are menacing the long-term productivity of marine resources. This makes the biocapacity deficit in this region which implies that the ecological footprint is much greater than its biocapacity. Thus, the ecological resources to support these activities exceeded the capacity to supply these resources (Ansari et al., 2020).

Discussion

Circular Economy

AFED (2012) reported that Arab countries' demand for natural products and services was more than double what their ecosystems could supply. Imbalances between domestic demand and supply of ecological services restrict future growth and well-being. The report stated that most Arab countries currently have an ecological debt, therefore Arab countries must take action collectively to move forward with a new vision for sustainable economics and ecology. To survive, the ways that can be reached among others are regional cooperation, resource efficiency, and balanced consumption.

To ensure the long-term well-being of the people and the planet, the European Environmental Bureau (EEB, 2020) campaigns for the need to transition to a new economy and modes of producing and consuming that respect planetary boundaries and global justice. One way is through a circular economy. The circular economy as a concept is meant to help achieving the goal of reducing resource consumption. Especially when broadening its definition, from pure recycling to smarter and more circular product design and alternative business models, the circular economy can indeed be a powerful tool in this regard. The objective of a true circular economy means reducing the absolute quantity of natural resources that enter the economy, and reducing the quantity of waste coming out.

Hence, realizing the circular economy in smart city toward circular smart city can lead to enormous environmental advantages, as well as economic and social benefits. Ellen MacArthur Foundation (2017, 2019) mentioned a city that embeds circular economy principles throughout its territory and builds a regenerative urban system as a Circular City. Cities have a high concentration of linear economic activity, take-make-waste, at once also having a high concentration of resources, data, talent, and capital which spread over a relatively small geographical territory. Cities are also centers of change and innovation. This makes cities become uniquely as well as strategically positioned to steer development.

Hence, (Islamic) smart city in GCC countries and its government have a crucial part in reducing ecological footprints by implementing circular economy. The circular economy offers cities three powerful principles which, when adopted together and underpinned by design, can help tackle these remaining emissions (Henderson, 2023), namely: (1) eliminate waste and pollution which can reduce and avoid emissions across the value chain; (2) circulate products and material which can enable embodied emissions to be retained within the economy; (3) regenerate nature and natural systems which can improve carbon sequestration.

Table 4 shows the perceptions of residents on structures pillar related to the existing infrastructure of the cities for circular economy facility and green spaces in (Islamic) smart cities in GCC countries on the top 100 smart cities in 2023. Each (Islamic) smart city provides recycling services with a satisfaction level of residents more than 0.6, especially Abu Dhabi and Dubai with a satisfaction level of more than 0.8. Related to green spaces, the satisfaction level of residents in Abu Dhabi and Dubai also has the highest score.

In the context of GCC countries, this region has moved toward a path of a more sustainable future and has initiated a circular paradigm shift in waste generation to reduce cost and enlarge resource efficiency, as well as support diversifying the economy away from using up the planet's finite resources while eliminating the bump on the environment (Bejjani et al., 2019; Rubeis et al., 2022). There are continual works to ameliorate waste recovery, particularly recovery for waste-to-energy. Through the Mesaieed waste-to-energy plant in 2014, Qatar became the first country to implement. With a high scale capacity 2,300 to 2,500 tonnes a day, the plant can generate an energy capacity of 42 MW. In recent times, the UAE declared several plants of waste-to-energy in Sharjah, Umm Al Quwain, and Dubai. There is a plan also for Abu Dhabi to build some plants. Some countries which adopt the waste-to-energy concept, include Jordan and Saudi Arabia (El Mir et al., 2021).

By adopting a CE model, World Government Summit reported that GCC countries can save nearly US$138 billion by 2030 if they correspond to almost 1% of the region's cumulative GDP between 2020 and 2030 (Bejjani et al., 2019; Bridge, 2019; Mubasher, 2019; Zawya, 2019). However, to augment circularity in line with the three principles of the CE in pivotal areas such as energy, biomass, and water, GCC countries have deficiencies in a holistic approach. By embracing a CE, it becomes an opportunity for GCC countries to reach GDP growth by extracting value from waste and mitigating wasteful consumption. Wider social and environmental benefits will also be created. Hence, this region should prioritize occasions in the built environment that would raise gains of approximately US$23 billion between 2020 and 2030 (Bejjani et al., 2019).

Table 4. Structures Pillar for Circular Economy and Green Space in (Islamic) Smart Cities in GCC Countries.

STRUCTURES	Abu Dhabi	Dubai	Riyadh	Mecca	Jeddah	Doha	Madinah	Muscat
	Score							
Recycling services are satisfactory	86.4	80.9	72.2	66.8	70.6	76.0	63.8	68.1
Green spaces are satisfactory	82.6	80.6	69.2	68.0	66.1	75.4	56.6	72.9

Source: IMD Smart City Index Report 2023, modified.

Technology and a Digitally-Enabled Community

Human capital will have an impact on society's understanding of and willingness to employ environmentally friendly and energy-saving technology (Amer et al., 2022). Several studies in United States (US), India, China, and G7 Countries, have proved that human capital are helpful in mitigating environmental deterioration and decreasing ecological footprint (Ahmed, Asghar, et al., 2020; Ahmed, Zafar, et al., 2020; Ahmed & Wang, 2019; Zafar et al., 2019). For this reason, technology and digitally-enabled community are two inseparable things in developing (Islamic) smart city.

Table 5 shows the perceptions of residents on issues related to technology pillar in (Islamic) smart cities in GCC countries. Technology pillar illustrates the technological provisions and services available to the residents. Technology pillar is evaluated under five key areas: health and safety, mobility, activities, opportunities, and governance. Each indicator presents the score for the city.

Related to circular economy, digitalization is guiding new routes of generating value (Moigne, 2021; Musari, 2021), for both established and emerging businesses. New technologies carry immense opportunities for companies to unlock the value of a regenerative recycling economy. The advancement of technologies has accelerated the expansion, growth, and execution of circular business models, heading newborn works, fresh connections canals, as well as modern and efficient operations that all combination of them opens the gateway to disrupt the existing industry models today (Lacy, 2017). This is disruptive circular sciences and technologies (DCST) which conceptually refers to innovations and methodologies that aim to create a circular economy toward a regenerative and sustainable economic system. The goal of DCST is to develop technologies and solutions that enable the transformation of waste into valuable resources, products, and services.

Technology is central to enabling and driving value in the circular economy (Moigne, 2021; Musari, 2021). Nevertheless, technology is an enabler, not a panacea. This is in line with the results of the study Böhmecke-Schwafert et al. (2022) which found that the ground-level experts performing blockchain innovations for the circular economy across the US and Europe do not consider blockchain as a panacea, but merely an infrastructural resource. However, no doubt, blockchain can contribute to the regeneration of nature. It can serve circulation tasks, identifying as well as monitoring materials and components right through the supply chain so that they can be either reused, remanufactured, or (when that is no longer possible) recycled or composted (Murphy, 2022).

The Implementation of (Islamic) Smart Cities in the GCC Countries

Cities are a vital machine for economic growth. Global GDP currently is generated as about 85% by cities. This speedy growth sets an expansive force on urban sources, capacities, and life quality. All of these components array cities with a considerable catalytic competency to navigate the circular economy agenda ahead and make them among the biggest recipients of this green transition (Vyas, 2018).

Table 5. Technology Pillar in (Islamic) Smart Cities in GCC Countries.

TECHNOLOGIES	Abu Dhabi	Dubai	Riyadh	Mecca	Jeddah	Doha	Madinah	Muscat
	Score							
Health & Safety								
Online reporting of city maintenance problems provides a speedy solution	80.8	74.8	79.5	79.1	71.9	77.5	63.2	69.9
A website or App allows residents to easily give away unwanted items	76.8	75.8	71.3	68.6	61.3	72.7	58.7	61.5
Free public wifi has improved access to city services	72.8	78.3	64.8	62.2	60.0	70.1	46.8	57.4
CCTV cameras has made residents feel safer	87.2	86.7	81.1	78.5	75.2	80.5	62.5	73.1
A website or App allows residents to effectively monitor air pollution	60.4	56.2	62.7	61.4	59.9	54.8	44.8	43.2
Arranging medical appointments online has improved access	86.2	84.2	88.2	83.6	87.0	75.8	76.2	69.8
Mobility								
Car-sharing Apps have reduced congestion	65.1	62.9	68.1	70.8	64.6	72.0	54.0	53.8
Apps that direct you to an available parking space have reduced journey time	73.6	65.7	71.8	71.4	74.1	63.6	64.9	55.7
Bicycle hiring has reduced congestion	68.4	63.9	55.7	50.9	59.1	60.9	41.8	40.0
Online scheduling and ticket sales has made public transport easier to use	81.8	77.4	75.7	77.1	73.6	75.1	68.0	72.5
The city provides information on traffic congestion through mobile phone	80.5	78.4	74.6	70.8	74.5	70.5	56.3	60.8

(*Continued*)

Table 5. (*Continued*)

TECHNOLOGIES	**Abu Dhabi**	**Dubai**	**Riyadh**	**Mecca**	**Jeddah**	**Doha**	**Madinah**	**Muscat**
	Score							
Activities								
Online purchasing of tickets to shows and museums has made it easier to attend	88.7	88.4	84.6	79.6	85.4	88.3	70.9	71.8
Opportunities (Work & School)								
Online access to job listings has made it easier to find work	78.7	76.5	78.8	74.0	76.1	72.3	63.5	69.8
IT skills are taught well in schools	77.0	72.8	74.1	72.7	67.1	71.8	61.9	63.0
Online services provided by the city have made it easier to start a new business	77.0	80.1	76.8	74.2	74.9	66.8	64.9	64.0
The current internet speed and reliability meet connectivity needs	85.9	83.5	82.1	82.9	75.5	83.5	68.9	75.2
Governance								
Online public access to city finances has reduced corruption	70.1	66.5	74.1	73.6	72.6	64.1	56.4	57.7
Online voting has increased participation	67.3	58.3	73.3	69.9	71.6	59.6	61.5	59.7
An online platform where residents can propose ideas has improved city life	73.3	69.7	76.8	70.8	74.3	62.9	62.4	56.1
Processing Identification Documents online has reduced waiting times	83.1	82.6	86.9	86.1	88.0	80.8	77.4	77.2

Source: IMD Smart City Index Report 2023, modified.

Likewise in GCC countries, the finest route is to pivot to urban areas. The economic development has run in line with a quickly increasing population, mainly centralized in cities. GCC cities currently accommodate 85% of the population in this region which is predicted to rise 90% by 2050. Their economies still depend heavily on natural resources, which at the same time the supply of which is finite. Therefore, GCC cities need to shift fast to an alternative model that would lead to efficiency and sustainability in the use of resources. To establish a robust framework for sustainable growth, this region must adopt the circular economy concept (Strategy&, 2019).

Circular economy principles applied to the city could support urban policymakers in achieving their objectives in a number of different ways (Ellen MacArthur Foundation, 2017), namely: (1) relieve pressures on municipal services and budgets; (2) increase disposable income; (3) encourage an innovation-rich urban economy; (4) reduce carbon emissions; (5) increase liveability; (6) potential for positive impact on employment opportunities in the city. Hence, by considering the development of GCC countries that exist in the Degenerative Development Zone within the Quadrant of HDI-Ecological Footprint as shown in Fig. 1, (Islamic) smart cities in this region will take a global strategic role in green transition to implement the circular economy as one of the sustainability paradigms.

Therefore, it is encouraged to focus on the untapped potential of *israf* (wastage) principle of Islamic economics and collaborate with "refuse" and "reduce" principles of circular economy to achieve the very much needed targets of reducing the unnecessary resource consumption and the carbon footprint (Sancar, 2023). The following is an overview of the implementation of circular economy in (Islamic) smart cities in the GCC countries.

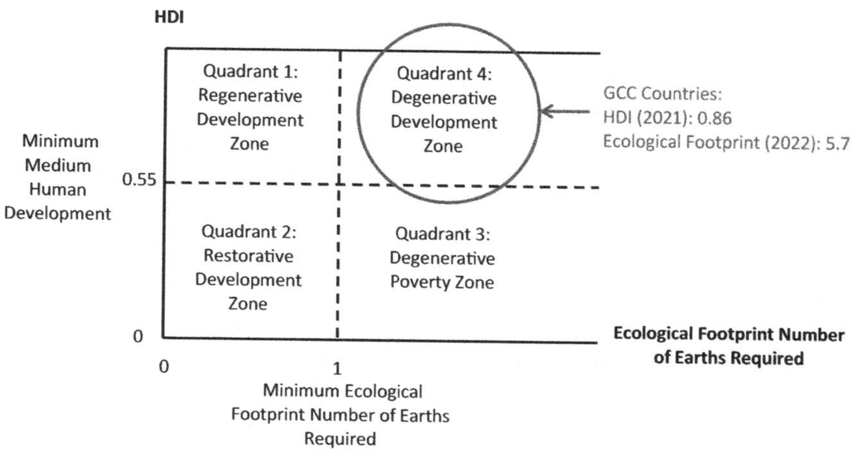

Fig. 1. Quadrant of HDI-Ecological Footprint. *Source*: Modified from Musari and Khan (2023), Khan and Tabet (2023), and Khan (2022).

UAE (Abu Dhabi & Dubai)

UAE environmental stewardship and sustainable development strategies are well known in the region. The major UAE environmental project is the National Climate Change Plan 2017–2050. The strategy envisions a climate-resilient, low-carbon economy that meets UAE development goals and Paris Agreement requirements. The plan covers energy, industry, transportation, infrastructure, waste, and land use. The strategy targets governance, research and innovation, education and awareness, funding and investment, and international cooperation (Musari & Khan, 2023).

Referring to the Government of UAE (2021), the UAE Circular Economy Policy 2021–2031 has been approved by the Cabinet of UAE in January 2021. The policy contains a comprehensive framework for assigning the approach of the state by moving away to a regenerative circular economy from a degenerative linear. A regenerative circular economy works within sustainable governance with the optimum natural resources to be used enclosed by the limits of the environment by embracing the way of production and consumption which secure the well-being of both present and future generations. The commitment to the Circular Economy as well as sustainable production and consumption is fully aligned with the Vision of UAE 2021 and the Green Agenda of UAE 2030.

Although the Government of UAE (2021) would like for all sectors to adopt circular economy principles, four priority sectors have been identified based on their role currently in the national economy and on their potential for stimulating and developing a UAE circular economy, namely: sustainable food production and consumption. sustainable manufacturing, sustainable transportation, as well as green infrastructure and development. Table 6 shows the main sector in which the circular economy principles can be implemented in the four priorities with the eight core sectors of focus that will support the circular economy transition to all sectors.

In Abu Dhabi, the Environment Agency – Abu Dhabi (EAD) is the entity responsible for environmental policies. The two main policy frameworks developed by EAD are the Abu Dhabi Environment Vision 2030 and Single Use Plastic Policy. Abu Dhabi Environment Vision 2030 aims to preserve and enhance Abu Dhabi's natural heritage in the efficient use of resources and contributing to a better quality of life for all. Environment Vision 2030 identifies the following five priority areas: (1) climate change; (2) clean air and noise pollution; (3) water resources; (4) biodiversity, habitats, and cultural heritage; (5) waste management.

Regarding the policy of Single Use Plastic, in collaboration with key stakeholders, EAD targets for collecting 20 million single-use plastic bottles a year for recycling by installing 70 reverse vending machines (RVMs) and 26 smart bins in areas with high footfall in the Emirate of Abu Dhabi such as in airport, malls, sport venues, universities, and in the Corniche. To nurture an awareness for recycling, RVMs and smart bins will allow the community to place their single-use plastic bottles and aluminum cans for recycling and receive rewards for doing so. This is the key step to build a circular system and reduce material flow (EAD, 2023).

Table 6. Four Priorities and Eight Core Sectors in UAE Circular Economy Policy 2021–2031.

Sustainable Food Production and Consumption:	**Overarching Priorities**:	**Green Infrastructure and Development**:
1. Reduce food loss and waste.	1. Set clear direction.	1. Sustainable and smart urban plans.
2. Input productivity.	2. Gather data.	2. Green building and construction.
3. Regenerative production.	3. Technology, circular economy, innovation, research and development (R&D).	3. Utilisation, renovation, and upgrade of buildings and infrastructures.
4. Upcycling biological waste.	4. Building awareness and capacity.	**Sustainable Transportation**:
Sustainable Manufacturing:	5. Partnership and collaboration.	1. Low-carbon vehicles.
1. Resource efficiency.	6. Access to sustainable finance.	2. Green integrated inter-modal plans.
2. Remanufacturing and reuse business models.	7. Circular procurement policy.	3. Public transport and sharing models.
3. Product and material design.	8. Waste management.	
4. Green sectoral clusters (Eco-industrial parks).		

Source: Government of UAE (2021).

To establish and strengthen leadership capacity among youth in order to address issues of environmental sustainability, inflame competitiveness amongst them to help degrade the environmental footprint, and drive endeavors toward establishing sustainable communities, EAD (2020) has organized the Sustainable Campus Initiative (SCI) since 2014. This youth community is a manifestation of the existence of a digitally-enabled community because those in this generation can be categorized as Generation Y and then Generation Z, which have been shaped by the digital era and tranquil to adopt and adapt technologies.

In Dubai, the Supreme Council of Energy (DSCE) has established the Circular Economy Committee to bring public and private entities together to support the circular economy in this smart city. The Committee purposes to: (1) develop and launch innovative tools, measuring means, initiatives, and policies to promote circularity; (2) support start-ups and develop the capabilities of entrepreneurs in the sector; (3) encourage the utilization of modern technologies capabilities; (4) enhance the efficiency of natural resources and promote sustainable best practices.

Dubai also has a nonprofit international circular economy network, namely Circular Economy Club (CEC). CEC Dubai voluntarily brings the circular economy activities to the city. The CEC Dubai network is open to anyone who is interested in learning more about and helping to support the transition toward a circular economy. As a networking arm of the Circular Economy Institute, CEC actually is the global circular economy network in 140 countries with 280 CEC locals which have the mission to bring circular economy to every corner of the world.

Moreover, there is also Dubai Electricity and Water Authority (DEWA) which by continuing to move along this pathway through a clear circular economy strategy, with the objective to shift away from the traditional linear business approach to a circular economy. DEWA has set a clear ambition to become a circular leader in the region by focusing on optimal resource use, creating social, economic and environmental value. The circular economy model of DEWA has five key circular principles: (1) circular design and use of circular material; (2) optimize asset management; (3) value retention and end-of-life treatment; (4) circular partnerships; (5) renewable energy, energy and water efficiency.

Saudi Arabia (Riyadh, Mecca, Jeddah, Madinah)

In order to realize the Saudi Vision 2030, the Ministry of Municipal, Rural Affairs and Housing (MOMRAH) established smart cities program to drive a transformation process toward urban smartness which selected 17 cities (Ministry of Municipal and Rural Affairs Saudi Arabia, 2017) to implement smart urban projects. Among 17 smart cities in Saudi, there are 4 cities that are included in the top 100 in 2023, i.e., Riyadh, Mecca, Jeddah, and Madinah. As smart cities in other countries, smart cities in Saudi also carry several dimensions in developing the cities, namely smart economy, smart living, smart governance, smart people, smart environment, and smart mobility.

Table 7 shows the dimensions of smart city development in Riyadh, Mecca, Jeddah, and Madinah. By calculating smart city indicators on a scale of values from zero as the lowest value to ten as the highest value through a questionnaire given to representatives of local institutions involved in smart city strategies, a study conducted by Aldegheishem (2023) found aggregate results for indicators of six dimensions for these four smart cities. The results show that Riyadh has the highest score, followed by Mecca, then Jeddah and Madinah. This is in line with the IMD Smart City Index Report 2023 which places the highest rankings and ratings in sequence in Riyadh, Mecca, Jeddah, and Madinah (see Table 2).

Table 7. Assessing the Dimensions of Smart City in Saudi Arabia

Dimensions	**Riyadh**	**Mecca**	**Jeddah**	**Madinah**
Smart economy	4.24	3.88	4.06	3.68
Smart living	4.16	3.85	4.02	3.38
Smart governance	4.11	3.29	3.25	3.18
Smart people	4.00	3.46	3.21	3.10
Smart environment	3.78	3.11	3.02	3.01
Smart mobility	3.26	2.74	2.25	2.21
Total	23.55	20.33	19.81	18.56

Source: Aldegheishem (2023).

Riyadh indeed is the largest hub for financial–economic operations and the highest population in Saudi Arabia. Riyadh is planned to provide 11,000 residential units for a population of 44,000 people. The 1,000-hectare development is designed to create 10,000 jobs in various sectors including green-tech industries to create a green circular economy for the city (Arab News, 2022; SmartCitiesWorld News Team, 2022). In transitioning toward becoming a smart urban center, Riyadh has main hurdles in achieving the smart city paradigm, such as modifying regulations, estimating infrastructure costs accurately, promoting public engagement in decision-making procedures, devising intelligent programs based on strategic resources, managing large-scale smart projects, transforming current practices, acquiring technical knowledge and proficiency in artificial intelligence, and establishing collaborative communication channels (Aldegheishem, 2023).

In Mecca, the comprehensive plan of this city was prepared by the Mecca and Madinah Development Authority and the High Authority for the Holy Areas for the Development of Mecca. Amanah Mecca is the municipality of the city of Mecca and is responsible for providing municipal services to the city and its residents and working to develop the city with all its facilities. The competent authorities embrace a major role in setting grounds for smart governance at the local level, leveraging a city's economic strengths, and implementing agenda for sustainable development (Al Sharif et al., 2021). One of the rising concerns in Saudi Arabia's Vision 2030 in Mecca is solid waste management (Algethami & Alhothali, 2023).

Regarding the circular economy, Amanah Mecca has worked to protect, recycle, and reuse all available fresh water by enabling renewable energy in a sustainable energy supply. Mecca has harnessed its potential from renewable energy, especially in solar and wind energy. Therefore, Mecca must invest in carbon saving technology and construct the advanced green building technology. Furthermore, the Ministries of Transport, and of Municipal and Rural Affairs have invested also in a smart transportation system and traffic management in Mecca to reduce the carbon footprint to make the city more livable by reducing mobility needs and increasing accessibility within the city (Al Sharif et al., 2021).

Then, as the second largest city in Saudi Arabia, Jeddah has implemented of smart environment through adapting the recycling system (Doheim et al., 2019). Jeddah has been proactive in mitigating and measuring food loss and waste. To recover the food, Jeddah places containers for food left-overs and requires restaurants and any ceremony services to contract with charities to redistribute the left-overs to those in need and sign contracts with companies to deliver the food to those in need (Saeedi et al., 2023).

For creating a digitally-enabled community, there is a King Abdullah University of Science and Technology (KAUST) in north Jeddah to be "an enduring model for advanced education and scientific research." KAUST is created to be a knowledge destination, a catalyst for innovation, economic development, and social prosperity in Saudi Arabia and the world. The talented minds from Saudi Arabia and the whole world are invited to KAUST for education, research, and conferences, where innovative works that address the current global challenges are encouraged by the latest research methods and all available resources. It is also worth mentioning that KAUST Campus is built on green specifications and has earned Leadership in Energy and Environmental Design Platinum rating (Doheim et al., 2019).

In Madinah, one of the sectors in this smart city is Environment & Agriculture. Several programs in this sector are Smart Green Buildings, Eco Digital Wallet, Sustainable Living Dashboard "Rasheed," "Farm-to-Table" Transformation Toolkit, Smart Waste Management, Interactive Connected Street. Smart Farming Solutions, Vertical & Community Agriculture, Central Farmers Community & Marketplace, and Zero Food Waste Platform "Ne3mah." Program of Zero Food Waste Platform "Ne3mah" is a form of implementing a circular economy. The activities in this program are a manifestation of the circular economy spirit.

Referring to https://smartmadinah.city, "Ne3mah" becomes a marketplace where private entities (i.e. homes and social gatherings) as well as commercial entities (namely hotels, supermarkets, restaurants, and bakeries) can offer their surplus of food to charities and people in need. Amid the high dependency on food imports, sensible consumption and mindful waste management are key for Madinah's city's food security given the poor adoption of sustainable habits across households and businesses. Therefore, Madinah needs to minimize city-level food waste by connecting sources that have excess food supply with segments in need. Through "Ne3mah" platform, users are connected to stores, restaurants, and bakeries in their proximity, from which extra food is shared free-of-charge or purchased at reduced rates. The platform is backed by a logistics network to help connect supply and demand seamlessly through automated schedules.

The big challenge for Madinah in establishing Environment & Agriculture sector is the unaware of sector stakeholders to sustainability commitments and environmental responsibility principles. Therefore, the sector strategic objectives in the program of this smart city is to manage effectively and with quality the city's priority resources such as water, air and agricultural crops. To realize them, Madinah infuses sustainability into the city day-to-day activities, drives efficient use of water and precious resources while effectively reducing waste, as well as improves agricultural crop yield and enables agricultural strategic plans and targets.

Qatar (Doha)

Qatar is effectively a city-state. Almost all of its population lives in Doha. The Ministry of Municipal Affairs was a de facto ministry of environment through urban planning and transportation policy. In 2021, the ministry was reshuffled and now is named the Ministry of Environment and Climate Change (Pal et al., 2022). The Qatar National Vision (QNV) 2030 and the United Nations (UN) Sustainable Development Goals (SDGs) place a strong emphasis on reducing waste, encouraging recycling, and moving toward a circular economy (Lundberg, 2022).

The Qatar National Environment and Climate Change Strategy (QNE) has identified biodiversity, air quality, water, greenhouse gas emissions, land use, as well as waste management and circular economy as prime concern areas with corresponding initiatives and goals. Waste management and circular economy are included in QNE as among all the priority areas in support of SDGs 9 and 12

(Al-Hababi, 2022). In the construction sector, Qatar has made significant strides toward attaining a circular economy by implementing recycling and reusing construction waste (Talballa & Gichuru, 2022). In the water industry, significant investment has taken place toward developing a sustainability where a transition toward renewable energy resources can support desalination, wastewater treatment, and reuse policies (Lawler et al., 2022).

Furthermore, targets for waste management and circular economy comprise: (1) 15% of municipal waste to be recycled; (2) 100% of unsanitary landfills to be closed and rehabilitated; (3) 35% of recycled materials in construction projects; (4) reaching 30% of circular procurement in public infrastructure; (5) reducing 50% of food waste per capita at consumer and retail levels, and (6) applying the principles of circular industry across industrial areas (Al-Hababi, 2022). To help reach these targets, referring to the Global Green Growth Institute (GGGI), the challenges and maximize opportunities of adopting a circular economy to contribute to climate action, pollution prevention, economic diversification, and resource efficiency have been addressed. It will also lay the foundation for integrating the principles of circularity and sustainable consumption and production into national development planning and implementation processes as well as corporate strategies and operations.

In Qatar, the Government indeed provides a supportive environment to enable the private sector to launch initiatives that help promote long-term sustainable growth. The companies, particularly energy companies, are also responding well to government efforts in public–private partnerships to adopt the circular economy model that would ensure that the four pillars of QNV 2030 will continue to be successfully addressed. One form of partnership is shown by the Abdullah Bin Hamad Al Attiyah International Foundation for Energy and Sustainable Development which has signed a MoU with the Qatar General Electricity and Water Corporation to tackle the circular economy (Dizon, 2019).

Oman (Muscat)

Implementing the smart city in Oman is vital for the following reasons: increased population; increased crime and accidents; security and privacy demand; better connectivity; social value; better quality of life; as well as efficiency and effectiveness for business operations. Smart cities are believed to be able to have a significant impact on the economic, social, political, technological, and environmental spheres (Oraba & Jalagat, Jr., 2023).

To bring a significant impact on the environmental sphere, Oman continues its commendable track record in environmental protection with the proposal of a smart city, Sultan Haitham City. The City promises a minimal ecological footprint, a focus on renewable energy, and employs technology to monitor environmental factors like water management and air quality, through incorporating solar power and water recycling as well as electric vehicles and waste-to-energy plants (The New Arab, 2023).

Indeed, through Oman Vision 2040, in line with the targets of the Paris Agreement, this country has announced its commitment to achieve net zero emissions by 2050. Therefore, the smart city must also in accordance with this commitment do an energy transition, ensuring environmental sustainability, ensuring a secure energy supply, encouraging job creation, minimizing economic costs to Omani citizens and industries, and optimizing the economic impact arising from the transition.

Generally, a study by Oraba & Jalagat, Jr. (2023) in Al-Irfan City, the first smart city in Oman which was launched in 2017, the main challenges faced by smart city implementation are: (1) the economic costs to construct the infrastructure based on technologies; (2) complexity and difficulty in the building permit, detailed planning and implementation, new technologies and applications, changing government policies and procedures, lack of accountability, financial complications, and complex systems; (3) technology and its implementation; (4) governance and coordination; (5) data privacy; and (6) social-public acceptance, sustainability issues for people, people awareness, and respect for others' rights or human rights issues.

Then, Oraba & Jalagat, Jr. (2023) proposed a couple of suggestions for a successful smart city implementation, namely: proper and systematic planning; concrete laws and regulations; sufficiency of needed resources and upgraded technologies; adequacy of collaborators (public–private partnerships); readiness of each governorate for smart city implementation; equal opportunity for SMEs to participate.

The GCC region has a high mass consumer society over the last five decades. It is indicated that the rise in per capita income, an increase in population growth and guest workers, and changes in technology, and marketing, as well as a linear economic paradigm, have generated significant waste in various sectors of the economy. Organized data is not available (Musari & Khan, 2023), but it is a reality that on one hand significant residual value is hidden in the piles of waste. On the other hand, this piling, up of waste has a negative footprint on the ecological environment of the GCC countries. This section presents a general picture of the potential of CE in the different sectors.

Therefore, to realize the circular economy in (Islamic) smart cities in GCC, there are some projects which can be implemented based on this paradigm, namely: circular economy for household and office furniture, clothing, carpets, curtains, furniture, and paper; circular economy of food waste; circular water economy; circular carbon economy; circular economy of heavy and light vehicles scrap; circular economy of construction and infrastructures scrap; circular economy of e-waste; circular economy of petrochemical and plastics; and circular economy for the industrial symbioses in GCC countries (Musari & Khan, 2023).

Conclusion

As urbanization is enlarging at a flying pace, the number of cities that engage the smart city concept is to rise, including in GCC countries. As a region with Very High Human Development, GCC countries have a high ecological footprint. Therefore, by implementing circular economy, (Islamic) smart cities in GCC countries and its government have a crucial part in reducing ecological footprints.

Then, technology is central to enabling and driving value in the circular economy. Nevertheless, technology is an enabler, not a panacea. So, (Islamic) smart cities in this region will take a global strategic role in green transition to implement the circular economy as one of the sustainability paradigms. Even the existence of (Islamic) smart cities in the GCC region can be an engine for creating ecological and environmental resilience to achieve the sustainable well-being.

Indeed, circular economy innovation is essential to tackle the imminent challenges of depleting resources and emerging environmental problems. Replace the current waste-driven linear economy paradigm with a new zero-waste circular economy paradigm consistent with the Islamic values which is adhered to by GCC countries. Therefore, to embrace a circular economy, (Islamic) smart cities in this region have a strategic role as a symbol of sustainable smart global Islamic civilization.

This chapter contributes to providing empirical evidence that (Islamic) smart cities in the GCC region have implemented the circular economy paradigm in their Smart Environment program. Theoretically, this chapter reinforces that the concept of a smart city symbolizes the integration of technology or digitalization with a community that cares about a sustainable lifestyle with one of the important components being ecological and environmental resilience through circular economy. A number of circular economy programs or projects discussed in this chapter can be replicated in other cities according to the level of technology they have to achieve sustainable well-being together.

For future research, it is highly recommended to: (1) re-investigate whether smart environment initiatives in (Islamic) smart cities in the GCC region, including the circular economy innovation, can reduce the ecological footprint of the city, even of its country in this region; (2) develop Islamic circularity management scorecard as Islamic performance management system of circular economy that enables businesses in (Islamic) smart city to drive strategies based on Shari'ah as well as measurement and follow-up; (3) develop the circular and regenerative business models and map DCST which can be adopted and adapted in (Islamic) smart cities.

References

AFED. (2012). Arab environment: Survival options - ecological footprint of Arab countries. In N. Saab (Ed.), *2012 Report of the Arab Forum for Environment and Development (AFED)*. Technical Publications and Environment & Development.

Ahmed, Z., Asghar, M. M., Malik, M. N., & Nawaz, K. (2020). Moving towards a sustainable environment: The dynamic linkage between natural resources, human capital, urbanization, economic growth, and ecological footprint in China. *Resources Policy*, *67*, 101677. https://doi.org/10.1016/j.resourpol.2020.101677

Ahmed, Z., & Wang, Z. (2019). Investigating the impact of human capital on the ecological footprint in India: An empirical analysis. *Environmental Science and Pollution Research*. https://doi.org/10.1007/s11356-019-05911-7

Ahmed, Z., Zafar, M. W., Ali, S., & Danish. (2020). Linking urbanization, human capital, and the ecological footprint in G7 countries: An empirical analysis. *Sustainable Cities and Society*, *55*, 102064. https://doi.org/10.1016/j.scs.2020.102064

Al Sharif, R. N., Aljuaid, R. M., Ramadan, N. M., & Alwetaishi, M. (2021). Makkah al-mukarramah-towards green and smart city. In *2021 International Conference of Women in Data Science at Taif University (WiDSTaif)*. https://doi.org/10.1109/widstaif52235.2021.94

Aldegheishem, A. (2023). Assessing the progress of smart cities in Saudi Arabia. *Smart Cities, 6*(4), 1958–1972. https://doi.org/10.3390/smartcities6040091

Algethami, H., & Alhothali, G. T. (2023). Waste collection optimisation: A path to a green and sustainable city of Makkah. *Logistics, 7*(3), 54. https://doi.org/10.3390/logistics7030054

Al-Hababi, R. (2022). The evolvement of Qatar's environmental sustainability policy: The strategies, regulations, and institutions. In L. Cochrane & R. Al-Hababi (Eds.), *Sustainable Qatar: Social, political and environmental perspectives* (pp. 17–36). Open Access, Springer.

Amer, E. A. A. A., Meyad, E. M. A., Gao, Y., Niu, X., Chen, N., Xu, H., & Zhang, D. (2022). Exploring the link between natural resources, urbanization, human capital, and ecological footprint: A case of GCC countries. *Ecological Indicators, 144*, 109556. https://doi.org/10.1016/j.ecolind.2022.109556

Ansari, M. A., Ahmad, M. R., Siddique, S., & Mansoor, K. (2020). An environment Kuznets curve for ecological footprint: Evidence from GCC countries. *Carbon Management, 11*(4), 355–368. https://doi.org/10.1080/17583004.2020.1790242

Arab News. (2022, July 1). *New Saudi smart city AlNama to be zero-carbon*. https://www.arabnews.com/node/2114701/business-economy

Asmyatullin, R. R., Tyrkba, Kh. V., & Ruzina, E. I. (2020). Smart cities in GCC: Comparative study of economic dimension. *IOP Conf. Series: Earth and Environmental Science, 459*(2020), 062045. https://doi.org/10.1088/1755-1315/459/6/062045

Bejjani, M., Anouti, Y., Klat, A., & Batal, J. (2019). *Putting GCC cities in the loop: Sustainable growth in a circular economy*. The Ideation Center, Strategy& Middle East and PricewaterhouseCoopers (PwC).

Böhmecke-Schwafert, M., Wehinger, M., & Teigland, R. (2022). Blockchain for the circular economy: Theorizing blockchain's role in the transition to a circular economy through an empirical investigation. *Business Strategy and the Environment*. http://dx.doi.org/10.1002/bse.3032

Bridge, S. (2019, February 12). How a circular economy could save the GCC $138bn by 2030. *Arabian Business*. https://www.arabianbusiness.com/politics-economics/413126-how-circular-economy-could-save-the-gcc-138bn-by-2030

Collins, A., & Flynn, A. (2015). *The ecological footprint: New developments in policy and practice*. Edward Elgar Publishing.

Damian, D., & Phan, T. (2022). Introduction to smart cities in Asia: Regulations, problems, and development. In T. Phan & D. Damian (Eds.), *Smart cities in Asia: Regulations, problems, and development* (pp. 1–7). Springer Nature. https://doi.org/10.1007/978-981-19-1701-1

Dizon, L. R. R. (2019, June 6). *Circular economy fits well with Qatar: Expert*. The Peninsula, Qatar Business. https://thepeninsulaqatar.com/article/06/06/2019/Circular-economy-fits-well-with-Qatar-Expert

Doagoo, B. C. (2022). Smart cities in Asia: An introductory note. In T. Phan & D. Damian (Eds.), *Smart cities in Asia: Regulations, problems, and development* (pp. 9–22). Springer Nature. https://doi.org/10.1007/978-981-19-1701-1

Doheim, R. M., Farag, A. A., & Badawi, S. (2019). Smart city vision and practices across the Kingdom of Saudi Arabia – A review. In A. Visvizi & M. D. Lytras (Eds.), *Smart cities: Issues and challenges, mapping political, social and economic risks and threats* (pp. 309–332). Elsevier. https://doi.org/10.1016/b978-0-12-816639-0.00017-x

EAD. (2020, November 29). Press release: Environment agency – Abu Dhabi (EAD) honours sustainable campus initiative youth for environmental work during covid-19. *Zawya*, https://www.zawya.com/en/press-release/environment-agency-abu-dhabi-honours-sustainable-campus-initiative-youth-for-environmental-work-during-covid-gf7470ov

EAD. (2023, July 13). By launching 70 reverse vending machines and 26 smart bins, EAD targets a reduction of 20 million single-use plastic bottles. *The Environment Agency – Abu Dhabi (EAD)*. https://www.ead.gov.ae/Media-Centre/News/launching-70-Reverse-Vending-Machines-2023

EEB. (2020). *A circular economy within ecological limits*. A discussion paper of European Environmental Bureau (EEB), February 14th. https://eeb.org/library/a-circular-economy-within-ecological-limits/

El Mir, J., Elgendy, K., & Khamlichi, H. (2021). Circular economy in cities of the MENA region: Prospects and challenges for material circularity. Friedrich-Ebert-Stiftung's Regional Climate and Energy Project in the Middle East and North Africa (MENA).

Ellen MacArthur Foundation. (2017). *Cities in the circular economy: An initial exploration.* https://ellenmacarthurfoundation.org/cities-in-the-circular-economy-an-initial-exploration

Ellen MacArthur Foundation. (2019). *City governments and their role in enabling a circular economy transition: An overview of urban policy levers.* https://ellenmacarthurfoundation.org/policy-levers

ElMassah, S., & Hassanein, E. A. (2023). Economic development and environmental sustainability in the GCC countries: New insights based on the economic complexity. *Sustainability, 15*, 7987. https://doi.org/10.3390/su15107987

Global Footprint Network. (2022). *National footprint and biocapacity accounts 2023 edition.* https://www.footprintnetwork.org/licenses/public-data-package-free/

Government of UAE. (2021). *UAE circular economy policy 2021-2031*. The Government of Uni Arab Emirate (GoUAE), in partnership with Ministry of Cabinet Affairs, Ministry of Climate Change & Environment, Ministry of Economy, and Ministry of State for Artificial Intelligence Office.

Henderson, P. (2023, September 18). The circular economy: A missing piece in city climate action plans? *Ellen MacArthur Foundation.* https://ellenmacarthurfoundation.org/articles/the-circular-economy-a-missing-piece-in-city-climate-action-plans

Hertog, S., & Luciani, G. (2009). Kuwait programme on development, governance and globalisation in the Gulf states energy and sustainability policies in the GCC. Centre for the Study of Global Governance, London School of Economics (LSE) - Kuwait Program Working Paper.

Joo, Y.-M., & Tan, T.-B. (2020). Smart cities in Asia: An introduction. In Y.-M. Joo & T.-B. Tan (Eds.), *Smart cities in Asia governing development in the era of hyper-connectivity* (pp. 1–17). Edward Elgar. https://doi.org/10.4337/9781788972888.00007

Khan, T. (2022). *Rethinking economics in the anthropogenic era*. A presentation material for Guest Lecture at Quaid-i-Azam University, Islamabad, October 13th.

Khan, T. (2023). *Externalities management from Islamic economic perspective: Rethinking economics and finance in the anthropogenic era*. A presentation material for Kulliyyah of Economics & Management Sciences (KENMS) of International Islamic University Malaysia (IIUM) Distinguished Intellectual Discourse Series, A Prelude to Key Themes in the 15th International Conference on Islamic Economics & Finance (ICIEF) 2023 Conference, May 26th.

Khan, T., & Tabet, I. (2023). Dynamic equilibrium, regenerative development, and designs of Islamic financial contracts. *Social Science Research Network (SSRN), October 2.* https://papers.ssrn.com/sol3/papers.cfm?abstract_id=4589599

Lacy, P. (2017, September 14). 5 disruptive technologies driving the circular economy. *World Economic (WE) Forum*. https://www.weforum.org/agenda/2017/09/new-tech-sustainable-circular-economy/

Lawler, J., Mazzoni, A., & Shannak, S. (2022). The domestic water sector in Qatar. In L. Cochrane & R. Al-Hababi (Eds.), *Sustainable Qatar: Social, political and environmental perspectives* (pp. 193–210). Open Access, Springer.

Ministry of Municipal and Rural Affairs Saudi Arabia. (2017). *Smart cities blueprint master document V 0.98 31-12-2017*.

Moigne, R. L. (2021, July 16). The power of digital technologies to enable the circular economy: Digitalising material flows. *Ellen MacArthur Foundation*. https://medium.com/circulatenews/the-power-of-digital-technologies-to-enable-the-circular-economy-5471d097ee7f

Mubasher. (2019, February 13). *GCC region can save $138bln by circular economy by 2030: Report. Zawya, Economy*. https://www.zawya.com/en/economy/gcc-region-can-save-138bln-by-circular-economy-by-2030-report-e6g33ync

Murphy, C. (2022, August 30). Blockchain can facilitate the transition to a circular economy – But scaling its use is a work in progress. *Tech Enablers Series of Ellen MacArthur Foundation, Part 2*. https://ellenmacarthurfoundation.org/tech-enablers-series/part-2

Musari, K. (2021). Circular economy for plastics and digitally-enabled community towards ASEAN halal hub in Asia. In P. O. de Pablos (Ed.), *Handbook of research on developing circular, digital, and green economies in Asia* (pp. 1–12). IGI Global. https://doi.org/10.4018/978-1-7998-8678-5.ch001

Musari, K., & Khan, T. (2023). *Sustainable development and circular economy in the GCC Countries*. A paper was presented at the Workshop "Circular Economy and Sustainable Development Model: It's Role in the GCC States' National Visions" held by Gulf Research Centre Cambridge (GRCC), which is an affiliate of the Gulf Research Center Foundation Geneva, and a partner institution of the Centre of Islamic Studies at the University of Cambridge, Cambridge – United Kingdom, July, 11th–13th.

Musari, K., & Naaz, S. (2023). How (Islamic) smart cities 2.0 are driving inclusive growth opportunities and fighting climate change? Evidence from Asia-9. In M. N. Almunawar, P. O. de Pablos, & M. Anshari (Eds.), *Sustainable development and the digital economy: Human-centricity, sustainability and resilience in Asia* (pp. 222–243). Routledge. https://doi.org/10.4324/9781003388753

Mutambik, I., Lee, J., Almuqrin, A., & Zhang, J. Z. (2023). Transitioning to smart cities in Gulf Cooperation Council countries: The role of leadership and organisational culture. *Sustainability*, *15*, 10490. https://doi.org/10.3390/su151310490

Oraba, R. M. B., & Jalagat, Jr., R. C. (2023). The challenges of smart city application in Oman: The al Irfan project experience. *European Academic Research*, *10*(12), 4291–4308.

Pal, L. A., Al-Hababi, R., & Cochrane, L. (2022). Pathways for a sustainable future. In L. Cochrane & R. Al-Hababi (Eds.), *Sustainable Qatar: Social, political and environmental perspectives* (pp. 365–379). Open Access, Springer.

Panda Media Centre. (2017, May 17). Asia leading the way: Pros and cons of future smart cities. *Panda Security*. https://www.pandasecurity.com/en/mediacenter/news/asia-leading-way-pros-cons-future-smart-cities/

Petrucci, A. L., & AlSheikh, A. (2022). *Resilience, growth, and GCC smart cities*. Konrad-Adenauer-Stiftung Policy Report No. 57 – January.

Power, D. (2017, March 30). Asia tipped to be the global smart cities leader. *ReadWrite*, https://readwrite.com/asia-tipped-global-smart-cities-leader-cl4/

Prakash, M. (2021, September 7). Unlocking citizen-centric smart cities in Asia. *Microsoft Stories Asia*. https://news.microsoft.com/apac/2021/09/07/unlocking-citizen-centric-smart-cities-in-asia/

Rees, W. (1986). *The regional capsule concept: An heuristic model for thinking about regional development and environmental policy*. Concept Note, School of Community and Regional Planning, University of British Columbia.

Rees, W., & Wackernagel, M. (1996). Urban ecological footprints: Why cities cannot be sustainable – and why they are a key to sustainability. *Environmental Impact Assessment Review, 16*(4–6), 223–248. https://doi.org/10.1016/S0195-9255(96)00022-4

Rubeis, M., Jedrzejewski, M., & Castoldi, S. (2022, December 15). Addressing barriers to circularity in the GCC. *The Gulf Petrochemicals and Chemicals Association (GPCA)*. https://www.gpca.org.ae/2022/12/15/addressing-barriers-to-circularity-in-the-gcc/

Saeedi, K., Visvizi, A., Alahmadi, D., & Babour, A. (2023). Smart cities and households' recyclable waste management: The case of Jeddah. *Sustainability, 15*(8), 6776. https://doi.org/10.3390/su15086776

Sancar, M. A. (2023). *The untapped potential of colloboration between "israf/wastage" principle of Islamic economics and refuse and reduce principles of circular economy*. A presentation material at the 11th Islamic Economics Workshop under the theme "Islamic Economics and Environment" held by Research Center for Islamic Economics (IKAM) in collaboration with the Turkish Entrepreneurship and Business Ethics Association (İGİAD) and Istanbul 29 Mayis University Economic and Financial Systems Application and Research Center (EFSAM), Istanbul, October 13th–15th.

Saxena, S., & Al-Tamimi, T. A. S. M. (2018). Visioning "smart city" across the Gulf Cooperation Council (GCC) countries. *Foresight, 20*(3), 237–251. https://doi.org/10.1108/fs-11-2017-0068

SmartCitiesWorld News Team. (2022, July 4). Plans unveiled for zero-carbon smart city in Riyadh. *SmartCitiesWorld*. https://www.smartcitiesworld.net/planning-and-redevelopment/planning-and-redevelopment/plans-unveiled-for-zero-carbon-smart-city-in-riyadh-7857

Strategy&. (2019). *Putting GCC cities in the loop sustainable growth in a circular economy*. Strategy& PricewaterhouseCoopers (PwC), Ideation Center insight.

Talballa, H. M., & Gichuru, J. (2022). Toward the circular Qatari zero-waste management sector. In L. Cochrane & R. Al-Hababi (Eds.), *Sustainable Qatar: Social, political and environmental perspectives* (pp. 305–328). Open Access, Springer.

Tekdoğan, Ö. F. (2023). *A critical approach to the green & sustainability movement*. A presentation material at the 11th Islamic Economics Workshop under the theme "Islamic Economics and Environment" held by Research Center for Islamic Economics (IKAM) in collaboration with the Turkish Entrepreneurship and Business Ethics Association (İGİAD) and Istanbul 29 Mayis University Economic and Financial Systems Application and Research Center (EFSAM), Istanbul, October 13th-15th.

The New Arab. (2023, September 12). *Is Oman's 'smart city' a model for sustainable development? The New Arab, Environment and Climate*. https://www.newarab.com/features/omans-smart-city-model-sustainable-development

UN – Department of Economic and Social Affairs, Population Division. (2019). *World urbanization prospects: The 2018 revision (ST/ESA/SER.A/420)*. United Nations (UN).

UNDP. (2022). *Human development report 2021/2022*. United Nations Development Programme (UNDP). http://hdr.undp.org/en/data

Vyas, S. (2018, February 13). *Smart cities – Powering a circular economy*. LinkedIn. https://www.linkedin.com/pulse/smart-cities-powering-circular-economy-sameer-vyas

Wackernagel, M., & Beyers, B. (2019). *Ecological footprint: Managing our biocapacity budget*. New Society Publishers.

Zafar, M. W., Zaidi, S. A. H., Khan, N. R., Mirza, F. M., Hou, F., & Kirmani, S. A. A. (2019). The impact of natural resources, human capital, and foreign direct investment on the ecological footprint: The case of the United States. *Resources Policy, Elsevier, 63*(C), https://doi.org/10.1016/j.resourpol.2019.101428

Zawya. (2019, February 12). Adopting a circular economic model could save GCC $138bln between 2020 and 2030. *Zawya*, Press Release. https://www.zawya.com/en/press-release/adopting-a-circular-economic-model-could-save-gcc-138bln-between-2020-and-2030-xh5i

Chapter 22

Understanding and Optimizing Hydrogen Transport Technologies for 21st Century Smart Cities: Innovation for Morocco's Green Interconnections in Hydrogen, Ammonia, and Kerosene Production

Ayat-Allah Bouramdane

Laboratory of Renewable Energies and Advanced Materials (LERMA), College of Engineering and Architecture, International University of Rabat (IUR), IUR Campus, Technopolis Park, Rocade Rabat-Salé, Sala Al Jadida, 11103, Morocco

Abstract

In smart cities striving for innovation, development, and prosperity, hydrogen offers a promising path for decarbonization. However, its effective integration into the evolving energy landscape requires understanding regional intricacies and identifying areas for improvement. This chapter examines hydrogen transport from production to utilization, evaluating technologies' pros, cons, and process equations and using Analytic Hierarchy Process (AHP) as a Multi-Criteria Decision-Making (MCDM) tool to assess these technologies based on multiple criteria. It also explores barriers and opportunities in hydrogen transport within the 21st-century energy transition, providing insights for overcoming challenges. Evaluation criteria for hydrogen transport technologies were ranked by relative importance, with energy efficiency topping the list, followed by energy density, infrastructure requirements, cost, range, and flexibility. Safety, technological maturity, scalability, and compatibility with existing infrastructure received lower weights. Hydrogen transport technologies were categorized into three performance levels: low, medium, and high. Hydrogen tube trailers ranked lowest, while chemical hydrides,

The Emerald Handbook of Smart Cities in the Gulf Region:
Innovation, Development, Transformation, and Prosperity for Vision 2040, 405–480

doi:10.1108/978-1-83608-292-720241030

hydrail, liquid organic hydrogen carriers, hydrogen pipelines, and hydrogen blending exhibited moderate performance. Compressed hydrogen gas, liquid hydrogen, ammonia carriers, and hydrogen fueling stations demonstrated the highest performance. The proposed framework is crucial for next-gen smart cities, cutting emissions, boosting growth, and speeding up development with a strong hydrogen infrastructure. This makes the region a sustainable tech leader, improving air quality and well-being. Aligned with Gulf Region goals, it is key for smart cities. Policymakers, industries, and researchers can use these insights to overcome barriers and seize hydrogen transport tech opportunities.

Keywords: 21st century smart cities; analytic hierarchy process; energy transition; hydrogen transport technologies; multi-criteria decision-making

1. Introduction

1.1. Research Motivation

1.1.1. Hydrogen for Decarbonization and Climate Change.

Currently, approximately one-third of global energy-related emissions are generated by economic sectors that lack viable alternatives to fossil fuels. These emissions predominantly arise from energy-intensive industries and freight transportation (Fig. 1). Hydrogen is considered a promising solution for decarbonizing various sectors of the economy, such as transportation, industry, and power generation. By using hydrogen produced from renewable sources, such as electrolysis powered by renewable energy (Kumar & Lim, 2022), the overall carbon footprint can be significantly reduced, helping to mitigate climate change and reduce greenhouse gas emissions (Haggi et al., 2022; Khokhar, 2023). The demand for hydrogen in Europe is projected to exceed 45 million tons by 2050. Various sectors, including transportation, heating, and heavy industry, are expected to embrace hydrogen as they strive for decarbonization in the coming decades. This growing demand has led to significant investments in hydrogen technology, indicating the increasing momentum behind its adoption (Fig. 2).

1.1.2. Hydrogen for Renewable Energy Integration

In order to meet the goals set in the Paris Agreement (Bouramdane, 2022a, 2023b, 2023c), a significant overhaul of the global energy system is required, transitioning from a heavy reliance on fossil fuels to a more sustainable and renewable low-carbon energy system. Extensive research conducted by the International Renewable Energy Agency (IRENA) suggests that more than 90% of the required reductions in global CO_2 emissions can be achieved through these transformative measures (Fig. 3). It is projected that renewable energy sources will directly contribute to a 41% reduction in emissions, with an additional 13% reduction

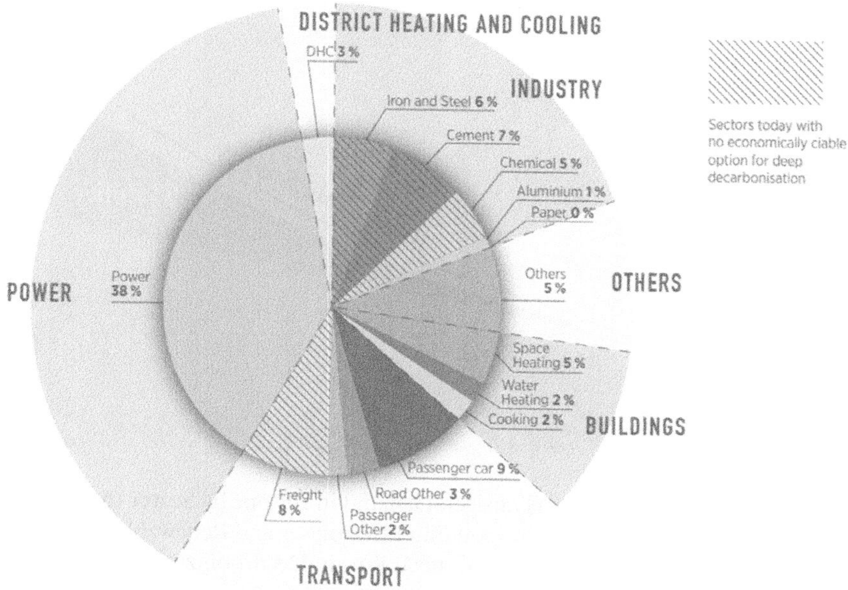

Fig. 1. Distribution of Global Energy-Related CO_2 Emissions by Sector in 2015. Approximately One-Third of These Emissions Currently Lack Economically Feasible Alternatives for Achieving Deep Decarbonization. *Source*: IRENA (2018).

achieved through the electrification of various sectors (Fig. 3). These findings highlight the crucial role of renewable energy in mitigating climate change and advancing towards a greener future.

To achieve this target, the global share of renewable energy in final energy consumption must rise from the current 18% to 65% by 2050 (left panel of Fig. 4). The power system will heavily rely on variable renewable energy sources, particularly wind and solar, which will constitute the majority of the generation capacity (right panel of the Fig. 4). To accommodate these high levels of variable generation economically, the power system must enhance its flexibility and adaptability.

Hydrogen can play a vital role in the energy transition by acting as a versatile energy carrier and facilitating the integration of renewable energy sources into the existing energy system. It can be produced during periods of excess renewable energy generation and stored for later use when demand exceeds supply, thus helping to balance the intermittency and seasonality of renewables and support grid stability (Bragatto et al., 2022; Elberry et al., 2021). In fact, hydrogen can serve as a form of energy storage, enabling the long-term storage and transportation of energy. It can be converted back into electricity through fuel cells or used in industrial processes, providing flexibility in energy supply and demand management.

There exist various energy storage technologies, each with its own set of advantages and disadvantages (Bouramdane, 2021, 2023a). When considering

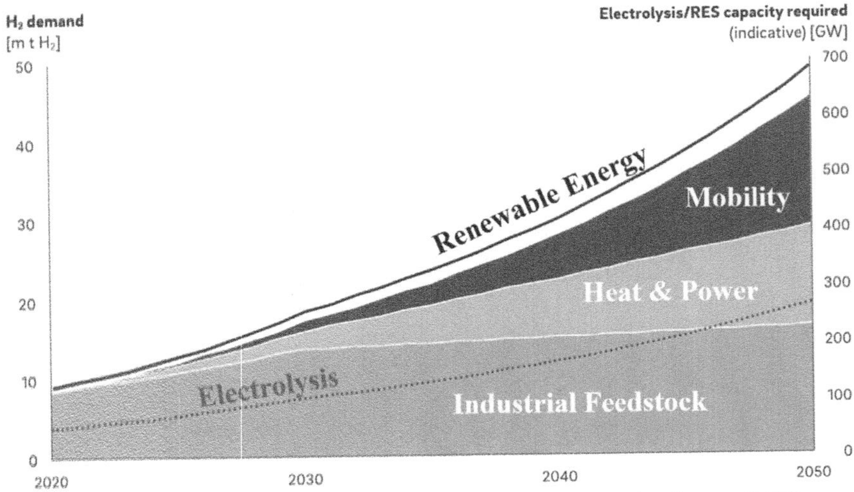

Fig. 2. Projected Growth of Hydrogen Demand in Europe by Sector (Million Metric Tons of H_2) and the Corresponding Electrolyser and Renewable Capacity (Gigawatts) Is Driven by Hydrogen's Crucial Role in Decarbonization. *Source*: Roland Berger (n.d.).

shorter discharge durations, typically below a few hours, options like compressed hydrogen and ammonia tend to be significantly more expensive compared to alternatives such as Pumped-Hydro Energy Storage (PHES) or lithium-ion battery storage (Fig. 5). However, as discharge durations extend, the attractiveness of compressed hydrogen and ammonia increases (Fig. 5). This is mainly due to their relatively lower capital costs for storing larger volumes of energy, as the investment costs involve the development of underground salt caverns or storage tanks. Among the range of storage technologies considered, compressed hydrogen emerges as the most economically viable choice for discharge durations exceeding 20–45 hours (Fig. 5).

Hydrogen enables "sector coupling" by connecting the electricity system with industry, buildings, and transportation, enhancing flexibility and facilitating the integration of variable renewable energies into the power system (Fig. 6).

1.1.3. Hydrogen for Energy Independence

Hydrogen can also contribute to enhancing energy independence and security by diversifying energy sources. It can be produced domestically, reducing dependence on fossil fuel imports and the geopolitical vulnerabilities associated with traditional energy supplies. This can enhance energy resilience and reduce the impact of supply disruptions (Groppi et al., 2018; Nikolova & Stoilov, 2022).

1.1.4. Hydrogen for Technological Innovation and Economic Opportunities

The development of the hydrogen value chain presents opportunities for technological innovation and economic growth. It can drive the development of new

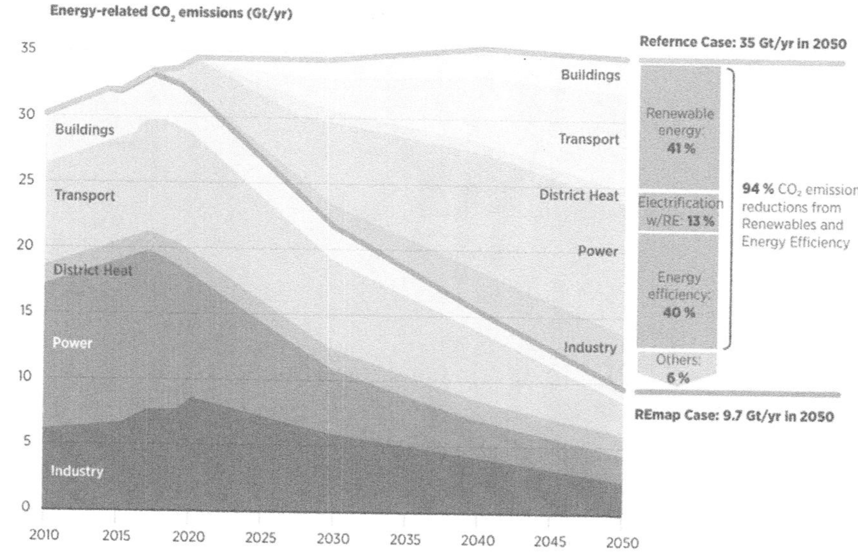

Notes: REmap refers to IRENA's roadmap to rapidly scale up renewables within the coming decades; the Reference Case refers to the path set by current plans and policies; CO_2 = carbon dioxide; Gt/yr = gigatonnes per year; w/RE = with renewable energy.
Source: IRENA (2018).

Fig. 3. Comparison of Energy-Related Carbon Dioxide Emissions Under Current Policies (Reference Case) and Accelerated Uptake of Renewables (REmap), 2010–2050 The Paris Agreement Sets a Target to Limit the Global Temperature Increase to "well below 2°C" Compared to Pre-Industrial Levels. To Achieve This Goal, Substantial Reductions in Emissions Across All Sectors Are Required. Renewable Energy and Energy Efficiency Measures Have the Potential to Contribute Over 90% of the Reduction in Energy-Related CO_2 Emissions. The Figure Highlights the Significant Role of Renewables in Driving Emissions Reductions and Transitioning Toward a More Sustainable and Low-Carbon Energy System. *Source*: IRENA (2018).

technologies, such as advanced electrolysis systems, fuel cells, and hydrogen storage solutions (Ashari et al., 2023; Wheeldon, 2022).

Additionally, it can create new industries, job opportunities, and economic sectors related to hydrogen production, infrastructure development, and hydrogen-powered applications (Bezdek, 2020; Mneimneh et al., 2023).

1.1.5. Hydrogen for Air Quality and Health Benefits

Hydrogen, when used in fuel cells or combustion processes, produces zero-or low-emission byproducts, such as water vapor. This can help improve air quality and reduce pollution, particularly in urban areas where air pollution poses significant health risks. By replacing conventional fossil fuel-based systems with hydrogen-based alternatives, we can achieve cleaner and healthier environments (Jacobson et al., 2005; Jensen et al., 2011).

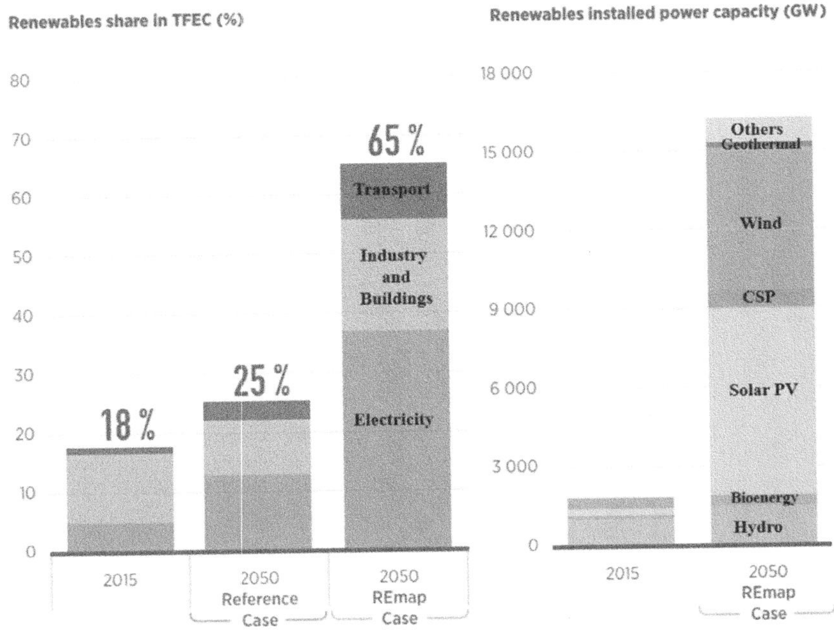

Fig. 4. Share of Renewables in Total Final Energy Consumption (TFEC) (Left) and Projected Renewable Power Capacity in 2050 (Right). The Left Panel Illustrates the Comparison Between the Reference Case and REmap Scenarios, Showcasing the Share of Renewables in Total Final Energy Consumption (TFEC). The Right Panel Displays the Projected Installed Renewable Power Capacity in 2050 Under the REmap Scenario. The Findings Emphasize the Significant Potential of Renewables to Contribute to the World's Energy Mix, with a Growing Dependence on Electricity and Substantial Growth in Installed Power Capacity by 2050. CSP stands for Concentrated Solar Power, GW Represents Gigawatt, PV Denotes Photovoltaic, and TFEC Refers to Total Final Energy Consumption. *Source*: IRENA (2018).

Given hydrogen's role in decarbonization, renewable energy integration, energy independence, technological innovation, economic opportunities, and air quality and health benefits, it is essential to comprehensively analyze and optimize the technical, economic, environmental, social, and political aspects of hydrogen transport technologies. This understanding is crucial for facilitating the successful transition to a sustainable energy system in the 21st century.

1.2. Existing Research

There is a significant amount of research being conducted on various aspects of hydrogen, including its production, storage, transportation, and utilization.

Research is focused on developing efficient and sustainable methods of hydrogen production (Dincer & Acar, 2015) from renewable and non-renewable sources (Acar & Dincer, 2014) for public transportation sector (Aydin & Dincer, 2022) using

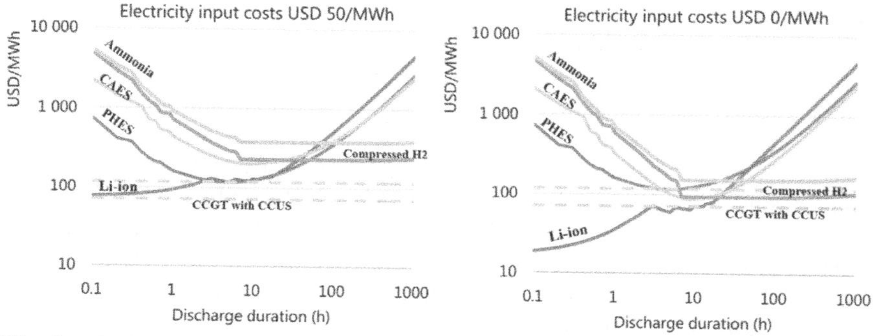

Fig. 5. Relationship between the Levelized Costs of Storage and the Duration of Discharge. It Provides Insights into How the Costs of Storage Vary Depending on the Time It Takes to Discharge the Stored Energy. PHES Stands for PHES, CAES Represents Compressed Air Energy Storage, and Li-Ion Indicates Lithium-Ion Battery Technology. Moreover, Compressed Hydrogen Storage Refers to Storing Hydrogen Gas Under Pressure, Typically in Salt Caverns, While Ammonia Storage Involves Storing Ammonia in Tanks. Additionally, CCGT with CCUS Refers to Combined Cycle Gas Turbine Power Plants Equipped with Carbon Capture, Utilization, and Storage Technology. *Source*: International Energy Agency (IEA, 2019).

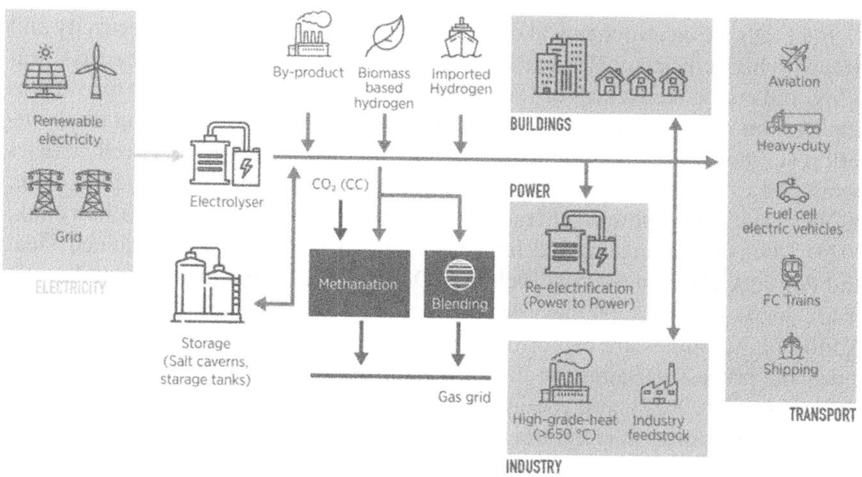

Fig. 6. Role of Hydrogen in Integrating Variable Renewable Energy into End Uses. Hydrogen Enables Coupling Between Electricity, Buildings, Transport, and Industry. It Has the Potential to Decarbonize Challenging Sectors Such as Industry, Where Renewable Hydrogen Can Replace Fossil Fuel-Based Hydrogen. In Buildings and Power, Hydrogen Injection into the Gas Grid Reduces Reliance on Natural Gas. In Transport, Hydrogen-Fueled Vehicles, Particularly Fuel Cell Electric Vehicles (FCEVs), Offer Low-Carbon Mobility Options for High-Utilization Sectors That Are Currently Limited by Battery Technology. Additionally, Hydrogen Provides Flexibility for Accommodating Higher Shares of Variable Renewable Energy (VRE) in the Power System. *Source*: IRENA (2018).

Multi-Criteria Decision-Making (MCDM) (Seker & Aydin, 2021), highlighting the factors affecting their scale-up (Agyekum et al., 2022) and other challenges (Çelik & Yildiz, 2017; Dash et al., 2023). This includes studying advanced electrolysis techniques, such as proton exchange membrane (PEM) electrolysis and solid oxide electrolysis cells (SOECs) (Kumar & Lim, 2022; Nami et al., 2022), as well as exploring novel pathways like solar-driven water splitting (Ekspong et al., 2021; Jiao et al., 2022), biomass conversion (Dou et al., 2019; Kumar et al., 2020), and methane reforming with carbon capture and storage (CCS) (Antonini et al., 2020; Ingale et al., 2022).

The storage and transportation of hydrogen are critical for its widespread use (Li et al., 2022; Suri & Siddique, 2019). Research is being conducted on different transportation methods (Li et al., 2023; Xiangmin, 2008) such as liquid hydrogen (Aziz, 2021) in maritime transportation (Ustolin et al., 2022), pipelines and truck trailers (Tayarani & Ramji, 2022) or different storage methods (Tarhan & Çil, 2021; Usman, 2022), such as compressed hydrogen gas (CHG) (Orlova et al., 2023), cryo-compression and cryo-adsorption (Petitpas et al., 2014), absorption and desorption processes (Li et al., 2022), liquid hydrogen (Ulucan et al., 2022; Zheng et al., 2021), and solid-state hydrogen storage materials (Zacharia & Ullah Rather, 2016) such as metal hydride materials (Sakintuna et al., 2007) and absorption (Kumar et al., 2022). Additionally, studies are investigating the design and optimization of hydrogen distribution networks, pipelines, and infrastructure to ensure safe and efficient transport with minimum energy consumption (Xiao, 2011) and minimal cost (Yeates et al., 2021).

Fuel cells are electrochemical devices that convert hydrogen into electricity and heat with high efficiency. Research is focused on improving the performance, durability, and cost-effectiveness of fuel cell technologies (Fan et al., 2021) for various applications (Peighambardoust et al., 2010; Sun et al., 2018), including transportation (Lemons, 1989; Ogungbemi et al., 2020) – e.g., hydrogen fuel cell vehicles compared to Diesel engine vehicles and battery electric vehicles (Cunanan et al., 2021) – , stationary power generation (Corigliano et al., 2022; Visvanathan et al., 2023) – considering ageing mechanisms (Song et al., 2021) – , Combined Heat and Power (CHP) generation (Tochigi & Dowaki, 2021), and microgrid (Akinyele et al., 2020), and portable devices (Chen et al., 2021; Kruusenberg et al., 2022). Additionally, studies are exploring the use of hydrogen in other sectors, such as industrial processes (Samiee et al., 2020; Zhao et al., 2018), heating (Dodds et al., 2015; Lee et al., 2022), and energy storage (Gan & Liu, 2017; Yufit et al., 2013).

Research is aimed at integrating hydrogen into existing energy systems and optimizing its use for maximum efficiency and effectiveness. This includes developing models and algorithms for energy management (Abomazid et al., 2022; Ferrara et al., 2021), demand-response strategies (Liu et al., 2022; Wu & Li, 2023) and microgrid (Alzahrani et al., 2022; Tooryan et al., 2022), nanogrid (Arsalis et al., 2022) or smart grids integration (Arnone et al., 2014; Speckmann et al., 2021). The goal is to understand how hydrogen can be integrated with renewable energy sources and other energy carriers to create a reliable and sustainable energy system.

Materials play a crucial role in various hydrogen technologies, including electrolysis, fuel cells, and hydrogen storage. Research is focused on developing advanced materials, such as catalysts, membranes, and electrode materials, to improve the performance and longevity of hydrogen-based systems. This involves exploring new materials, nanostructured materials, and composite materials with enhanced properties (Allendorf, 2015; Filippov & Yaroslavtsev, 2021).

Research is being conducted to analyze the policy frameworks, economics, and market dynamics of hydrogen. This includes evaluating the potential for cost reduction (Sun et al., 2019; Talebian et al., 2020), assessing the environmental and socio-economic impacts (Cader et al., 2021; Delima & Vezirolu, 2001), and identifying the barriers and opportunities for hydrogen deployment (Melaina et al., 2012; Trencher & Edianto, 2021). Additionally, studies are examining the role of government policies, regulations, and incentives in promoting hydrogen adoption (Glushchenko, 2020; Jahangiri et al., 2021).

The research work conducted in Bouramdane's thesis (Bouramdane, 2021) (SMRY (Bouramdane, 2022b, 2023a) proposes different scenarios for the largescale integration of onshore wind with photovoltaic (PV) and concentrated solar power (CSP), without or with various capacities of battery and thermal storage. The work evaluates different assumptions of renewable penetration, the impact of different storage configurations, and the effect of investment, operation, and maintenance costs on this integration. It analyzes how these energy sources can complement each other spatially and temporally in the current and future climate characterized by global warming. The thesis also identifies geographically favorable areas for each energy source. Such an analysis could help make an informed decision in the context of green hydrogen production in Morocco.

The field is rapidly evolving, and researchers worldwide are continuously contributing to advancing our understanding of hydrogen technologies and their potential applications.

1.3. Knowledge Gap

While significant research has been conducted on hydrogen transport technologies, there are still several knowledge gaps that warrant further investigation.

While there are studies discussing the general steps involved in hydrogen transport (Section 1.2), there is a need for more comprehensive and detailed research that explores the entire supply chain from production to utilization. This includes examining aspects such as hydrogen production methods, storage technologies, distribution networks, and end-use applications, with a focus on efficiency, safety, and environmental impacts.

Although there are studies that highlight the advantages and disadvantages of specific hydrogen transport technologies, there is a lack of comprehensive comparative analyses that cover a wide range of technologies. Furthermore, there is a need for more research focusing on the process equations and modeling of different transport technologies, which can provide insights into their efficiency, energy requirements, and environmental impacts.

While MCDM-AHP has been utilized in various decision-making contexts, there is limited research specifically applying AHP to assess hydrogen transport technologies. Further studies are needed to explore the application of AHP in evaluating different technologies based on multiple criteria, such as cost, efficiency, environmental impact, safety, and scalability. This research can contribute to more informed decision-making and the prioritization of hydrogen transport options.

While there is recognition of the barriers and opportunities associated with hydrogen transport technologies in the context of the 21st-century energy transition, further investigation is needed to understand these factors in greater detail. Research should focus on identifying and addressing barriers such as infrastructure development, cost competitiveness, regulatory frameworks, public acceptance, and technological advancements. Additionally, exploring the potential opportunities, such as integration with renewable energy sources, decarbonization of sectors, and economic growth, can provide valuable insights into overcoming barriers and maximizing the benefits of hydrogen transport technologies.

Addressing these knowledge gaps will contribute to a more comprehensive understanding of hydrogen transport technologies, enabling informed decision-making and fostering their effective integration into the evolving energy landscape.

1.4. Research Questions

This chapter seeks to address the existing knowledge gap (Section 1.3) by examining the following research questions:

1. What are the general steps involved in hydrogen transport, from production to utilization?
2. What are the advantages, disadvantages, and process equations associated with various hydrogen transport technologies?
3. How can the AHP be employed to assess these technologies based on multiple criteria?
4. What are the barriers and opportunities associated with hydrogen transport technologies in the context of the 21st century energy transition, and how can these barriers be overcome while taking advantage of the opportunities?

1.5. Methodology

The methodology involves a comprehensive literature review to gather information on the general steps, advantages, disadvantages, and process equations of hydrogen transport technologies, as well as the barriers and opportunities of hydrogen transport technologies in the context of the 21st century energy transition.

The AHP is employed as a MCDM approach to establish relative weights for the evaluation criteria (i.e., energy efficiency, energy density, safety, infrastructure requirements, cost, range, and flexibility, environmental impact, technological maturity, scalability, and compatibility with existing infrastructures) and rank the technologies (i.e., CHG, Liquid Hydrogen (LH_2), hydrogen pipelines, hydrogen tube trailers, ammonia (NH_2) carriers, chemical hydrides, hydrail (hydrogen rail), hydrogen blending, Liquid Organic Hydrogen Carriers (LOHC), and hydrogen fueling stations) accordingly.

1.6. Practical Implications

The practical implications of this chapter are significant for stakeholders involved in the energy transition. It provides insights into the complexities of hydrogen transport and facilitates informed decision-making for policymakers and industry professionals. The examination of the advantages, disadvantages, and process equations of hydrogen transport technologies assists in technology selection and deployment. The utilization of the AHP enables systematic evaluation based on multiple criteria. Identifying barriers and opportunities helps stakeholders overcome challenges and leverage benefits. Overall, this chapter informs policy formulation, guides technology selection, and promotes supportive frameworks for the successful integration of hydrogen transport technologies in the energy transition.

This research chapter on hydrogen transport technologies provides a comprehensive analysis of various methods, considering factors such as energy efficiency, safety, cost, and environmental impact. In the context of hydrogen transport between Morocco and other countries (e.g., Mauritania, southern Africa, and Egypt, as depicted in Fig. 7), the practical implications of this research are significant, given the potential for hydrogen to play a crucial role in sustainable energy solutions and international collaboration.

In fact, Morocco has developed a National Hydrogen Strategy to guide its vision for 2050. The strategy involves the use of hydrogen as a feedstock in the local production of green ammonia, the export of green hydrogen and its derivatives, and the use of hydrogen as an electricity storage vector and as fuel in the transport sector. In the short term (2020–2030), the focus is on the local production of green ammonia and the export of hydrogen and its derivatives. In the middle term (2030–2040), the strategy envisions the reduction in the cost of producing hydrogen and hydrogen-based fuels, and the use of hydrogen in various sectors. In the long term (2040–2050), the strategy sees an expansion of the hydrogen trade at a global level, as well as its use in industry, residential heating, and transport in the country (Green Hydrogen Organisation (GH2), n.d.; IEA, 2023; M. de la Transition Énergétique et du Développement Durable, 2023). The roadmap is powered by wind and solar energy, and it sets out plans to develop Morocco's hydrogen production and to create an export market. The country aims to create an economic and industrial sector around green molecules – particularly hydrogen, ammonia, and methanol – to contribute to reducing greenhouse gas emissions. An investment of €300 million has already been pledged, and the country is expected to become one of the most advanced African countries in the green hydrogen industry (Climate Policy Database, 2021; Green Hydrogen Organisation (GH2), n.d.). Morocco is increasing its investment in research and development for hydrogen production through electrolysis, water splitting, and solar thermolysis. The country aims to be a global leader in the transition to green hydrogen and is set to produce approximately 10,000 tonnes of hydrogen per year starting from 2025, for both domestic and export markets in Morocco and Europe (Mission Innovation, n.d.).

Solar potential around the world varies across regions, with distinct levels of sunlight availability (Fig. 8). High solar potential regions are characterized by

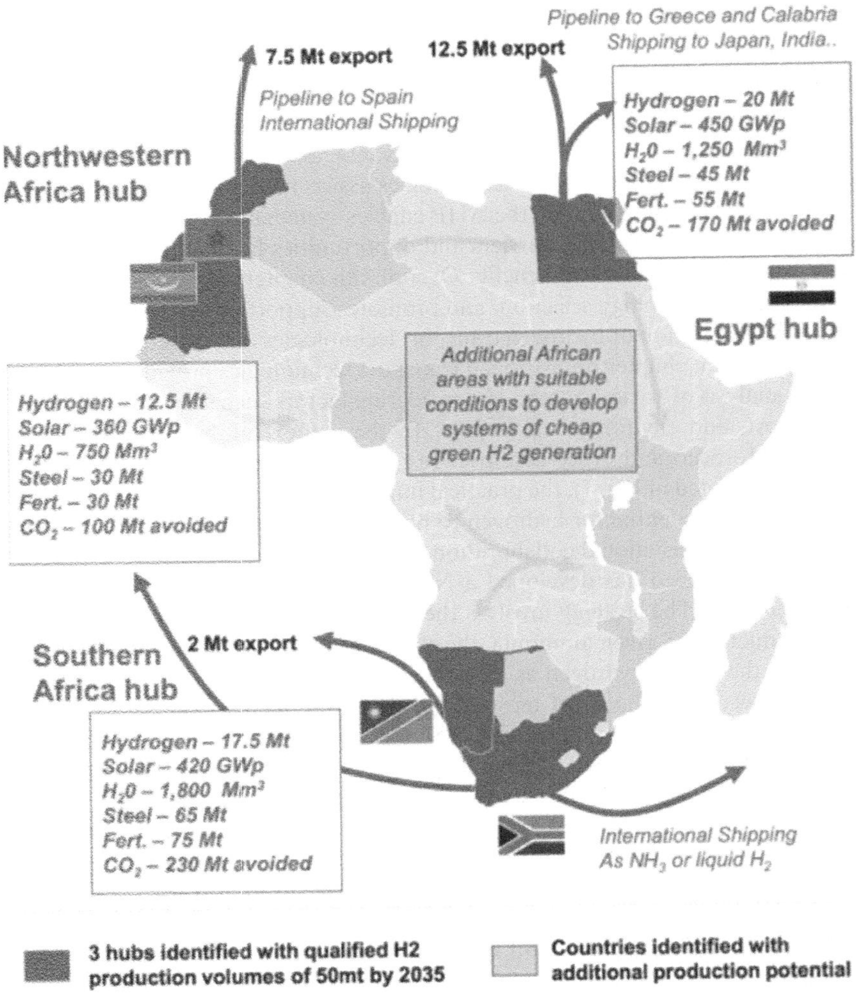

Fig. 7. The Benefits of Using Solar Energy to Produce Green Hydrogen Across Four African Hubs: Mauritania, Morocco, Southern Africa, and Egypt. *Source*: Green Car Congress (2022).

abundant sunlight throughout the year, often situated near the equator or in arid environments. The Sahara Desert in Africa, parts of the Middle East, and specific areas in Australia and South America exemplify such high-potential zones. These regions offer an advantageous environment for large-scale solar energy generation. Moving to medium solar potential regions, these areas receive a moderate amount of sunlight, albeit less than their high-potential counterparts. Typically surrounding the latitudes of high-potential zones, they include regions in Europe, parts of North America, and certain areas in Asia. Assessing the economic viability of solar projects becomes crucial in such regions, considering installation

Fig. 8. Solar Potential Worldwide Varies by Region, with High-Potential Areas, Like the Sahara Desert, the Middle East, Australia, and South America, Offering Abundant Sunlight for Large-Scale Solar Energy Generation. Medium-Potential Regions, Including Europe, Parts of North America, and Specific Areas in Asia, Receive a Moderate Amount of Sunlight. Conversely, Low-Potential Regions, such as Northern Canada, Scandinavia, and Certain Parts of Russia, Receive Comparatively Less Sunlight. Feasibility in These Areas Relies on Optimizing Technology Efficiency for Less Solar-Intensive Environments. The Analysis Highlights the Potential for High-Solar Regions to Export Surplus Solar Energy for Hydrogen Production, Enabling the Synthesis of Valuable Energy Carriers Like Ammonia and Kerosene, Thus Fostering a More Balanced and Efficient Global Energy Landscape. *Source*: Global Solar Atlas (n.d.).

and maintenance costs in relation to energy output. Conversely, low solar potential regions receive comparatively lower levels of sunlight. These areas are often found in higher latitudes, polar regions, and places with frequent cloud cover. Examples encompass the northern parts of Canada, Scandinavia, and certain regions in Russia. In these areas, the feasibility of solar energy projects relies on employing technologies that maximize efficiency in less solar-intensive environments (Bouramdane, 2021, Chapitre 1; 2023d). An interesting aspect of this solar potential analysis is the potential for regions with high solar potential to export solar-generated energy to regions with lower potential. This holds significance in the context of hydrogen production and subsequent synthesis into ammonia and kerosene. High-potential regions can leverage their surplus solar energy to produce hydrogen, which can then be transported and used in regions with lower solar potential. This interconnection facilitates the creation of valuable energy carriers, contributing to a more balanced and efficient global energy landscape.

Wind potential across the world exhibits diverse characteristics across regions, with varying levels of wind energy availability (Fig. 9). High wind potential regions are distinguished by consistently strong and reliable winds, commonly found in coastal areas, mountain passes, and open plains. Notable examples include the coastal areas of Northern Europe, the Great Plains in the United States, and specific parts of South America, where these conditions offer an ideal

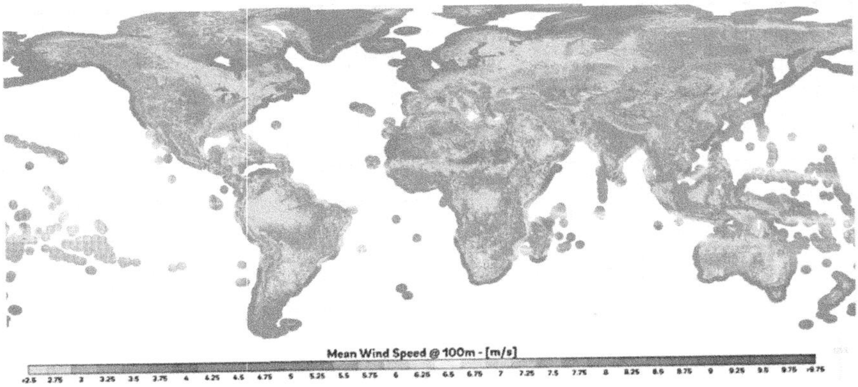

Fig. 9. Global Wind Potential Varies Across Regions, with High-Potential Areas, Such as the Coastal Regions of Northern Europe and the Great Plains in the United States, Characterized by Strong and Reliable Winds Ideal for Large-Scale Wind Energy Projects. Medium-Potential Regions, Found Inland in North America, Europe, and Parts of Asia, Offer Moderate Wind Resources That Can Be Economically Viable with Careful Planning. Conversely, Low-Potential Regions, Including Densely Forested Areas and Urban Environments, Pose Challenges for Wind Energy Generation. An Intriguing Aspect Arises as High-Wind Regions Can Export Surplus Wind-Generated Energy to Low-Wind Regions, Holding Significance for Hydrogen, Ammonia, And Kerosene Production. Regions with High Wind Potential Can Leverage Their Resources to Produce Hydrogen, Which Can Be Transported and Used in Areas with Lower Wind Potential, Contributing to a More Balanced and Efficient Global Energy Landscape. *Source*: Global Wind Atlas (n.d.).

environment for the development of large-scale wind energy projects. Medium wind potential regions, on the other hand, possess moderate wind resources that are suitable for wind energy generation. These areas typically experience moderately windy conditions and are often situated away from the most favorable wind zones. Examples encompass various inland regions in North America, Europe, and parts of Asia, where wind energy projects can still be economically viable but may require careful planning and optimization. Conversely, low wind potential regions present challenges for wind energy generation due to limited wind resources, making the economic feasibility of such projects more challenging. These areas are characterized by consistently low wind speeds or obstructed terrain, including densely forested regions, urban environments, and certain parts of deserts, where alternative energy sources may need consideration (Bouramdane, 2021, Chapitre 1). An intriguing aspect emerges where regions endowed with high wind potential can play a pivotal role in exporting surplus wind-generated energy to areas characterized by lower wind potential. This exploration holds particular significance in the context of hydrogen production and subsequent synthesis into valuable energy carriers like ammonia and kerosene. High-potential wind regions have the capacity to leverage their abundant wind resources to produce hydrogen through electrolysis or other green methods (Bouramdane, 2023e, 2023f). The produced hydrogen can then be efficiently transported and used in regions with lower wind potential. This interconnection not only facilitates the creation of valuable energy carriers but also contributes to establishing a more balanced and efficient global energy landscape.

The Levelized Cost of Hydrogen (LCOH) can vary significantly depending on several factors, including technology, local resource availability, infrastructure, and government policies. Some regions may have a higher LCOH due to the use of more expensive or less efficient hydrogen production technologies. For instance, certain areas heavily reliant on fossil fuel-based hydrogen production methods, such as steam methane reforming (SMR) without carbon capture (Bouramdane, 2023e), could experience higher costs. Regions with limited access to low-cost renewable energy sources (Figs. 8 and 9) may also experience higher LCOH (Fig. 10). Hydrogen production through electrolysis using renewable energy tends to be more cost-effective in areas with abundant and cheap renewable power. Regions with a mix of renewable and non-renewable energy sources may experience medium-range LCOH (Fig. 10). The cost of hydrogen production in these areas can be influenced by the balance conventional and renewable energy use. The availability and development of hydrogen infrastructure, such as pipelines and storage facilities, can impact LCOH. Regions with moderate infrastructure development may fall into the medium-cost category. Areas with abundant and affordable renewable energy resources, such as solar and wind power (Figs. 8 and 9), are likely to have lower

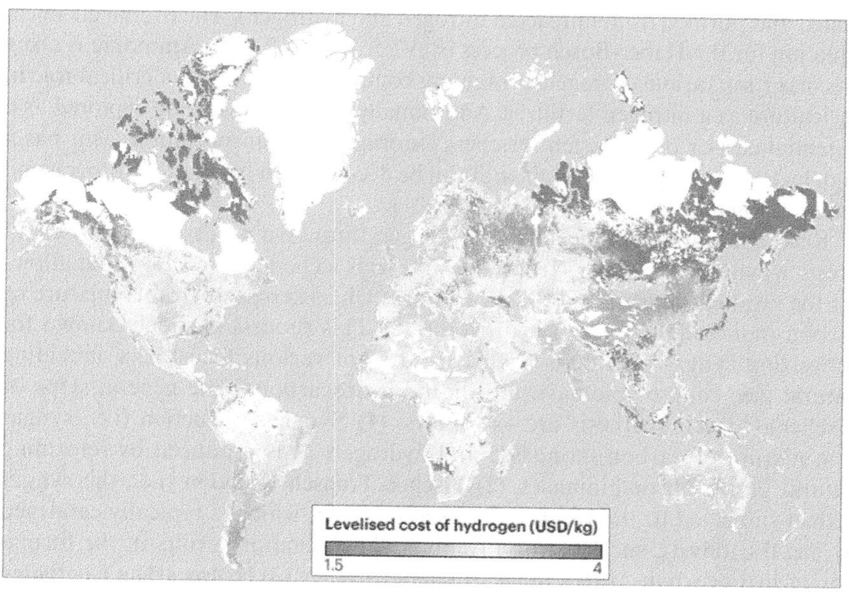

Fig. 10. The LCOH, Representing the Average Cost of Hydrogen Production Over Its Lifetime Inclusive of Capital, Operating, and Fuel Costs, Is Computed Globally with a Generation Mix Comprising 25% Solar Photovoltaic (PV), 25% Wind, and 25% Electrolysis. *Source*: The International Energy Agency (IEA) (n.d.).

LCOH (Fig. 10). Electrolysis of water using renewable energy sources can be a cost-effective method for hydrogen production in these regions. Regions with strong government support, favorable policies, and financial incentives for hydrogen production can experience lower LCOH (Fig. 10). Supportive policies may include subsidies, tax incentives, and regulations promoting the use of green hydrogen. It is essential to note that advancements in technology, economies of scale, and evolving market dynamics can influence the LCOH distribution over time. Additionally, the global push towards decarbonization and the growing interest in green hydrogen could impact the cost landscape.

Ammonia (NH_3) can be produced from hydrogen. The most common method for ammonia production is the Haber-Bosch process, which involves the reaction of nitrogen (N_2) and hydrogen (H_2) under high pressure and temperature. This process is crucial for the large-scale production of ammonia, which is widely used in fertilizers, as well as in various industrial applications. The Haber–Bosch process steps are as follows: **(1) Nitrogen Fixation** (i.e., nitrogen is obtained from the air through a process called nitrogen fixation. Typically, air is compressed and passed over a catalyst to convert a smart portion of the nitrogen into ammonia); **(2) Hydrogen Production** (i.e., hydrogen is produced by processes such as SMR, partial oxidation (POX), or electrolysis of water); **(3) Haber-Bosch Reaction** (i.e., nitrogen and hydrogen are combined in a reactor under high pressure, typically around 200–300 atmospheres, and elevated temperature, around 400–500°C, using an iron or iron-based catalyst. This results in the formation of ammonia); **(4) Ammonia Separation and Purification** (i.e., the produced ammonia is then separated and purified from unreacted nitrogen and hydrogen). The overall chemical equation for the Haber-Bosch process is: $N2 + 3H2 \rightarrow 2NH3$. Ammonia is a key precursor for various nitrogen-containing compounds and plays a critical role in agriculture as a nitrogen fertilizer. Additionally, ammonia is being explored as a potential carrier of hydrogen for energy storage and transportation, as it has a high hydrogen content by weight and can be decomposed back into hydrogen and nitrogen when needed (Ghavam et al., 2021).

Kerosene can be produced from hydrogen through a process called Fischer–Tropsch synthesis. Fischer–Tropsch synthesis is a chemical reaction that allows for the production of liquid hydrocarbons, such as kerosene, from a mixture of carbon monoxide (CO) and hydrogen (H_2). This process is widely known for converting syngas (synthesis gas) derived from various feedstocks, including natural gas, coal, or biomass, into liquid hydrocarbons. The general steps of Fischer–Tropsch synthesis are as follows: **(1) Syngas Production** (i.e., syngas is a mixture of carbon monoxide and hydrogen. It is produced by reforming natural gas, coal, or biomass); **(2) Fischer–Tropsch Reaction** (i.e., the syngas is then subjected to the Fischer–Tropsch reaction, which is typically catalyzed by metal catalysts, such as iron or cobalt. This reaction results in the formation of hydrocarbons with a range of chain lengths); **(3) Hydrocarbon Upgrading** (i.e., the hydrocarbons produced in the Fischer–Tropsch reaction are often in the form of a waxy substance. These hydrocarbons need to be further processed to obtain liquid fuels like kerosene. This may involve hydrocracking and refining steps). This process has been historically used in the production of synthetic fuels, including kerosene, particularly during times when conventional petroleum sources were scarce or expensive. It is worth noting that the Fischer–Tropsch process requires a source of hydrogen, and the hydrogen itself can be produced through various methods, including electrolysis or reforming of natural gas. The overall environmental impact and sustainability of the process depend on the source of hydrogen used in conjunction with Fischer–Tropsch synthesis (Medrano-García et al., 2022; Meurer & Kern, 2021).

1.7. Smart Cities Integration in the Gulf Region

The concept of smart cities aligns seamlessly with the findings and implications of our research on hydrogen transport technologies. In the Gulf Region, where urbanization and technological advancements are rapidly shaping city landscapes, the adoption of smart city initiatives becomes imperative for sustainable development. The efficiency and sustainability criteria assessed in our study directly correlate with the core principles of smart cities, emphasizing the need for intelligent, data-driven solutions in urban planning. The deployment of hydrogen transport technologies, identified and assessed in our research, can play a pivotal role in creating smarter, greener cities in the Gulf Region. These technologies, ranging from CHG to LOHC, provide the foundation for sustainable and clean energy systems essential for smart urban environments.

The integration of the MCDM-AHP methodology into our assessment aligns with the data-centric approach of smart cities. This methodology ensures that decisions related to adopting specific hydrogen transport technologies are based on a comprehensive evaluation of key criteria, promoting informed and strategic choices in urban planning. The smart city vision in the Gulf Region can leverage our research to optimize energy efficiency, enhance safety, and reduce environmental impact, fostering a technologically advanced and sustainable urban ecosystem.

Furthermore, considering the specific context of Morocco's green interconnections explored in our study, the insights gained can be directly applied to shape smart city initiatives in the Gulf Region. The practical implications of our research extend beyond the theoretical realm, offering tangible guidance for integrating hydrogen, ammonia, and kerosene production into the green energy landscape of Gulf cities. In conclusion, our research not only aligns with the principles of smart cities but also provides actionable insights for implementing sustainable and intelligent solutions in the urban development of the Gulf Region.

1.8. Outline

This chapter sets out to explore the realm of hydrogen transport technologies, recognizing their pivotal role in the 21st century energy transition. The introduction (Section 1) provides a comprehensive overview of the research's motivation (Section 1.1), existing studies (Section 1.2), identified knowledge gaps (Section 1.3), research questions (Section 1.4), methodology (Section 1.5), and the practical implications (Sections 1.6 and 1.7) of the investigation.

The methodology section (Section 2) provides a detailed overview of the research approach employed, including data collection methods, analytical techniques, and the application of the AHP for assessing hydrogen transport technologies.

Delving into the fundamental aspects, Section 3.1 outlines the key steps involved in hydrogen transport, commencing from production to eventual utilization. The chapter covers different hydrogen production methods, storage options, distribution, and diverse applications in different industries.

In Section 3.2, the chapter presents a thorough evaluation of various hydrogen transport technologies, detailing the advantages, disadvantages, and process equations associated with each method.

To comprehensively assess the hydrogen transport technologies based on multiple criteria, Section 3.3 introduces the AHP as a robust evaluation framework. It expounds on how AHP can effectively weigh the pros and cons of each technology, offering a systematic approach to decision-making (Section 2.1).

Recognizing the significance of hydrogen transport in the context of the energy transition, Section 3.4 addresses the barriers that impede the widespread adoption of these technologies. Simultaneously, it highlights the potential opportunities these technologies offer to expedite decarbonization and enhance energy sustainability. Solutions to overcome the barriers while capitalizing on the opportunities are also discussed.

Section 4 explores the implications of optimizing hydrogen transport technologies in the Gulf Region's smart cities, providing valuable insights for diverse stakeholders and guiding strategic decisions in urban planning, infrastructure development, and policy-making.

The chapter concludes by summarizing the key findings from the investigation and outlines the implications for policymakers, industries, and stakeholders (Section 5).

2. Methodology

To answer the research questions of this study (Section 1.4), we first conduct a comprehensive review of existing literature and research studies on hydrogen transport, focusing on the general steps involved in the process from production to utilization. This includes understanding the different stages, such as hydrogen production, storage, transportation, and end-use applications (Section 3.1).

Second, we gather data on various hydrogen transport technologies, including CHG, Liquid Hydrogen (LH_2), hydrogen pipelines, hydrogen tube trailers, ammonia (NH_2) carriers, chemical hydrides, hydrail (hydrogen rail), hydrogen blending, LOHC, and hydrogen fueling stations. We analyze the advantages, disadvantages, and process equations associated with each technology based on available literature, industry reports, and expert opinions (Section 3.2).

Third, we apply the AHP as a MCDM method (Section 2.1) to assess the different hydrogen transport technologies. We define a set of evaluation criteria, such as energy efficiency, energy density, safety, infrastructure requirements, cost, range, and flexibility, environmental impact, technological maturity, scalability, and compatibility with existing infrastructure. Assign relative weights to each criterion based on expert judgments or stakeholder opinions. Then, we use AHP to compare and rank the technologies based on these criteria, taking into account the importance of each criterion (Section 3.3).

Fourth, we analyze the barriers and opportunities associated with hydrogen transport technologies in the context of the 21st century energy transition. We identify the key challenges and obstacles, such as infrastructure development, cost and economics, safety concerns, technological maturity and efficiency, limited range and infrastructure coverage, public acceptance and awareness, policy and regulation, durability and degradation, and availability of critical materials. We explore potential strategies and solutions to overcome these barriers and leverage the opportunities for the widespread adoption of hydrogen transport technologies (Section 3.4).

2.1. MCDM-AHP

The MCDM framework is a systematic approach used to evaluate and compare multiple criteria or factors in decision-making processes (French, 1996).

MCDM types include techniques such as:

- **AHP**: A widely used MCDM method that structures decision problems hierarchically and derives priority weights through pairwise comparisons (Saaty, 1990).
- **Technique for Order of Preference by Similarity to Ideal Solution (TOPSIS)**: Ranks alternatives based on their distances to ideal and anti-ideal solutions, providing a compromise solution (Hwang & Yoon, 1981).
- **Preference Ranking Organization Method for Enrichment Evaluations (PROMETHEE)**: Evaluates alternatives by pairwise comparisons and generates rankings considering multiple criteria (Brans, 2006).
- **Weighted Sum Model (WSM)**: Aggregates criteria values using weighted sums to rank alternatives based on their overall scores (Zeleny, 2004).
- **ELECTRE (Elimination and Choice Translating Reality)**: Eliminates alternatives based on certain thresholds and assigns rankings using concordance and discordance measures (Roy, 1968).
- **VIKOR (VIseKriterijumska Optimizacija I Kompromisno Resenje)**: Finds the compromise solution by determining the maximum "group utility" and the minimum "individual regret" simultaneously (Opricovic & Tzeng, 2004).
- **Analytic Network Process (ANP)**: Handles dependencies and feedback among criteria and alternatives by constructing a network structure for decision-making (Saaty, 2005).
- **Simple Additive Weighting (SAW)**: Computes overall scores for alternatives by summing the weighted values of each criterion (Yoon & Hwang, 1995).
- **Grey Relational Analysis (GRA)**: Measures the relationships between alternatives and criteria using grey relational coefficients (Ju-long, 1982).

- **Fuzzy Analytic Hierarchy Process (FAHP)**: Incorporates fuzzy sets theory into AHP to handle imprecise and uncertain decision information (Buckley et al., 1999).

MCDM applications encompass a diverse range of fields, such as environmental management (i.e., sustainable resource allocation and environmental impact assessment) (Thanh, 2022), financial decision making (e.g., making investment decisions and portfolio selection) (Biagi et al., 2021; Hashemizadeh & Ju, 2019), healthcare management (e.g., for medical diagnosis, hospital ranking, and healthcare resource allocation) (Bharsakade et al., 2021), project management (e.g., for project selection, scheduling, and risk analysis) (Chen et al., 2018), supplier selection (i.e., to evaluate and rank potential suppliers) (Chan et al., 2008), and risk of hydrogen energy storage and transportation in cities (Sun et al., 2023).

As stated previously, the AHP is a decision-making method that quantifies subjective judgments to evaluate and prioritize alternatives based on multiple criteria (Saaty, 1990).

The AHP has found various applications in the energy sector, particularly in hydrogen-related studies (Xu et al., 2022; Zhang et al., 2022), and renewable energy planning (i.e., PV and CSP (Bouramdane, 2023d, 2023g), onshore wind (Bouramdane, 2023h), offshore wind (Bouramdane, 2023i), and offshore floating PV plants (Bouramdane, 2023j, 2023k).

Moreover, the MCDM-AHP approach finds application in the domain of cybersecurity solutions within smart grid environments, demonstrating its seamless integration with artificial intelligence, as highlighted by Bouramdane's research (Bouramdane, 2023l, 2023m). This integration is exemplified in a recent investigation conducted by Bouramdane (2023n). Furthermore, the same methodology has been employed by the researcher to evaluate water management strategies in smart cities (Bouramdane, 2023o), encompassing applications in water desalination (Bouramdane, 2023p).

In a comprehensive analysis of hydrogen production technologies in Morocco, Bouramdane et al. (2023f) utilized the MCDM-AHP methodology. Their evaluation considered factors such as technological feasibility, economic viability, environmental impact, and social acceptance. The study identified high-performing technologies, such as Autothermal Reforming (ATR) with Carbon Capture and Storage, deemed suitable for hydrogen production in Morocco. Moderate-performing technologies like PV and CSP exhibited promise, while low-performing technologies may encounter challenges in meeting specified criteria. The study emphasizes the significance of stakeholder perspectives, particularly in scenarios involving renewable energy penetration, in shaping the suitability of technologies. For a more comprehensive understanding of hydrogen technologies, readers are encouraged to explore Bouramdane et al.'s extensive body of work (Bouramdane, 2023c, 2023e, 2023q, 2023x, 2023y).

This study aims to fill a knowledge gap as, to the best of the author's knowledge, no previous research has utilized AHP to prioritize hydrogen transport technologies.

2.1.1. AHP Theoretical Framework

The AHP is a decision-making method that involves pairwise comparisons to derive priority weights for criteria and alternatives (Saaty, 1990). The AHP process can be represented mathematically as follows:

1. **Hierarchical Structure**: Break down the decision-making problem into its key elements, such as the objective, criteria, sub-criteria, alternatives, and more. Next, construct a linear hierarchy of these components, keeping the levels limited in number.
2. **Pairwise Comparison Matrix**: Construct the judgment matrix (C) using Saaty's scale, typically ranging from 1 to 9 (Table 1), where the elements represent the decision maker's preferences for one criterion over another. The formation of this matrix follows the equation (1):

$$C = \begin{bmatrix} 1 & c_{12} & c_{13} & \cdots & c_{1n} \\ \dfrac{1}{c_{12}} & 1 & c_{23} & \cdots & c_{2n} \\ \dfrac{1}{c_{13}} & \dfrac{1}{c_{23}} & 1 & \cdots & c_{3n} \\ \vdots & \vdots & \vdots & \ddots & \vdots \\ \dfrac{1}{c_{1n}} & \dfrac{1}{c_{2n}} & \dfrac{1}{c_{3n}} & \cdots & 1 \end{bmatrix} \tag{1}$$

where c_{ij} represents the pairwise comparison of criterion i with criterion j based on the decision maker's preference.

3. **Normalization of the Pairwise Comparison Matrix**: Normalize the matrix (C) by dividing each element by the sum of the corresponding column (2). This step ensures that the column-wise sums of the matrix are equal to 1, representing the relative weights

$$W_{ij} = \frac{c_{ij}}{\sum_{j=1}^{n} c_{ij}} \tag{2}$$

where W_{ij} represents the normalized value of the element in the pairwise comparison matrix at position (i, j) and c_{ij} is the original value of the element.

4. **Calculation of Criteria Weights**: Calculate the criteria weights (3) by taking the column-wise average of the normalized matrix. This provides the relative importance of each criterion in the decision-making process.

$$w_j = \frac{1}{n}\sum_{i=1}^{n} W_{ij} \tag{3}$$

Table 1. The 1–9 Importance Scale (Saaty, 2008).

Rating Scale	Explanation
1	Equally important
3	Moderately important
5	Strongly important
7	Very strongly important
9	Certainly important
2, 4, 6, 8	Intermediate values

where w_j is the weight of criterion j, n is the number of criteria, and W_{ij} represents the element in the normalized matrix corresponding to criterion i and alternative j.

By computing w_j for each criterion, we obtain the relative importance of each criterion in the decision-making process. The larger the weight, the more significant the criterion is in influencing the overall decision.

5. **Calculation of Consistency Index (CI) and Consistency Ratio (CR)** (4) (5) to check the consistency of the judgments.

$$CI = \frac{\lambda_{\max} - n}{(n-1)} \quad (4)$$

$$CR = \frac{CI}{RI} \quad (5)$$

where: CI is the Consistency Index, λ_{max} is the maximum eigenvalue of the matrix, n is the number of criteria, CR is the Consistency Ratio, RI is the Consistency Index based on the matrix size, which can be found in Saaty's random consistency (Table 2).

Table 2. Values of Random Consistency Index (RI) (110).

Matrix Size *n*	1	2	3	4	5	6	7	8	9	10	11
RI	0	0	0.58	0.90	1.12	1.24	1.32	1.41	1.45	1.49	1.51

During this stage, any inconsistencies in the pairwise comparison matrix are identified. In order to ensure the consistency of the pairwise comparisons, the CR should be less than 10%. If the CR exceeds this threshold, it indicates inconsistencies in the matrix, and the element values should be re-evaluated and the matrix updated accordingly (Fig. 11).

3. Results

This section presents a comprehensive analysis of hydrogen transport technologies, encompassing several key aspects. Firstly, it examines the general steps involved

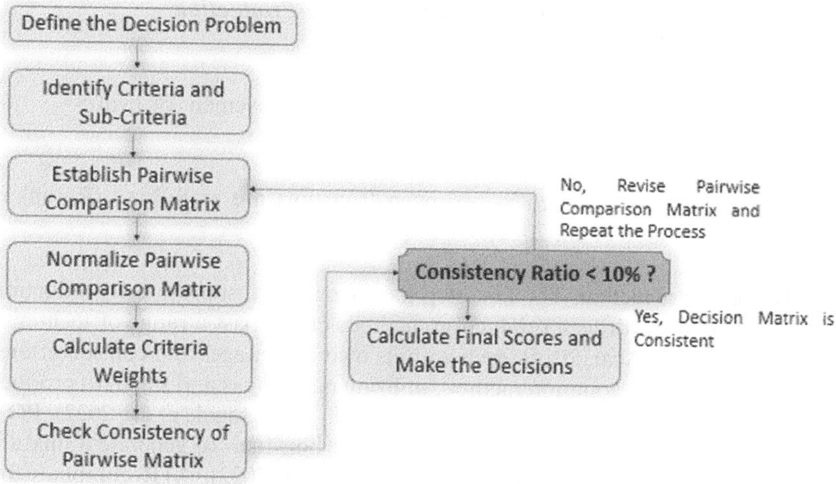

Fig. 11. AHP Algorithm: A Systematic Approach for Evaluating and Prioritizing Hydrogen Transport Technologies Based on Multiple Criteria. *Source*: Own Elaboration.

in hydrogen transport, starting from production to utilization (Section 3.1). This analysis provides insights into the complexities and requirements of implementing hydrogen-based systems, enabling policymakers and industry professionals to better understand the necessary infrastructure and processes involved.

Secondly, Section 3.2 presents an in-depth exploration of the advantages, disadvantages, and process equations associated with various hydrogen transport technologies. By evaluating these factors, policymakers and industry stakeholders can make informed decisions about technology selection and deployment. This assessment takes into account criteria such as efficiency, safety, cost, and environmental impact, allowing for a comprehensive comparison of the different technologies.

Thirdly, Section 3.3 employs the MCDM approach, specifically the AHP, to assess hydrogen transport technologies based on multiple criteria. The key criteria for evaluation are identified (Section 3.3.1), and a pairwise comparison matrix is constructed to quantify the relative importance of each criterion. The criteria weights are determined using the AHP methodology, which provides a systematic and structured approach to decision-making. The weighted sum evaluation method is then applied to analyze the optimal hydrogen transport technology based on the established criteria and their respective weights (Section 3.3.2).

Lastly, Section 3.4 addresses the barriers and opportunities associated with hydrogen transport technologies in the context of the 21st century energy transition. By identifying these challenges and potential benefits, stakeholders can develop strategies to overcome the barriers and take advantage of the opportunities presented by hydrogen transport technologies. This analysis will contribute to the successful integration of hydrogen transport technologies into the energy system, supporting the transition towards a sustainable and decarbonized future.

3.1. General Steps in Hydrogen Transport: From Production to Utilization

> Hydrogen transport technologies involve the movement of hydrogen from its production source to the point of use.

The general steps involved in hydrogen transport are as follows (Ratnakar et al., 2021; Wickham et al., 2022):

1. **Hydrogen Production**: The first step in hydrogen transport is the production of hydrogen. Hydrogen production methods encompass a range of technologies (Dincer & Acar, 2015), each with its own advantages and considerations. SMR is the most common method, utilizing natural gas and steam to produce hydrogen with high efficiency (Antonini et al., 2020; Ingale et al., 2022). POX involves the controlled oxidation of hydrocarbon fuels to generate a mixture of hydrogen and carbon monoxide (Zhu & Dai, 2022). ATR combines the principles of SMR and POX to achieve a balanced hydrogen-to-carbon monoxide ratio (Zhang et al., 2022). Biomass gasification converts organic materials into syngas, which can be further processed to obtain hydrogen (Chang et al., 2011). Solar water splitting utilizes solar energy to split water into hydrogen and oxygen through various techniques, including thermochemical and photoelectrochemical (PEC) water splitting (Ekspong et al., 2021; Jiao et al., 2022). Electrolysis is another significant method involving the use of electricity to split water molecules into hydrogen and oxygen, with different types such as PEM, Alkaline (AE), SOEC, and advanced electrolyzers like high-temperature polymer electrolyte membrane electrolyzers (HT-PEME) and direct current (DC) electrolyzers (Kumar & Lim, 2022; Nami et al., 2022). Each method offers unique features and challenges, making it essential to consider factors such as energy efficiency, cost, scalability, environmental impact, and technological readiness in selecting the most suitable hydrogen production technology for specific applications (Bouramdane, 2023z).
2. **Purification and Compression**: After hydrogen is produced, it needs to be purified to remove impurities such as moisture, carbon dioxide, and other contaminants. Purification is important to ensure the quality and safety of the hydrogen. Once purified, hydrogen is typically compressed to increase its density and facilitate transportation (Orlova et al., 2023; Wickham et al., 2022).
3. **Storage and Transportation**: Hydrogen can be stored and transported in different forms depending on the specific application and distance (Li et al., 2022; Suri & Siddique, 2019). Various storage techniques have been developed, each with its own characteristics and considerations (Tarhan & Çil, 2021; Usman, 2022). Compressed Gas Storage involves compressing hydrogen gas at high pressures and storing it in specially designed containers. This method offers high storage capacity but requires robust infrastructure and poses challenges related to compression and safety (Orlova et al., 2023). Liquid Hydrogen Storage involves cooling hydrogen gas to extremely low temperatures, transforming it into a liquid state for storage. This method

provides high energy density but requires cryogenic systems and insulation to maintain low temperatures (Ulucan et al., 2022; Zheng et al., 2021). Metal Hydride Storage utilizes the ability of certain metals to absorb and release hydrogen, providing a safe and compact storage solution. However, it may have limitations in terms of storage capacity and cycling efficiency (Sakintuna et al., 2007). Chemical Hydride Storage involves chemically bonding hydrogen with specific materials, releasing it through controlled reactions when needed. This method offers high energy density but requires additional processing steps and may have limitations in terms of hydrogen release kinetics (Okada, 2019). Solid-state Storage, such as hydrogen absorbed into porous materials or hydrogen stored in carbon nanotubes, shows promise for high storage capacity and reversible hydrogen release. However, it requires further research and development to overcome technical challenges (Zacharia & Ullah Rather, 2016). Each hydrogen storage method has its advantages, limitations, and suitability for specific applications, highlighting the need for comprehensive evaluations considering factors such as energy density, safety, storage capacity, cost, and system compatibility. Common methods of transportation include CHG, Liquid Hydrogen (LH_2), hydrogen pipelines, hydrogen tube trailers, ammonia (NH_3) carriers, chemical hydrides, hydrail, hydrogen blending, LOHC, and hydrogen fueling stations. Each method has its own advantages and considerations in terms of energy efficiency, safety, infrastructure requirements, and cost (Sections 3.2 and 3.3).

4. **Distribution and Infrastructure**: The infrastructure required for hydrogen transport includes storage facilities, pipelines, tube trailers, and refueling stations. These infrastructure elements need to be strategically located to ensure efficient distribution and accessibility of hydrogen to end-users (Xiao, 2011; Yeates et al., 2021). The development of a robust hydrogen infrastructure is crucial for the widespread adoption of hydrogen as an energy carrier (Section 3.4).
5. **Utilization**: Once hydrogen reaches its destination, it can be utilized in various applications, such as fuel cells for power generation, hydrogen combustion for heat production, or as a feedstock in industrial processes. The utilization of hydrogen contributes to decarbonization efforts by replacing fossil fuels and reducing greenhouse gas emissions (Akinyele et al., 2020; Yufit et al., 2013) (Sections 1.1 and 1.2).
6. **Safety and Regulations**: Throughout the entire hydrogen transport process, safety is a paramount concern (Sections 3.2, 3.3, and 3.4). Handling and transporting hydrogen require strict safety measures to prevent accidents or leaks. Regulatory frameworks and standards are in place to ensure the safe production, storage, and utilization of hydrogen.
7. **Monitoring and Maintenance**: Continuous monitoring and maintenance of the hydrogen transport infrastructure are necessary to ensure optimal performance, safety, and reliability. Regular inspections, leak detection systems, and preventive maintenance activities help identify and address any potential issues in a timely manner (Guy et al., 2021; Jung et al., 2019).

The general steps involved in hydrogen transport form a comprehensive process that encompasses production, purification, storage, transportation, distribution, utilization, safety, and maintenance. Each step plays a critical role in the efficient and sustainable transport of hydrogen for various energy applications.

3.2. Hydrogen Transport Technologies: Advantages, Disadvantages, and Process Equations

In this section, we will explore the advantages and disadvantages of various hydrogen transport technologies, including CHG, Liquid Hydrogen (LH_2), hydrogen pipelines, hydrogen tube trailers, ammonia (NH_2) carriers, chemical hydrides, hydrail (hydrogen rail), hydrogen blending, LOHC, and hydrogen fueling stations, and examine the equations involved in the process of each technology.

There are several hydrogen transport options available today (Li et al., 2022; Suri & Siddique, 2019). Each of these options has its own advantages and challenges, and their suitability depends on factors such as the distance of transportation, infrastructure availability, safety considerations, and local regulations. The choice of hydrogen transport technology often depends on the specific use case and regional requirements (Tayarani & Ramji, 2022; Xiangmin, 2008).

Here are some of the main ones.

3.2.1. CHG

Hydrogen can be compressed and transported in gaseous form using high-pressure cylinders or tanks. This method requires strong containers and proper safety measures. CHG has both advantages and disadvantages (Suri & Siddique, 2019).

CHG offers several advantages as a hydrogen transport technology, including the simplicity of the infrastructure required (i.e., CHG is a relatively straightforward technology that involves compressing hydrogen gas into storage tanks or cylinders. The infrastructure required for CHG is simpler compared to other hydrogen transport methods), and the technology is mature (i.e., compressed gas cylinders and tanks are widely used and familiar in various industries. The technology for manufacturing, transporting, and handling compressed gas is well-developed and mature), flexible distribution (i.e., CHG allows for flexible distribution of hydrogen to various destinations, including industrial facilities, refueling stations, and other end-users. It can be transported by trucks, trailers, or other means over short to medium distances), and it has a high energy density (i.e., CHG can achieve a high energy density, enabling the storage of a significant amount of hydrogen in a relatively small volume. This makes it suitable for applications that require portable or space-constrained storage), making it a promising solution for efficient and convenient hydrogen-powered transportation.

However, it is important to consider the disadvantages of CHG as a hydrogen transport technology, which include its low energy density compared to liquid hydrogen (i.e., while CHG offers a higher energy density than hydrogen gas at ambient conditions, it is still lower than that of liquid hydrogen. This means that larger storage tanks or cylinders are required to transport the same amount of hydrogen compared to liquid hydrogen), the high compression energy requirement (i.e., compressing hydrogen gas to high pressures, such as 350 bar or 700 bar, requires significant energy input. The compression process can be energy-intensive and may reduce the overall energy efficiency of the hydrogen transportation system), the safety considerations (i.e., handling and storing compressed gas can pose safety risks. The high pressures involved require robust containers and stringent safety protocols to prevent leaks, ruptures, or accidents. Safety measures and regulations must be followed throughout the entire CHG transport and storage process), the limited range (i.e., CHG is suitable for short to medium distances due to its lower energy density and the need for frequent refilling or replacement of cylinders. For long-distance transport, alternative methods like pipelines or liquid hydrogen may be more efficient), and the infrastructure requirements (i.e., CHG transportation may require a dedicated infrastructure for high-pressure storage, compression, and distribution. Building and maintaining this infrastructure can be costly and may require significant investment), which can hinder widespread adoption and accessibility.

The properties of compressed hydrogen can be described using the Ideal Gas Law, which relates pressure, volume, temperature, and the number of moles of gas. The equation for the Ideal Gas Law is (6) (Tenny & Cooper, 2020):

$$PV = nRT \tag{6}$$

where: P is the pressure of the gas (in units of force per unit area, such as pascals or bar), V is the volume of the gas (in units of cubic meters or liters), n is the number of moles of gas, R is the ideal gas constant (8.314 J/mol.K or 0.0831 bar.L/mol.K), T is the temperature of the gas (in units of Kelvin).

This equation describes the relationship between these variables when the gas behaves ideally, meaning there are no intermolecular forces or volume occupied by the gas molecules themselves.

The compression ratio (CR) is the ratio of the final compressed gas pressure to the initial gas pressure (Ashkezari, 2022). It is calculated as (7):

$$CR = \frac{P_2}{P_1} \tag{7}$$

where $P2$ is the final compressed gas pressure and $P1$ is the initial gas pressure.

Boyle's Law describes the relationship between the pressure and volume of a gas at a constant temperature (Tenny & Cooper, 2020). It can be expressed as (8):

$$P_1 * V_1 = P_2 * V_2 \tag{8}$$

where $P1$ and $V1$ are the initial pressure and volume, and $P2$ and $V2$ are the final pressure and volume.

For an adiabatic compression/expansion process (no heat exchange with the surroundings), the relationship between pressure and volume can be described using the adiabatic equation (9):

$$P_1 * V_1 * \gamma = P_2 * V_2 * \gamma \tag{9}$$

where $P1$ and $V1$ are the initial pressure and volume, and $P2$ and $V2$ are the final pressure and volume, and γ is the specific heat ratio (ratio of specific heat capacities) of the gas.

Compression work (CW) refers to the energy input required to compress the hydrogen gas to the desired pressure. It can be calculated as the product of the pressure difference and the change in volume (10):

$$CW = P_2 * (V_2 - V_1) \tag{10}$$

where $P2$ is the final compressed gas pressure, $V2$ is the final volume, and $V1$ is the initial volume.

These equations and concepts are relevant to the understanding, analysis, and design of the compression process in CHG systems. They help in calculating the relationship between pressure, volume, temperature, work, and efficiency involved in compressing hydrogen gas for transportation.

3.2.2. Liquid Hydrogen (LH_2)

Hydrogen can be cooled to extremely low temperatures (−253°C) to convert it into a liquid state. Liquid hydrogen offers a higher energy density compared to compressed gas, but it requires specialized cryogenic containers and insulation (Aziz, 2021; Ustolin et al., 2022).

LH_2 has its own set of advantages and disadvantages.

The advantages include the high energy density (i.e., LH_2 has a significantly higher energy density compared to CHG. This means that a larger amount of hydrogen can be stored and transported in a given volume, making LH_2 advantageous for applications that require higher energy storage capacity), the longer-range capability (i.e., the higher energy density of LH_2 enables longer-range transport compared to CHG. LH_2 allows for more efficient long-distance transportation, reducing the need for frequent refilling or replacement of storage containers), the reduced compression energy (i.e., LH_2 eliminates the need for high-pressure compression equipment used in CHG systems, reducing the energy required for compression. This can improve the overall energy efficiency of the hydrogen transportation process), the versatility (i.e., LH_2 can be transported using various means, including specialized cryogenic tankers, insulated containers, or dewars. This versatility allows for different transportation options based on the specific requirements and infrastructure available), the rapid refueling (i.e., LH_2 refueling is relatively quick, similar to refueling with conventional liquid fuels. This can help reduce downtime for vehicles and improve operational efficiency), and the co-production with other industries (i.e., LH_2 can be co-produced as a byproduct of industrial processes such as ammonia production or natural gas liquefaction, reducing the overall cost and environmental impact of hydrogen production).

However, it is important to consider the disadvantages of LH_2 as a hydrogen transport technology, such as the fact that specialized equipment and infrastructure are needed for cryogenic storage and handling (i.e., handling and storing liquid hydrogen require specialized equipment and infrastructure due to its extremely low boiling point of −253°C. Maintaining the necessary cryogenic temperatures during transport and storage can be technically challenging and costly), boil-off and venting losses (i.e., due to the nature of LH_2, there is a constant loss of hydrogen through boil-off and venting. This loss can occur during transport and storage, reducing the overall efficiency of the LH_2 system and potentially impacting the economics and environmental sustainability of the technology), safety considerations (i.e., LH_2 presents safety challenges due to its extremely low temperature and potential for rapid phase change. Specialized safety measures and protocols, including proper insulation, venting systems, and leak detection, are required to ensure safe handling and transportation), infrastructure requirements (i.e., establishing LH_2 transportation infrastructure requires significant investment and specialized facilities for handling, storage, and distribution. Cryogenic storage tanks, filling stations, and appropriate handling equipment need to be implemented, which can be costly and limit the availability of LH_2 transport infrastructure), and high cost (i.e., the production, storage, and transportation of LH_2 involve complex and energy-intensive processes, which can contribute to higher costs compared to other hydrogen storage and transport technologies).

It is important to assess these advantages and disadvantages in the context of specific applications, distances, and requirements to determine the most suitable hydrogen transport technology. The choice between LH_2 and CHG depends on factors such as energy density requirements, range, infrastructure availability, and economic considerations.

The properties of LH_2 can be described using various thermodynamic equations and properties (Aziz, 2021; Eckroll, 2017).

The Clausius–Clapeyron equation relates the vapor pressure of a substance to its temperature and enthalpy of vaporization (Brown, 1951; Knözinger, 2020). For liquid hydrogen, the Clausius–Clapeyron equation can be used to calculate the vapor pressure as a function of temperature (11):

$$\ln\left(\frac{P_2}{P_1}\right) = \frac{-\Delta \mathrm{H}_{\mathrm{vap}}}{\mathrm{R}} * \left(\frac{1}{T_2} - \frac{1}{T_1}\right) \tag{11}$$

where: P_1 and P_2 are the initial and final vapor pressures, T_1 and T_2 are the initial and final temperatures (in Kelvin), ΔH_{vap} is the enthalpy of vaporization, R is the ideal gas constant.

The enthalpy of vaporization (ΔH_{vap}) represents the energy required to convert liquid hydrogen into hydrogen gas at a given temperature. It can be experimentally determined or estimated using empirical relationships.

During the transport of liquid hydrogen, heat transfer equations are used to calculate the amount of heat transfer between the liquid hydrogen and its surroundings. These equations consider factors such as surface area, temperature gradients, and thermal conductivity.

Thermodynamic properties, such as specific heat capacity, density, and viscosity, are essential for modeling and understanding the behavior of LH_2 during transport. These properties are typically obtained from experimental data or calculated using equations specific to hydrogen.

It is important to note that the equations and models used for LH_2 transport can be complex and depend on various factors, including the specific equipment, storage vessels, and environmental conditions. The equations provided here are simplified representations to give an overview of the relevant principles involved in liquid hydrogen transport. Detailed analysis and design of LH_2 transport systems require specialized knowledge and considerations (Aziz, 2021; Eckroll, 2017).

3.2.3. Hydrogen Pipelines

Similar to natural gas pipelines, hydrogen pipelines can transport large volumes of hydrogen over long distances. Dedicated hydrogen pipelines are already in use in some regions, especially where hydrogen production and consumption are concentrated (Tayarani & Ramji, 2022; Yeates et al., 2021).

Hydrogen pipelines offer several advantages as a hydrogen transport technology, such as large-scale and high-volume transport (i.e., hydrogen pipelines have the potential to transport large volumes of hydrogen over long distances. This makes them suitable for applications that require a continuous and substantial supply of hydrogen, such as industrial processes or large-scale power generation), efficient and continuous supply (i.e., once established, hydrogen pipelines can provide a continuous supply of hydrogen, eliminating the need for frequent refilling or replacement of storage containers. This can improve the operational efficiency of hydrogen-dependent industries and applications), reduced energy losses (i.e., compared to other transport methods, such as trucking or shipping, pipelines typically have lower energy losses during transportation. This is because pipelines operate at a constant pressure, minimizing the need for compression or decompression along the route), lower environmental impact (i.e., hydrogen pipelines can contribute to a lower environmental impact compared to other transport methods that rely on fossil fuels. They enable the transportation of clean hydrogen, which can help reduce greenhouse gas emissions and contribute to a more sustainable energy system), and integration with existing infrastructure (i.e., in regions with existing natural gas pipelines networks, repurposing or retrofitting these pipelines for hydrogen transport can be a cost-effective and efficient solution. This allows for the integration of hydrogen transport into the existing energy infrastructure, minimizing the need for extensive new infrastructure development), facilitating cost-effective deployment and utilization.

However, it is important to consider the disadvantages of hydrogen pipelines as a hydrogen transport technology, which include the limited infrastructure availability (i.e., currently, hydrogen pipeline infrastructure is relatively limited and localized. Expanding the pipeline network to cover larger geographical areas can be costly and time-consuming, requiring significant investment and coordination), the potential hydrogen embrittlement (i.e., hydrogen can cause embrittlement in certain materials, which can be a concern for pipeline integrity and safety. The selection of suitable materials and the implementation of appropriate safety measures are critical to mitigate the risk of hydrogen embrittlement), the limited flexibility (i.e., once a hydrogen pipeline is installed, it may have limited flexibility in terms of routing and adaptability to changing demand patterns. This can be a disadvantage in regions or industries where demand for hydrogen fluctuates or requires dynamic supply routes), the high upfront capital investment (i.e., building a hydrogen pipeline infrastructure involves significant upfront capital investment. The cost associated with pipeline construction, maintenance, and monitoring can be substantial, particularly when establishing new pipeline networks), and the need for safety considerations (i.e., the safe operation of hydrogen pipelines is of utmost importance. Strict safety regulations and monitoring systems are necessary to prevent leaks, manage hydrogen concentrations, and mitigate potential hazards associated with hydrogen transport).

It is important to evaluate these advantages and disadvantages in the context of specific applications, geographical locations, and infrastructure availability. The suitability of hydrogen pipelines as a transport technology depends on factors such as distance, volume requirements, safety considerations, and the availability of existing pipeline networks.

The process of hydrogen transport through pipelines involves various considerations and equations (156).

The Darcy–Weisbach equation is commonly used to calculate the pressure drop or head loss in a pipeline due to friction (Brown, 2002, 2007). It can be expressed as (12):

$$\Delta P = \frac{f * \left(\frac{L}{D}\right) * (\rho * V^2)}{2} \tag{12}$$

where: ΔP is the pressure drop, f is the friction factor, L is the length of the pipeline, D is the diameter of the pipeline, ρ is the density of the hydrogen gas, and V is the velocity of the hydrogen gas in the pipeline.

The Hazen–Williams equation is another method to estimate the pressure drop in a pipeline, particularly for water or liquids (Liou, 1998; Maschinenbau, 2012). Although it may not be directly applicable to hydrogen, it can provide an initial estimate in the absence of specific hydrogen-related data. The equation is (13):

$$\Delta P = 10.67 * C * \left(\frac{Q}{C}\right)^{1.852} * (D^{4.87}) * (L^{0.54}) \tag{13}$$

where: ΔP is the pressure drop, C is the Hazen–Williams coefficient, Q is the flow rate of hydrogen, D is the diameter of the pipeline, and L is the length of the pipeline.

The compressibility factor accounts for the change in gas density due to variations in pressure and temperature (Wang et al., 2023). It is often represented by the symbol *Z*. Accurate determination of the compressibility factor is crucial for calculating the flow properties of hydrogen in pipelines.

The Ideal Gas Law can be applied to calculate the properties of hydrogen gas within the pipeline. It relates pressure, volume, temperature, and the number of moles of gas (Tenny & Cooper, 2020). The equation is (14):

$$PV = nRT \tag{14}$$

where: *P* is the pressure of hydrogen gas in the pipeline, *V* is the volume of the hydrogen gas, *n* is the number of moles of hydrogen gas, *R* is the ideal gas constant, *T* is the temperature of the hydrogen gas.

These equations, among others, are used to analyze the flow behavior, pressure drop, and other characteristics of hydrogen gas within the pipeline during the transport processes. They help in designing, optimizing, and evaluating the performance of hydrogen pipeline systems.

3.2.4. Hydrogen Tube Trailers

Hydrogen tube trailers refer to trailers or trucks that carry CHG or liquid hydrogen. They are used for short to medium-distance transportation, making them suitable for regional distribution (Bonner, 2018; Suri & Siddique, 2019).

Hydrogen tube trailers offer distinct advantages as a hydrogen transport technology, including their high capacity (i.e., hydrogen tube trailers have the capacity to transport a significant amount of hydrogen in a single shipment. They can carry a large volume of CHG, allowing for efficient transport and delivery to the desired location), their flexibility and mobility (i.e., tube trailers offer flexibility in terms of transportation routes and destinations. They can be easily transported on highways and roads, enabling hydrogen delivery to various locations, including remote areas or sites without dedicated hydrogen infrastructure), their versatility (i.e., hydrogen tube trailers can be used for both short-distance and long-distance transport. They provide a flexible solution for delivering hydrogen to different end-users, including industrial facilities, hydrogen refueling stations, or research and development centers), their reliability (i.e., tube trailers are designed to meet safety standards and regulations for hydrogen transport. They are built with robust construction and safety features, such as pressure relief devices and monitoring systems, ensuring reliable and secure transportation of hydrogen), and their rapid refilling (i.e., tube trailers can be quickly refilled with hydrogen gas, minimizing downtime, and ensuring prompt delivery. This allows for efficient logistical operations and reduces delays in supplying hydrogen to customers).

However, it is important to consider the disadvantages of hydrogen tube trailers as a hydrogen transport technology, which include the limited storage capacity (i.e., compared to other hydrogen transport methods, such as liquid hydrogen or pipelines, tube trailers have a relatively lower storage capacity. This means that a larger number of tube trailer shipments may be required to transport the same volume of hydrogen), the infrastructure requirements (i.e., hydrogen tube trailers require dedicated loading and unloading facilities, which can be costly to establish and maintain. Additionally, the availability of tube trailer refilling stations may be limited, requiring careful logistical planning), the energy requirements (i.e., compressing hydrogen gas to high pressures for storage in tube trailers requires energy, which adds to the overall energy consumption of the hydrogen transport process. This energy requirement should be taken into account when assessing the overall efficiency and environmental impact of the technology), the transport limitations (i.e., tube trailers are subject to transportation limitations, such as weight restrictions and regulations on hazardous materials. These limitations may impact the distance, speed, or routes for hydrogen transport, potentially affecting the accessibility and efficiency of the delivery process), and the safety considerations (i.e., the safe handling, loading, and unloading of hydrogen from tube trailers require specialized training and equipment. Proper safety protocols, such as leak detection systems and emergency response plans, must be in place to minimize the risks associated with hydrogen transportation).

It is important to evaluate these advantages and disadvantages in the context of specific applications, distances, and requirements to determine the most suitable hydrogen transport technology. The choice between hydrogen tube trailers and other options depends on factors such as transport distance, storage capacity requirements, infrastructure availability, safety considerations, and economic viability.

The process of hydrogen transport using tube trailers involves considerations such as pressure, volume, and energy (Bonner, 2018; Suri & Siddique, 2019). While there are no specific equations unique to tube trailers, there are some general equations that are relevant to the overall process of hydrogen transport.

The ideal gas law (6) can be used to calculate the relationship between pressure, volume, temperature, and the number of moles of hydrogen gas (Tenny & Cooper, 2020).

The process of compressing hydrogen gas for storage in tube trailers requires energy. The energy required E for compression can be calculated using the equation (15):

$$E = P * V * \frac{(P_2 - P_1)}{\eta} \tag{15}$$

where: P is the operating pressure of the hydrogen gas, V is the volume of the hydrogen gas, η is the efficiency of the hydrogen gas, P_2 is the final pressure after compression, P_1 is the initial pressure before the compression process.

This equation helps estimate the energy required to compress hydrogen gas into tube trailers for transport.

During the transport of hydrogen gas in tube trailers, there may be pressure drops due to factors such as friction. The Darcy–Weisbach equation or other pressure drop equations can be used to estimate the pressure drop in the pipeline or tubes (16) (Brown, 2002, 2007):

$$\Delta P = \frac{f * \left(\frac{L}{D}\right) * (\rho * V^2)}{2} \tag{16}$$

where: ΔP is the pressure drop, f is the friction factor, L is the length of the tube trailer, D is the diameter of the tube, ρ is the density of the hydrogen gas, and V is the velocity of the hydrogen gas.

It is important to note that these equations are not specific to tube trailers but are relevant to the overall process of hydrogen transport. The specific design and operation of tube trailers may involve additional considerations and equations depending on factors such as materials, safety standards, and specific operating conditions.

3.2.5. Ammonia (NH_3) Carriers

Hydrogen can be combined with nitrogen to produce ammonia. Ammonia has a higher energy density than liquid hydrogen and can be transported using existing infrastructure, such as ships and pipelines. At the destination, ammonia can be converted back to hydrogen (Ishimoto et al., 2020; Tullo, 2022).

NH_3 carriers offer significant advantages as a hydrogen transport technology, including their high hydrogen density (i.e., ammonia has a high hydrogen density, which means a large amount of hydrogen can be stored and transported in a relatively smaller volume compared to other hydrogen carriers. This can lead to more efficient transport and reduced infrastructure requirements), their established infrastructure (i.e., Ammonia is a well-known and widely used chemical, and there is already an existing infrastructure for its production, storage, and transportation. This can be leveraged for the transport of hydrogen, reducing the need for extensive new infrastructure development), their existing trade and handling expertise (i.e., Ammonia is already traded and handled globally as a commodity chemical, and there is a well-established industry with expertise in its production, storage, and transport. This can provide a smoother transition and facilitate the integration of ammonia carriers into existing logistics and supply chain networks), their international trade opportunities (i.e., Ammonia carriers can enable the international trade of hydrogen, as ammonia can be easily shipped and handled globally. This can open up new markets for hydrogen and support the development of a global hydrogen economy), and the existing ammonia utilization (i.e., ammonia has existing industrial uses, such as fertilizers and chemical production. By using ammonia as a hydrogen carrier, it allows for the potential for dual-use applications, where ammonia can be further processed or used for other purposes, providing additional value and economic benefits) (Inal et al., 2022).

However, it is important to consider the disadvantages of NH_3 carriers as a hydrogen transport technology, including the challenges related to ammonia production and conversion (i.e., ammonia production typically involves the use of fossil fuels, which can result in greenhouse gas emissions. However, efforts are underway to develop renewable and low-carbon ammonia production methods. Additionally, converting ammonia back into hydrogen requires an additional step, which adds complexity and energy requirements to the process), safety concerns (i.e., ammonia is toxic and can be hazardous to human health and the environment. Special precautions and safety measures are required during handling, storage, and transport to ensure the safe management of ammonia carriers), ammonia infrastructure adaptation (i.e., while existing infrastructure for ammonia production, storage, and transport is available, it may require adaptation and modifications to meet the specific requirements for hydrogen transport. This can involve additional costs and logistical challenges) (Zawacki 2018, Economic Analysis of NH_3 and NH_3-H2 Supply Chains for Hydrogen Delivery), energy efficiency and losses (i.e., the conversion of hydrogen into ammonia and back into hydrogen involves energy losses. The overall energy efficiency of the process needs to be considered to assess the feasibility and viability of using ammonia carriers for hydrogen transport), system complexity (i.e., the use of ammonia carriers adds complexity to the hydrogen transport system, requiring additional processes and infrastructure for the conversion of hydrogen to ammonia, storage, and conversion back to hydrogen. This complexity can impact the overall efficiency and cost-effectiveness of the technology), which requires proper mitigation measures and environmental considerations.

It is important to note that the advantages and disadvantages may vary depending on specific applications, geographical location, and the availability of infrastructure and expertise. The suitability of ammonia carriers as a hydrogen transport technology depends on various factors, including the desired scale of transport, safety considerations, environmental impact, and the overall energy system context (Li et al., 2023).

The process of using ammonia (NH_2) carriers as a hydrogen transport technology involves several steps, including the production of ammonia, its storage, transport, and conversion back to hydrogen (Ishimoto et al., 2020; Tullo, 2022).

Here are some equations that are relevant to these processes:

- Ammonia Synthesis: The production of ammonia involves the Haber-Bosch process, which is represented by the following equation: $N_2 + 3H_2 \rightleftharpoons 2NH_3$. This equation represents the reaction of nitrogen gas (N_2) and hydrogen gas (H_2) to form ammonia (NH_3) under high pressure and temperature conditions. The equilibrium constant of the reaction is influenced by temperature and pressure.
- Hydrogen Storage in Ammonia: Ammonia can be used as a carrier for hydrogen storage. The equation for the storage of hydrogen in ammonia can be written as: $NH_3 + H_2 \rightleftharpoons NH_3H$. In this equation, ammonia (NH_3) reacts with hydrogen gas (H_2) to form the ammonium compound (NH_3H), which acts as a storage medium for hydrogen.
- Hydrogen Release from Ammonia: To release hydrogen from ammonia, a process called ammonia cracking or decomposition is employed. The equation for this process is: $NH_3H \rightleftharpoons NH_3 + H_2$. The ammonium compound NH_3H decomposes into ammonia (NH_3) and hydrogen gas (H_2), releasing the stored hydrogen.
- Ammonia-to-Hydrogen Conversion Efficiency: The efficiency of converting ammonia back into hydrogen can be calculated using the following equation (17):

$$\eta = \frac{H_{2released}}{H_{2stored}} * 100 \tag{17}$$

This equation measures the efficiency η of the ammonia-to-hydrogen conversion process and indicates the amount of hydrogen successfully recovered from the stored ammonia.

It is important to note that these equations represent the fundamental reactions and processes involved in the use of ammonia carriers for hydrogen transport. The actual implementation and optimization of these processes may involve additional considerations, such as reaction kinetics, temperature, pressure, catalysts, and energy requirements. Additionally, the equations do not account for the energy losses or practical considerations associated with the overall system design, such as heat management, separation techniques, and safety measures (Ishimoto et al., 2020; Tullo, 2022).

3.2.6. Chemical Hydrides

Hydrogen can be chemically bound to certain solid materials, such as metal hydrides. These materials can store and release hydrogen when needed. Chemical hydrides offer a higher energy density, but the release and regeneration of hydrogen require additional processes (Biniwale et al., 2008; Okada, 2019).

Chemical hydrides present notable advantages as a hydrogen transport technology. These include their high hydrogen density (i.e., chemical hydrides have a high hydrogen storage capacity, which means a significant amount of hydrogen can be stored in a relatively small volume or mass. This allows for efficient storage and transport of hydrogen), safe and stable (i.e., chemical hydrides are often more stable and safer to handle compared to other hydrogen storage methods, such as compressed gas or liquid hydrogen. They have lower flammability and reduced risk of explosion, making them potentially suitable for a wider range of transportation and storage applications), long storage lifetimes (i.e., chemical hydrides can have long storage lifetimes, allowing for extended periods of hydrogen storage without significant degradation or loss. This makes them suitable for applications where long-term storage is required, such as remote locations or intermittent energy supply), high energy density (i.e., chemical hydrides offer high energy density, meaning they can store a significant amount of energy per unit of weight or volume. This is advantageous for applications where space and wright are critical factors, such as portable or mobile hydrogen storage systems), and compatibility with existing infrastructure (i.e., chemical hydrides can be designed to be compatible with existing infrastructure, such as pipelines, tanks, or transportation networks. This allows for easier integration into the existing hydrogen infrastructure and reduces the need for extensive infrastructure modifications).

However, it is important to acknowledge the disadvantages of chemical hydrides as a hydrogen transport technology, which include the challenges associated with the hydrogen release kinetics (i.e., the release of hydrogen from chemical hydrides can be relatively slow, requiring specific conditions, such as elevated temperature or the use of catalysts, to accelerate the reaction. This limits the rate at which hydrogen can be delivered or used) (Mori 2018, Hydrogen Release and Uptake of Complex Hydrides for Hydrogen Storage), the system complexity (i.e., the use of chemical hydrides involves additional system complexity, such as the need for reactors, heat management systems, and recycling or regeneration processes. This complexity can increase the overall cost and energy requirements of the hydrogen storage and transport system), the limited reversibility (i.e., some chemical hydrides have limited reversibility, meaning that once the hydrogen is released, the hydride material needs to be regenerated or replaced before it can be used for hydrogen storage again. This adds to the overall cost and logistics of the system), the high cost and challenging scalability (i.e., the cost of chemical hydrides can be relatively high compared to other hydrogen storage methods. Additionally, scaling up the production and availability of chemical hydrides to meet large-scale hydrogen transport requirements can be challenging and may require significant investment), and the environmental considerations (i.e., the production, recycling, and disposal of chemical hydrides involve the use of specific chemicals or materials that have environmental implications. The sustainability and environmental impact of the entire lifecycle of chemical hydrides need to be carefully considered).

It is important to note that the advantages and disadvantages of chemical hydrides may vary depending on the specific material used and the intended application. The suitability of chemical hydrides as a hydrogen transport technology depends on factors such as the required storage capacity, desired release kinetics, safety considerations, cost constraints, and the availability of infrastructure and expertise.

The process of using chemical hydrides as a hydrogen transport technology involves the storage, release, and regeneration of hydrogen through chemical reactions (Biniwale et al., 2008; Okada, 2019). Here are some general equations that are relevant to these processes:

- Hydrogen Storage in Chemical Hydrides: The storage of hydrogen in chemical hydrides involves the absorption of hydrogen by the hydride material. The reaction can be represented as: $H_2 + MH \rightarrow M + H_2$, where H_2 is the hydrogen gas, MH is the chemical hydride compound, M is the metal or element in the hydride material, and $M + H_2$ represents the stored hydrogen.
- Hydrogen Release from Chemical Hydrides: The release of hydrogen from chemical hydrides typically requires the application of heat, pressure, or a combination of both. The reaction can be represented as: $M + H_2 \rightarrow MH +$ Heat, where $M + H_2$ represents the stored hydrogen, MH is the chemical hydride compound, and Heat represents the energy input required for hydrogen release.
- Regeneration of Chemical Hydrides: After the hydrogen is released from the chemical hydride, the material needs to be regenerated or restored for subsequent use. The regeneration reaction can be represented as: $M + Heat \rightarrow MH$, where M is the metal or element in the chemical hydride, Heat represents the energy input required for regeneration, and MH is the regenerated hydride material.

It is important to note that the specific equations and reactions involved in chemical hydrides as hydrogen storage materials can vary depending on the type of hydride used. Different chemical compounds and materials have different reaction mechanisms, temperature and pressure requirements, and regeneration processes. Additionally, the kinetics and efficiency of these reactions can also vary based on the specific material and conditions.

Furthermore, it is worth mentioning that the equations provided above are simplified representations of the reactions involved and may not capture all the intricacies and details of the actual processes. The actual implementation and optimization of chemical hydrides as a hydrogen transport technology may involve additional considerations, such as the use of catalysts, reaction kinetics, and temperature and pressure control, among others.

3.2.7. Hydrail (Hydrogen Rail)

Hydrail, also known as hydrogen rail, is a form of hydrogen transport technology that involves using hydrogen fuel cells to power trains, eliminating the need for electrified infrastructure. Hydrogen can be transported to refueling stations near rail lines, enabling long-distance and clean rail transportation (Ehrhart et al., 2021; Westbroek, 2021).

Hydrail offers numerous advantages as a hydrogen transport technology, including zero emissions (i.e., hydrail offers the advantage of being a zero-emission technology. Hydrogen fuel cells produce electricity through the electrochemical reaction of hydrogen and oxygen, generating only water vapor as a byproduct. This eliminates harmful emissions such as greenhouse gases and pollutants, contributing to improved air quality and reduced environmental impact), high energy efficiency (i.e., hydrogen fuel cells are highly efficient in converting hydrogen fuel into electricity. Compared to conventional internal combustion engines, fuel cells can achieve higher energy conversion efficiencies, resulting in more effective use of the fuel and reduced energy waste), long-range capabilities (i.e., hydrail systems can provide long-range capabilities, making them suitable for regional and intercity rail transport. Hydrogen fuel cells offer a consistent power output, enabling trains to maintain high speeds and performance over extended distances without the need for frequent refueling), quiet and smooth operation (i.e., hydrail systems operate quietly and provide smooth acceleration and deceleration, enhancing passenger comfort and reducing noise pollution. This makes them particularly suitable for urban areas and regions with strict noise regulations), and the ability to store excess energy through regenerative braking (i.e., hydrogen fuel cells can efficiently store excess energy produced during regenerative braking, where the kinetic energy of the train is converted back into electricity. This stored energy can be reused to power the train during acceleration, improving overall energy efficiency and reducing energy consumption).

Despite its advantages, hydrail faces certain challenges as a hydrogen transport technology, including the need for extensive infrastructure development (i.e., the implementation of hydrail systems requires the development of dedicated hydrogen refueling infrastructure along rail lines. This involves installing hydrogen storage and dispensing facilities, which can be costly and time-consuming, particularly in areas with limited existing hydrogen infrastructure), hydrogen production and distribution complexities (i.e., the production and distribution of hydrogen for hydrail systems involves specific challenges. Hydrogen can be produced through various methods, such as electrolysis, SMR, or biomass gasification, each with its own energy requirements, costs, and environmental considerations. Establishing a reliable and sustainable hydrogen supply chain for hydrail applications can be complex), storage and safety considerations (i.e., storing and handling hydrogen onboard trains and at refueling stations require adherence to strict safety protocols. Hydrogen is highly flammable and has specific requirements for safe storage and transportation. Implementing adequate safety measures and ensuring compliance with regulations is essential), higher initial costs compared to traditional rail technologies (i.e., the initial investment and operational costs of hydrail systems can be higher compared to traditional diesel or electric rail technologies. The cost of hydrogen fuel cell technology, infrastructure development, and maintenance can pose financial challenges, although costs are expected to decrease with technological advancements and economies of scale), and limited availability and standardization of hydrail systems (i.e., the availability and standardization of hydrail systems are currently limited. The technology is still in the early stages of development and adoption, which can present challenges for procurement, maintenance, and interoperability of hydrail trains and associated infrastructure).

It is important to note that the advantages and disadvantages of hydrail can vary depending on specific contexts, such as geographical location, energy sources, policy support, and market conditions. The viability and suitability of hydrail as a hydrogen transport technology depend on careful consideration of these factors and the specific requirements of the rail transport sector (Ehrhart et al., 2021; Westbroek, 2021).

The operation of hydrail, or hydrogen rail, involves the use of hydrogen fuel cells to generate electricity to power trains (Ehrhart et al., 2021; Westbroek, 2021). Here are some equations relevant to the process of hydrail as a hydrogen transport technology:

- Hydrogen Fuel Cell Operation: The hydrogen fuel cell is the key component of hydrail technology. It involves the following electrochemical reactions:

  - Anode Reaction: $2H_2 \rightarrow 4H^+ + 4e^-$
  - Cathode Reaction: $O_2 + 4H^+ + 4e^- \rightarrow 2H_2O$
  - Overall Cell Reaction: $2H_2 + O_2 \rightarrow 2H_2O$

In these reactions, hydrogen gas (H_2) is supplied to the anode, where it is oxidized to release protons (H_2) and electrons (e-). The protons travel through an electrolyte to the cathode, while the electrons flow through an external circuit, creating an electrical current. At the cathode, oxygen gas (O_2) combines with the protons and electrons to form water (H_2O). This overall process generates electricity to power the train.

- Hydrogen Production: The production of hydrogen for hydrail systems can involve various methods, such as electrolysis or SMR. The specific equations depend on the chosen production method. For example, the equation for water electrolysis is: $2H_2O \rightarrow 2H_2 + O_2$. This equation represents the splitting of water molecules (H_2O) into hydrogen gas (H_2) and oxygen gas (O_2) using electricity.

- Hydrogen Compression and Storage: The compression and storage of hydrogen for hydrail systems typically involve high-pressure tanks. The process does not involve specific chemical reactions but focuses on maintaining hydrogen gas at high pressures, often around 350–700 bar, to increase storage density and range.

It is important to note that the equations provided above represent simplified version of the reactions involved in hydrail technology and may not capture all the intricacies and details of the actual processes. The implementation of hydrail systems may involve additional considerations, such as thermal management, system efficiency, and control strategies. Additionally, the equations related to hydrogen production methods, such as electrolysis or SMR, vary depending on the specific process, catalysts, and conditions used. The equations provided here serve as general representations of the reactions involved but may not reflect the specific details of each method. It is also worth mentioning that hydrail technology primarily focuses on the use of hydrogen fuel cells to generate electricity for the trains. The equations provided above pertain to the operation of hydrogen fuel cells, which enable the conversion of hydrogen into electrical energy.

3.2.8. Hydrogen Blending

Hydrogen can be blended with natural gas in small percentages (up to 20%) and transported through exiting natural gas pipelines. This approach allows for the gradual introduction of hydrogen into existing infrastructure without major modifications (Cavana et al., 2021; Erdener et al., 2022; Mahajan et al., 2022; Zhou et al., 2022).

Hydrogen blending offers several advantages, including carbon emission reduction (i.e., hydrogen blending can contribute to reducing carbon emissions when hydrogen, a clean-burning fuel, is substituted for natural gas. This process helps to lower greenhouse gas emissions and mitigate climate change), the utilization of existing infrastructure (i.e., one of the key advantages of hydrogen blending is the utilization of the existing natural gas infrastructure. This allows for a relatively seamless integration of hydrogen without the need for major infrastructure modifications or the establishment of dedicated hydrogen transport systems), flexibility and scalability for gradual integration (i.e., hydrogen blending offers flexibility and scalability as it allows for incremental increases in the hydrogen content over time. This gradual approach helps to manage the transition to a hydrogen-based energy system and enables adjustments based on infrastructure and market readiness), and the promotion of fuel diversity by incorporating hydrogen into the natural gas supply (i.e., hydrogen blending promotes fuel diversity by incorporating hydrogen into the natural gas supply. This diversification can enhance energy security, reduce dependence on fossil fuels, and provide a pathway for the utilization of renewable hydrogen sources).

Despite its advantages, hydrogen blending faces limitations such as the limited hydrogen content that can be blended (i.e., the maximum hydrogen content that can be blended into natural gas is typically limited to a certain percentage, often around 10-20% due to safety and operational considerations. This limitation may not achieve the full benefits of hydrogen as a clean energy source), the need for compatibility and equipment modifications (i.e., blending hydrogen into natural gas may require equipment modifications to ensure compatibility. Certain components of the gas distribution infrastructure, such as pipelines and appliances, may need to be upgraded or replaced to handle higher hydrogen concentrations, adding potential costs and logistical challenges), potential safety considerations (i.e., hydrogen has different safety characteristics compared to natural gas, such as its flammability range and leak detection methods. Adequate safety measures and protocols must be in place to ensure the safe blending and distribution of hydrogen in the existing infrastructure), as well as the associated costs and potential challenges in achieving higher concentrations of hydrogen in the natural gas distribution system (i.e., the costs associated with hydrogen blending include the production, purification, and distribution of hydrogen, as well as any necessary infrastructure modifications. These costs may vary depending on factors such as the hydrogen source, scale of blending, and existing infrastructure conditions).

It is important to note that the advantages and disadvantages of hydrogen blending can be influenced by factors such as the specific blending ratio, infrastructure conditions, regulatory frameworks, and the availability of hydrogen from renewable sources. Additionally, hydrogen blending is often considered as an interim solution to transition towards higher levels of pure hydrogen use in dedicated hydrogen transport systems (Mahajan et al., 2022; Zhou et al., 2022).

The process of hydrogen blending in the natural gas distribution system does not involve specific chemical equations. Instead, it primarily involves the mixing of hydrogen gas (H_2) with natural gas (mainly methane, CH_2) in predetermined proportions (Cavana et al., 2021; Erdener et al., 2022). The blending process can be represented mathematically using the following equation (18):

$$H_{2blend} = \frac{V_{H_2}}{(V_{H_{2tot}} + V_{NG_{tot}})} * 100 \qquad (18)$$

where H_{2blend} is the hydrogen in blend, VH_2 is the volume of hydrogen, $VH_{2\,tot}$ is the total volume of hydrogen, and VNG_{tot} is the total volume of natural gas.

In this equation, the percentage of hydrogen in the blend is calculated by dividing the volume of hydrogen by the total volume of hydrogen and natural gas combined, then multiplying by 100 to express it as a percentage.

For example, if 10 cubic meters of hydrogen is blended with 90 cubic meters of natural gas, the percentage of hydrogen in the blend would be 10%.

This equation is a simplified representation, and in practice, blending hydrogen into the natural gas distribution system requires careful consideration of factors such as safety, compatibility, and infrastructure modifications. The specific blending ratios and procedures may vary depending on the intended applications and the regulations in place. Additionally, other equations related to the costs, energy content, and combustion characteristics of the blended gas may also be relevant in assessing the feasibility and performance of hydrogen blending as a hydrogen transport technology (Cavana et al., 2021; Erdener et al., 2022).

3.2.9. LOHC

Hydrogen can be chemically bonded to a liquid carrier, such as an organic compound. This allows for easy and safe transportation of hydrogen in liquid form, which can be later released and used as needed (Niermann et al., 2021; Southall & Lukashuk, 2022).

LOHC offer several advantages, including high hydrogen density for efficient storage and transportation (i.e., LOHC systems enable high hydrogen density, allowing for efficient storage and transportation of hydrogen. The liquid organic compounds can carry a significant amount of hydrogen, enhancing the energy storage capacity), safety and stability with non-toxic and non-flammable characteristics (i.e., LOHCs are typically non-toxic, non-flammable, and chemically stable. This makes them safer to handle and transport compared to gaseous or cryogenic hydrogen. They can be stored and transported under ambient conditions without the need for high-pressure or low-temperature infrastructure), reversible hydrogen release for on-demand usage (i.e., LOHC systems enable reversible hydrogen storage and release. Hydrogen can be chemically bound to the organic compound under specific conditions, and later released when needed for various applications, such as fuel cells or industrial processes), and compatibility with existing infrastructure, facilitating seamless integration into current transportation and distribution networks (i.e., LOHC technology can use existing infrastructure, such as pipelines and tankers, that are already in place for conventional liquid fuels. This compatibility facilitates the integration of LOHC systems into existing transportation and distribution networks).

Despite its advantages, LOHC face challenges such as energy-intensive processes for hydrogenation and dehydrogenation (i.e., the process of hydrogenation and dehydrogenation in LOHC systems requires energy input. Energy is needed to release and rebind hydrogen from the organic carrier, which can reduce the overall energy efficiency of the system), system complexity due to multiple steps (i.e., LOHC systems involve multiple steps, including hydrogenation, dehydrogenation, and catalyst regeneration. These steps require additional infrastructure, equipment, and control systems, adding complexity and potential operational challenges), chemical regeneration requirement (i.e., the process of regenerating the liquid organic carrier after hydrogen release involve additional chemical reactions and purification steps. These processes may require additional resources and can introduce complexities in terms of cost, waste management, and environmental considerations), and limited commercialization compared to more established hydrogen transport technologies (i.e., while LOHC technology shows promise, it is still at a relatively early stage of development and commercialization. The availability of commercially viable LOHC systems and the associated infrastructure may be limited compared to more established hydrogen transport technologies).

It is worth noting that the advantages and disadvantages of LOHC as a hydrogen transport technology can vary depending on the specific liquid organic compound used, the system design, and the intended application. Further research and technological advancements are needed to optimize LOHC systems for widespread adoption and overcome some of the current limitations.

The process of LOHC as a hydrogen transport technology involves several chemical reactions for the hydrogenation and dehydrogenation of the liquid organic compounds (Niermann et al., 2021; Southall & Lukashuk, 2022). The equations below represent the general process:

- Hydrogenation Reaction: $LOHC + H_2 \rightleftharpoons LOHC - H_2$. In this equation, the LOHC reacts with hydrogen gas (H_2) under specific conditions and with the presence of a catalyst to form a hydrogenated LOHC compound (LOHC-H_2). This reaction involves the binding of hydrogen to the organic compound.
- Dehydrogenation Reaction: $LOHC - H_2 \rightleftharpoons LOHC + H_2$ The dehydrogenation reaction is the reverse of the hydrogenation reaction. The hydrogenated LOHC compound (LOHC-H_2) releases hydrogen gas (H_2) when the conditions and catalyst are appropriately adjusted. This reaction involves the breaking of the chemical bond between hydrogen and the organic compound, releasing hydrogen for various applications.

It is important to note that the specific organic compounds used as LOHCs may vary, and the equations involved can be more complex and depend on the specific chemical properties and reaction mechanisms of the chosen LOHC. Additionally, the hydrogenation and dehydrogenation reactions may require the use of specific catalysts and operating conditions to optimize the efficiency and reversibility of the process.

3.2.10. Hydrogen Fueling Stations

Hydrogen can be produced onsite at hydrogen fueling stations using various methods such as electrolysis or reforming processes. This eliminates the need for long-distance transport and enables local distribution to hydrogen-powered vehicles (Sun et al., 2019; Xiangmin, 2008).

Hydrogen fueling stations serve as a crucial infrastructure for hydrogen transport technology, offering the advantages of fast refueling times (i.e., hydrogen fueling stations offer fast refueling times, similar to traditional gasoline stations, enabling quick turnaround times for hydrogen-powered vehicles), extended driving range (i.e., hydrogen fuel cell vehicles have a longer driving range compared to battery electric vehicles, allowing for extended travel without the need for frequent recharging), zero emissions (i.e., hydrogen fueling stations enable the use of hydrogen as a clean fuel source, resulting in zero tailpipe emissions when hydrogen is used in fuel cell vehicles. This helps reduce air pollution and mitigate climate change), and scalability and energy storage (i.e., hydrogen fueling stations can be scalable to meet increasing demand, and they also serve as a means of energy storage. Excess renewable energy can be used to produce hydrogen through electrolysis and stored at fueling stations for later use, contributing to grid stability).

One of the disadvantages of hydrogen fueling stations as hydrogen transport technology is the limited infrastructure (i.e., the infrastructure for hydrogen fueling stations is still relatively limited compared to traditional gasoline stations or electric charging infrastructure. This can restrict the availability of hydrogen fuel and limit the adoption of hydrogenpowered vehicles), the high cost and investment (i.e., the establishment of hydrogen fueling stations requires significant upfront investment and ongoing operational costs. The high costs associated with infrastructure installation, hydrogen production, and storage can be a barrier to widespread deployment), the challenging hydrogen production (i.e., the production of hydrogen for fueling stations may still heavily rely on fossil fuel-based processes, such as SMR, which can generate carbon emissions. The transition to renewable and sustainable hydrogen production methods is necessary to maximize the environmental benefits of hydrogen fueling stations), and the safety considerations (i.e., handling and storing hydrogen at fueling stations require adherence to strict safety protocols. While hydrogen is non-toxic, it is highly flammable and requires proper safety measures and training to ensure safe operations).

It is important to note that the advantages and disadvantages of hydrogen fueling stations can vary based on regional factors, infrastructure development, and the overall hydrogen supply chain. Ongoing research and advancements in hydrogen production, storage, and infrastructure will play a crucial role in addressing these limitations and further enhancing the viability and adoption of hydrogen fueling stations.

The process of hydrogen fueling stations primarily involves the compression and dispensing of gaseous hydrogen for refueling hydrogen-powered vehicles, especially those equipped with fuel cell systems. While there are no specific chemical reactions like in some other hydrogen production methods, the overall process can be represented by the following simplified equations:

- Hydrogen Compression: H_2 gas $\rightleftharpoons$ compressed hydrogen. In this equation, hydrogen gas (H_2) is compressed to a high pressure suitable for refueling vehicles. The compression process typically involves a compressor that pressurizes the hydrogen gas to the desired level.
- Hydrogen Dispensing: Compressed hydrogen $\rightleftharpoons$ hydrogen refueling nozzle $\rightleftharpoons$ hydrogen tank of vehicle. In this equation, the compressed hydrogen is dispensed through a refueling nozzle into the hydrogen tank of the vehicle. The nozzle is designed to connect securely to the vehicle's tank and facilitate the transfer of hydrogen.

It is important to note that the specific equipment and systems used in hydrogen fueling stations can vary, and the actual process may involve additional steps, such as cooling to remove heat generated during compression and safety measures to ensure proper handling of hydrogen. Additionally, some fueling stations may use liquid hydrogen (LH_2) rather than gaseous hydrogen, in which case, additional steps for liquefaction and storage of hydrogen are involved. Overall, hydrogen fueling stations play a critical role in enabling the practical use of hydrogen as a clean transportation fuel, allowing vehicles equipped with fuel cells to access hydrogen quickly and efficiently for extended driving ranges with zero-emissions.

The most common routes for large-scale hydrogen transportation, which include hydrogen pipelines, ammonia, liquefied hydrogen, and LOHC, are depicted in Fig. 12.

Fig. 13 illustrates the impact of transport distance on the cost of hydrogen delivery from a single production and delivery point. It demonstrates that for shorter distances (up to 3,000 km), CHG, especially through pipelines, tends to be the most cost-effective option. However, for longer distances, both liquefied hydrogen and LOHC offer lower costs, with liquefied hydrogen performing slightly better. It is important to note that as the distance increases, the boil-off losses of liquefied hydrogen reduce its competitiveness. For distances exceeding approximately 16,000 km, chemical carriers like LOHC and ammonia become the preferred choices for hydrogen delivery.

Fig. 14 illustrates the breakdown of total costs for various hydrogen delivery options in a case study. The study focuses on delivering 1 million tons of renewable hydrogen annually to a single industrial customer through a dedicated pipeline or shipping route with a transport distance of 2,500 km. The costs are categorized into packing costs, unpacking costs, and costs associated with storage

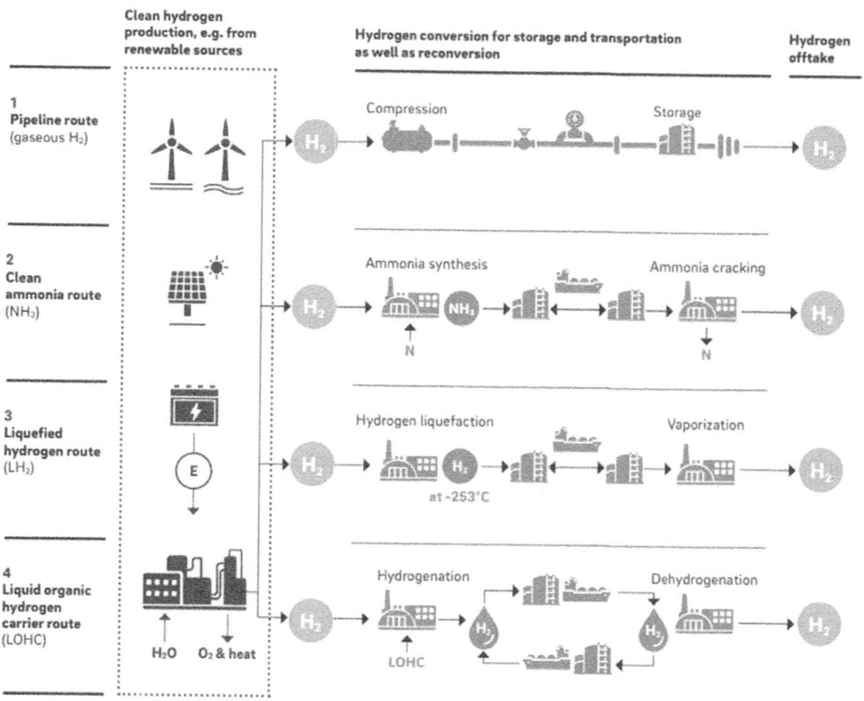

Fig. 12. Comparison of Large-Scale Hydrogen Transportation Routes: (1) Hydrogen Pipelines (CHG), (2) Ammonia (NH_3), Liquefied Hydrogen (LH_2), and Liquid Organic Hydrogen Carriers (LOHC). *Source*: Roland Berger (n.d.).

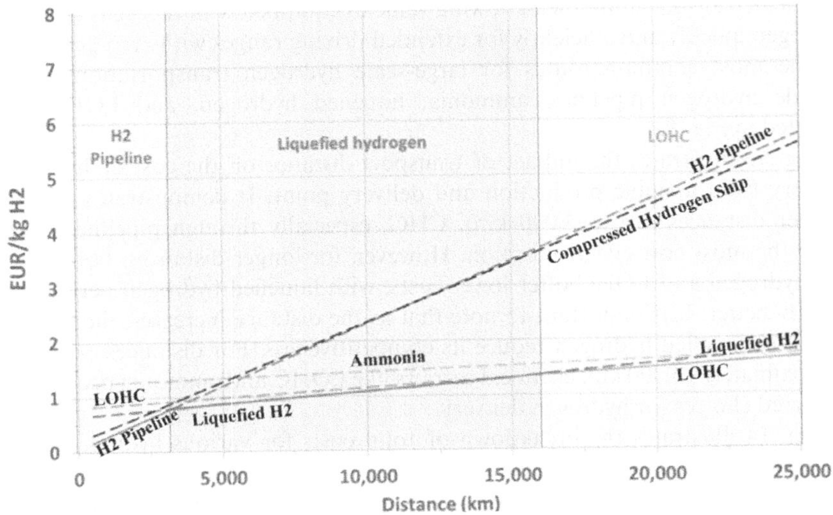

Fig. 13. The Cost of Delivering 1 MT of Hydrogen Through a Simple Point-to-Point Transport Route Under a Low Electricity Cost Scenario. *Source*: Joint Research Centre (JRC) (2023).

and the utilization of a transport fleet. It is important to note that the availability and suitability of these options may vary depending on the infrastructure, regulations, and market development in a given region. Additionally, research and development in hydrogen transportation are ongoing, and new technologies may emerge in the future.

3.3. Employing MCDM-AHP for Hydrogen Transport Technologies Assessment

In this section, we will employ the AHP to address the MCDM problem with hydrogen transport technologies (Section 2.1).

3.3.1. Key Criteria for Evaluation

When assessing hydrogen transport technologies, several relevant criteria need to be considered (Section). These criteria include:

1. **"C1" Energy Efficiency**: The energy efficiency of the transport technology, including the efficiency of compression, liquefaction, or other conversion processes involved.
2. **"C2" Energy Density**: The energy density of the hydrogen carrier, which determines the volume or mass of hydrogen that can be transported per unit of storage or transport capacity.
3. **"C3" Safety**: The safety considerations associated with handling, storage, and transport of the hydrogen carrier, including factors such as flammability, toxicity, and the potential for leaks or accidents.

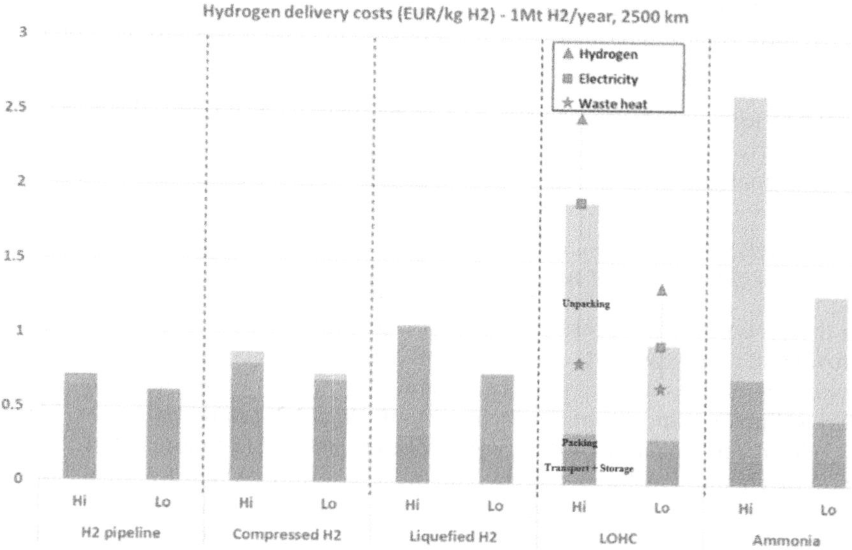

Fig. 14. The Hydrogen Delivery Costs for a Case Study Scenario Involving the Transportation of 1 Million Tons of Renewable Hydrogen Per Year to a Single Industrial Customer. The Transport Takes Place Via a Dedicated Pipeline or Shipping Route, with a Transport Distance of 2,500 km. The Figure Showcases the Cost Breakdown, Highlighting the Packing Costs, Unpacking Costs, and Costs Associated with Storage and the Use of a Transport Fleet. It is Worth Noting That, apart from Pipelines, All Transport Options Exclusively Rely on Shipping as the Mode of Transportation. *Source*: Joint Research Centre (JRC) (2023).

4. **"C4" Infrastructure Requirements**: The infrastructure requirements for each transport technology, including the need for storage facilities, pipelines, tube trailers, or other specialized infrastructure.
5. **"C5" Cost**: The overall cost of the transport technology, including capital costs, operational costs, and maintenance costs.
6. **"C6" Range and Flexibility**: The range or distance over which the transport technology can effectively transport hydrogen, as well as the flexibility to accommodate different transportation needs and routes.
7. **"C7" Environmental Impact**: The environmental impact of the transport technology, including factors such as greenhouse gas emissions, air pollution, and water consumption.
8. **"C8" Technological Maturity**: The level of technological maturity and commercial readiness of the transport technology, including the availability of equipment, infrastructure, and expertise.
9. **"C9" Scalability**: The scalability of the transport technology to meet the growing demand for hydrogen transportation, including its ability to handle large volumes of hydrogen and support a widespread hydrogen infrastructure.

10. **"C10" Compatibility with Existing Infrastructure**: The compatibility of the transport technology with existing infrastructure, such as transportation networks, storage facilities, and refueling stations.

When assessing hydrogen transport technologies, considering these criteria will help evaluate their suitability for specific applications and facilitate informed decision-making.

3.3.2. Analyzing the Optimal Hydrogen Transport Technology: Pairwise Comparison, Criteria Weights, and Weighted Sum Evaluation

In this section, we will present the pairwise comparison matrix between criteria, provide the relative weights of criteria, and calculate the weighted sum of alternatives. This analysis will help evaluate and determine the optimal hydrogen transport technology among options such as (A1) CHG, (A2) Liquid Hydrogen (LH_2), (A3) hydrogen pipelines, (A4) hydrogen tube trailers, (A5) ammonia (NH_2) carriers, (A6) chemical hydrides, (A7) hydrail (hydrogen rail), (A8) hydrogen blending, (A9) LOHC, and (A10) hydrogen fueling stations. The evaluation will be based on multiple criteria, including (C1) energy efficiency, (C2) energy density, (C3) safety, (C4) infrastructure requirements, (C5) cost, (C6) range and flexibility, (C7) environmental impact, (C8) technological maturity, (C9) scalability, and (C10) compatibility with existing infrastructure.

The hierarchical structure for evaluating hydrogen transport technologies is as follows (Fig. 15).

The pairwise comparison matrix between criteria is presented in (19), while Table 3 displays the relative weights assigned to each criterion. Finally, the weighted sum of each hydrogen transport technology is illustrated in Table 4, providing a comprehensive evaluation of their performance based on the given criteria.

The pairwise comparison matrix (19) allows us to assess the relative importance of each criterion with respect to the others. The values in the matrix represent the judgments made on the importance or priority of one criterion over another. The matrix enables us to quantify the degree of preference or significance assigned to each criterion based on the decisionmaker's perception. By analyzing the pairwise

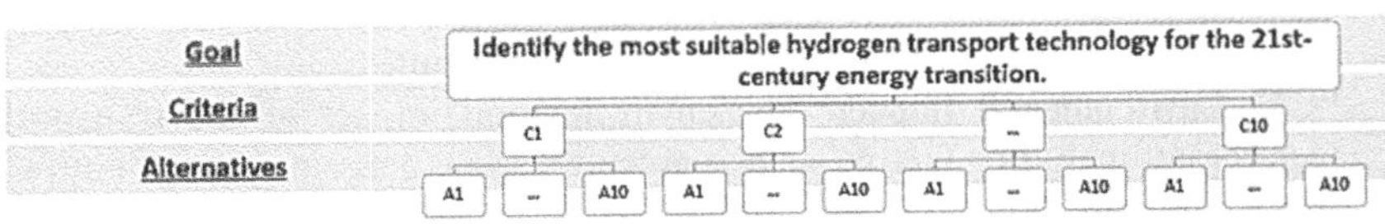

Fig. 15. The Hierarchical Structure for Evaluating Hydrogen Transport Technologies. It Includes the Overall Goal at the Top Level, Followed by the Evaluation Criteria at the Second Level, and the Different Hydrogen Transport Technologies at the Third Level. This Structure Allows for a Systematic Assessment of Each Technology Based on the Specified Criteria, Enabling a Comprehensive Analysis and Comparison of Their Performance and Suitability for the 21st-Century Energy Transition. *Source*: Own Elaboration.

comparison matrix, we can identify the criteria that are considered more influential or critical in the evaluation process.

$$C = \begin{bmatrix} 1 & 3 & 5 & 3 & 4 & 4 & 3 & 4 & 3 & 2 \\ 1/3 & 1 & 3 & 3 & 4 & 4 & 3 & 4 & 2 & 2 \\ 1/5 & 1/3 & 1 & 3 & 2 & 2 & 2 & 3 & 2 & 1 \\ 1/3 & 1/3 & 1/3 & 1 & 3 & 3 & 2 & 3 & 2 & 2 \\ 1/4 & 1/4 & 1/2 & 1/3 & 1 & 3 & 2 & 3 & 2 & 2 \\ 1/4 & 1/4 & 1/2 & 1/3 & 1/3 & 1 & 1 & 2 & 2 & 2 \\ 1/3 & 1/3 & 1/2 & 1/2 & 1/2 & 1 & 1 & 2 & 2 & 1 \\ 1/4 & 1/4 & 1/3 & 1/3 & 1/3 & 1/2 & 1/2 & 1 & 2 & 2 \\ 1/3 & 1/2 & 1/2 & 1/2 & 1/3 & 1/2 & 1/2 & 1/2 & 1 & 1 \\ 1/2 & 1/2 & 1 & 1/2 & 1/2 & 1/2 & 1 & 1/2 & 1 & 1 \end{bmatrix} \quad (19)$$

The maximum eigenvalue λ_{max} of the pairwise comparison matrix (19) is determined as 9.0127.

The number of criteria n is 10. $CI == \frac{9.0127 - 10}{10 - 1} = 0.0025$

The RI for a matrix of size 10 is 1.49 (Table 2). $CR = \frac{0.0025}{1.49} \approx 0.0017$

Since the CR value (0.0017) is less than 0.1 (10%), the pairwise comparison matrix is considered consistent.

Table 3. Relative Weights of Criteria (i.e., Their Importance or Priority in the Evaluation of Hydrogen Transport Technologies). Higher Relative Weights Suggest Greater Significance or Influence of a Particular Criterion in the Decision-Making Process. These Weights Serve as a Guide to Emphasize the Significance of Certain Factors and Consider Them Appropriately in the Decision-Making Process. *Source*: Own Elaboration.

Criteria	**Relative Weights (RWs)**
C1 "Energy Efficiency"	0.196
C2 "Energy Density"	0.130
C3 "Safety"	0.043
C4 "Infrastructure Requirements"	0.130
C5 "Cost"	0.086
C6 "Range and Flexibility"	0.086
C7 "Environmental Impact"	0.086
C8 "Technological Maturity"	0.065
C9 "Scalability"	0.054
C10 "Compatibility with Existing Infrastructure"	0.054

Table 4. Weighted Sum of Each Hydrogen Transport Technology. A Higher Weighted Sum Indicates That a Technology Is More Favorable and Suitable According to the Evaluated Criteria and Their Relative Importance. Conversely, a Lower Weighted Sum Suggests That a Technology May Have Limitations or Lower Performance in Relation to the Assessed Criteria. *Source*: Own Elaboration.

Alternatives	Weighted Sum
A1 "Compressed Hydrogen Gas (CHG)"	0.168
A2 "Liquid Hydrogen (LH_2)"	0.150
A3 "Hydrogen Pipelines"	0.116
A4 "Hydrogen Tube Trailers"	0.096
A5 "Ammonia (NH_2) Carriers"	0.135
A6 "Chemical Hydrides"	0.121
A7 "Hydrail (Hydrogen Rail)"	0.120
A8 "Hydrogen Blending"	0.113
A9 "Liquid Organic Hydrogen Carriers (LOHC)"	0.120
A10 "Hydrogen Fueling Stations"	0.129

The relative weights of the criteria provide an indication of their importance in the evaluation of hydrogen transport technologies. A criterion with a higher relative weight is considered more crucial in determining the optimal hydrogen transport technology. On the other hand, a criterion with a lower relative weight has a relatively smaller impact on the decision. By analyzing the relative weights (Table 3), decision-makers can identify the key criteria that should be given more attention when evaluating and selecting the optimal hydrogen transport technology.

The interpretation of these relative weights can be categorized as follows:

- High relative weights: Energy Efficiency (C1): With a relative weight of 0.196, energy efficiency is considered highly important in the evaluation process. It indicates a strong preference for hydrogen transport technologies that maximize energy efficiency.

- Medium Relative Weights:
  - Energy Density (C2): Energy density has a relative weight of 0.130, reflecting its significance in the evaluation. It suggests a preference for technologies with higher energy density, but it is not as critical as energy efficiency.
  - Infrastructure Requirements (C4): This criterion also shares a relative weight of 0.130 with energy density, indicating its importance. It suggests a preference for technologies with lower infrastructure requirements or those that can leverage existing infrastructure.

- Medium-Low Relative Weights:
  - Cost (C5): With a relative weight of 0.086, cost is considered moderately important. While it is a consideration, it is not as critical as energy efficiency and energy density.
  - Range and Flexibility (C6): Range and flexibility also have a relative weight of 0.086, indicating their moderate importance. They are considered significant but not as critical as energy efficiency, energy density, and infrastructure requirements.
  - Environmental Impact (C7): Environmental impact shares the same relative weight of 0.086 with cost and range/flexibility, suggesting its significance in the evaluation process. It indicates a consideration for technologies with a lower environmental impact.

- Low Relative Weights:
  - Safety (C3): Safety has a relatively low relative weight of 0.043, indicating it is considered less important compared to other criteria.
  - Technological Maturity (C8): With a relative weight of 0.065, technological maturity is considered relatively less important in the evaluation process.
  - Scalability (C9) and Compatibility with Existing Infrastructure (C10): Both criteria share a relative weight of 0.054, indicating their relatively lower importance in the evaluation.

By categorizing the relative weights into low, medium, and high, we can see which criteria carry more weight in the evaluation process. The high relative weights represent criteria of utmost importance, while the medium and medium-low relative weights indicate criteria that are significant but not as critical. The low relative weights suggest criteria that have a relatively lower impact on the evaluation.

Based on the resulted weighted sum of alternatives (Table 4), we can interpret the performance of each hydrogen transport technology as follows:

- Low weighted sum: hydrogen tube trailers “A4” have the lowest weighted sum of 0.096, indicating a relatively lower overall performance compared to other options.
- Medium weighted sum: chemical hydrides “A6” (0.121), hydrail (hydrogen rail) “A7” (0.120), Liquid Organic Hydrogen Carriers (LOHC) “A9”, hydrogen pipelines “A3” (0.116), hydrogen blending “A8” (0.113) have medium weighted sums, suggesting a moderate overall performance.
- High weighted sum: Compressed Hydrogen Gas (CHG) “A1” (0.168), Liquid Hydrogen (LH_2) “A2” (0.15), Ammonia (NH_3) carriers “A5” (0.135), hydrogen fueling stations “A10” (0.129) have high weighted sums, indicating a relatively higher overall performance compared to other alternatives.

The interpretation of these weighted sums depends on the specific context and criteria used for evaluation. A technology with a high weighted sum is considered more favorable and suitable based on the criteria and their relative importance in the analysis. Conversely, a technology with a low weighted sum may have certain limitations or lower performance in relation to the evaluated criteria.

To explore the comprehensive analysis of criteria-based assessment for hydrogen transport technologies, please refer to the supplementary material accompanying this chapter.

3.4. Barriers and Opportunities of Hydrogen Transport Technologies for the 21st Century Energy Transition

This section explores the various barriers and opportunities associated with the adoption and implementation of hydrogen transport technologies, including CHG, Liquid Hydrogen (LH_2), hydrogen pipelines, hydrogen tube trailers, ammonia (NH_2) carriers, chemical hydrides, hydrail (hydrogen rail), hydrogen blending, LOHC, and hydrogen fueling stations. It examines the challenges and obstacles that hinder the widespread deployment of these technologies. Additionally, it highlights the potential opportunities and benefits these technologies offer in achieving the energy transition goals of the 21st century.

3.4.1. Overcoming Barriers

In the 21st-century energy transition, hydrogen transport technologies face several barriers that can hinder their widespread adoption and integration. Addressing the barriers to hydrogen transport technologies for the 21st century energy transition requires a comprehensive approach involving various strategies and actions. These barriers and associated mitigation actions include (Hosseini & Butler, 2020; Singla et al., 2021):

1. **Infrastructure development** for hydrogen transportation is a complex and demanding task that involves establishing an interconnected network of pipelines, storage facilities, and fueling stations. It requires substantial investments, coordination among stakeholders, and addressing regulatory challenges. To overcome this hurdle, collaboration among governments, industry players, and research institutions is crucial. Long-term planning, coordination, and investment are necessary to ensure the availability of robust infrastructure that can support the widespread adoption of hydrogen as an energy carrier.
2. **Cost and economics** of hydrogen transport technologies present a significant challenge due to their high upfront expenses. The costs associated with infrastructure development, storage systems, and fuel cell vehicles can be higher compared to conventional fossil fuel technologies. The scale of hydrogen production and demand also impacts cost competitiveness. To address this challenge, it is important to implement policies and incentives that promote costeffectiveness. Financial support for research and development, private

sector investment incentives, and innovative business models can play a role in reducing costs. Strategies such as scaling up production, improving efficiency, and optimizing supply chains can further contribute to cost reduction in hydrogen transport technologies.

3. **Safety concerns** surrounding hydrogen transport technologies stem from their high flammability and wide flammability range. These concerns necessitate specialized safety measures, infrastructure, and training to ensure safe handling, storage, and transportation. To overcome this barrier, it is crucial to establish robust safety standards and regulations. This includes conducting thorough risk assessments, developing safety protocols and guidelines, and implementing stringent monitoring and inspection procedures. Public awareness campaigns can also contribute by educating the public about the safety measures implemented in hydrogen transport technologies. By addressing safety concerns through comprehensive measures, the adoption of hydrogen as an energy carrier can be facilitated.
4. **Technological maturity and efficiency** of hydrogen transport technologies are important considerations for their widespread adoption. While some technologies are still in the early stages of development, efforts should be directed toward improving their efficiency and advancing their technological maturity. This involves enhancing the energy conversion and utilization of hydrogen throughout the production, storage, and distribution processes. Continued research and development efforts are necessary to drive progress in fuel cell technology, storage and compression methods, and production processes. Collaboration between academia, industry, and research institutions can play a vital role in accelerating technological advancements and fostering innovation in the field of hydrogen transport technologies.
5. **Limited range and infrastructure coverage** pose challenges to the widespread adoption of hydrogen transport technologies. The limited range of hydrogen vehicles and the sparse availability of refueling infrastructure hinder the practicality and convenience of using hydrogen as a transportation fuel. To address this, strategic deployment plans should be implemented to expand the coverage of hydrogen infrastructure. This includes identifying priority areas for infrastructure development, establishing refueling stations along important transportation routes, and fostering partnerships to accelerate infrastructure expansion. Collaboration between governments, private sector entities, and research institutions is crucial for achieving a comprehensive and accessible hydrogen infrastructure network.
6. **Public acceptance and awareness** play a crucial role in the successful adoption of hydrogen transport technologies. Building trust, addressing misconceptions, and promoting the benefits of hydrogen as a clean and sustainable energy source are essential. Educating the general public through awareness campaigns, demonstration projects, and showcasing the advantages of hydrogen transport technologies can help increase acceptance. Engaging with communities, addressing concerns, and promoting transparency are key to building public trust and support. Efforts should be focused on enhancing public awareness and understanding of the potential of hydrogen as an

alternative fuel, emphasizing its positive impact on the environment and energy transition.

7. **Supportive policies and regulations** are vital for the development and deployment of hydrogen transport technologies. The absence of clear and consistent policies can impede progress in this field. Governments should prioritize the development of supportive policies and regulatory frameworks that incentivize the adoption of hydrogen technologies. This includes financial incentives, well-defined regulations, and setting targets for hydrogen utilization. Collaboration between governments, industry stakeholders, and international organizations can help harmonize policies and regulations across regions, creating a favorable environment for the growth of hydrogen transport technologies.
8. **Durability and degradation** pose potential barriers to the widespread adoption of hydrogen transport technologies. As these technologies age, they may experience performance degradation and reduced efficiency. Factors such as material compatibility, exposure to harsh environments, and the presence of impurities can contribute to system degradation. It is essential to address these challenges to ensure the long-term reliability and effectiveness of hydrogen transport technologies. Research efforts should focus on improving durability and reliability through the development of robust materials, performance testing, and preventive maintenance practices. Collaboration between industry and research institutions is crucial to advancing materials science and engineering, effectively tackling durability and degradation concerns.
9. **Availability and reliance on critical materials** present a barrier to the widespread deployment of hydrogen transport technologies. These technologies often require specific materials, such as catalysts, membranes, or storage materials, which may have limited global supply or be concentrated in specific regions. This reliance on critical materials raises concerns regarding cost, accessibility, and potential geopolitical implications (Bouramdane, 2023r). To overcome these barriers, it is important to develop alternative materials and optimize the use of existing ones. Strategies such as diversifying the supply chain, promoting recycling and circular economy practices, and exploring alternative materials and technologies can help ensure the availability of critical materials for hydrogen transport technologies. International collaboration, resource mapping, and strategic partnerships are key to mitigating the risks associated with the availability of critical materials.

By addressing these barriers through a combination of policy measures, research and development efforts, public engagement, and strategic collaborations, the widespread adoption and integration of hydrogen transport technologies can be facilitated, contributing to the successful energy transition in the 21st century.

3.4.2. Unlocking Opportunities

The opportunities presented by hydrogen transport technologies for the 21st-century energy transition are significant and promising. To take advantage of the opportunities presented by hydrogen transport technologies, several key strategies can be implemented (Hosseini & Butler, 2020; Singla et al., 2021):

1. **Harnessing Clean Energy Sources**: Hydrogen is a clean fuel that produces only water vapor when used in fuel cells or combustion engines, contributing to reduced greenhouse gas emissions and improved air quality. Governments and industry stakeholders should invest in renewable energy sources such as wind, solar, and hydroelectric power to produce clean hydrogen. By coupling hydrogen production with renewable energy, the carbon footprint of hydrogen can be minimized, contributing to a more sustainable energy system.
2. **Leveraging Energy Storage Capabilities**: Hydrogen serves as a versatile energy carrier, facilitating efficient energy storage and balancing the intermittency of renewable energy sources like solar and wind. Hydrogen can play a significant role in energy storage by converting excess renewable energy into hydrogen through electrolysis. This stored hydrogen can be used for power generation during periods of high demand or when renewable energy supply is limited. By integrating hydrogen as an energy storage solution, the stability and reliability of the energy grid can be enhanced.
3. **Decarbonizing Transportation**: Hydrogen-powered vehicles, such as fuel cell vehicles and hydrogen trains (hydrail), have the potential to decarbonize various transportation sectors, reducing reliance on fossil fuels. Governments can provide incentives for the adoption of hydrogen vehicles, develop refueling infrastructure, and support research and development efforts to improve the efficiency and affordability of hydrogen-powered transportation.
4. **Long Range and Quick Refueling**: Hydrogen-powered vehicles offer longer driving ranges and shorter refueling times compared to some electric vehicles, providing convenience and flexibility to users. Governments and industry stakeholders can promote the expansion of hydrogen refueling infrastructure, ensuring convenient access for consumers and supporting long-distance travel.
5. **Compatibility with Existing Infrastructure**: Certain hydrogen transport technologies, like hydrogen blending and ammonia carriers, can be integrated into existing infrastructure, reducing the need for extensive modifications. The compatibility of hydrogen transport technologies with existing infrastructure, such as natural gas pipelines, can facilitate their integration into the energy system. Governments and industry stakeholders can explore retrofitting options, develop guidelines for safe integration, and incentivize the utilization of existing infrastructure for hydrogen transport.
6. **Industrial Applications**: Hydrogen has various industrial applications, such as in manufacturing processes and the production of chemicals and fertilizers. Governments and industry stakeholders can promote the use of hydrogen in industrial sectors by providing incentives, supporting research and development, and facilitating collaborations between industry players.

7. **Scalability**: Hydrogen technologies are scalable and can be deployed at different levels, from small fueling stations to large industrial applications, catering to diverse energy demands.
8. **Hydrogen Economy**: The development of a hydrogen economy can foster economic growth, job creation, and technological innovation in the energy sector. Governments can support the development of a hydrogen economy by fostering a favorable regulatory environment, promoting private sector investments, and encouraging the scaling up of hydrogen production, distribution, and utilization. Collaboration between different sectors and stakeholders can drive the growth and maturity of the hydrogen economy.
9. **Global Energy Transition**: Hydrogen can facilitate a global transition to sustainable energy, enabling countries to cooperate on renewable energy projects and reduce dependency on fossil fuels.
10. **Diversification of Energy Sources**: Embracing hydrogen technologies offers an opportunity to diversify energy sources and reduce geopolitical risks associated with traditional energy supplies. Governments can collaborate on international initiatives to develop hydrogen infrastructure, share best practices, and establish common standards and regulations. This will facilitate the global energy transition towards a more sustainable and diversified energy mix.

Overall, hydrogen transport technologies present an encouraging outlook for the 21st-century energy transition, with the potential to play a crucial role in achieving a sustainable, low-carbon future. However, addressing the barriers and challenges is essential to fully harnessing these opportunities and realizing the transformative potential of hydrogen in the global energy landscape.

4. Discussion

In this section, we delve into the implications of our research for the smart cities agenda in the Gulf Region, aiming to provide actionable insights that resonate with diverse communities and stakeholders. Urban planners can leverage our findings to integrate sustainable hydrogen transport technologies seamlessly into the urban fabric, fostering efficient and environmentally conscious city planning. Architects, informed by our assessment of various hydrogen transport options, can contribute to the development of infrastructure that aligns with the evolving energy landscape. Policy makers gain valuable perspectives on crafting regulations that support the integration of hydrogen technologies, ensuring safety and efficiency in smart cities. Faculty and students find inspiration for research endeavors and academic pursuits in the realm of clean energy and urban development. Education administrators can adapt curricula to equip the workforce with skills tailored to the demands of a hydrogen-powered future. Lastly, government officers are presented with nuanced recommendations and initiatives to guide policy-making, facilitating the broader impact of smart cities in the Gulf Region.

4.1. Implications of Optimizing Hydrogen Transport Technologies in the Gulf Region's Smart Cities

The findings of this research hold significant implications for the smart cities agenda in the Gulf region. The adoption of hydrogen transport technologies aligns with the broader objectives of smart urban development, offering a pathway towards sustainable, efficient, and technologically advanced transportation systems. In this section, we discuss the various facets of the implications:

- **Environmental Sustainability**: Hydrogen, as a clean energy carrier, contributes to the reduction of carbon emissions, aligning with the Gulf region's commitment to environmental sustainability in smart cities. Embracing hydrogen transport can lead to cleaner air, mitigating the impact of urban pollution and enhancing the overall quality of life.
- **Technological Innovation and Integration**: The diverse range of hydrogen transport technologies discussed in the research presents an opportunity for technological innovation and integration within the smart cities framework. The deployment of these technologies can be seen as a catalyst for advancing the technological landscape in the Gulf region, fostering a culture of innovation and cutting-edge solutions in the urban infrastructure.
- **Energy Efficiency and Security**: The use of MCDM-AHP methodology in assessing hydrogen transport technologies ensures that energy efficiency is a key criterion. Implementing energy-efficient systems in transport aligns with the smart cities agenda's focus on optimizing resource use and enhancing energy security.
- **Infrastructure Development**: The research highlights the importance of infrastructure requirements as a key criterion for assessing hydrogen transport technologies. The Gulf region can leverage this insight to plan and invest in the development of a robust and adaptive infrastructure that supports the deployment of hydrogen transport systems, aligning with the smart cities goal of integrated and smart infrastructure.
- **Economic Viability**: Understanding the cost implications of different hydrogen transport technologies is crucial for smart cities in the Gulf region. The research's emphasis on cost as a criterion provides valuable insights for decisionmakers, allowing them to make informed choices that balance economic viability with environmental sustainability.
- **Global Energy Transition**: The research addresses the barriers and opportunities associated with hydrogen transport technologies in the context of the 21st-century energy transition. The Gulf region's alignment with this global shift positions it as a participant in a larger movement towards sustainable and renewable energy, reinforcing its commitment to smart cities as hubs of progressive energy practices.
- **International Collaboration**: The practical implications of the study for Morocco's green interconnections in hydrogen, ammonia, and kerosene production also offer an avenue for international collaboration within the Gulf region. Smart cities can serve as focal points for cross-border initiatives, fostering collaboration in research, technology exchange, and the implementation of sustainable energy solutions.

In conclusion, the integration of hydrogen transport technologies into the smart cities agenda in the Gulf region presents an opportunity to lead in sustainable urban development, technological innovation, and global energy transition. The research provides a comprehensive framework for decision-makers to navigate the complexities of adopting hydrogen transport, ensuring that smart cities in the Gulf region remain at the forefront of environmentally conscious and technologically advanced urban development.

4.2. Interpretation of Key Findings for Diverse Communities and Stakeholders in Smart Cities

The analysis of various hydrogen transport technologies and their assessment using the MCDM-AHP methodology provides crucial insights for diverse communities and stakeholders in smart cities. Urban planners can benefit from understanding the feasibility and implications of adopting specific hydrogen transport technologies within urban infrastructure. The assessment criteria, such as energy efficiency, safety, and environmental impact, offer valuable information for architects and policy makers in designing and implementing sustainable urban systems.

For faculty and students, this research opens avenues for academic exploration and innovative research projects in the field of clean energy and sustainable transportation. Educational administrators can leverage these findings to tailor curricula that align with emerging technologies, ensuring that the workforce is equipped with the necessary skills for the evolving energy landscape. Government officers, particularly those involved in energy and transportation sectors, gain insights into the most viable and efficient hydrogen transport technologies, aiding in strategic decision-making for urban development.

Some propositions for initiatives related to smart cities include:

- **Integrated Hydrogen Infrastructure Planning**: Encourage the integration of hydrogen transport infrastructure in urban planning. This includes the development of dedicated hydrogen pipelines, fueling stations, and storage facilities to support a seamless transition to hydrogen-based transportation systems.
- **Public Awareness and Acceptance Programs**: Initiate public awareness programs to inform citizens about the benefits and safety measures associated with hydrogen transport technologies. Building public trust and acceptance is essential for the successful implementation of these technologies in smart cities.
- **Collaborative Research and Innovation Hubs**: Establish collaborative research and innovation hubs that bring together researchers, industry experts, and policymakers. These hubs can serve as incubators for new technologies, fostering cross-disciplinary collaboration to address challenges and accelerate the adoption of sustainable transport solutions.

Some policy recommendations include:

- **Incentivize Hydrogen Infrastructure Development**: Implement policies that provide incentives for the development of hydrogen infrastructure, including tax credits, grants, and subsidies. This will attract private investments and accelerate the deployment of hydrogen technologies in smart cities.

- **Regulatory Framework for Safety Standards**: Develop and enforce robust regulatory frameworks that ensure the safety and security of hydrogen transport technologies. Clear standards and guidelines will build confidence among stakeholders and facilitate the widespread adoption of these technologies.
- **Investment in Research and Development**: Allocate resources for research and development initiatives focused on advancing hydrogen transport technologies. Government funding and collaboration with private entities can drive innovation, making these technologies more cost-effective and scalable for smart cities.

In conclusion, the findings of this research offer valuable insights and recommendations for stakeholders in smart cities, emphasizing the importance of strategic planning, public engagement, and supportive policies to facilitate the successful integration of hydrogen transport technologies in urban environments.

5. Conclusion

5.1. Research Motivation

The motivation to study hydrogen transport technologies stems from the urgent need for sustainable and clean energy solutions in the face of climate change (Bouramdane, 2022c, 2022d, 2023s, 2023t, 2023u, 2023v, 2023w, 2024a) and the transition to a low-carbon economy (Bouramdane, 2022e, 2024b). Hydrogen has emerged as a promising energy carrier due to its potential to decarbonize various sectors, including transportation, industry, and power generation. Therefore, understanding and optimizing hydrogen transport technologies is crucial for the widespread adoption of hydrogen as a clean and sustainable energy carrier.

5.2. Research Questions and Methodology

This research chapter aims to comprehensively examine various aspects of hydrogen transport technologies and provide insights into their advantages, disadvantages, process equations, and barriers and opportunities for the 21st century energy transition.

To achieve the research objectives, the chapter first explored the general steps involved in hydrogen transport, from production to utilization. This provided a holistic perspective on the overall process and highlighted the key stages and considerations involved in transporting hydrogen.

Next, the advantages, disadvantages, and process equations of different hydrogen technologies are investigated. These included CHG, Liquid Hydrogen (LH_2), hydrogen pipelines, hydrogen tube trailers, ammonia (NH_3) carriers, chemical hydrides, hydrail (hydrogen rail), hydrogen blending, LOHC, and hydrogen fueling stations. By analyzing these technologies, the chapter provides insights into their characteristics.

To support decision-making in selecting the most suitable hydrogen transport technology, the AHP was employed as a MCDM tool. The research applied AHP to assess and rank the hydrogen transport technologies based on multiple criteria, including energy efficiency, energy density, safety, infrastructure requirements, cost, range, flexibility, environmental impact, technological maturity, scalability, and compatibility with existing infrastructure. This approach provides a systematic and quantitative framework for evaluating and comparing the technologies, enabling informed decision-making, and facilitating the identification of optimal solutions.

Lastly, the chapter examines the barriers and opportunities of hydrogen transport technologies for the 21st century energy transition, investigate strategies to overcome these barriers, and explores how to effectively capitalize on the available opportunities.

5.3. General Findings

Each hydrogen transport technology offers a distinct set of advantages and disadvantages:

- CHG is an efficient, flexible, and scalable hydrogen transport technology with a mature and well-developed infrastructure. It can easily integrate with existing natural gas infrastructure, making it a convenient option for deployment. However, CHG has a lower energy density compared to other alternatives, necessitating larger volumes for transportation. It also requires stringent safety measures and may incur higher transportation costs due to its larger volume. The environmental impact of CHG depends on the source of hydrogen production. Overall, CHG is a reliable and established solution for hydrogen transport.
- Liquid Hydrogen (LH_2) is a hydrogen transport technology that offers several advantages. It boasts high energy efficiency and a significantly higher energy density compared to other options, allowing for longer-range transportation. Safety measures are crucial due to the low temperature and potential for leaks or explosions. LH_2 requires specialized infrastructure for storage and handling, which may increase the initial investment costs. The environmental impact of LH_2 depends on the source of hydrogen production, but it generally produces low emissions during use. Technologically, LH_2 is considered mature and has been used in various applications. However, scalability and compatibility with existing infrastructure can be challenging due to the need for cryogenic storage and handling systems. Overall, LH_2 provides a promising solution for long-range hydrogen transportation, but careful consideration of safety, infrastructure requirements, and compatibility is necessary.
- Hydrogen pipelines are a highly efficient and reliable option for transporting hydrogen. They have low energy losses and can accommodate large volumes of hydrogen, compensating for its lower energy density. Safety measures, such as leak detection systems, ensure the secure transportation of hydrogen. While the upfront infrastructure costs are significant, the operational costs are

relatively low. Pipelines offer a long-range and fixed-route transportation solution, but additional infrastructure may be needed for distribution. They have a low environmental impact and are a mature technology. Scalability is a strength, allowing for future expansion, and they can be compatible with existing natural gas infrastructure. Overall, hydrogen pipelines are a well-established and promising option for large-scale hydrogen transport.

- Hydrogen tube trailers are a reliable and established hydrogen transport technology. They offer moderate energy efficiency and require proper safety protocols during transportation. While they have a lower energy density compared to other options, they are flexible in terms of range and can be used for regional transport. Hydrogen tube trailers require specialized infrastructure for loading and unloading, which may incur higher costs. The environmental impact depends on the source of hydrogen production. Overall, hydrogen tube trailers provide a viable solution for regional hydrogen transport, but their scalability and compatibility with existing infrastructure should be considered.
- Ammonia (NH_3) carriers are a hydrogen transport technology that offers a higher energy density and greater range and flexibility for long-distance transportation. However, they have lower energy efficiency and require specific safety measures due to the toxicity and flammability of ammonia. Infrastructure requirements include specialized storage and handling facilities, which can impact initial costs. The cost of ammonia transportation depends on factors such as distance and storage needs. The environmental impact is influenced by the production process, which may result in greenhouse gas emissions. Technologically, ammonia carriers are relatively mature, and their scalability depends on infrastructure development and ammonia production availability. Compatibility with existing infrastructure, such as ships and storage facilities, can be advantageous. Overall, careful evaluation of factors like energy efficiency, safety, cost, environmental impact, and compatibility is necessary to assess the suitability of ammonia carriers for hydrogen transport.
- Chemical hydrides are a hydrogen transport technology that offers high energy density, making them suitable for long-range transportation. However, they have lower energy efficiency due to the energy required for hydrogen release and regeneration. Safety considerations are essential due to the potential for hazardous chemical reactions. Infrastructure requirements include specialized storage and processing facilities, which can impact initial costs. The cost of chemical hydrides depends on the type of hydride and the regeneration process. The environmental impact varies depending on the hydride used and its life cycle. Technological maturity is moderate, and further research is ongoing for improved systems. Scalability is feasible but may require additional infrastructure development. Compatibility with existing infrastructure may be challenging, and dedicated facilities are often required. In conclusion, chemical hydrides have advantages in energy density and range but pose challenges in terms of efficiency, safety, and infrastructure requirements for hydrogen transport.
- Hydrail, also known as Hydrogen Rail, is a promising hydrogen transport technology that utilizes fuel cells to power trains. It offers high energy efficiency, zero emissions at the point of use when using green hydrogen, and the potential

for scalability as technology advances and infrastructure expands. However, challenges include the lower energy density of hydrogen, which affects storage and range, as well as the need for extensive hydrogen refueling infrastructure and strict safety measures. With further advancements and investment, hydrail could become a viable zero-emission solution for the rail industry, contributing to a more sustainable transportation system.

- Hydrogen Blending involves mixing hydrogen with natural gas or other fuels to reduce carbon emissions from existing natural gas infrastructure. It offers potential benefits in terms of energy efficiency and reduced emissions, especially when blending hydrogen with low-carbon gas sources. However, its energy density is lower than that of pure hydrogen, which may impact storage and range in certain applications. Safety considerations are generally well managed due to existing gas infrastructure standards. Hydrogen Blending is relatively cost-effective compared to some other hydrogen transport technologies since it utilizes existing infrastructure. However, its scalability may be limited by the availability of low-carbon hydrogen sources and the compatibility of blended gas with existing appliances. Overall, Hydrogen Blending represents a transitional approach to decarbonizing existing gas networks and could play a role in the early stages of hydrogen adoption.
- LOHC are an advanced hydrogen transport technology that offers numerous benefits. LOHCs exhibit high energy efficiency, enabling efficient storage and release of hydrogen with minimal energy losses. They have a high energy density, allowing for the storage of large quantities of hydrogen in a compact liquid form. LOHCs also excel in safety, as they can be stored and transported under ambient conditions, reducing the need for high-pressure vessels and enhancing overall safety. Moreover, LOHCs can leverage existing infrastructure for liquid fuels, making their implementation more feasible and cost-effective. In terms of environmental impact, LOHCs contribute to reducing greenhouse gas emissions and air pollution by facilitating the storage and transportation of clean hydrogen. Although LOHC technology is still in the early stages of development, it shows great promise for scalability and can be a key player in the future of hydrogen transport.
- Hydrogen Fueling Stations are essential for enabling the widespread use of hydrogen as a clean energy source. They offer high energy efficiency in storing and dispensing hydrogen, although there are some losses in the production and distribution processes. The stations provide efficient storage and transportation of hydrogen through high-pressure or cryogenic methods. Safety protocols are strictly followed during dispensing. Infrastructure requirements vary, but integration with existing fueling station networks is possible. The initial costs of establishing fueling stations are high, but advancements and economies of scale are expected to reduce costs. Hydrogen fueling stations offer comparable range and flexibility to conventional vehicles, with quick refueling times. They have a positive environmental impact when sourced from renewable energy. The technology is advancing rapidly, with various types of stations being developed. Scalability is a key advantage, and compatibility with existing infrastructure facilitates adoption. Overall, hydrogen fueling stations are

a promising solution for clean energy, supporting transportation and energy storage applications.

The relative weights assigned to the evaluation criteria for hydrogen transport technologies can be categorized as follows:

- High relative weights: "Energy Efficiency" has a significant relative weight of 0.196, indicating its strong importance in the evaluation process. Maximizing energy efficiency is a key preference.
- Medium relative weights: "Energy Density" and "Infrastructure Requirements" share a relative weight of 0.130, reflecting their importance in the evaluation. Higher energy density and lower infrastructure requirements are preferred.
- Medium-low relative weights: "Cost," "Range and Flexibility," and "Environmental Impact" all share a relative weight of 0.086, suggesting their moderate importance. While they are considered significant, they are not as critical as energy efficiency and density.
- Low relative weights: "Safety" has a relatively low relative weight of 0.043, indicating its lower importance compared to other criteria. "Technological Maturity", "Scalability", and "Compatibility with Existing Infrastructure" have relative weights of 0.065 and 0.054, indicating their relatively lesser importance in the evaluation process.

These relative weights provide insight into the priority placed on each criterion, guiding the evaluation and selection of hydrogen transport technologies based on their performance in these areas.

Based on the weighted sum of alternatives, the performance of each hydrogen transport technology can be summarized as follows:

- Low weighted sum: Hydrogen tube trailers have the lowest weighted sum, indicating a relatively lower overall performance compared to other options.
- Medium weighted sum: Chemical hydrides, hydrail (hydrogen rail), LOHC, hydrogen pipelines, and hydrogen blending have medium weighted sums, suggesting a moderate overall performance.
- High weighted sum: CHG, Liquid Hydrogen (LH_2), Ammonia (NH_3) carriers, and hydrogen fueling stations have high weighted sums, indicating a relatively higher overall performance compared to other alternatives.

These findings provide insights into the relative performance of each technology based on the weighted analysis, with certain options standing out as having higher overall performance compared to others.

Hydrogen transport technologies face various barriers, including infrastructure development, cost and economics, safety concerns, technological maturity and efficiency, limited range and infrastructure coverage, public acceptance and awareness, durability and degradation, and the availability of critical materials. Despite these challenges, they also offer numerous opportunities, such as utilizing clean energy sources, enabling energy storage, decarbonizing transportation, providing long-range and quick refueling capabilities, compatibility with existing infrastructure, industrial applications, scalability, facilitating the global energy transition, and diversifying energy sources. By addressing these barriers and leveraging these opportunities, hydrogen transport technologies can play a significant role in the 21st-century energy transition towards a sustainable and low-carbon future.

5.4. Practical Implications

The findings of this research have practical implications for policymakers, industry stakeholders, and researchers involved in the energy transition. By understanding the general steps involved in hydrogen transport as well as the advantages, disadvantages, and process equations associated with various hydrogen transport technologies, decision-makers can make informed choices that align with their specific goals and requirements. The AHP methodology provides a structured approach to assessing and prioritizing technologies based on various criteria, aiding in the selection and implementation of the most suitable options. Additionally, the identified barriers and opportunities highlight the areas that need attention and further development to overcome challenges and maximize the potential of hydrogen transport technologies.

5.5. Transformative Impact of Hydrogen Transport Technology Study on Smart Cities Design and Governance

This comprehensive study on hydrogen transport technologies presents a pivotal contribution to the design, implementation, and governance of smart cities. By analyzing the general steps in hydrogen transport and meticulously examining a spectrum of technologies, from CHG to LOHC, the research provides a robust foundation for integrating sustainable energy solutions into the urban fabric. The utilization of the MCDMAHP methodology for technology assessment, based on key criteria such as energy efficiency, safety, and environmental impact, offers a quantitative and informed basis for decision-making in smart cities. The identification of barriers and opportunities for hydrogen transport technologies further equips city planners and policymakers with strategic insights for navigating the complexities of the 21st-century energy transition. Moreover, the discussion of practical implications for Morocco's green interconnections emphasizes the real-world applicability of the study's findings, transcending theoretical frameworks and directly impacting the sustainable development and energy strategies of smart cities worldwide.

5.6. Future Research Directions and Integration Opportunities in Sustainable Energy

This study on hydrogen transport technologies lays a solid foundation for future research endeavors and offers multiple avenues for integration with other works in the field of sustainable energy. One potential avenue for further exploration is a detailed examination of the economic implications and feasibility of implementing the identified hydrogen transport technologies in various global contexts. Additionally, future research could delve into advancements in materials science and engineering to optimize the efficiency and safety of these technologies. Integrating the findings of this study with research focused on renewable energy sources and storage technologies could present a holistic approach to shaping the future of clean energy systems. Furthermore, collaborative efforts with experts in policy analysis and regulatory frameworks could provide valuable insights into overcoming the barriers identified in this study, facilitating a smoother transition towards a hydrogen-based energy landscape. Overall, by fostering interdisciplinary collaboration and addressing evolving challenges, future research can build upon the present study's insights and contribute to a more sustainable and integrated energy ecosystem.

Declaration

Competing Interests:

The author declares that she has no known competing financial interests or personal relationships that could have appeared to influence the work reported in this paper.

Authors' Contribution:

Not applicable.

Availability of Data and Material:

The data used in this study, as well as the detailed methodology adapted, are comprehensively described in the methodology section (Section 2) of this chapter.

Funding:

The publication of this book chapter in Emerald Publishing was made possible without incurring processing charges. The author would like to express her appreciation to the editorial board and staff of Emerald Publishing, – particularly Dr Miltiadis D. Lytras, Dr Afnan Alkhaldi, and Dr Sawsan Malik, – for their commitment to promoting open access and facilitating the sharing of knowledge within the academic community.

Acknowledgments

The author would like to extend her sincere appreciation to the editorial board and staff of Emerald Publishing, – particularly Dr Miltiadis D. Lytras, Dr Afnan Alkhaldi, and Dr Sawsan Malik – for the esteemed invitation to contribute to the "Emerald Handbook of Smart Cities in the Gulf Region: Innovation, Development, Transformation, and Prosperity for Vision 2040."

Ethical Considerations:

This research study does not involve human subjects, animals, or sensitive data. As such, no formal ethical approval was required. The study adheres to the principles of research integrity and responsible conduct, ensuring accurate reporting of findings and appropriate citation of sources.

References

Abomazid, A. M., El-Taweel, N. A., & Farag, H. E. Z. (2022). Optimal energy management of hydrogen energy facility using integrated battery energy storage and solar photovoltaic systems. *IEEE Transactions on Sustainable Energy*, *13*, 1457–1468.

Acar, C., & Dincer, I. (2014). Comparative assessment of hydrogen production methods from renewable and non-renewable sources. *International Journal of Hydrogen Energy*, *39*, 1–12.

Agyekum, E. B., Nutakor, C., Agwa, A. M., & Kamel, S. (2022). A critical review of renewable hydrogen production methods: Factors affecting their scale-up and its role in future energy generation. *Membranes*, *12*, 173. https://doi.org/10.3390/membranes12020173

Akinyele, D., Olabode, E., & Amole, A. (2020). Review of fuel cell technologies and applications for sustainable microgrid systems. *Inventions*, *5*, 42. https://doi.org/10.3390/inventions5030042

Allendorf, M. D. (2015). Hydrogen materials advanced research consortium (PowerPoint).

Alzahrani, A., Ramu, S. K., Devarajan, G., Vairavasundaram, I., & Vairavasundaram, S. (2022). A review on hydrogen-based hybrid microgrid system: Topologies for hydrogen energy storage, integration, and energy management with solar and wind energy. *Energies*, *15*(21), 7979. https://doi.org/10.3390/en15217979

Antonini, C., Treyer, K., Streb, A. R., van der Spek, M., Bauer, C., & Mazzotti, M. (2020). Hydrogen production from natural gas and biomethane with carbon capture and storage – A techno-environmental analysis. *Sustainable Energy and Fuels*, *4*, 2967–2986.

Arnone, D., Rossi, A., & Bertoncini, M. (2014). Energy management in a smart grid-integrated hydrogen-based storage electric grid balancing when integrating large-capacity renewable energy sources. In *Proceedings of the 2014 IEEE international conference on renewable energy research and applications (ICRERA)* (pp. 1–6).

Arsalis, A., Papanastasiou, P., & Georghiou, G. E. (2022). Integration of a green hydrogen subsystem in a photovoltaic-battery nanogrid system. In *2022 2nd International Conference on Energy Transition in the Mediterranean Area (SyNERGY MED)*, pp. 1–5.

Ashari, P. A., Blind, K., & Koch, C. (2023). Knowledge and technology transfer via publications, patents, standards: Exploring the hydrogen technological innovation system. *Technological Forecasting and Social Change*, *186*, 122179. https://doi.org/10.1016/j.techfore.2023.122179

Ashkezari, A. Z. (2022). Numerical analysis of performance and emissions behavior of a Bi-Fuel engine with compressed natural gas enriched with hydrogen using variable compression ratio strategy. *International Journal of Hydrogen Energy*, *47*, 28585–28595. https://doi.org/10.1016/j.ijhydene.2022.06.232

Aydin, M. I., & Dincer, I. (2022). A life cycle impact analysis of various hydrogen production methods for public transportation sector. *International Journal of Hydrogen Energy*, *47*, 28596–28608. https://doi.org/10.1016/j.ijhydene.2022.06.234

Aziz, M. (2021). Liquid hydrogen: A review on liquefaction, storage, transportation, and safety. *Energies, 14*(18), 5917. https://doi.org/10.3390/en14185917

Berger, R. (n.d.). *Hydrogen transportation | The key to unlocking the clean hydrogen economy*. Roland Berger GmbH.

Bezdek, R. H. (2020) The hydrogen economy and jobs of the future. *ECS Transactions, 98*, 3–16. https://doi.org/10.1149/09801.0003ecst

Bharsakade, R. S., Acharya, P., Ganapathy, L., & Tiwari, M. K. (2021). A lean approach to healthcare management using multi criteria decision making. *OPSEARCH, 58*, 610–635.

Biagi, V., Bollati, M., & Gravio, G. D. (2021). Decision making and project selection: An innovative MCDM Methodology for a technology company. In *2021 The 2nd International Conference on Industrial Engineering and Industrial Management*. Gijón, Spain.

Biniwale, R. B., Rayalu, S. S., Devotta, S., & Ichikawa, M. (2008). Chemical hydrides: A Solution to high capacity hydrogen storage and supply. *International Journal of Hydrogen Energy, 33*, 360–365.

Bonner, B. B. (2018). Advanced hydrogen fueling station supply: Tube trailers. In *Proceedings of the 2018 DOE hydrogen and fuel cells program annual merit review and peer evaluation meeting*.

Bouramdane, A.-A. (2021, October). *Scenarios of large-scale solar integration with wind in Morocco: Impact of storage, cost, spatio- temporal complementarity and climate Change*. [Theses, Institut Polytechnique de Paris]. https://tel.archives-ouvertes.fr/tel-03518906

Bouramdane, A.-A. (2022a). *Morocco's current trajectory to mitigate and adapt to global and African warming*. https://doi.org/10.13140/RG.2.2.16687.00161/1

Bouramdane, A.-A. (2022b). PV, CSP et éolien au maroc: Intégration à géométrie variable. *énergie/mines & carrières*. https://doi.org/10.5281/zenodo.7594221

Bouramdane, A.-A. (2022c). Sécheresse: L'extrême Va-t-il progressivement devenir la norme? *énergie/mines & carrières*. https://doi.org/10.5281/zenodo.7594311

Bouramdane, A.-A. (2022d). Chaleur caniculaire, Incendies gigantesques à répétition: Des signes du changement climatique? *énergie/mines & carrières*. https://doi.org/10.5281/zenodo.7594264

Bouramdane, A.-A. (2022e). Pourquoi l'Atténuation et l'Adaptation aux changements climatiques sont complémentaires? *énergie/mines & carrières*. https://doi.org/10.5281/zenodo.7594404

Bouramdane, A.-A. (2023a). Ma Thèse En Une Page: « Choix Optimal de Technologies Renouvelables au Maroc en Fonction des Scénarios de Pénétration et du Climat», La Revue de l'Énergie N°668, éditée par le Conseil Français de l'Énergie. ISSN: 0303-240X. https://www.larevuedelenergie.com/choix-optimal-de-technologies-renouvelables-au-maroc-en-fonction-des-scenarios-de-penetration-et-du-clima https://www.larevuedelenergie.com/abonnement/la-revue-de-lenergie-668/

Bouramdane, A.-A. (2023b). *Climate resilience: Insights from global negotiations and Morocco's path to sustainability*. Lambert Academic Publishing (LAP). ISBN: 978-620-6-75083-3. https://www.morebooks.shop/shop-ui/shop/product/9786206750833, https://www.morebooks.shop/shop-ui/shop/translation-bundle/a0559d0d940

Bouramdane, A.-A. (2023c). Africa's vulnerability and the road ahead from COP27 to COP28. *énergie/mines & carrières*. https://energiemines.ma/africas-vulnerability-and-the-road-ahead-from-cop27-to-cop28/

Bouramdane, A.-A. (2023d). *Unlocking photovoltaic potential in climate action – Challenged countries*. Lambert Academic Publishing (LAP). ISBN: 978-620-6-84437-2. https://www. morebooks.shop/shop-ui/shop/product/9786206844372, https://www.morebooks.shop/shop-ui/shop/translation-bundle/e1751605240

Bouramdane, A.-A. (2023e). *Hydrogen production technologies: Modeling, Pros and cons, applications, suitable regions, and unveiling sustainability and economics through LCA and LCC*. Lambert Academic Publishing (LAP), ISBN: 978-620-6-78174-5. https://www.morebooks.shop/shop-ui/shop/product/9786206781745

Bouramdane, A.-A. (2023f). Crafting an optimal portfolio for sustainable hydrogen production choices in Morocco. *Fuel, 358*, 130292. https://doi.org/10.1016/j.fuel.2023.130292

Bouramdane, A.-A. (2023g). Identifying large-scale photovoltaic and concentrated solar power hot spots: Multi-criteria decision-making framework, world academy of science. *Engineering and Technology International Journal of Energy and Power Engineering, 17*(02). https://doi.org/10.5281/zenodo.7324107. https://publications.waset.org/abstracts/157186/pdf

Bouramdane, A.-A. (2023h). Spatial suitability assessment of onshore wind systems using the analytic hierarchy process, world academy of science. *Engineering and Technology International Journal of Energy and Power Engineering, 17*(07). https://doi.org/10.5281/zenodo.7324223; https://publications.waset.org/abstracts/157232/pdf

Bouramdane, A.-A. (2023i). Site suitability of offshore wind energy: A Combination of geographic referenced information and analytic hierarchy process, World academy of science. *Engineering and Technology International Journal of Energy and Power Engineering, 17*(02). https://doi.org/10.5281/zenodo.7324191; https://publications.waset.org/abstracts/157240/pdf

Bouramdane, A.-A. (2023j). *Potential site for offshore floating photovoltaic systems in morocco: Evaluation criteria required considering climate change effects to achieve the energy trilemma*. Lambert Academic Publishing (LAP). ISBN: 978-620-6-15964-3, https://www.morebooks.shop/shop-ui/shop/product/9786206159643, https://www.morebooks.shop/shop-ui/shop/translation-bundle/48db9fb8e40

Bouramdane, A.-A. (2023k). *Préservation des ressources d'Eau et transition énergétique: Point sur le photovoltaïque flottant, énergie/mines & carrières*. https://doi.org/10.5281/zenodo.8021774

Bouramdane, A.-A. (2023l). *How to manage Vulnerabilities in the renewable energy environment?* Leadvent Group, Renewable Energy Cyber Security Forum. https://doi.org/10.5281/zenodo.7730062

Bouramdane, A.-A. (2023m). Role of smart grids in achieving Paris agreement goals and determining optimal cybersecurity measures. *énergie/mines & carrières*. https://doi.org/10.5281/zenodo.8418765

Bouramdane, A.-A. (2023n). Cyberattacks in smart grids: Challenges and solving the multi-criteria decision-making for cybersecurity options, including ones that incorporate artificial intelligence, using an analytical hierarchy process. *Journal of Cybersecurity and Privacy (JCP), 3*, 4. https://doi.org/10.3390/jcp3040031

Bouramdane, A.-A. (2023o). Optimal water management strategies: Paving the way for sustainability in smart cities. *Smart Cities, 6*, 5. https://doi.org/10.3390/smartcities6050128

Bouramdane, A.-A. (2023p). *Un avenir hydrogéné au maroc: Dessalement de l'Eau – Défis et promesses, énergie/mines & carrières*. https://energiemines.ma/un-avenir-hydrogene-au-maroc-dessalement-de-leau-defis-et-promesses/

Bouramdane, A.-A. (2023q). Assessing the environmental impact of hydrogen fuel cell technologies (PEMFCs, SOFCs, AFCs): A Cradle-To-Gate attributional life cycle analysis. *Green Economics, 1*(02). http://jomardpublishing.com/UploadFiles/Files/journals/GE/V1N2/BouramdaneA.pdf

Bouramdane, A.-A. (2023r). Minéraux de la Transition Énergétique: Criticité Géologique, Géostratégique et Environnementale. *énergie/mines & carrières*. https://doi.org/10.5281/zenodo.7594617

Bouramdane, A.-A. (2023s). Assessment of CMIP6 Multi-Model projections worldwide: Which regions are getting warmer and are going through a drought in Africa and Morocco? What changes from CMIP5 to CMIP6? *Sustainability, 15*, 690. https://doi.org/10.3390/su15010690

Bouramdane, A.-A. (2023t). Determining vulnerable areas to warming and drought in Africa and Morocco based on CMIP6 Projections: Towards the implementation of mitigation and adaptation measures. *GU General Assembly 2023, Vienna, Austria, 24–28* Apr 2023, EGU23-2456. https://doi.org/10.5194/egusphere-egu23-2456

Bouramdane, A.-A. (2023u). *Climate risks and energy transition in morocco: Vulnerability to climate losses and damages and uncertainty in the renewable electricity mix under different penetration.* Lambert Academic Publishing (LAP). ISBN: 978-620-6-17980-1. https://www.morebooks.shop/shop-ui/shop/product/9786206179801, https://www.morebooks.shop/gb/translation_bundle_cb314d0c040

Bouramdane, A.-A. (2023v). Lieux les plus sensibles Au changement climatique nécessitant des mesures d'Atténuation et d'Adaptation. *énergie/mines & carrières*. https://doi.org/10.5281/zenodo.7937556

Bouramdane, A.-A. (2023w). Quelle est la relation entre l'Agriculture et le changement climatique? *énergie/mines & carrières*. https://doi.org/10.5281/zenodo.7730008

Bouramdane, A.-A. (2023x). Production d'hydrogène vert au maroc: Quelle technologie est la plus adaptée à différents niveaux de pénétration renouvelable? *énergie/mines & carrières*. https://doi.org/10.5281/zenodo.8144588

Bouramdane, A.-A. (2023y). L'Oasis d'Hydrogène vert pour une agriculture marocaine durable, La jaune et la rouge N°790. *Le Magazine des Alumni de Polytechnique, Dossier: Environnement et Société, 12*, ISSN: 0021–5554. https://www.lajauneetlarouge.com/loasis-dhydrogene-vert-pour-une-agriculture-marocaine-durable/

Bouramdane, A.-A. (2023z). Hydrogène, Captage et Stockage du CO2 et Sobriété Énergétique: Tour d'Horizon. *énergie/mines & carrières*. https://doi.org/10.5281/zenodo.7774592.

Bouramdane, A. A. (2024a). Shaping resilient buildings and cities: Climate change impacts, Metrics, and strategies for mitigation and adaptation. *Information System and Smart City*. https://api.semanticscholar.org/CorpusID:266853661

Bouramdane, A. A. (2024b). Morocco's path to a climate-resilient energy transition: identifying emission drivers, proposing solutions, and addressing barriers. *DP Sciences-Science and Technology for Energy Transition (STET), Formerly Oil Gas Science and Technology-Revue d'IFP Energies Nouvelles*. https://doi.org/10.2516/stet/2024021

Bragatto, T., Carere, F., Cresta, M., Gatta, F. M., Geri, A., Lanza, V., Maccioni, M., & Paulucci, M. (2022). Location and sizing of hydrogen based systems in distribution network for renewable energy integration. *Electric Power Systems Research, 204*, 107919. https://doi.org/10.1016/j.epsr.2022.107919

Brans, J. P. (2006). Chapter 5 PROMETHEE Methods. In *Multiple Criteria Decision Analysis: State of the Art Surveys* (pp. 163–196, Volume 78). Springer: New York, NY, USA. https://doi.org/10.1007/0-387-23081-5_5

Brown, O. L. I. (1951). The Clausius-Clapeyron equation. *Journal of Chemical Education, 28*, 428.

Brown, G. O. (2002). The history of the Darcy-Weisbach equation for pipe flow resistance. In *Proceedings of the environmental and water resources history session at the world water and environmental resources congress*. ASCE: Reston, VA, USA. https://doi.org/10.1061/40650(2003)4

Brown, G. O. (2007). The history of the Darcy-Weisbach equation.

Buckley, J. J., Feuring, T., & Hayashi, Y. (1999). Fuzzy hierarchical analysis, FUZZ-IEEE'99. 1999 IEEE International Fuzzy Systems. *Conference Proceedings (Cat. No.99CH36315), 2*, 1009–1013.

Cader, J., Koneczna, R., & Olczak, P. (2021). The impact of economic, energy, and environmental factors on the development of the hydrogen economy. *Energies, 14*(24), 8231. https://doi.org/10.3390/en14248231

Cavana, M., Mazza, A., Chicco, G., & Leone, P. (2021). Electrical and gas networks coupling through hydrogen blending under increasing distributed photovoltaic generation. *Applied Energy*, *290*, 116764.

Çelik, D., & Yildiz, M. (2017). Investigation of hydrogen production methods in accordance with green chemistry principles. *International Journal of Hydrogen Energy*, *42*, 23395–23401.

Chan, F. T. S., Kumar, N., Tiwari, M. K., Lau, H. C. W., & Choy, K. L. (2008). Global supplier selection: A Fuzzy-AHP approach. *International Journal of Production Research*, *46*, 3825–3857.

Chang, A. C.-C., Chang, H. F., Lin, F.-J., Lin, K.-H., & Chen, C.-H. (2011). Biomass gasification for hydrogen production. *Fuel and Energy Abstracts*, *3*, 1–6.

Chen, Y.-S., Chuang, H.-M., Sangaiah, A. K., Lin, C.-K., & Huang, W. (2018). A Study for project risk management using an advanced MCDM-Based DEMATEL-ANP approach. *Journal of Ambient Intelligence and Humanized Computing*, *10*, 2669–2681.

Chen, Y., Jian, Q., Huang, Z., Zhao, J., Bai, X., & Li, D. (2021). Improvement of thermal management of proton exchange membrane fuel cell stack used for portable devices by integrating the ultrathin vapor chamber. *International Journal of Hydrogen Energy*, *46*, 27832–27844. https://doi.org/10.1016/j.ijhydene.2021.06.012

Climate Policy Database. (2021). *National Roadmap for Green Hydrogen Morocco*. https://climatepolicydatabase.org/policies/national-roadmap-green-hydrogen

Corigliano, O., Pagnotta, L., & Fragiacomo, P. (2022). On the technology of solid oxide fuel cell (SOFC) energy systems for stationary power generation: A review. *Sustainability*, *14*, 12345. https://doi.org/10.3390/su140212345

Cunanan, C., Tran, M., Lee, Y., Kwok, S., Leung, V., & Fowler, M. W. (2021). A Review of heavy-duty vehicle powertrain technologies: Diesel engine vehicles, battery electric vehicles, and hydrogen fuel cell electric vehicles. *Clean Technologies*, *3*, 1–14. https://doi.org/10.3390/cleantechnol3010001

Dash, S. K., Chakraborty, S., & Elangovan, D. (2023). A brief review of hydrogen production methods and their challenges. *Energies*, *16*(3), 1141. https://doi.org/10.3390/en16031141

Delima, L., & Vezirolu, T. (2001). Long-term environmental and socio-economic impact of a hydrogen energy program in Brazil. *International Journal of Hydrogen Energy*, *26*, 39–45.

Dincer, I., & Acar, C. (2015). Review and evaluation of hydrogen production methods for better sustainability. *International Journal of Hydrogen Energy*, *40*, 11094–11111.

Dodds, P. E., Staffell, I., Hawkes, A. D., Li, F. G., Grünewald, P., McDowall, W., & Ekins, P. (2015). Hydrogen and fuel cell technologies for heating: A review. *International Journal of Hydrogen Energy*, *40*, 2065–2083.

Dou, B., Zhang, H., Song, Y., Zhao, L., Jiang, B., Mingxing, H., Ruan, C., Chen, H., & Xu, Y. (2019). Hydrogen production from the thermochemical conversion of biomass: Issues and challenges. *Sustainable Energy & Fuels*, *3*, 1–15. https://doi.org/10.1039/C8SE00535D

Eckroll, J. (2017). Concepts for large scale hydrogen liquefaction plants. In *Proceedings of the 2017 Norwegian University of Science and Technology (NTNU)*.

Ehrhart, B. D., Anleu, G. A. B., Mohmand, J., Klebanoff, L. E., & Baird, A. (2021). *Current efforts in hydrogen for rail*. Proposed for presentation at the 16th Annual International Hydrail Conference held June 29-29, 2021.

Ekspong, J., Larsen, C., Stenberg, J., Kwong, W. L., Wang, J., Zhang, J., Johansson, E. M. J., Messinger, J., Edman, L., & Wågberg, T. (2021). Solar-driven water splitting at 13.8% solar-to-hydrogen efficiency by an earth-abundant electrolyzer. *ACS Sustainable Chemistry & Engineering*, *9*, 12345–12355. https://doi.org/10.1021/acssuschemeng.1c01234

Elberry, A. M., Thakur, J., & Veysey, J. (2021). Seasonal hydrogen storage for sustainable renewable energy integration in the electricity sector: A case study of Finland. *Journal of Energy Storage*, *33*, 102038. https://doi.org/10.1016/j.est.2020.102038

Erdener, B. C., Sergi, B., Guerra, O. J., Chueca, A. L., Pambour, K. A., Brancucci, C., & Hodge, B.-M. S. (2022). A review of technical and regulatory limits for hydrogen blending in natural gas pipelines. *International Journal of Hydrogen Energy*, *47*, 12345–12356. https://doi.org/10.1016/j.ijhydene.2022.01.001

Fan, L., Tu, Z., & Chan, S. H. (2021). Recent development of hydrogen and fuel cell technologies: A review. *Energy Reports*, *7*, 1–18. https://doi.org/10.1016/j.egyr.2021.05.001

Ferrara, A., Jakubek, S., & Hametner, C. (2021). Energy management of heavy-duty fuel cell vehicles in real-world driving scenarios: Robust design of strategies to maximize the hydrogen economy and system lifetime. *Energy Conversion and Management*, *232*, 113795.

Filippov, S. P., & Yaroslavtsev, A. B. (2021). Hydrogen energy: Development prospects and materials. *Russian Chemical Reviews*, *90*, 627–643.

French, S. (1996). Multicriteria methodology for decision aiding. *Journal of the Operational Research Society*, *48*, 1257–1258.

Gan, L., & Liu, M. (2017). Progress of heat management of fuel cell for energy storage.

Ghavam, S., Vahdati, M., Wilson, I. A. G., & Styring, P. (2021). Sustainable ammonia production processes. *Frontiers in Energy Research*. https://doi.org/10.3389/fenrg.2021.618401 https://api.semanticscholar.org/CorpusID:232413387

Global Solar Atlas. (n.d.). https://globalsolaratlas.info/

Global Wind Atlas. (n.d.). https://globalwindatlas.info/fr

Glushchenko, A. M. (2020). Decarbonization of the steel industry: The role of state economic policy. *The Problems of Economy*, *1*, 1–10.

Green Car Congress. (2022). *Energy, technologies, issues and policies for sustainable mobility*. https://www.greencarcongress.com/2022/12/20221222-h2africa.html

Green Hydrogen Organisation (GH2). (n.d.) *GH2 Country Portal – Morocco*. https://gh2.org/countries/morocco

Groppi, D., Garcia, D. A., Basso, G. L., Cumo, F., & de Santoli, L. (2018). Analysing economic and environmental sustainability related to the use of battery and hydrogen energy storages for increasing the energy independence of small islands. *Energy Conversion and Management*, *157*, 150–160. https://doi.org/10.1016/j.enconman.2017.11.033

Guy, P., László, T., & Julien, C. (2021). Effects of hydrogen addition on design, Maintenance and surveillance of gas networks. *Processes*, *9*, 1234. https://doi.org/10.3390/pr9011234

Haggi, H., Fenton, J. M., Brooker, P., & Sun, W. (2022). (Invited) Renewable hydrogen systems enable deep energy decarbonization of power and transportation sectors. *ECS Meeting Abstracts*, *MA2022-01*, 1–2.

Hashemizadeh, A., & Ju, Y. (2019). Project portfolio selection for construction contractors by MCDM–GIS approach. *International Journal of Environmental Science and Technology*, *16*, 1–14. https://doi.org/10.1007/s13762-019-02300-z

Hosseini, S. E., & Butler, B. (2020). An overview of development and challenges in hydrogen powered vehicles. *International Journal of Green Energy*, *17*, 13–37.

Hwang, C.-L., & Yoon, K. (1981). Multiple attribute decision making: Methods and applications a State-of-the-Art survey. *Lecture Notes in Economics and Mathematical Systems*. (Volume 48). Springer, Berlin, Germany.

IEA. (2019). *The future of hydrogen: Seizing today's opportunities*. Retrieved July 4, 2023, from https://www.hydrogenexpo.com/media/9370/the_future_of_hydrogen_iea.pdf

IEA. (2023). *National hydrogen strategy*. Retrieved October 11, 2023, from https://www.iea.org/policies/16971-national-hydrogen-strategy

Inal, O. B., Zincir, B., & Deniz, C. (2022). Investigation on the decarbonization of shipping: An approach to hydrogen and ammonia. *International Journal of Hydrogen Energy*, *47*, 28605–28614. https://doi.org/10.1016/j.ijhydene.2022.06.235

Ingale, G. U., Kwon, H. M., Jeong, S., Park, D., Kim, W. I., Bang, B., il Lim, Y., Kim, S. W., Kang, Y.-B., Mun, J., Jun, S., & Lee, U. (2022). Assessment of greenhouse gas emissions from hydrogen production processes: Turquoise hydrogen vs. Steam methane reforming. *Energies*, *15*, 1234. https://doi.org/10.3390/en15041234

International Energy Agency (IEA). (n.d.). *Levelised cost of hydrogen maps*. https://www.iea.org/data-and-statistics/data-tools/levelised-cost-of-hydrogen-maps

IRENA. (2018). *Hydrogen from renewable power: Technology outlook for the energy transition*. Retrieved July 4, 2023, from https://www.irena.org/publications/2018/Sep/Hydrogen-from-renewable-power

Ishimoto, Y., Voldsund, M., Nekså, P., Roussanaly, S., Berstad, D. O., & Gardarsdottir, S. (2020). Large-Scale production and transport of hydrogen from Norway to Europe and Japan: Value chain analysis and comparison of liquid hydrogen and ammonia as energy carriers. *International Journal of Hydrogen Energy*, *45*, 32865–32883.

Jacobson, M. Z., Colella, W. G., & Golden, D. M. (2005). Cleaning the air and improving health with hydrogen fuel-cell vehicles. *Science*, *308*, 1901–1905.

Jahangiri, M., Mostafaeipour, A., Habib, H. U. R., Saghaei, H., & Waqar, A. (2021). Effect of emission penalty and annual interest rate on cogeneration of electricity, heat, and hydrogen in Karachi: 3E Assessment and sensitivity analysis. *The Journal of Engineering*, 1–16.

Jensen, S. S., Ketzel, M., Brandt, J., Frohn, L. M., Winther, M., Nielsen, O. K., Jørgensen, K. A., & Karlsson, K. (2011). Impacts of large-scale introduction of hydrogen in the road transport sector on urban air pollution and human exposure in Copenhagen.

Jiao, F., Lu, B., Chen, C., Dai, F., Liu, T., & Liu, Q. (2022). In-depth energy and irreversibility analysis in the solar driven two-step thermochemical water splitting cycle for hydrogen production. *Journal of Thermal Science*, *32*, 1–16.

JRC. (2023). European commission, Science for policy briefs, Assessment of hydrogen delivery options. Retrieved July 4, 2023, from https://joint-research-centre.ec.europa.eu/system/files/2021-06/jrc124206_assessment_of_hydrogen_delivery_options.pdf

Ju-long, D. (1982). Control problems of grey systems. *Systems & Control Letters*, *1*, 288–294.

Jung, C., Senholdt, M., Spenke, C., Schmidt, T. J., & Ulmer, S. (2019). Hydrogen monitoring in the heat transfer fluid of parabolic trough plants. SOLARPACES 2018: International Conference on Concentrating Solar Power and Chemical Energy Systems.

Khokhar, H. N. (2023). Decarbonisation of global economies; is net zero emission achievable? The case for hydrogen and fuel cell technology for innovative futures. *Journal of Entrepreneurship and Innovation in Emerging Economies*, *9*, 92–130.

Knözinger, H. (2020). Clausius-Clapeyron Equation, Catalysis from A to Z.

Kruusenberg, I., Chadha, K., & Atal, T. (2022). High power density fuel cell systems for portable electric generators. *ECS Meeting Abstracts*, *MA2022-01*, 1–2.

Kumar, G., Eswari, A. P., Kavitha, S., Kumar, M. D., Kannah, R. Y., How, L. C., Muthukaruppan, G., & Jeyakumar, R. B. (2020). Thermochemical conversion routes of hydrogen production from organic biomass: Processes, challenges and limitations. *Biomass Conversion and Biorefinery*, *13*, 8509–8534.

Kumar, A., Muthukumar, P., Sharma, P., & Kumar, E. A. (2022). Absorption based solid state hydrogen storage system: A review. *Sustainable Energy Technologies and Assessments*, *50*, 102204. https://doi.org/10.1016/j.seta.2022.102204

Kumar, S. S., & Lim, H. (2022). An overview of water electrolysis technologies for green hydrogen production. *Energy Reports*, *8*, 13793–13813.

Lee, H., Lee, D., & Kim, Y. (2022). Heating performance of a coolant-Source heat pump using waste heat from stack and electric devices in fuel cell electric vehicles under cold conditions. *Energy Conversion and Management*, *252*, 115117. https://doi.org/10.1016/j.enconman.2022.115117

Lemons, R. A. (1989). Fuel cell for transportation. In *Proceedings of the 1989 international conference on fuel cells*.

Li, H., Cao, X., Liu, Y., Shao, Y., Nan, Z., Teng, L., Peng, W., & Bian, J. (2022). Safety of hydrogen storage and transportation: An overview on mechanisms, Techniques, and challenges. *Energy Reports, 15*, 1234. https://doi.org/10.3390/en15041234

Li, Q., Lin, X., Luo, Q., Chen, Y., Wang, J., Jiang, B., & Sheng Pan, F. (2022). Kinetics of the hydrogen absorption and desorption processes of hydrogen storage alloys: A review. *International Journal of Minerals, Metallurgy and Materials, 29*, 32–48.

Li, M., Ming, P., Huo, R., & Mu, H. (2023). Economic assessment and comparative analysis of hydrogen transportation with various technical process. *Journal of Renewable and Sustainable Energy, 15*, 025904. https://doi.org/10.1063/5.0134603

Liou, C. P. (1998). Limitations and proper use of the Hazen-Williams equation. *Journal of Hydraulic Engineering, 124*, 951–954.

Liu, Z., Yang, M., Jia, W., & Ding, T. (2022). Low-carbon economic dispatch of integrated electricity-heat-hydrogen systems considering integrated demand response. In *2022 IEEE/IAS Industrial and Commercial Power System Asia (I&CPS Asia)*, pp. 1173–1177.

Lutostansky, E. M., Creitz, L., Jung, S., Schork, J. M., Worthington, D. R. E., & Xu, Y., (2013) Modeling of underground hydrogen pipelines. *Process Safety Progress, 32*. https://doi.org/10.1002/prs.11588

M. de la Transition Énergétique et du Développement Durable. (2023). Feuille de route hydrogène vert vecteur de transition énergétique et de croissance durable. Retrieved October 11, 2023, from https://www.mem.gov.ma/Lists/Lst_rapports/Attachments/36/Feuille%20de%20route%20de%20hydrog%C3%A8ne%20vert.pdf

Mahajan, D., Tan, K., Venkatesh, T., Kileti, P., & Clayton, C. R. (2022). Hydrogen blending in gas pipeline networks – A Review. *Energies, 15*, 3582. https://doi.org/10.3390/en15103582

Maschinenbau, F. (2012). Hazen–Williams equation.

Medrano-García, J. D., Charalambous, M. A., & Guillén-Gosálbez, G. (2022). Economic and environmental barriers of CO_2-based Fischer–Tropsch electro-Diesel. *ACS Sustainable Chemistry & Engineering, 10*, 11751–11759. https://api.semanticscholar.org/CorpusID:251862224

Melaina, M. W., Steward, D., Penev, M., McQueen, S., Jaffe, S., & Talón, C. (2012). Hydrogen infrastructure market readiness: Opportunities and potential for near-term cost reductions. In *Proceedings of the Hydrogen Infrastructure Market Readiness Workshop and Summary of Feedback Provided through the Hydrogen Station Cost Calculator*. NREL/BK-5600-55961; National Renewable Energy Laboratory: Golden, CO, USA. https://www.nrel.gov/docs/fy12osti/55961.pdf

Meurer, A., & Kern, J. (2021). Fischer–Tropsch synthesis as the key for decentralized sustainable kerosene production. *Energies, 14*, 1836. https://api.semanticscholar.org/CorpusID:233698429

Mission Innovation. (n.d.). *National Innovation Pathway*. https://www.mission-innovation.net/our-members/morocco/pathway/

Mneimneh, F., Ghazzawi, H., Hejjeh, M. A., Manganelli, M., & Ramakrishna, S. (2023). Roadmap to achieving sustainable development via green hydrogen. *Energies, 16*(3), 1368. https://doi.org/10.3390/en16031368

Nami, H., Rizvandi, O. B., Chatzichristodoulou, C., Hendriksen, P. V., & Frandsen, H. L. (2022). Techno-economic analysis of current and emerging electrolysis technologies for green hydrogen production. *Energy Conversion and Management, 264*, 115668. https://doi.org/10.1016/j.enconman.2022.115668

Niermann, M., Timmerberg, S., Drünert, S., & Kaltschmitt, M. (2021). Liquid organic hydrogen carriers and alternatives for international transport of renewable hydrogen. *Renewable & Sustainable Energy Reviews, 135*, 110171.

Nikolova, D., & Stoilov, D. (2022). Household energy efficiency, Diversification of suppliers and renewable hydrogen production the pillars of European energy independence. *2022 14th electrical engineering faculty conference (BulEF)*, pp. 1–3.

Ogungbemi, E., Wilberforce, T., Ijaodola, O., Thompson, J., & Olabi, A. G. (2020). Selection of proton exchange membrane fuel cell for transportation. *International Journal of Hydrogen Energy*, *45*, 12345–12356. https://doi.org/10.1016/j.ijhydene.2019.12.034

Okada, Y. (2019). Innovative system for hydrogen storage and transportation by organic chemical hydride method. *Journal of the Japan Welding Society*, *88*, 123–130.

Opricovic, S., & Tzeng, G.-H. (2004). Compromise solution by MCDM Methods: A comparative analysis of VIKOR and TOPSIS. *European Journal of Operational Research*, *156*, 445–455.

Orlova, S., Mezeckis, N., & Vasudev, V. (2023). Compression of hydrogen gas for energy storage: A review. *Latvian Journal of Physics and Technical Sciences*, *60*, 4–16.

Peighambardoust, S. J., Rowshanzamir, S., & Amjadi, M. (2010). Review of the proton exchange membranes for fuel cell applications. *International Journal of Hydrogen Energy*, *35*, 9349–9384.

Petitpas, G., Bénard, P., Klebanoff, L. E., Xiao, J., & Aceves, S. M. (2014). A comparative analysis of the cryo-compression and cryo-adsorption hydrogen storage methods. *International Journal of Hydrogen Energy*, *39*, 10564–10584.

Ratnakar, R. R., Gupta, N., Zhang, K., van Doorne, C., Fesmire, J. E., Dindoruk, B., & Balakotaiah, V. (2021). Hydrogen supply chain and challenges in large-scale LH2 storage and transportation. *International Journal of Hydrogen Energy*, *46*, 34927–34940. https://doi.org/10.1016/j.ijhydene.2021.08.021

Roy, B. (1968). Classement et Choix en Présence de Points de Vue Multiples.

Saaty, T. L. (1990). *The analytic hierarchy process: Planning, priority setting, resource allocation.*

Saaty, T. L. (2005). Theory and applications of the analytic network process: Decision making with benefits, opportunities, costs, and risks. In *Proceedings of the 2005 International series in operations research & management science* (vol. 95). Springer, Boston, MA. https://doi.org/10.1007/0-387-33987-6_1

Saaty, T. L. (2008). Decision making with the analytic hierarchy process. *International Journal of Services Sciences*, *1*, 83–98.

Sakintuna, B., Lamari-Darkrim, F., & Hirscher, M. (2007). Metal hydride materials for solid hydrogen storage: A review. *International Journal of Hydrogen Energy*, *32*, 1121–1140.

Samiee, L., Goodarzvand-Chegini, F., GhasemiKafrudi, E., & Kashefi, K. (2020). *Hydrogen recovery in an industrial Chlor-Alkali plant using alkaline Fuel cell and hydrogen boiler Techniques: Techno-Economic Assessment and Emission Estimation.*

Seker, S., & Aydin, N. (2021). Assessment of hydrogen production methods via integrated MCDM approach under uncertainty. *International Journal of Hydrogen Energy*, *46*, 28208–28223. https://doi.org/10.1016/j.ijhydene.2021.06.131

Singla, M. K., Nijhawan, P., & Oberoi, A. S. (2021). Hydrogen fuel and fuel cell technology for cleaner future: A review. *Environmental Science and Pollution Research*, *28*, 15607–15626.

Song, P., Qiao, G., Song, J., Ye, Q., Xu, G., & Deng, Z. (2021). Study on aging mechanism of membrane electrode assembly in stationary power generation fuel cell with variable load condition. In *2021 International Conference on Power System Technology (POWERCON)*, pp. 209–212.

Southall, E., & Lukashuk, L. V. (2022). Hydrogen storage and transportation technologies to enable the hydrogen economy: Liquid organic hydrogen carriers. *Johnson Matthey Technology Review*, *66*, 2–15. https://doi.org/10.1595/205651322X16207857531246

Speckmann, F.-W., Weeber, M., & Birke, K. P. (2021). Techno-economic assessment of hydrogen usage in a smart grid, employing a staggered optimization algorithm. *E3S Web of Conferences*, *328*, 02001. https://doi.org/10.1051/e3sconf/202132802001

Sun, D., Guo, D., & Xie, D. (2023). Using multicriteria decision making to evaluate the risk of hydrogen energy storage and transportation in cities. *Sustainability*, *15*, 1234. https://doi.org/10.3390/su1501234

Sun, H., He, C., Yu, X., Wu, M., & Ling, Y. (2019). Optimal siting and sizing of hydrogen refueling stations considering distributed hydrogen production and cost reduction for regional consumers. *International Journal of Energy Research*, *43*, 4184–4200.

Sun, Z., Lin, B., & Yan, F. (2018). Anion-exchange membranes for alkaline fuel-cell applications: The effects of cations. *Chem-SusChem*, *11*(1), 58–70.

Suri, S. K., & Siddique, M. (2019). Novel and optimized techniques for storage and transportation of hydrogen: Perspectives and challenges. *Journal of Applied and Emerging Sciences*, *9*, 1–12.

Talebian, H., Herrera, O. E., & Mérida, W. (2020). Policy effectiveness on emissions and cost reduction for hydrogen supply chains: The case for British Columbia. *International Journal of Hydrogen Energy*, *45*, 34927–34940. https://doi.org/10.1016/j.ijhydene.2020.07.222

Tarhan, C., & Çil, M. A. (2021). A Study on hydrogen, the clean energy of the future: Hydrogen storage methods. *Journal of Energy Storage*, *40*, 102676.

Tayarani, H., & Ramji, A. (2022). Life cycle assessment of hydrogen transportation pathways via pipelines and truck trailers: Implications as a low carbon fuel. *Sustainability*.

Tenny, K. M., & Cooper, J. (2020). Ideal Gas Behavior.

Thanh, N. V. (2022). Optimal waste-to-energy strategy assisted by fuzzy MCDM model for sustainable solid waste management. *Sustainability*, *14*, 12345. https://doi.org/10.3390/su14012345

Tochigi, S., & Dowaki, K. (2021). Environmental impacts analysis of stationary fuel cell combined heat and power generation systems. *Journal of the Japan Institute of Energy*, *100*, 1234–1243. https://doi.org/10.3775/jie.100.1234

Tooryan, F., Hassanzadehfard, H., Dargahi, V., & Jin, S. (2022). A cost-effective approach for optimal energy management of a hybrid CCHP Microgrid with different hydrogen production considering load growth analysis. *International Journal of Hydrogen Energy*, *47*, 12345–12356. https://doi.org/10.1016/j.ijhydene.2022.01.123

Trencher, G., & Edianto, A. S. (2021). Drivers and barriers to the adoption of fuel cell passenger vehicles and buses in Germany. *Energies*, *14*, 833. https://doi.org/10.3390/en14040833

Tullo, A. H. (2022). Organics challenge ammonia as hydrogen carriers. *ACS Central Science*, *8*, 1471–1473.

Ulucan, T., Akhade, S. A., Ambalakatte, A., Autrey, T. S., Cairns, A., Chen, P., Cho, Y. W., F. Gallucci, W. Gao, J. B. Grinderslev, Grubel, K., Jensen, T. R., de Jongh, P. E., Kothandaraman, J., Lamb, K. E., Lee, Y.-S., Makhloufi, C., Ngene, P., Olivier, P., Webb, C. J., Wegman, B., Wood, B. K., & Weidenthaler, C. (2022). Hydrogen storage in liquid hydrogen carriers: Recent activities and new trends. *Progress in Energy*, *5*.

Usman, M. R. (2022). Hydrogen storage methods: Review and current status. *Renewable and Sustainable Energy Reviews*, *161*, 112303. https://doi.org/10.1016/j.rser.2022.112303

Ustolin, F., Campari, A., & Taccani, R. (2022). An extensive review of liquid hydrogen in transportation with focus on the maritime sector. *Journal of Marine Science and Engineering*, *10*(1), 123. https://doi.org/10.3390/jmse100100123

Visvanathan, V. K., Palaniswamy, K., Ponnaiyan, D., Chandran, M., Kumaresan, T., Ramasamy, J., & Sundaram, S. (2023). Fuel cell products for sustainable transportation and stationary power generation: Review on market perspective. *Energies*, *16*(3), 1234. https://doi.org/10.3390/en16031234

Wang, X., Shen, Y., Tao, C., & Wu, F. (2023). Review on computation models of high-pressure hydrogen compressibility factor. In *Proceedings of the 19th International Flow Measurement Conference 2022 on Flow Measurement FLOMEKO 2022*.

Westbroek, W. (2021). Laying the tracks for hydrail. *Electric and Hybrid Rail Technology*, *1*, 12–15.

Wheeldon, M. (2022). The business of growing a hydrogen business: Lessons from disruptive innovation theory. *The APPEA Journal*, *62*, 123–130.

Wickham, D. J., Hawkes, A. D., & Jalil-Vega, F. (2022). Hydrogen supply chain optimisation for the transport sector – Focus on hydrogen purity and purification requirements. *Applied Energy*, *305*, 117740. https://doi.org/10.1016/j.apenergy.2021.117740

Wu, Q., & Li, C. (2023). Dynamic pricing and energy management of hydrogen-based integrated energy service provider considering integrated demand response with a bi-level approach. *Journal of Energy Storage*, *61*, 106830. https://doi.org/10.1016/j.est.2023.106830

Xiangmin, P. (2008). Comparison of hydrogen transportation methods for hydrogen refueling station. *Journal of Tongji University*, *36*, 123–128.

Xiao, F. (2011). Hydrogen distribution networks targeting minimum exergy consumption. *CIESC Journal*, *62*, 1234–1240.

Xu, J. Z., Wang, M., & Guo, P. (2022). Construction of hydrogen safety evaluation model based on analytic hierarchy process (AHP). *Trends in Renewable Energy*, *8*(2), 140–151. https://doi.org/10.17737/tre.2022.8.2.00150

Yeates, C., Schmidt-Hattenberger, C., & Bruhn, D. (2021). *A method for simultaneous minimal-cost supply node location and network design in pipelined infrastructure*. In *Proceedings of the 11th Trondheim CCS Conference (TCCS-11)*, Trondheim, Norway, 21–23 June 2021.

Yoon, K., & Hwang, C.-L. (1995). *Multiple attribute decision making: An introduction*. SAGE Publications, Inc.: Thousand Oaks, CA, USA. https://dx.doi.org/10.4135/9781412985161

Yufit, V., Hale, B., Matian, M., Mazur, P., & Brandon, N. P. (2013). Development of a regenerative hydrogen-vanadium fuel cell for energy storage applications. *Journal of The Electrochemical Society*, *160*, A856–A861.

Zacharia, R., & Ullah Rather, S. (2016). Review of solid state hydrogen storage methods adopting different kinds of novel materials. *Journal of Nanomaterials*, *2015*, 1–4. https://doi.org/10.1155/2015/723120

Zeleny, M. (2004). Multiple Criteria Decision Making (MCDM). In *Proceedings of the 2004 international conference on multiple criteria decision making*.

Zhang, S., Xue, J., & Sun, X. (2022). Suitability evaluation of hydrogen production and hydrogenation station based on AHPFuzzy comprehensive evaluation method. *Energy Engineering*, *8*, 123–135. https://doi.org/10.1016/j.energyeng.2022.123

Zhao, S., Liu, P., Niu, Y., Chen, Z., Khan, A. U., Zhang, P., & Li, X. (2018). A Novel early warning system based on a sediment microbial fuel cell for In situ and real time hexavalent chromium detection in industrial wastewater. *Sensors (Basel, Switzerland)*, *18*, 1234. https://doi.org/10.3390/s18041234

Zheng, J., Zhou, H., Wang, C.-G., Ye, E., Xu, J., Loh, X. J., & Li, Z. (2021). Current research progress and perspectives on liquid hydrogen rich molecules in sustainable hydrogen storage. *Energy Storage Materials*, *35*, 695–722.

Zhou, D., Yan, S., Huang, D., Shao, T., Xiao, W., Hao, J., Wang, C., & Yu, T. (2022). Modeling and simulation of the hydrogen blended gas-electricity integrated energy system and influence analysis of hydrogen blending modes. *Energy*, *239*, 121629.

Zhu, H., & Dai, H. (2022). Enhancement of partial oxidation reformer by the free-section addition for hydrogen production. *Renewable Energy*, *193*, 1234–1243. https://doi.org/10.1016/j.renene.2022.05.123

Chapter 23

Smart City Framework for Nusantara Capital City Development

Bambang Susantono[a], *Mohammed Ali Berawi*[b] *and Mustika Sari*[b, c]

[a]*Department of Urban and Regional Planning, Diponegoro University, Semarang, Indonesia*
[b]*Department of Civil Engineering, Faculty of Engineering, Universitas Indonesia, Indonesia*
[c]*Center for Sustainable Infrastructure Development, Faculty of Engineering, Universitas Indonesia, Indonesia*

Abstract

Nusantara, Indonesia's currently developed new capital city, aims to lead the nation toward a more efficient and sustainable future by fostering an inclusive, sustainable, and prosperous development for all. Envisioned to be built as a smart city, Nusantara leverages the latest technological advancements across various domains as the basis for this cutting-edge urban development. This chapter discusses the advanced technologies implemented to realize the smart city concept in the development of Nusantara. The Nusantara smart city framework encompasses six domains: smart governance, transportation and mobility, smart living, natural resources and energy, smart industry, and human resources, along with the smart built environment and infrastructure, aiming to transform Nusantara into a smart city that epitomizes efficiency, sustainability, and inclusivity. This framework outlines integrating advanced technologies to foster a resilient economy, a sustainable environment, and an enhanced quality of life for its citizens through improved administrative procedures, transportation systems, public safety, healthcare access, resource management, and infrastructure development. Implementing this framework can provide insight into the future development of smart cities in the Gulf Region, poised to significantly impact societal well-being and economic

The Emerald Handbook of Smart Cities in the Gulf Region:
Innovation, Development, Transformation, and Prosperity for Vision 2040, 481–514

doi:10.1108/978-1-83608-292-720241032

resilience significantly, demonstrating a model for future urban development that harmonizes community engagement, technological innovation, and environmental preservation.

Keywords: Smart city; Nusantara New Capital City; smart governance; smart transportation and mobility; smart living; natural resources and energy; industry and human resources; built-environment and infrastructure

1. Introduction

Indonesia is currently developing a new capital city on Kalimantan Island, called Nusantara, motivated by many urgent factors. First, this initiative is driven by critically significant future challenges. Following the Indonesia 2045 vision, which commemorates the centennial of its independence, the Indonesian economy is projected to achieve a position among the top five globally by 2045 (Muhyiddin, 2019). To achieve this goal, it is crucial to undergo an economic transformation supported by human development efforts and science and technology advancement. Additionally, sustainable economic development, equitable development, as well as national resilience and governance are essential components that need to be addressed.

Secondly, the development of the new capital city assumes a vital role in propelling inclusive and equitable economic growth across the Indonesian archipelago, encompassing the formerly underdeveloped Eastern Indonesia region. Throughout the past, Jakarta and its surrounding areas have served as the epicenter for governance, politics, industry, trade, investment, technology, and culture. Nevertheless, the prevailing dominance has resulted in noteworthy regional disparities in development and well-being across Indonesia (Leasiwal et al., 2022). The lack of alignment between this incongruity and the objectives of attaining sustainable economic growth has resulted in unexplored regional potential, disparities across various regions, and the potential for conflicts that could jeopardize national unity (Yusuf et al., 2023). Nusantara, the new capital city in Kalimantan, is planned to become a prominent economic focal point strategically located in the central region of Indonesia. This development aims to stimulate the emergence of further economic centers and leverage the abundant resources available in the area.

Thirdly, Jakarta encounters many obstacles, making it no longer unsuited to serve as the nation's capital. Jakarta grapples with the burden of a high population density, surpassing the national population density average with a density of 16,704 people per square kilometer compared to the national average of approximately 141 people per square kilometer. Moreover, Jakarta grapples with significant traffic congestion, as seen by its 29th position among 389 cities worldwide in 2022 (TomTom. 2022). Moreover, Jakarta is faced with significant environmental and geological challenges, consistently plagued by recurring floods and subsidence, resulting in certain regions residing at an elevation lower than the sea level

(Takagi et al., 2016). Given these limitations, it is evident that Jakarta's suitability as the capital city of Indonesia is more apparent, emphasizing the pressing need to relocate the capital to a more appropriate setting.

To help Nusantara achieve its objectives as Indonesia's new capital city, it is planned with three urban development guidelines safeguarding its future developments, including Nusantara as a forest, sponge, and smart city. As a forest city, Nusantara is dedicated to creating a harmonious balance between urban development and protecting its natural environment. As a sponge city, Nusantara will have extensive blue and green spaces as well as city operations, ensuring resiliency toward numerous ecological threats. Furthermore, Nusantara is also developed as a smart city, with a new approach utilizing smart technologies to improve its operations and provide better public services, improving the residents' quality of life (Susantono & Berawi, 2023).

The smart city concept has been widely recognized as a technology-centric urban paradigm that utilizes innovative solutions to integrate people, information, and various city elements. It has also blossomed as an integration between communication technologies and urban infrastructure, aiming to optimize the efficiency of city operations and elevate the overall quality of citizen's life, thus boosting productivity and competitiveness (Berawi, 2022b; Lin et al., 2019; Phung et al., 2022). Implementing a smart city incorporates a range of smart components and exchanging intelligent information that flows between various city aspects, collectively contributing to the city's smartness (Caragliu et al., 2011; Zygiaris, 2013).

Combining technology and data in city management helps increase the efficiency of basic service delivery and sustainable resource usage. Furthermore, with the smart reporting technology that speeds responses to urban problems, smart cities also improve the government's provision to societal needs (Berawi et al., 2022). Various studies have indicated that with the development of smart cities, the quality of life of its residents can be improved, resource efficiency can be optimized, new digital-based sectors can be promoted, and other social, economic, and environmental benefits can also be obtained.

To operate properly, smart city development needs a framework that guides governments and society to navigate the complexities of urban digital transformation efficiently. A smart city framework facilitates efficient decision-making in city management, ensuring the smart city is implemented with a clear structure and measurable results. This framework holds significance in guiding the city's digital transformation process and optimizing the outcomes of digital transformation initiatives. In turn, it facilitates efficient and cost-effective urban development by establishing distinct priorities and objectives for resource allocation.

Various smart city frameworks exist today, one is the European Cities' Smart City Model, which encompasses six essential domains: Environment, Governance, Economy, People, Life, and Mobility (Giffinger & Gudrun, 2010). Furthermore, the Smart Dubai Framework is structured around six domains: Smart Economy, Smart Living, Smart Governance, Smart Environment, Smart People, and Smart Mobility. Meanwhile, London follows the Smart London Plan, focusing on four key domains: Smart Infrastructure, Smart Environment, Smart Mobility, and Smart Society.

For the development of Nusantara, a smart city framework has been formulated to cover all aspects of life in Nusantara, following its principles of being a green, sustainable, inclusive, and resilient city. This chapter discusses the Nusantara smart city framework from the government's perspective, its formulation, the domains included, and the smart city features determined to be presented in each domain. This chapter is written in five (5) sections, followed by the description of Nusantara in Section 2, methods used to formulate the smart city framework in Section 3, discussion on the Nusantara Smart City Framework in Section 4, and conclusion in Section 5.

2. Nusantara New Capital City Development

2.1. Nusantara New Capital City's 2045 Vision

The objective of Nusantara, as stated in Indonesia's Presidential Regulation No. 63 of 2022, is to become a World Class City for All. This vision underscores the commitment to enhancing domestic value chains across Eastern Indonesia, promoting economic growth in the Kalimantan region, and facilitating the implementation of globally recognized governance standards. The core of the Nusantara vision revolves around three principal objectives.

To begin with, through the supervision of spatial planning, land utilization, water resources, energy management, waste disposal, sanitation, public transportation, and the development of secure, comfortable, and globally sustainable settlement environments, Nusantara endeavors to transform into a global sustainable city. Additionally, Nusantara will guide Indonesia's forthcoming economic environment by promoting progressive, competitive, and innovative socio-economic changes. This objective will be achieved by optimizing technological advancements, architectural and urban planning, and the diverse socio-cultural fabric complemented by the growth of superhub economics. Moreover, Nusantara is expected to also serve as the archetypal representation of Indonesia's national identity, embodying the nation's grandeur, unity, diversity, and socio-cultural identity (see Fig. 1).

2.2. Project Location and Development Area

Nestled in the heart of Indonesia, Nusantara is positioned as the catalyst for economic growth stimulation across Indonesia, and it is a global city employing international standards of governance and urban development. Nusantara is situated in the Eastern part of Kalimantan Island, between two existing cities of Balikpapan and Samarinda. It has a total land and water area of around 324,332 hectares, with waters covering 68,188 hectares and land covering 256,142 hectares, divided into two regions: an urban area of 56,180 hectares called KIKN and a development area of 199,962 hectares called KPIKN. Fig. 2 provides an intricate overview of Nusantara within the wider geographical context.

There are multiple distinct zones in the KIKN region, which include KIPP or the Central Government Core Area, West IKN, North IKN, East IKN 1, East IKN 2, Muara Jawa, and Simpang Samboja. Each region has been determined to strategically serve as a pivotal site for distinctive economic initiatives, resulting in various development strategies.

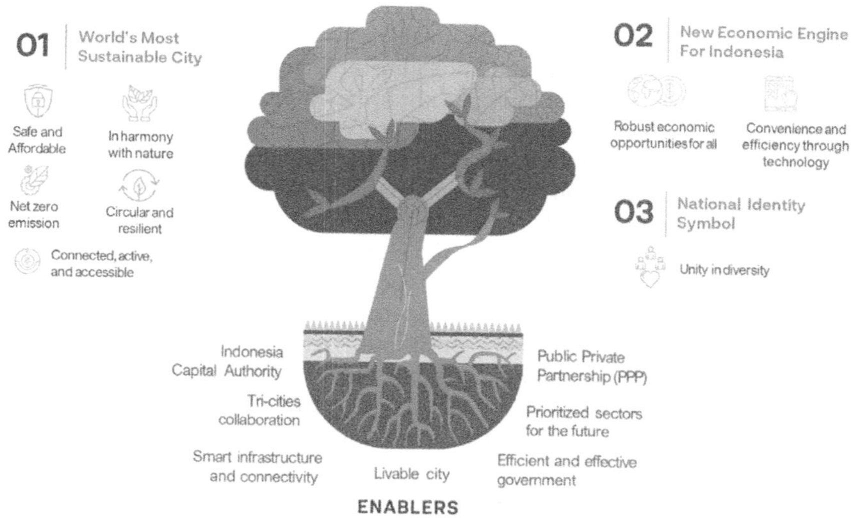

Fig. 1. Three Principles of Nusantara Development (*Source*: Nusantara Capital City, 2022).

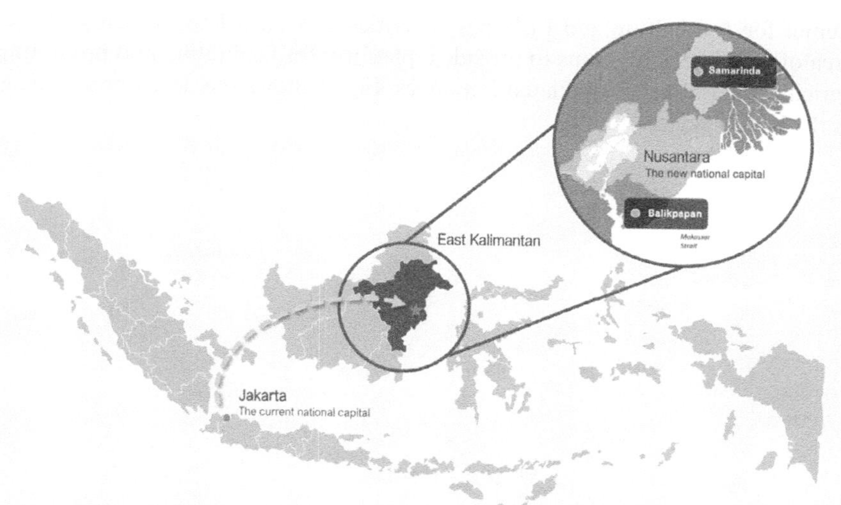

Fig. 2. Nusantara New Capital City: A Context Within Indonesia and Among Two Neighboring Cities (*Source*: Nusantara Capital Authority, 2022).

The KIPP area, which will function as the national government center accommodating essential government departments tasked with pivotal national decision-making responsibilities, has an estimated 6,671 hectares. The proposed facility within the expansive 17,206-hectare region referred to as the West IKN area, the Center for Economics, Business, and Finance, is scheduled to be constructed. Furthermore, the region of South IKN, comprising 9,751 hectares, has been designated to build the New Renewable Energy Center, aiming to promote the progress and integration of sustainable energy sources, thereby reducing dependence on finite resources and mitigating ecological impacts.

An entertainment and sports center are being considered for construction in the designated East IKN 1 area, where a proposed development plan calls for 9,761 hectares of land development. The objective of the proposed facility is to function as a pivotal location for sporting, entertainment, and recreational activities, accommodating the needs of both the nearby inhabitants and tourists. Educational services will be established in the North IKN region, which provides an expansive expanse of 12,067 hectares. On the other hand, the region identified as East IKN 2, which encompasses 3,720 hectares, has been planned to establish the research and innovation center developed to drive technological progress, foster industry expansion, and facilitate the exchange of knowledge.

A total of 2,986 hectares will be designated by the Simpang Samboja area to construct a trade and distribution center for commodities, aiming to establish a logistics center that is both strategically located and operates at a high level of efficiency, thereby enabling the streamlined transportation of merchandise on a national and global scale. Furthermore, the Kuala Samboja region has designated a land area of 4,299 hectares to establish a food and agro-industrial hub to facilitate the growth of the sectors involving many stages, such as the production, processing, and sale of agricultural and food commodities. Last, establishing the Center for Agriculture and Fisheries Activities in Muara Jawa, which spans an area of 9,084 hectares, aims to provide a platform for facilitating and promoting agricultural and fisheries-related initiatives. Fig. 3 shows the development areas in the KIKN region.

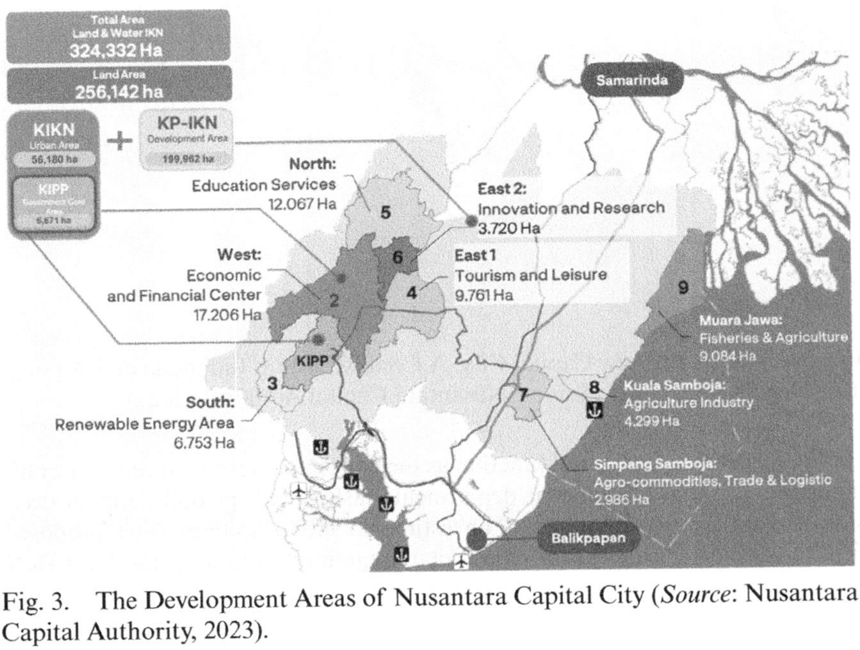

Fig. 3. The Development Areas of Nusantara Capital City (*Source*: Nusantara Capital Authority, 2023).

3. Developing Nusantara Smart City Framework: A Comprehensive Approach

The smart city concept is at the heart of Nusantara's development, as stated and defined in Law No. 3/2022 and Presidential Regulation No. 63/2022. As an approach that utilizes ICT advancement, urban data management, and digital technologies to plan and manage urban functions in an efficient, innovative, inclusive, and resilient way, it has been considered a comprehensive way of fostering the development of Nusantara that is dynamic, inclusive, ready to face the future, and connected. Therefore, to realize a robust smart city for Nusantara, there is a need to envision, plan, and design it. This section is divided into three subsections, describing each method in detail.

3.1. Envisioning Nusantara Smart City

Following the legal provisions stipulated in the Law of the Republic of Indonesia No. 3 of 2022 concerning the Capital City, the strategic development and management of the Nusantara New Capital City represent a forward-thinking approach directed toward positioning the National Capital as a world city for all with the main aim of realizing an ideal city that not only serves as a prototype for urban development and management within Indonesia but also extends its influence on a global scale. This vision is geared toward positioning the Nusantara as (a) a globally sustainable city fostering comfort, environmental harmony, resilience through resource efficiency, and a commitment to low-carbon initiatives; (b) a catalyst for Indonesia's future economic growth, providing inclusive economic opportunities by nurturing potential, innovation, and technological advancements; and (c) a symbol of the nation's identity, representing unity in diversity as accordance with Bhinneka Tunggal Ika principle.

The law also mandates that the foundational framework governing the construction and evolution of this capital city pivots on fundamental principles encompassing equality, ecological equilibrium, resilience, sustainable development, livability, connectivity, and the integration of smart city initiatives. Therefore, the strategic vision of the Nusantara smart city was formulated based on the guiding principles and urban design directives of the Nusantara, alongside the establishment of Key Performance Indicators (KPIs) specific to the Nusantara, adopting a results-focused evaluation methodology, fostering an environment conducive to innovative, adaptable, and agile management practices, as the foundation and orientation of smart city master plan.

KPIs serve as benchmarks for the Nusantara Capital Authority (NCA) as the stakeholder responsible for preparing, developing, and relocating the national capital and administrating the Special Capital Region of Nusantara. KPIs are instrumental in measuring the success of realizing the vision and achieving the goals and objectives of Nusantara's KPI, assessing the efficient and effective implementation of the Nusantara Master Plan, evaluating the success of collaborations with third parties in Nusantara's development and management, providing early warning information on obstacles and mitigation steps during Nusantara's development and management, and furnishing information for controlling and evaluating Nusantara's development and management implementation. Fig. 4 summarizes the KPIs of Nusantara city.

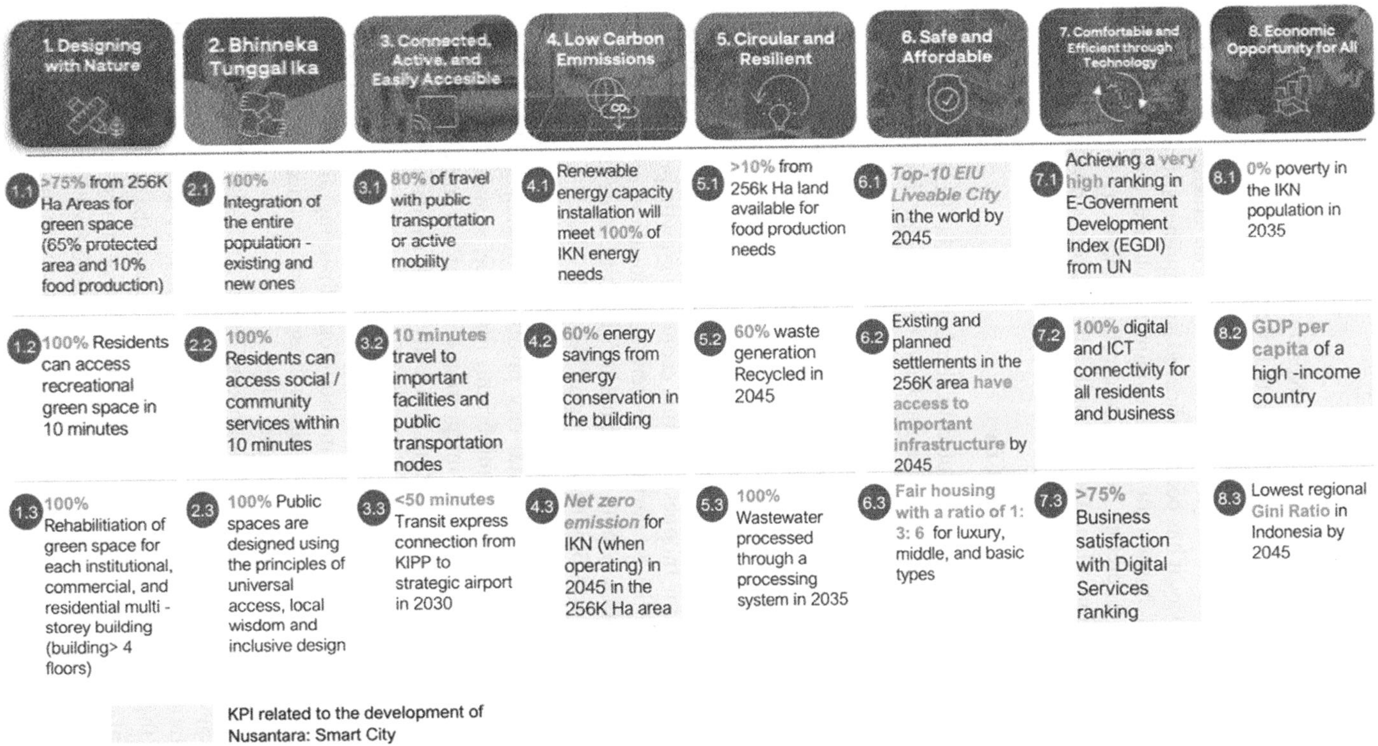

Fig. 4. Nusantara's 24 KPIs and relations to Nusantara Smart City (*Source*: Nusantara Capital Authority, 2022).

Nusantara's vision also aligns with the fundamental elements required for a smart city, which includes all aspects for economic growth, sustainability, and improved quality of life. In the economic growth element, investments should promote sustainable economic growth while creating jobs in the digital, circular economy. Furthermore, the city development should be able to attract and retain companies and prepare the workforce for the future, where entrepreneurship and innovation are encouraged. In the sustainability element, smart city development should be able to manage energy and natural resources wisely, recycle and reuse assets, reduce carbon footprint, and drive toward having cleaner air and less noise. Moreover, for the quality-of-life element, Nusantara smart city should be able to foster diversity and inclusivity, improve public education, health, and safety, and enable constituent engagement.

3.2. Defining User Experiences

User experience encompasses all interactions, emotions, perceptions, and responses experienced by users during engagement with an ecosystem, involving the aspects of functionality, usability, visual esthetics, responsiveness, and alignment with user needs. User experience research aims to deliver substantial, effective, user-friendly products or services with added value. To create a good public service experience, the design of user experience requires in-depth understanding through data collection, user research, user needs analysis, prototyping, and testing of products or services with real users.

Through a meticulous analysis of demographic data and the comprehensive NUSANTARA development plan outlined in Presidential Regulation No. 63 of 2022, A total of ten user groups have been defined to effectively capture the diverse experiences of individuals residing, working, and visiting the Nusantara, each with specific characteristics, needs, motivations, and challenges. Ten user personas, created based on data and information collected from user research, interviews, surveys, and direct observations, include (1) government workers relocating from Jakarta to Nusantara; (2) blue collar workers; (3) business communities; (4) local, existing residents; (5) inter-city commuters; (6) tourists; (7) non-government work; (8) children; (9) older people; and (10) differently abled, such as wheelchair and visually impaired.

These personas describe users' demographic characteristics, needs, motivations, preferences, and goals. By visualizing users in the form of personas, differences in the use of smart features or functions by different groups of users can be identified and informed. Each user persona has a background, demographics, description, and narrative explaining their characteristics and needs. This persona can include information such as their age, gender, education, occupation, interests, habits, and challenges they face. In addition, the user persona also consists of the user's goals, problems to be solved, preferences in using the product, style, content, and interaction preferences. These user personas and their journeys are systematically crafted for every user segment. The identities and experiences serve as the foundation for the design of the Nusantara smart city.

3.3. Determining Smart Domains in Nusantara Smart City

Functions of smart cities identified from customer journeys cover enhanced working and living experiences. Along with the experience-centric features, the Presidential Regulation No. 63 of 2022 and the IESE Cities in Motion Index (CIMI) that measures the sustainability and smartness of key global cities are also utilized to determine the Nusantara smart city's domains, services, features, and technical requirements.

IESE Cities in Motion is a research platform jointly launched by the Center for Globalization and Strategy and the Strategy Department of IESE Business School (Berrone & Ricart, 2022). In this context, the IESE CIMI has been designed to create next-level indicators in terms of comprehensiveness, attributes, comparability, quality, and included information, which is aimed at measuring the future sustainability of leading global cities and the quality of life of their residents.

These indicators can serve as the foundation for the smart city indicators in Nusantara, as the CIMI Model aims to help citizens and the government understand a city's performance across ten (10) key dimensions: governance, urban planning, public management, technology, environment, international outreach, social cohesion, mobility and transport, human capital and economy. These dimensions were utilized to help define the components and areas of focus within the concept of Nusantara smart city.

4. Nusantara Smart City Initiatives Framework

Drawing from the visions of the Nusantara smart city, the characteristics and needs of the users, and the guidance from the IESE Cities in Motion Model, the framework for the Nusantara smart city was formulated, encompassing six (6) key domains: (1) smart governance; (2) smart transportation and mobility; (3) smart living; (4) natural resources and energy; (5) industry and human resources; and (6) built-environment and infrastructure.

4.1. Smart Governance Domain

Governance serves as the conducive atmosphere that necessitates effective procedures and sufficient legal structures to permit the government to be responsive to the demands of its constituents (UN-HABITAT, 2012). Garcia Alonso and Lippez-De Castro (2016) and Albino et al. (2015) defined governance as the collaboration and interaction of various stakeholders during decision-making. The concept is commonly used to describe the action or manner of governing a state, an organization, or another constellation of actors. It demonstrates that governance and government are related but distinct concepts. According to Scholl and Alawadhi (2016), smart governance utilizes intelligent and adaptive actions and activities to oversee and make decisions regarding a particular matter. Smart governance, according to Scholl and Scholl (2014), serves as the foundation for an intelligent, transparent, and participatory government.

Smart governance represents a philosophy centered around leveraging information and communication technology (ICT) to enhance governance and promote active public engagement in governmental affairs (Pereira et al., 2018). Several key factors underpin this approach. Firstly, the increasing accessibility and prevalence of ICT enable governments to employ these tools to enhance governance practices and encourage community involvement. Moreover, global advancements in quality of life and economic development have amplified societal expectations for a more transparent and responsive government. Citizens are increasingly inclined toward greater involvement and seek a deeper understanding of governmental processes and resource allocation (De Guimarães et al., 2020). Efficiency is another critical driver; effective governance ensures the reasonable and equitable use of available resources. Governments can significantly enhance their governance frameworks by amalgamating technology with citizen engagement, fostering a more robust and participatory approach.

In the smart governance domain of Nusantara smart city framework, there are four (4) subdomains developed, including (a) City Management and Planning Service; (b) Digital Service for Citizen; (c) Smart Business Permit; and (d) IKNOW App.

a. City Management and Planning Service

The subdomain of City Management and Planning Service comprises several smart features, including the Integrated Command Control Center (ICCC), Centralized Citizen Reporting Center, and Urban Planning System. The ICCC oversees numerous systems, such as infrastructure monitoring, public facility monitoring, safety and supervision, emergency and crisis management, and analytical purposes. On the other hand, the Centralized Citizen Reporting Center facilitates communication between citizens and the city management, particularly reporting issues and concerns.

The Urban Planning System provides government officers with advanced tools to enhance their decision-making capabilities with the statistical and predictive analytics technology facilitating data-driven evaluations. Furthermore, this system allows authorities to compare Nusantara's urban master plan and satellite imagery, precisely depicting the urban landscape and promoting the development of the urban areas. In addition, the system enables rigorous planning for land use, spatial zoning, incorporation of green spaces, and implementation of building codes, informed by comprehensive information.

b. Digital Service for Citizen

The Nusantara Smart City App is a platform that incorporates a wide range of essential services to meet the different demands of citizens. This platform merges procedures of Resident Administration, methods for Public Feedback, updates on news and events, facilities for space rental, emergency services, and a variety of service links encompassing education, health, utilities, and transportation. Significantly, the system integrates environmental monitoring capabilities crucial for implementing sustainable urban management practices.

Moreover, the one-stop public services system embodies a comprehensive platform that provides citizens with efficient and simplified access to a wide range of important services. The comprehensive system encompasses Transfer Registration, Tax Payments, and Service Integration, thereby ensuring convenience and efficiency in addressing the administrative requirements of citizens.

The digital identity is a centralized repository that stores vital information for digital identification. The feature offers users a digitally protected virtual representation, facilitating the efficient organization and control of crucial personal information.

Furthermore, the Help Desk feature incorporated inside the Nusantara Smart City App is crucial in assisting citizens. The system functions as a comprehensive service center, catering to requests regarding government services, providing technical support, and spreading information on important domains, such as housing, jobs, education, and healthcare services. Moreover, it enables the assistance of visitors by offering necessary recommendations and gathers significant user input via surveys, ensuring continuous improvement and responsiveness to the visitors' requirements.

c. Smart Business Permit

The Smart Permit is a framework incorporating various essential components, including business permit applications, utility registration, tax payments, employment, and government assistance. Its primary objective is to enhance operational efficiency for businesses by streamlining these processes. The business permit module in this feature facilitates applying for permits for entities. It allows them to submit applications, obtain recommendations for suitable locations based on spatial inspections, report on the progress of construction, apply for building permits, and monitor the licensing status.

In addition, the system's utility registration component enables the efficient registration of building utilities, enhancing the management of important services. Moreover, the tax payment feature allows organizations to submit tailored files and promotes the seamless settlement of taxes and levies, ensuring adherence to regulatory responsibilities.

Aligned to promote employment opportunities, it facilitates the advertisement of job vacancies by entities, supporting the creation of jobs and contributing to economic progress. Concurrently, it enhances vital governmental support by providing datasets to facilitate informed decision-making and streamline the registration process for government programs. These initiatives are designed to aid and create a conducive environment for the growth and long-term viability of startups and Small and Medium-sized Enterprises (SMEs) within the business ecosystem.

d. IKNOW App

The IKNOW App is an internal application that incorporates multiple essential functions to enhance the efficiency of administrative operations within the NCA framework. The Data Sharing Platform is pivotal in facilitating secure intergovernmental data sharing, which is crucial to developing informed policymaking and maintaining regulatory consistency. The e-office for NCA provides a comprehensive range of functionalities, including digital correspondence management, cashless administration, budget planning, financial control, and rigorous asset inventory keeping.

Moreover, the Integrated Portal is a consolidated platform for NCA officials, facilitating access to e-learning resources, secure online meeting spaces, and the Government Data Sharing Platform. Furthermore, the Smart App provides NCA officials with effective tools for managing tasks, enabling them to monitor calendars, coordinate meetings, secure premises, and oversee job assignments.

In addition, the Smart Payroll and Benefits Service enhances efficiency in administrative operations by automating payroll calculations, expense reimbursements, and benefits management specifically designed for NCA officials. The integrated system effectively communicates with the Ministry of Finance tax site, guaranteeing precision and effectiveness in financial operations. In general, integrating these interconnected features within the IKNOW app assists in improving operational efficiency and strengthening administrative capacities, hence fostering effective governance within the Nusantara smart city.

4.2. Smart Transportation and Mobility Domain

The key objectives of the smart city initiatives are to improve the residents' quality of life by implementing seamless user experiences and providing easy and efficient modes of mobility, enabling the movement from one place to another place, which requires efficient, effective, safe, comfortable, and sustainable transportation (Singh & Edla, 2022). Mobility and transportation play a crucial role in the context of smart cities, constituting more than a quarter of the overall energy consumption associated with such urban environments (Nguyen et al., 2020). Hence, implementing smart transportation and mobility systems constitutes a pivotal basis for developing smart cities. Its primary objective is to mitigate issues such as traffic congestion, environmental pollution, and road accidents while also enhancing accessibility and mitigating the negative impacts associated with transportation (Savin, 2020).

The Nusantara smart city framework is set to implement a comprehensive smart transportation initiative across four key subdomains: (a) Intelligent Transportation System (ITS), (b) Mobility-as-a-Services, (c) Smart Vehicle Ecosystem, and (d) Smart Logistics System.

a. Intelligent Transportation System

The ITS is an advanced system providing innovative services related to different modes of transport and traffic management that combines information technology (IT), data communication, electronic sensing, and geospatial technology to enhance traffic management, optimize transport efficiency, and provide safety services within the transportation sector, by enabling safer, more coordinated, and smarter use of transport networks (Li et al., 2018; Turner & Uludag, 2016).

ITS has emerged as an influential factor in reshaping urban environments. It incorporates various components, including sensors, cameras, traffic management systems, and data analytics tools, into transportation infrastructures to enhance the ability to monitor and manage traffic in real time, allowing timely responses to any issues. The ITS developed within the Nusantara smart city encompasses several interlinked systems designed to optimize the efficiency and safety aspects of transportation and mobility systems.

The Autonomous Driving System (ADS) is a vehicle control system facilitating driverless operation using sensors such as cameras, radar, as well as Light Detection and Ranging (LiDAR) (Shafiee et al., 2021). It is responsible for enabling autonomous driving by utilizing a collection of functional blocks with functions and attributes crucial in determining the characteristics of the surrounding environment and objects (Iclodean et al., 2022). In addition, Advanced Traffic and Parking Management Systems (ATPMS) incorporate components such as traffic control and dynamic parking management. These systems aim to identify traffic-related problems and offer crucial data regarding parking availability and traffic conditions, utilizing digital twins to simulate and analyze data.

The Advanced Public Transportation System (APTS) is a transportation system that incorporates many modes of transportation by utilizing real-time data obtained from sensors and Global Positioning System (GPS) technology. This system dynamically adapts routes and schedules in response to demand, encouraging greater utilization of public transportation services. On the other hand, the Incident Management System (IMS) is a comprehensive framework that facilitates the management and occurrences of traffic events. It leverages advanced technologies, including big data, the Internet of Things (IoT), and Closed-Circuit Television (CCTV) to enable real-time detection of incidents, coordination of emergency responses, and dissemination of information through city channels and smart applications.

Furthermore, the Advanced Traveler Information System (ATIS) provides up-to-date data regarding the conditions of transportation system and public transportation timetables to assist travelers in making informed decisions. Commercial Vehicle Operations Management (CVOS) optimizes the operations of commercial vehicles to improve efficiency and safety within the supply chain. The Electronic Payment System (EPS) for Transport facilitates the adoption of cashless transactions within the public transportation sector. ITS aims to improve mobility, safety, and efficiency in the Nusantara smart city and promote a technologically advanced and networked transportation ecosystem.

b. Mobility-as-as-Service

Mobility-as-a-Service (MaaS) is a crucial component of smart mobility plans, as it combines the commercialization of transportation services with flexible coordination of microservices (Melis et al., 2018). The primary objective of this innovative approach is to develop customized applications that appeal to a wide range of user demographics to provide a smooth and personalized transportation experience for persons with varying mobility needs and preferences (Melis et al., 2018).

MaaS operates by integrating diverse transportation modes into a unified digital service platform, which includes alternatives such as taxis, bicycles, and buses. It encompasses the utilization of ITS to enhance the accessibility and integration of public transportation through a range of characteristics. This comprehensive service serves as a centralized platform for users' cross-modal transportation requirements, providing integrated features including the ability to purchase tickets for different modes of transportation, receive real-time updates on traffic and transportation, access vehicle and bicycle rental services, and utilize tracking capabilities that are advantageous for individuals using public transport (Koźlak & Pawłowska, 2019).

The worldwide adoption of MaaS encompasses a wide range of geographical areas and contains integrated services that enable seamless transit across different modes. The services provided include extensive traffic condition information, integrated ticketing systems, and surveillance features for monitoring family members utilizing public transportation. On the other hand, the development of MaaS in the Nusantara is predominantly centered around endeavors such as the MaaS app, vehicle rentals, and urban air mobility (UAM), with a particular emphasis on customizing the service to cater to the distinct transportation requirements within the locality effectively.

Global MaaS implementation spans various geographical regions and encompasses integrated services facilitating seamless transit across diverse transportation modes (Kupriyanovsky et al., 2017). These provisions encompass a spectrum of services, including comprehensive traffic information, integrated ticketing systems, and even surveillance features catering to the monitoring needs of individuals utilizing public transportation. In contrast, the development of MaaS within the Nusantara predominantly focuses on initiatives such as the MaaS app, vehicle rentals, and UAM. This approach aims to effectively customize services to suit the distinctive transportation requirements of the region.

Moreover, Nusantara's MaaS implementation involves features like vehicle rentals centered around electric vehicles (EVs), catering to users' diverse mobility needs for short, long-term, or on-demand trips. Integrating vehicle rental services with other transportation options, such as public transit or taxis, enhances flexibility and convenience. Additionally, the emphasis on UAM involves deploying electric vertical take-off and landing (eVTOL) aircraft and drones to revolutionize urban transportation. UAM strives to mitigate traffic congestion, reduce travel times, and offer efficient transportation alternatives for passengers and logistics within urban landscapes.

c. Electric Vehicle Ecosystem

The increasing emphasis on smart cities that promote efficiency, reliability, sustainability, and connectivity has led to the widespread adoption of EVs as a predominant means of future transportation (Aymen & Mahmoudi, 2019). EVs have emerged worldwide as a potent and eco-conscious initiative. When integrated into urban settings, they can symbolize a city's dedication to sustainable mobility, representing a pivotal element within the broader framework of the smart city concept.

Consequently, there is a growing need for a strong and reliable EV charging infrastructure (Antoun et al., 2020).

The Electric Vehicle Ecosystem (EVE) stands as a pivotal driver in fostering environmentally sustainable mobility within the developmental scope of the Nusantara Capital City. Its framework comprises essential infrastructure featuring Public Electric Vehicle Charging Stations and Public Electric Vehicle Battery Exchange Stations. These strategic initiatives are designed to increase the efficiency and availability of EVs, particularly in public transportation, amplifying the city's commitment to promoting environmentally friendly commuting alternatives.

The EV Charging for Public Commercial infrastructure offers charging facilities tailored for public transportation strategically located across residential and commercial areas in Nusantara. These stations are integrated with smart city applications, enabling monitoring and managing EV charging points, thus contributing to the city's widespread adoption of electric public transportation. Moreover, Battery Swapping for Public Commercial serves as an infrastructure designed specifically for smaller EVs like scooters, personal mobility devices, and electric bicycles. Following Presidential Regulation No. 59 of 2019 concerning implementing Battery Electric Vehicles, these stations are designed to enable efficient battery swapping, often in conjunction with Public Electric Vehicle Charging Stations. The integration of monitoring applications becomes essential in ensuring the availability of batteries for efficient swapping services, thereby optimizing the utilization of smaller EVs within urban environments.

d. Smart Logistics

The conceptualization of smart logistics exhibits a wide range of interpretations, resulting in a lack of consensus within the academic domain (Ding et al., 2021; Uckelmann, 2008). Smart logistics is a comprehensive field that involves several logistical operations, including transportation, warehousing, and customer service. These operations are carried out utilizing advanced planning, management, and control mechanisms that exceed traditional approaches in terms of intelligence and efficiency. The level and characteristics of this intelligence reflect variation across different applications and domains of emphasis. Several ideas have emerged within the realm of smart logistics, encompassing autonomous logistics (Windt & Hülsmann, 2007), ITSs (EU, 2019), the Physical Internet (PI, π) (Montreuil et al., 2013), smart freight (Sternberg et al., 2010), and customer-centric intelligent logistics (McFarlane et al., 2016). These concepts encompass numerous aspects of smart logistics, each targeting certain elements and employing varying degrees of intelligence in their methodologies.

Smart logistics is a highly efficient approach for addressing the demands of rapidly evolving customer expectations, capitalizing on the advantages presented by emerging technology, and enabling the implementation of novel business models. Integrating these emerging technologies in smart logistics aims to increase flexibility, adaptability, and proactivity inside dynamic supply chains.

Two smart features are included in Nusantara smart logistics, including smart delivery and smart warehouse. Smart delivery enhances the efficiency of the last-mile delivery process by integrating technology-based solutions. It encompasses effective shipment tracking methods, flexible storage options such as lockers, and customized item collection systems. Smart delivery optimizes the concluding phase of the delivery procedure using these advanced technologies, enhancing effectiveness and ease for end users.

In addition, Smart Warehouse plays a crucial role by prioritizing the improvement of storage capacity for goods in the first stages of the supply chain. This feature of smart logistics enhances multiple components, such as sorting mechanisms, preparation procedures, inventory management systems, packaging processes, operational analysis tools, and fleet monitoring mechanisms. Through these advancements, Smart Warehouse demonstrates a substantial enhancement in the efficiency and effectiveness of product handling, commencing from the early phases of the supply chain.

4.3. Smart Living Domain

The concept of smart living can be defined in various theoretical ways. It is defined as an environment customized to meet its inhabitants' needs and preferences by integrating sensing, actuation, verbal and non-verbal interactions, and computation functionalities to facilitate inhabitants' attainment of comfort and autonomy (Cicirelli et al., 2016). On the contrary, Giffinger and Gudrun (2010) defined spaces with Smart Living concept houses as facilities that consider social cohesion in the neighborhood, occupants' health conditions, individual safety, housing quality, and educational facilities. Furthermore, Madakam and Ramaswamy (2016) stated that Smart Living comprises smart appliances and smart homes. In this context, smart appliances connect devices in technological ways to the smart home. In contrast, smart homes have the capability of a residential facility to monitor or adjust the indoor environmental systems.

In the context of a smart city, smart living refers to an intelligent urban setting that prioritizes creating a positive environment for its occupants. This setting is distinguished by its extensive availability of cultural and entertainment options, strong technical and social infrastructure, and broad access to public services. Furthermore, it underscores a dedication to conserving the environment and creating sufficient green areas. In essence, the fundamental objective is to foster a comprehensive and intuitive environment that elevates the standard of living for the city residents (Dewalska–Opitek, 2014).

To improve the quality of life in smart cities, smart living for Nusantara capital city development involves data analysis to aggregate areas at risk of crime, allowing for more efficient deployment of police and emergency response resources (Berawi et al., 2021). Smart city initiatives can also include using real-time data to improve emergency response time and leveraging sensors and drones to assess damage and aid disaster response efforts. The smart living domain in Nusantara comprises Health and Welfare and Public Safety, as well as Space Management.

a. Health and Welfare

The Health and Welfare subdomain entails multifaceted initiatives to improve residents' well-being. An Integrated Health App is a comprehensive platform enabling convenient access to health services, including monitoring health metrics, scheduling consultations, and facilitating medication orders. Moreover, Smart Health Facilities incorporate intelligent management systems and devices within healthcare settings to streamline patient data management, optimize workflows, and enhance operational efficiency through real-time data provided by sensors and wearable devices. Furthering the welfare agenda, Smart Occupational Safety integrates smart technologies to monitor workplace conditions, collect employee data, issue early warnings, and promote safety awareness among workers.

Other than Artificial Intelligence (AI) and advanced hardware technologies, 5G network technology has also been recognized for its wide-ranging benefits in the healthcare sector, offering transformative advantages (Shilvya et al., 2022). It enables pre-hospital care by allowing remote monitoring of patient's vital signs through wearables, sensors, and video systems, facilitating early interventions. Additionally, 5G supports virtual consultations between healthcare providers and patients, reducing the need for physical visits and cutting down travel costs.

b. Public Safety and Space Management

The Public Safety and Space Management subdomain focuses on ensuring the security and efficient management of public spaces within the smart city framework. Utilizing AI technology and advanced sensors, Urban Safety and Mobility features continuously monitor urban areas, detecting suspicious activities and patterns, aiming to create a secure environment for residents. Emergency Response Management systems also integrate AI and robust communication channels to predict, identify, and respond effectively to emergencies by coordinating responses from various teams and disseminating crucial information promptly. Complementing these efforts, the Early Warning System (EWS) feature harnesses smart technology to detect and alert about potential threats or hazardous events, such as natural disasters or accidents, enabling swift notifications and efficient countermeasures. Moreover, the Smart Spaces feature combines IT, AI, and smart devices to create environments within buildings or rooms that monitor, manage, and automate various aspects of daily life, encompassing comfort, energy efficiency, security, and overall user experience.

4.4. Natural Resources and Energy Domain

The domain of natural resources and energy within the context of smart city initiatives refers to the utilization of advanced technologies and strategies to effectively manage and enhance the utilization of resources, including but not limited to power, water, and other renewable resources (Neirotti et al., 2014). It includes the deployment of smart grids, which are electrical distribution systems capable of effectively providing sustainable, cost-effective, and reliable electricity services by considering the actions and patterns of all interconnected consumers. Additionally, it encompasses the administration of public lighting infrastructure and the stewardship of natural resources, including sustainable sources such as thermal energy, solar power, cooling systems, water resources, and wind energy.

This domain aims to mitigate expenses, enhance the dependability and transparency of energy provision systems, and foster sustainability principles. In the Nusantara smart city framework, there are three (3) subdomains developed, including (a) Resource Management; (b) Energy Management and Digitalization; and (c) Smart Forest and Green Management.

a. Resource Management

Due to increasing urbanization and population growth, Indonesia faces significant obstacles in waste management, clean water supply, and wastewater treatment. The annual waste generated in the country is projected to reach approximately 150 million tonnes by 2025 (Hoornweg & Bhada-Tata, 2012). One of the primary challenges is the insufficiency of infrastructure for waste and water management, accompanied by a lack of public knowledge regarding the importance of reducing trash and conserving water (Masia & van der Poll, 2021). To effectively tackle these concerns, it is essential to implement technological interventions alongside more stringent legislative enforcement measures.

The concept of smart waste management encompasses a novel approach that utilizes advanced technology, innovation, and data analysis to optimize operational efficiency, reduce waste management expenses, and lessen waste disposal's environmental impacts (Das et al., 2019). In contrast to conventional approaches characterized by higher costs and potential environmental hazards, smart waste management incorporates sensors, data analytics, and technology to enhance the efficiency of waste collection and processing. Within the context of this framework, the functional requirements consist of three components: Smart Trash, Smart Waste Fleet, and The Material Recovery Facility (3R MRF).

Simultaneously, smart water management facilitates the advancement of water supply sustainability by utilizing sensor technology, Supervisory Control and Data Acquisition (SCADA), water quality monitoring, non-potable management, and river pollution monitoring. The process of real-time water quality monitoring entails the assessment of key parameters such as dissolved oxygen, pH, conductivity, and temperature to promptly detect instances of pollution and the presence of harmful substances. The utilization of SCADA systems enables the instantaneous acquisition and examination of data within water distribution networks, allowing operators to oversee variables such as flow, pressure, energy usage, water quality, device conditions, and notifications regarding leaks and low pressure.

In addition, Smart Wastewater Management utilizes IoT sensors to continuously monitor and assess real-time wastewater flow, water quality, and potential leaks. This technology facilitates the timely identification of problems, enables efficient repairs, and aids in planning network maintenance activities. Predictive data analysis is employed to enhance the capacity of wastewater treatment plants. At the same time, smart stormwater management techniques are utilized to gather and store rainwater to preserve water ecosystems effectively. Incorporating greywater recycling into sensor-based allocation systems is valuable in managing sustainable water resources. Additionally, using Geographical Information

Systems (GIS) for monitoring enables continuous real-time water quality assessment, facilitating compliance with environmental regulations.

b. Energy Management and Digitalization

Digitalization in the energy sector refers to the application of technologies to improve business, policy, or decisions in general to increase overall efficiency, cost, security, and sustainability. It involves the use of information, computing, communication, and connectivity technologies to transform and integrate local, independently produce energy, and enable the management of a decentralized energy system (Światowiec-Szczepańska & Stępień, 2022). The integration of effective energy management practices with digitalization presents an opportunity to attain the objective of promoting sustainable energy utilization, hence enhancing efficiency and mitigating adverse environmental consequences.

This subdomain encompasses several smart features such as Smart Grid, Smart Energy Market, and Vehicle-to-X (V2X). The purpose of advanced technology is to support the design and management of the forthcoming smart energy systems characterized by significant integration of renewable energy sources and a decentralized organizational framework.

Smart Grids play a pivotal role in Nusantara's electricity system, efficiently distributing power from decentralized generators and swiftly responding to fluctuations in supply and demand. Integration through smart meters and automated distribution systems enables real-time energy trading among consumers and prosumers, fostering sustainable, reliable, and locally balanced generation-consumption relationships while reducing energy transportation costs. Key requisites for smart grid features encompass advanced automation, real-time monitoring, and bidirectional communication capabilities.

The Smart Energy Market integrates clean energy production and infrastructure, offering customers a variety of energy contracts to enhance overall efficiency. Customers transition into prosumers, trading energy within their communities and integrating distributed generators. Peer-to-peer (P2P) energy markets facilitate direct exchange without intermediaries, encouraging renewable energy use and yielding economic benefits. The crucial attributes of smart energy market features revolve around adaptable energy trading platforms and user-friendly interfaces. Meanwhile, V2X denotes a vehicle's ability to interact with its surroundings, including road infrastructure, other vehicles, energy systems, and communication systems. The "X" encompasses various elements like Vehicle-to-Grid (V2G), Vehicle-to-Home (V2H), and Vehicle-to-Building (V2B), showcasing the integration potential of vehicles within the broader energy ecosystem

c. Smart Forest and Green Management

Forests encompass extensive regions characterized by abundant trees and bushes fulfilling crucial ecological functions such as serving as habitats for various species, controlling climatic conditions, and effectively sequestering carbon. In the region of East Kalimantan, approximately 66% of the land is occupied by forests. However, it is important to note that these forests comprise productive areas characterized by monoculture stands. Production forests are subject to intense management practices to extract wood, pulp, and paper. These forests exhibit distinct structural and compositional characteristics and harbor reduced species diversity. To facilitate the restoration of natural forests, Nusantara employs a comprehensive approach that integrates reforestation efforts with the establishment of the Mentawir Nursery in North Penajam Paser Regency, located in East Kalimantan, which possesses the capacity to produce an impressive annual output of 15 million seedlings.

Implementing smart forestry and green management in Nusantara is pivotal in supporting the sustainability of forest cities and ensuring food security. Wilson and Lovell (2016) highlight smart agroforestry as a land-use strategy that effectively achieves the dual objectives of enhancing food security and reducing environmental harm. This system successfully amalgamates agricultural and forestry methodologies by harnessing cutting-edge technologies such as sensors, drones, and satellite imagery, resulting in improved agricultural output and managed forest resources.

Agroforestry, which integrates agricultural and forestry methods, provides an extensive array of advantages, including but not limited to improved soil quality, enhanced carbon sequestration, decreased reliance on fossil fuels, increased resilience to agricultural uncertainties, and the establishment of diverse wildlife habitats (Baah-Acheamfour et al., 2014; Davis et al., 2012; Wilson & Lovell, 2016; Winans et al., 2015). Sensors are of the utmost importance in monitoring soil moisture levels to optimize irrigation, whereas soil analysis assists farmers in choosing the most suitable crops. In addition, the assistance provided by satellite imagery monitoring in the form of efficient crop management expedites farmers' financial gains.

Four (4) smart features are included in Nusantara's smart forest and green management, including Carbon Stock and Emission Monitoring, Smart Forest Biodiversity Monitoring, Precision Farming, and Smart Forest Fire Management.

Carbon Stock and Emission Monitoring focuses on quantifying and utilizing carbon stores in forests and replanting to counterbalance emissions. By employing advanced technologies, such as spatial carbon calculators and carbon dioxide (CO_2) emission monitoring sensors, this endeavor makes a valuable contribution to the comprehensive assessment of carbon stocks.

The Smart Forest Biodiversity Monitoring focuses on systematically monitoring forest biodiversity and assisting in reforestation efforts by utilizing IoT sensors, drones, and video traps. These technologies enable real-time monitoring and facilitate the building of a comprehensive database. This technology allows for the monitoring of events as they occur and assists in promptly reporting emergencies related to the presence of wildlife in residential areas, including a wide range of functional and technical requirements.

Precision farming is a crucial component of Nusantara's agroforestry and forestry systems, as it incorporates advanced technologies such as GIS and climate and soil sensors. These tools are of utmost importance in formulating cultivation strategies and managing agricultural land, as they contribute significantly to the precision of farming techniques and the fulfillment of food demands within the region of Nusantara. The objectives can be efficiently achieved by outlining comprehensive functional and technological criteria.

4.5. Industry and Human Resources Domain

The Industry and Human Resources domain highlights the role of smart technology in fostering sustainable learning environments across diverse settings such as schools, homes, and offices. By leveraging digital payment platforms and intelligent applications, Nusantara can effectively advance toward its smart city aspirations. The adoption of smart industry in Nusantara is vital for several reasons. Firstly, technological advancements and industrial innovation enhance operational efficiency and productivity while reducing costs and time. Secondly, it fosters increased competitiveness, positioning Indonesian cities as more appealing to international investors and companies. Moreover, its implementation bolsters economic growth, generating new employment opportunities within the region. Lastly, smart industries contribute to an environment conducive to innovation and the development of cutting-edge technologies, further propelling progress and sustainability.

There are six subdomains in Nusantara's Industry and Human Resources domain: Smart Education, Smart Job Training and Matching, Smart Entertainment and Tourism, Digital Payment System, Urban Citizen Living Lab, and Digital Platform for SMEs.

a. Smart Education

Education is considered a critical human element within smart cities and a fundamental service (Anthopoulos et al., 2016). It is regarded as an essential element in promoting social progress and cultivating development within the context of smart city frameworks, as the smart people domain was significantly highlighted as a key indicator of a smart city (Letaifa, 2015). According to Boes et al. (2016), smart education in Barcelona has established a campus system that promotes collaboration and synergy among diverse organizations – including local, national, and international businesses, universities, research institutions, SMEs, and local entrepreneurs – at the local, national, and international levels. In addition, it has been demonstrated that tele-education is efficacious in transmitting knowledge and enhancing public consciousness regarding critical matters such as energy conservation and recycling (Alexopoulos et al., 2021). Furthermore, it has played a pivotal role in increasing senior citizens' access to ICT education, enhancing their connectivity and outreach (Molnar, 2014).

The smart education domain within the smart city framework encompasses four key features. Firstly, the digital library is an online repository housing various digital resources, such as books, articles, journals, manuscripts, and audiovisual materials, accessible via electronic media. It functions as a digital alternative to traditional libraries, enabling internet-based access, search, and retrieval of various content across diverse academic disciplines. Secondly, digital classrooms and e-learning platforms leverage technology to create interactive online learning environments, facilitating access to educational materials and enabling virtual interactions and remote learning. Thirdly, a Student Admission Platform is a digital system designed to streamline the student admission process in educational institutions. Finally, the Integrated Education Platform integrates various educational components – from learning modules to student and teacher management, school administration, and collaborative tools – into a unified entity. This platform's primary objective is to provide a comprehensive solution to enhance learning and administrative processes within educational institutions.

b. Smart Job Training and Matching

The Smart Job Training and Matching feature enhances employees' professional competencies and skills. Fundamentally, this notion aims to strengthen capabilities and skill sets via a comprehensive strategy incorporating training, advanced education, and formal certification procedures. Professional training, a systematic endeavor, is intended to enhance job-related competencies, including technical mastery, managerial insight, and interpersonal abilities. Typically, these programs are provided via a multitude of channels, including specialized training agencies, academic institutions, or easily accessible online platforms. In addition, skill development initiatives play a crucial role in enhancing performance in the workplace by emphasizing the improvement of technical expertise, leadership attributes, and proficient communication capabilities. This can be accomplished through various means, including work exposure, mentoring programs, training sessions, and practical experiences. Moreover, in the capacity of validation mechanisms, certification platforms conduct exhaustive tests, evaluations, or assessments to ascertain the competence of candidates. These platforms provide concrete proof of expertise and understanding, frequently demonstrated via online assessments, portfolios, or direct assessments and accredited by industry-specific organizations.

c. Smart Entertainment and Tourism

In the Smart Entertainment and Tourism subdomain, innovative approaches revolutionize tourist experiences through several smart features: Visitor Application, Smart Digital Museum, and Entertainment and Creativity Program. Visitor Apps stand as a prime example, leveraging smartphones and smart devices to elevate the quality of tourist experiences at destinations. These apps offer personalized information and interactive features, catering to the preferences and needs of individual travelers, thereby enhancing their overall engagement. Moreover, the emergence of Smart Digital Museums marks a transformative stride in this subdomain, integrating technological advancements within museum settings to deliver interactive and informative experiences. The Entertainment and Creativity Program is another vital component geared toward providing captivating and interactive entertainment experiences within the tourism industry. This initiative aims to enhance the allure of tourist destinations by introducing innovative and inspiring entertainment facets, thereby enriching the overall visitor experience and boosting the destinations' appeal.

d. Digital Payment System

The Integrated Digital Payment System seamlessly connects electronic payment methods, including credit cards, bank transfers, digital wallets, and online payment options. Its primary objective is simplifying and streamlining the payment process. Their implementation will promote the accessibility of payment services and the growth of financial and digital literacy by developing a knowledge society based on innovative digital technologies (Korobeynikova et al., 2018). It allows users to swiftly and effortlessly conduct transactions via electronic devices such as smartphones, tablets, or computers using linked applications or platforms, eliminating the need for cash or physical payment tools.

e. Urban Citizen Living Lab

In Nusantara's capital city development context, the Urban Citizen Living Lab embodies an essential hub to catalyze urban innovation and progress through several initiatives, including the Technology Demonstration Center, Chemical and Pharmacy Center, Equipment and Machinery Center, and Social Science Center.

The principal aim of the Technology Demonstration Center is to function as a hub for Research and Development (R&D), showcase state-of-the-art technologies, and provide extensive educational opportunities. With an emphasis on improving urban life, this center offers practical experiences to stakeholders and residents of urban areas. The primary elements of this gathering consist of designated areas for research and development, trade shows, training workshops, and vendor exhibitions. Furthermore, the Chemical and Pharmacy Center is crucial in advancing research and innovation in urban environments, particularly in the chemical sector. This center promotes collaborations between academia and industry and facilitates various functions, including dedicated R&D facilities, spaces for product evaluation, and workforce development programs.

Furthermore, emphasizing advanced technology, energy, transportation, and urban infrastructure, the Equipment and Machinery Center grants users access to an extensive selection of state-of-the-art machinery and equipment to encourage experimentation and innovation in urban environments. The Social Science Center is concurrently committed to the examination and resolution of urban social issues. This facility facilitates collaborations, data analysis, and comprehensive studies among scholars, researchers, and urban residents, particularly in social science research and development. The primary objective of this initiative is to produce novel insights, develop viable public policies, and invent inventive resolutions that improve the quality of life in urban areas, foster sustainability, and empower local communities.

f. Digital Platform for SMEs

A digital platform for SMEs is a system or technological infrastructure designed to assist SMEs, particularly in digital marketing, sales, and management activities. This platform provides various digital services and applications that can help SMEs increase their efficiency, productivity, and competitiveness in the market. Digital platforms contribute to the innovation of the SME business model, creating new distribution channels and new ways to develop and deliver value to customer segments (Matarazzo et al., 2021).

4.6. Built Environment and Infrastructure Domain

The smart built environment within smart cities revolves around digitalizing the physical environment, thereby influencing private enterprises, public services, and society (Heaton & Parlikad, 2019). It involves integrating the urban built environment with the natural environment and urban services, improving the efficiency and security of urban infrastructures (Shahrour et al., 2017). The development strategy for the capital city of Nusantara prioritizes environmental sustainability and infrastructure efficiency through various initiatives. The fundamental elements of this framework are the smart built environment and infrastructure, which consist of two essential components: smart buildings and smart infrastructure.

Smart buildings integrate advanced technology and systems that enhance sustainability, security, comfort, and overall performance. Similarly, smart infrastructure incorporates cutting-edge technology and digital systems into designing, constructing, and operating infrastructure. It encompasses a range of essential infrastructure development such as roads, bridges, water treatment facilities, and other public works. This innovative approach utilizes cutting-edge technologies such as the IoT, AI, and other digital technologies to establish a more effective, environmentally friendly, and flexible infrastructure that addresses the changing requirements of society.

a. Smart Building

The fundamental components essential to smart buildings include the Access Control System, Communication System, Energy System, Safety System, Heating, Ventilation, and Air Conditioning (HVAC) System, Lighting System, Mobility System, Security System, and Resource System. Nevertheless, successfully deploying intelligent buildings requires essential prerequisites, including an Integrated Building Management System, Control Room and Data Center, Fiber to the Room (FTTR) network, and Digital Twin system.

The Access Control System comprises Touchless Access Control and streamlined Visitor Management, enhancing security and productivity. The communication system includes Intercom Systems as well as Digital and Audio Visual Boards, facilitating uninterrupted communication for safety and the dissemination of information. The energy system utilizes automated and sub-meter readers to monitor and optimize energy consumption. Furthermore, the Safety System integrates several features, such as Active Disaster Response, Smart Fire Suppression, Emergency Buttons, and Wildlife Hazard Protection, thereby augmenting emergency safety protocols.

In addition, the HVAC system is responsible for regulating temperature and air quality, while the Lighting system is designed to optimize illumination to achieve maximum efficiency. Additionally, the Mobility system aims to boost transportation and improve accessibility. The Security System utilizes many advanced technologies, such as Intelligent Video Surveillance, Locking Systems, Virtual Gates, and Occupancy Monitoring, to enhance security measures in diverse environments. Finally, the Resource System prioritizes implementing strategies such as Smart Water Management, Drinking Water Dispensers, Smart Waste Chute, Smart Bins Cans, and Smart Toilets to enhance water use and optimize waste management.

b. Smart Infrastructure

The fundamental components of smart infrastructure systems are centered on real-time monitoring and managing diverse critical infrastructure systems, including but not limited to traffic flow, water treatment facilities, and the maintenance of bridges and roads. The utilization of data-driven analysis provides infrastructure owners and operators with actionable insights that can be used to make informed decisions in the management and maintenance of infrastructure, further enhancing the infrastructure's efficiency and long-term sustainability. The Digital Twin is a crucial feature in the subdomain of smart infrastructure.

A Digital Twin is a virtual representation of a physical object or operation, which leverages both real-time and past data to facilitate decision-making, optimize asset management strategies, and monitor current conditions (Cureton & Dunn, 2020). This technology harnesses the power of virtual and augmented reality to construct a full representation of the real asset or system (El Saddik, 2018). This approach significantly improves the effectiveness of planning and implementation of smart city initiatives, leading to improvements in financial outcomes, resource allocation, environmental sustainability, and the overall well-being of citizens (Agostinelli, 2021; Khallaf et al., 2022).

Digital Twin technology is crucial in various aspects of smart city development, including Planning, Scenario Planning and Design, Operational Planning and Scheduling, and Operational Decision Support. The initiative follows the Indonesian Digital Government idea, which emphasizes the application of ICT to enhance the efficiency of public services, as stated in Presidential Regulation No. 95 of 2018. Incorporating Digital Twin technology into Nusantara's development framework will assist authorities and operators in meticulous building and facility management.

Fig. 5 summarizes the components of Nusantara's smart city initiatives, including the domains, subdomains, and smart features.

4.7. Nusantara Smart City Initiatives Framework

The framework for Nusantara smart city initiatives was developed by considering several layers, starting from the development principles, smart city domains, and smart city enablers to development goals. The development principles that direct the capital city of Nusantara encompass a holistic and interconnected vision that integrates technological progress, environmental awareness, resilience, sustainability, and inclusivity. At the core of this vision lies a profound alignment to

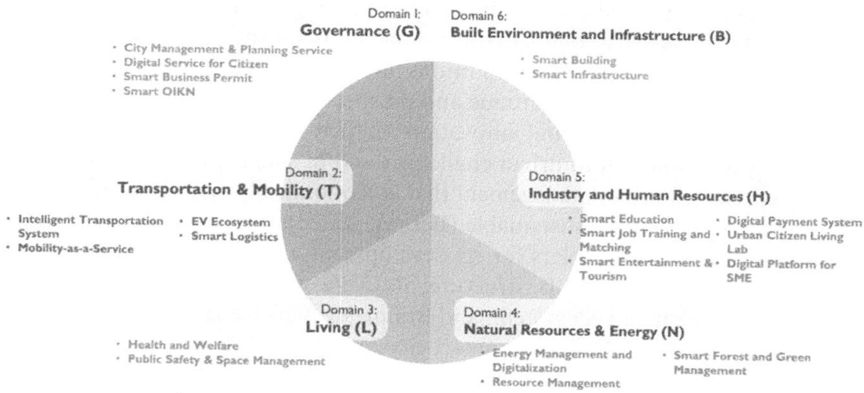

Fig. 5. Smart City Components: Domains, Smart Services, and Smart Features (*Source*: Nusantara Capital Authority, 2023).

environmental stewardship reflected in the city's blueprint, which aims to minimize ecological impacts while fostering a strong bond with the environment. Rooted in this ecological philosophy are ubiquitous components, including lush green areas, the incorporation of sustainable infrastructure, and a commitment to utilizing renewable energy resources; these comprise the fundamental pillars of its environmentally conscious urban planning.

The fundamental principle underlying Nusantara's developmental framework is the concept of resilience, which acknowledges the critical necessity to confront and recover from complex challenges, including those of an economic, environmental, and societal nature. The municipality prioritizes the development of resilient infrastructure that can effectively endure natural disasters and adjust to changing and dynamic conditions; thus, this strengthens the foundation of its resilience strategy. On the other hand, ingrained as a guiding principle, sustainability permeates the fabric of urban planning and daily life within the city. Prioritizing resource efficiency, waste reduction, and the pursuit of carbon neutrality, Nusantara endeavors to foster a circular economy that minimizes waste while regenerating or repurposing resources.

Inclusivity serves as a fundamental cornerstone of the social fabric of Nusantara, fostering a deep-seated sense of belonging and fairness among its heterogeneous inhabitants. The city endeavors to design environments and provide opportunities that are universally accessible to all inhabitants, regardless of their socioeconomic status, ensuring equitable access to education, healthcare, housing, and employment for each individual.

The city's commitment to smart city development through its six domains incorporates innovation and cutting-edge technology to improve productivity, connectedness, and overall quality of life. By utilizing data-driven solutions, smart infrastructure implementation, and digital connectivity, Nusantara aims to create an interconnected urban environment that fosters sustainable development while empowering its citizens. All of these principles integrate to steer the capital city of Nusantara toward a future in which technological innovation, resilience, sustainability, inclusivity, and environmental consciousness coexist amicably. This coming together creates the conditions for developing a thriving, integrated, and progressive city that will continue and prosper for many years to come.

Technology development and innovation will play a crucial role in establishing a smart city as a solution to urban challenges and in shaping a future model for digitally connected urban management that is economically viable, socially desirable, and environmentally sustainable (Berawi, 2022a). Nevertheless, Nusantara identifies three critical enablers for its development as a smart city: Hardware, which refers to the tangible infrastructure; Software, which encompasses the digital operating system and ecosystem; and Brainware, which concerns the development of human capital. Nusantara smart city consists of different interconnected layers of those three aspects.

Additionally, telecommunication infrastructure is crucial since it serves as the backbone of connectivity, enabling various technologies to be implemented in the city and a catalyst for innovation and businesses to flourish. Moreover, as a requirement for basic telecommunication providers, the fifth generation (5G) network would act as a driver for advanced technology. By mandating 5G networks for basic telecommunications providers and advancements in technology, the overall enhancement in quality of life could be facilitated. A multitude of city services are optimized and managed at the data center and city command center, which serve as integrators for all smart city domains. The conceptual framework for Nusantara smart city initiatives that include the five development goals, six smart city domains, three key enables, and three development goals are illustrated in Fig. 6.

The implications of this research contribution to the Smart Cities Agenda highlight the potential of Nusantara's model to greatly impact and influence the global smart city movement, offering insights and frameworks that can be adapted by cities worldwide to navigate the complex interplay of technology, economy, and society toward achieving sustainable urban futures. It can also provide valuable insights for various stakeholders in smart cities, including city planners, architects, policymakers, lecturers, students, educational administrators, and government officials.

To achieve the goal of Nusantara as a pioneering smart city, the key initiative is to establish an integrated smart city management and operations center. This center will improve urban operations by utilizing data-driven insights to enhance service delivery and decision-making. Improving smart mobility and public transport with technology-driven solutions aims to make transportation more efficient, decrease traffic congestion, and encourage sustainable travel choices. The city's sustainable growth will be guaranteed through a focus on renewable energy and resource efficiency projects and the establishment of collaborative governance and citizen engagement platforms. This approach aims to promote active participation from residents in shaping their urban environment.

The government and policymakers should encourage investment in digital infrastructure and related technologies to assist in the transformation of smart cities. Enhancing rules and standards is crucial to ensuring environmental sustainability

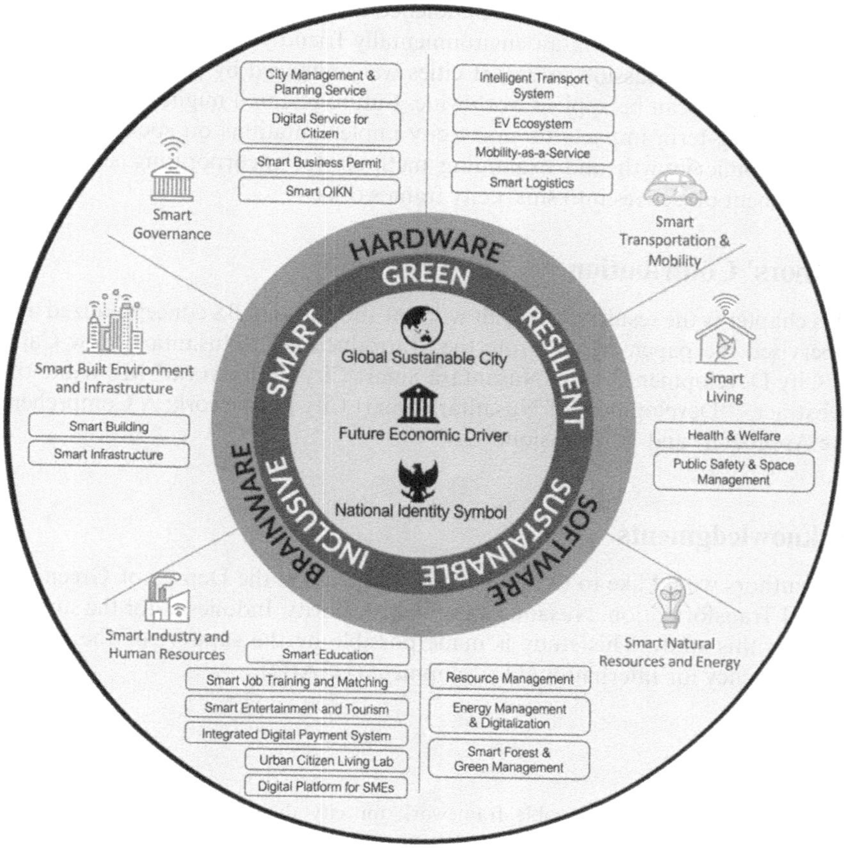

Fig. 6. Nusantara Smart City Initiatives Conceptual Framework (*Source*: Nusantara Capital Authority, 2023).

and social inclusivity in the smart city framework. Furthermore, promoting collaboration across Gulf countries to exchange knowledge and resources would be essential for successfully implementing the smart city agenda. Thus, these recommendations are hoped to set a standard for urban development that effectively combines technology, community, and environmental conservation.

5. Conclusion

This chapter discusses the development of Nusantara, Indonesia's future capital city, focusing on transforming it into a smart city. The method used to create the framework for the smart city projects includes defining Nusantara's goals, evaluating user feedback, and utilizing the CIMI for benchmarking. This chapter explain how Nusantara is using innovative technologies to improve the living standards of its citizens and efficiently manage resources and city operations.

It provides a crucial guide for a wide audience, including policymakers and urban developers, on establishing an environmentally friendly, resilient, and inclusive smart city. The discussion on smart cities was enhanced by providing a strong framework that can be applied worldwide. Future research might focus on studying the long-term impacts of smart city implementations on social dynamics and economic growth and examining methods for incorporating sustainable development objectives into smart city frameworks.

Authors' Contribution

This chapter is the result of the joint work of the authors. BS conceptualized and supervised the paper. MAB wrote the "Introduction," "Nusantara New Capital City Development," and "Nusantara Smart City Framework." MS wrote the "Abstract," "Development of Nusantara Smart City Framework: A Comprehensive Approach" and "Conclusion."

Acknowledgments

The authors would like to thank the technical team at the Deputy of Green and Digital Transformation, Nusantara Capital Authority, Indonesia, for the support given to this study. This study is made possible by the support of the United States Agency for International Development (USAID).

References

Agostinelli, S. (2021). Actionable framework for city digital twin-enabled predictive maintenance and security management systems. *WIT Transactions on the Built Environment*, *205*, 223–233. https://doi.org/10.2495/BIM210191

Albino, V., Berardi, U., & Dangelico, R. M. (2015). Smart cities: Definitions, dimensions, performance, and initiatives. *Journal of Urban Technology*, *22*(1), 3–21. https://doi.org/10.1080/10630732.2014.942092

Alexopoulos, C., Euripidis, L., & Charalabidis, Y. (2021). A methodology for participatory planning of smart city interventions. In *Public administration and information technology* (Vol. 37). https://doi.org/10.1007/978-3-030-61033-3_8

Anthopoulos, L., Janssen, M., & Weerakkody, V. (2016). A Unified Smart City Model (USCM) for smart city conceptualization and benchmarking. *International Journal of Electronic Government Research*, *12*(2). https://doi.org/10.4018/IJEGR.2016040105

Antoun, J., Kabir, M. E., Moussa, B., Atallah, R., & Assi, C. (2020). A detailed security assessment of the EV charging ecosystem. *IEEE Network*, *34*(3), 200–207. https://doi.org/10.1109/MNET.001.1900348

Aymen, F., & Mahmoudi, C. (2019). A novel energy optimization approach for electrical vehicles in a smart city. *Energies*, *12*(5). https://doi.org/10.3390/en12050929

Baah-Acheamfour, M., Carlyle, C. N., Bork, E. W., & Chang, S. X. (2014). Trees increase soil carbon and its stability in three agroforestry systems in central Alberta, Canada. *Forest Ecology and Management*, *328*, 131–139.

Berawi, M. A. (2022a). City of tomorrow: The new capital city of Indonesia. *International Journal of Technology*, *13*(4), 690–694. https://doi.org/10.14716/ijtech.v13i4.6011

Berawi, M. A. (2022b). Fostering smart city development to enhance quality of life. *International Journal of Technology*, *13*(3), 454–457. https://doi.org/10.14716/ijtech.v13i3.5733

Berawi, M. A., Leviäkangas, P., Siahaan, S. A. O., Hafidza, A., Sari, M., Miraj, P., Harwahyu, R., & Saroji, G. (2021). Increasing disaster victim survival rate: SaveMyLife Mobile Application development. *International Journal of Disaster Risk Reduction*, *60*. https://doi.org/10.1016/j.ijdrr.2021.102290

Berawi, M. A., Susantono, B., & Sari, M. (2022). Fostering resilient and smart cities to enhance sustainable urban development. In *The Routledge handbook of sustainable cities and landscapes in the Pacific Rim* (pp. 704–719). Routledge. https://doi.org/10.4324/9781003033530-60

Berrone, P., & Ricart, J. E. (2022). IESE Cities in Motion Index. In *IESE Business School*. IESE Business School, University of Navarra. https://media.iese.edu/research/pdfs/ST-0633-E.pdf

Boes, K., Buhalis, D., & Inversini, A. (2016). Smart tourism destinations: Ecosystems for tourism destination competitiveness. *International Journal of Tourism Cities*, *2*(2). https://doi.org/10.1108/IJTC-12-2015-0032

Caragliu, A., del Bo, C., & Nijkamp, P. (2011). Smart cities in Europe. *Journal of Urban Technology*, *18*(2), 65–82. https://doi.org/10.1080/10630732.2011.601117

Cicirelli, F., Fortino, G., Giordano, A., Guerrieri, A., Spezzano, G., & Vinci, A. (2016). On the design of smart homes: A framework for activity recognition in home environment. *Journal of Medical Systems*, *40*(9). https://doi.org/10.1007/s10916-016-0549-7

Cureton, P., & Dunn, N. (2020). Digital twins of cities and evasive futures. In *Shaping smart for better cities: Rethinking and shaping relationships between urban space and digital technologies*. https://doi.org/10.1016/B978-0-12-818636-7.00017-2

Das, S., Lee, S. H., Kumar, P., Kim, K. H., Lee, S. S., & Bhattacharya, S. S. (2019). Solid waste management: Scope and the challenge of sustainability. *Journal of Cleaner Production*, *228*. https://doi.org/10.1016/j.jclepro.2019.04.323

Davis, A. S., Hill, J. D., Chase, C. A., Johanns, A. M., & Liebman, M. (2012). *Increasing cropping system diversity balances productivity, profitability and environmental health*.

De Guimarães, J. C. F., Severo, E. A., Felix Júnior, L. A., Da Costa, W. P. L. B., & Salmoria, F. T. (2020). Governance and quality of life in smart cities: Towards sustainable development goals. *Journal of Cleaner Production*, *253*, 119926. https://doi.org/10.1016/j.jclepro.2019.119926

Dewalska–Opitek, A. (2014). Smart city concept – The citizens' perspective. *Communications in Computer and Information Science*, *471*. https://doi.org/10.1007/978-3-662-45317-9_35

Ding, Y., Jin, M., Li, S., & Feng, D. (2021). Smart logistics based on the Internet of Things technology: An overview. *International Journal of Logistics Research and Applications*, *24*(4). https://doi.org/10.1080/13675567.2020.1757053

El Saddik, A. (2018). Digital twins: The convergence of multimedia technologies. *IEEE MultiMedia*, *25*(2), 87–92. https://doi.org/10.1109/MMUL.2018.023121167

EU. (2019). *Intelligent transport systems*. https://ec.europa.eu/transport/themes/its_en

Garcia Alonso, R., & Lippez-De Castro, S. (2016). Technology helps, people make: A smart city governance framework grounded in deliberative democracy. In *Public administration and information technology* (Vol. 11). https://doi.org/10.1007/978-3-319-17620-8_18

Giffinger, R., & Gudrun, H. (2010). Smart cities ranking: An effective instrument for the positioning of the cities? *ACE: Architecture, City and Environment*, *4*(12), 7–26. https://doi.org/10.5821/ace.v4i12.2483

Heaton, J., & Parlikad, A. K. (2019). A conceptual framework for the alignment of infrastructure assets to citizen requirements within a smart cities framework. *Cities*, *90*, 32–41. https://doi.org/10.1016/j.cities.2019.01.041

Hoornweg, D., & Bhada-Tata, P. (2012). What a waste: A global review of solid waste management. *World Bank*, *15*(November), 1–2. http://hdl.handle.net/10986/17388

Iclodean, C., Varga, B. O., & Cordoş, N. (2022). *Autonomous driving systems* (pp. 69–138). https://doi.org/10.1007/978-3-031-14678-7_3

Khallaf, R., Khallaf, L., Anumba, C. J., & Madubuike, O. C. (2022). Review of digital twins for constructed facilities. *Buildings*, *12*(11). https://doi.org/10.3390/buildings 12112029

Korobeynikova, O., Korobeynikov, D., & Popova, L. (2018). Problem solving payment innovations in the digital economy. *Proceedings of the International Scientific Conference "Competitive, Sustainable and Secure Development of the Regional Economy: Response to Global Challenges" (CSSDRE 2018)*. https://doi.org/10.2991/cssdre-18.2018.7

Koźlak, A., & Pawłowska, B. (2019). Mobility-as-a-service for improving mobility in smart cities – A comparative analysis of selected cities. *Prace Naukowe Uniwersytetu Ekonomicznego We Wrocławiu*, *63*(10), 31–46. https://doi.org/10.15611/pn.2019.10.03

Kupriyanovsky, V., Akimov, A., Pokusaev, O., Alenkov, V., Namiot, D., & Sinyagov, S. (2017). Intellectual mobility and mobility as a service in Smart Cities. *International Journal of Open Information Technologies*, *5*, 77–122.

Letaifa, S. (2015). How to strategize smart cities: Revealing the SMART model. *Journal of Business Research*, *68*, 1414–1419. https://doi.org/10.1016/J.JBUSRES.2015.01.024

Leasiwal, T. C., Payapo, R. W., Oppier, H., & Setya, D. (2022). Analysis of income disparities between regions in the western region of Indonesia and the eastern region of Indonesia. *International Journal of Science and Society*, *4*(4), 638–656. https://doi.org/10.54783/ijsoc.v4i4.612

Li, D., Deng, L., Cai, Z., Franks, B., & Yao, X. (2018). Notice of retraction: Intelligent transportation system in Macao based on deep self-coding learning. *IEEE Transactions on Industrial Informatics*, *14*(7). https://doi.org/10.1109/TII.2018.2810291

Lin, C., Zhao, G., Yu, C., & Wu, Y. (2019). Smart city development and residents' well-being. *Sustainability*, *11*(3), 676. https://doi.org/10.3390/su11030676

Madakam, S., & Ramaswamy, R. (2016, January). Smart cities – Six dimensions. *Conference on Advances in Computing and Information Technology*.

Masia, O. A., & van der Poll, J. A. (2021). A framework for agile project management for the water industry in developing economies. *Journal for Global Business Advancement*, *14*(1). https://doi.org/10.1504/JGBA.2021.114320

Matarazzo, M., Penco, L., Profumo, G., & Quaglia, R. (2021). Digital transformation and customer value creation in Made in Italy SMEs: A dynamic capabilities perspective. *Journal of Business Research*, *123*, 642–656. https://doi.org/10.1016/j.jbusres.2020.10.033

McFarlane, D., Giannikas, V., & Lu, W. (2016). Intelligent logistics: Involving the customer. *Computers in Industry*, *81*. https://doi.org/10.1016/j.compind.2015.10.002

Melis, A., Mirri, S., Prandi, C., Prandini, M., Salomoni, P., & Callegati, F. (2018). Integrating personalized and accessible itineraries in MaaS ecosystems through microservices. *Mobile Networks and Applications*, *23*(1). https://doi.org/10.1007/s11036-017-0831-z

Molnar, A. (2014). On better understanding the usage of mobile phones for learning purposes. *Bulletin of the Technical Committee on Learning Technology*, *16*(2–3), 18–20.

Montreuil, B., Meller, R. D., & Ballot, E. (2013). Physical Internet foundations. *Studies in Computational Intelligence*, *472*. https://doi.org/10.1007/978-3-642-35852-4_10

Muhyiddin, M. (2019). Future challenges on Indonesia's Vision 2045. *Jurnal Perencanaan Pembangunan: The Indonesian Journal of Development Planning*, *3*(1). https://doi.org/10.36574/jpp.v3i1.63

Neirotti, P., De Marco, A., Cagliano, A. C., Mangano, G., & Scorrano, F. (2014). Current trends in smart city initiatives: Some stylised facts. *Cities*, *38*, 25–36. https://doi.org/10.1016/j.cities.2013.12.010

Nguyen, D. D., Rohács, J., Rohács, D., & Boros, A. (2020). Intelligent total transportation management system for future smart cities. *Applied Sciences (Switzerland)*, *10*(24). https://doi.org/10.3390/app10248933

Pereira, G. V., Parycek, P., Falco, E., & Kleinhans, R. (2018). Smart governance in the context of smart cities: A literature review. *Information Polity*, *23*(2), 143–162. https://doi.org/10.3233/IP-170067

Phung, S., Richards, A. C., & Mohammadzadeh, M. (2022). Smart cities in the Pacific Rim: A mapping of urban evolution in the Pacific. In *The Routledge handbook of sustainable cities and landscapes in the Pacific Rim* (pp. 670–686). Taylor and Francis. https://doi.org/10.4324/9781003033530-58

Savin, G. V. (2020). Quality indicator of the smart city transportation and logistics system. *Economic Analysis: Theory and Practice*, *19*(11), 2116–2135. https://doi.org/10.24891/ea.19.11.2116

Scholl, H. J., & Alawadhi, S. (2016). Creating Smart Governance: The key to radical ICT overhaul at the City of Munich. *Information Polity*, *21*(1). https://doi.org/10.3233/IP-150369

Scholl, H. J., & Scholl, M. C. (2014). *Smart Governance: A Roadmap for Research and Practice*. https://doi.org/10.9776/14060

Shafiee, M. J., Jeddi, A., Nazemi, A., Fieguth, P., & Wong, A. (2021). Deep neural network perception models and robust autonomous driving systems: Practical solutions for mitigation and improvement. *IEEE Signal Processing Magazine*, *38*(1), 22–30. https://doi.org/10.1109/MSP.2020.2982820

Shahrour, I., Alileche, L., & Alfurjani, A. (2017). Smart cities: System and tools used for the digital modelling of physical urban systems. *2017 Sensors Networks Smart and Emerging Technologies, SENSET 2017*, 1–4. https://doi.org/10.1109/SENSET.2017.8125056

Shilvya, J. A., George, S. T., Subathra, M. S. P., Manimegalai, P., Mohammed, M. A., Jaber, M. M., Kazemzadeh, A., & Al-Andoli, M. N. (2022). Home based monitoring for smart health-care systems: A survey. In *Wireless communications and mobile computing* (Vol. 2022, pp. 1–10). https://doi.org/10.1155/2022/1829876

Singh, A. K., & Edla, D. R. (2022). Smart transportation – A futuristic intelligent mobility approach towards smarter cities. *Proceedings of 4th International Conference on Cybernetics, Cognition and Machine Learning Applications, ICCCMLA 2022*. https://doi.org/10.1109/ICCCMLA56841.2022.9989297

Sternberg, H., Hagen, A., Paganelli, P., & Lumsden, K. (2010). Intelligent cargo - Enabling future's sustainable and accountable transportation system. *World Journal of Science, Technology and Sustainable Development*, *7*(3), 253–262. https://doi.org/10.1108/20425945201000016

Susantono, B., & Berawi, M. A. (2023). Nusantara smart city: Improving the quality of life of its citizens - Academia - The Jakarta Post. *Jakarta Post*. https://www.thejakartapost.com/opinion/2023/06/09/nusantara-smart-city-improving-the-quality-of-life-of-its-citizens.html

Świątowiec-Szczepańska, J., & Stępień, B. (2022). Drivers of digitalization in the energy sector – The managerial perspective from the catching up economy. *Energies*, *15*(4). https://doi.org/10.3390/en15041437

Takagi, H., Esteban, M., Mikami, T., & Fujii, D. (2016). Projection of coastal floods in 2050 Jakarta. *Urban Climate*, *17*, 135–145. https://doi.org/10.1016/j.uclim.2016.05.003

TomTom. (2022). *Jakarta traffic report: TomTom Traffic Index*. TomTom International BV. https://www.tomtom.com/traffic-index/jakarta-traffic/

Turner, S. W., & Uludag, S. (2016). Intelligent transportation as the key enabler of smart cities. *Proceedings of the NOMS 2016 - 2016 IEEE/IFIP Network Operations and Management Symposium*. https://doi.org/10.1109/NOMS.2016.7502999

Uckelmann, D. (2008). A definition approach to smart logistics. *Lecture Notes in Computer Science (Including Subseries Lecture Notes in Artificial Intelligence and Lecture Notes in Bioinformatics), 5174 LNCS*. https://doi.org/10.1007/978-3-540-85500-2_28

UN-HABITAT. (2012). State of the world's cities 2008/2009: Harmonious cities. In *State of the World's Cities 2008/2009: Harmonious Cities* (Vol. 9781849772). https://doi.org/10.4324/9781849772624

Wilson, M., & Lovell, S. (2016). Agroforestry – The next step in sustainable and resilient agriculture. *Sustainability*, *8*(6), 574. https://doi.org/10.3390/su8060574

Winans, K. S., Tardif, A.-S., Lteif, A. E., & Whalen, J. K. (2015). Carbon sequestration potential and cost-benefit analysis of hybrid poplar, grain corn and hay cultivation in southern Quebec, Canada. *Agroforestry Systems*, *89*, 421–433.

Windt, K., & Hülsmann, M. (2007). Changing paradigms in logistics-understanding the shift from conventional control to autonomous cooperation and control. In *Understanding autonomous cooperation and control in logistics: The impact of autonomy on management, information, communication and material flow*. https://doi.org/10.1007/978-3-540-47450-0_1

Yusuf, A. A., Roos, E. L., Horridge, J. M., & Hartono, D. (2023). Indonesian capital city relocation and regional economy's transition toward less carbon-intensive economy: An inter-regional CGE analysis. *Japan and the World Economy*, *68*. https://doi.org/10.1016/j.japwor.2023.101212

Zygiaris, S. (2013). Smart city reference model: Assisting planners to conceptualize the building of smart city innovation ecosystems. *Journal of the Knowledge Economy*, *4*(2). https://doi.org/10.1007/s13132-012-0089-4

Chapter 24

Artificial Intelligence for Smart City Vision 2040 in Gulf Region

Miltiadis D. Lytras[a], *Afnan Alkhaldi*[b], *Sawsan Malik*[b] *and Andreea Claudia Serban*[c]

[a]*College of Engineering, Efrat University, Saudi Arabia*
[b]*Arab Open University, Kuwait*
[c]*Bucharest University of Economic Studies, Romania*

Abstract

This chapter delves into the transformative potential of Artificial Intelligence (AI) in spearheading the smart city vision by 2040 in the Gulf Region. It emphasizes the integration of AI with smart city governance, strategy, and infrastructure, underlining the pivotal role AI plays in digital transformation. The discourse navigates through strategic areas such as the exploitation of Large Language Models for enhancing smart city services, the utilization of AI in analyzing social network content and crowdsourcing applications, and leveraging AI's reasoning capabilities to bolster predictive analytics and smart city services innovation. The chapter also showcases how AI-driven innovation fosters the development of new markets and industries within the smart city ecosystem. Highlighting the Gulf Region's strategic investment in human capital and technology, the authors present a vision where smart cities serve as hubs of sustainability, innovation, and economic growth. Through the lens of the Smart City GSUTI Framework, the narrative illustrates the comprehensive integration of governance, strategy, utilization frameworks, technologies, and infrastructural capabilities. This holistic approach ensures that AI not only enhances the quality of life and economic prosperity in the Gulf Region but also positions it as a global leader in smart city development. The chapter is a call

The Emerald Handbook of Smart Cities in the Gulf Region:
Innovation, Development, Transformation, and Prosperity for Vision 2040, 515–525

Published under exclusive licence by Emerald Publishing Limited
doi:10.1108/978-1-83608-292-720241034

to action for leveraging AI's potential to transform the Gulf Region into a model for sustainable, innovative, and smart urban development by 2040.

Keywords: Smart cities; artificial intelligence; digital transformation; smart cities governance; Gulf Region; Gulf Cooperation Council Countries; Vision 2040

1. Introduction

The latest developments in the domain of Artificial Intelligence (AI) have convinced everyone that the impact of this phenomenon in humanity at a global scale will be bold. In the context of smart cities research, particularly in the Gulf Region, this prediction holds true. Here, funds and venture capitals are actively investing in both well-established AI companies and start-ups. Simultaneously, investments in human capital and the attraction of top talent from around the globe are transforming the Gulf region into a sustainable hub of innovation for smart cities and AI research.

In this concluding chapter of the volume, we provide our initial thoughts on some transformative applications of AI in the context of next-generation smart cities research. In Fig. 1, we summarize four selective areas indicating the potential impact of AI on smart cities research namely:

- Strategic Area #1: Exploitation of Large Language Models (LLMs) on the provision of smart cities services
- Strategic Area #2: Utilization of AI insights over social networks content and crowdsourcing applications
- Strategic Area #3: Robust AI reasoning capabilities, for the promotion of prescriptive and predictive analytics and smart city services
- Strategic Area #4: AI-driven innovation, including disrupting AI services that have the capacity to develop new markets and industries of smart city services.

Especially when it comes to national or regional contexts, like the countries in the Gulf Region or the Gulf region as a collective urban space it is important to elaborate on the variety of new smart city business models powered by AI.

As discussed in Chapter 1, the Smart City GSUTI Framework offers a context for further elaboration. The five areas of the framework include the following pillars:

- Smart cities governance
- Smart cities strategy
- Smart city utilization framework
- Enabling smart city technologies
- Enabling infrastructures & capabilities.

The integration of the five pillars of the framework for the holistic smart city governance with the four strategic areas of AI deployment generates diverse use

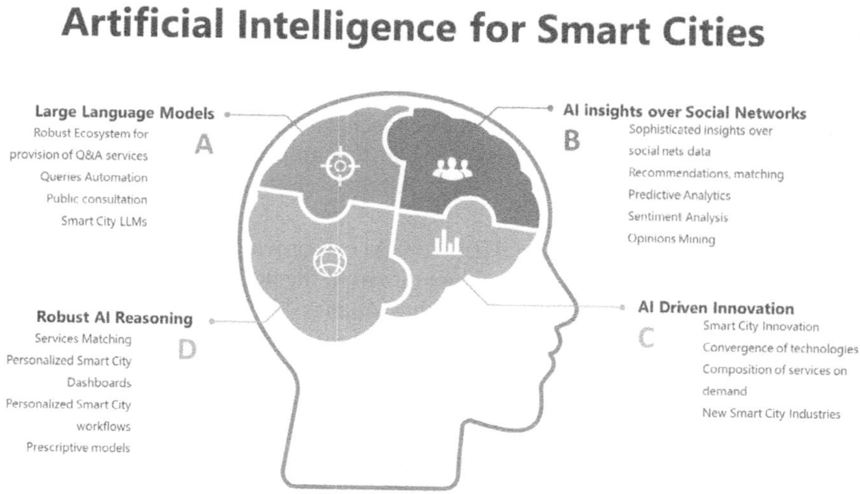

Fig. 1. Indicative Innovative and Disruptive Application Domains of AI for Smart Cities (*Source*: The Authors).

cases and scenarios with significant impact on the sustainable economic and social development in the Gulf Region. In the second section, we proceed further with additional clarifications, while in the remainder of this section, we provide some key direction for projects in each of the strategic areas highlighted before.

1.1. Large Language Models

The current evolving LLMs such as ChatGPT, Claude, Gemini, PaLM2, are absolute transformers. They can be utilized in the context of global and national smart city projects in diverse directions such as:

- Robust Ecosystem for Provision of Q&A Services: A new generation of question and answering systems or dialectic systems will offer to smart city services users the capacity to exploit the collective wisdom and to enjoy personalized services, that can be matched to needs and specifications.
- Queries Automation: LLMs also offer the capability for the design of integrated workflows aiming to respond effectively and to automate queries fulfillment. This is an amazing capability especially in the smart city utilization areas that discussed in Chapter 1, such as energy, tourism, economy, environment, healthcare, etc.
- Public Consultation: LLMs can also facilitate targeted public consultation procedures by eliciting input and feedback and revising policy documents and procedures.
- Smart City LLMs: It is more than evident that in the near future LLMs will also customized for specialized needs. An LLMs submodule related to smart cities, innovation, and technology can serve multiple purposes in the smart city context.

1.2. AI Insights Over Social Networks

The deployment of AI models over large social media data sets, can also train and offer new services and applications. For sure significant considerations related to data privacy and protection or consent of social media users have to be secured, but the overall capacity of AI in this area to offer new services is commendable (Alkhammash et al., 2019; Alhalabi et al., 2021).

- Sophisticated insights over social net data: AI can provide insights and reasoning over social media posts of significant items in the political agenda of smart cities, including well-being, quality of life, quality of health, climate change, social services, etc. It can also serve a new generation of fully customized dashboards and analytics ecosystems for the promotion of strategic objectives in the smart city agenda.
- Recommendations, matching: AI can also promote trusted algorithms on profiles, services, and applications matching allowing users and citizens to exploit unique paths for the exploration of services and content (Lytras & Serban, 2020; Lytras et al., 2020, 2021a, 2021b).
- Predictive analytics: This kind of analytics can also be exploited in the context of smart city megaprojects or national projects for enhanced decision-making and for improving the efficiency and the capacity of services.
- Sentiment analysis: Social media content can be also exploited with sophisticated AI methods such as sentiment analysis in order to meet diverse and complicated smart city challenges such as the detection of bullying, harassment over social media as well as for promoting the mental health of citizens.
- Opinions mining & fake news detection: Social media content and AI methods, also can promote a new era of opinions mining services and fake news detection for smart city community applications and civic engagement platforms.

1.3. Robust AI Reasoning

The amazing reasoning capabilities of AI can also serve diverse strategic orientations of the smart city framework. The following are just a few ideas that can be implemented with the support of the AI:

- Services Matching: Dynamic composition of smart cities services utilizing profile info of users or recommendations.
- Personalized Smart City Dashboards: Utilization of personal or business dashboards for smart cities exploiting diverse areas of smart cities utilization, e.g., energy, education, healthcare, governance, economy, tourism, etc.
- Personalized Smart City Workflows: Integrated business process automation in the context of smart city national or mega projects.
- Prescriptive Models: Robust optimization AI-driven techniques for different aspects of economic and social sustainability on integrated smart city framework.

1.4. AI-Driven Innovation

The AI's disruptive character can also lead to a new thread of innovation wave. In this direction, the following are indicative use cases and scenarios for the adoption of AI toward a next-generation AI-driven innovation (Nikiforova et al., 2023).

- Smart city innovation
- Convergence of technologies: Exploitation of novel services with integration of IoT, AI, Metaverse, Augmented Reality, etc.
- Composition of services on demand
- New smart city industries: A new direction of futuristic scenarios for smart city utilization including Metaverse or Industries of AI-enabled services.

2. Application Scenarios for the Deployment of AI in Future Smart Cities Research and Services

It seems that there are unlimited possibilities and use cases for the deployment of the AI for smart city applications. In Fig. 2, and in the next paragraphs we discuss indicative scenarios for future AI-enabled services anchored to the Smart City GSUTI framework proposed by the editors (see Chapter 1 of this Handbook). The purpose of this listing is just to highlight the unique value proposition of AI.

2.1. Smart City Use Cases Attached to the Smart City

2.1.1. Environment

In this context, AI can play a robust role in the utilization of ML/AI models for managing critical environmental challenges such as:

- Pollution monitoring
- Environmental protection apps
- Ecosystem health assessment
- Waste management
- IoT-enabled scheduling.

2.1.2. Energy

Smart energy, is a bold pillar of any smart city utilization framework and diverse strategies can play a critical role in addressing the following challenges:

- Demand forecasting
- Grid stability monitoring
- Renewable energy utilization
- Energy analytics
- Dynamic pricing.

Especially in the context of the Gulf Region, the promotion of new models for sustainable economic development detached from oil production, will boost the adoption of AI for managing smart energy grids, solar power grids, and many other innovative applications. In the same direction, AI-enabled management of international energy networks will also be a bold area of research and innovation in the Gulf Region.

2.1.3. Education

AI services will also affect significantly the provision of educational services in the Gulf Region. Diverse AI technologies and models will power several of the following learning contexts:

- Learning assistants
- Personalized learning environments
- Collaborative learning platforms
- AI-enabled skills building
- Transformative learning.

AI will provide unique opportunities, for learning content packaging, dynamic customization of learning environments and contexts, and robust capacity for skills building and growth. Beyond this AI, LLMs will offer unique student and faculty experiences, and provide new communication challenges to the higher education administration for the co-design of educational strategies (Lytras et al., 2022; Sairete et al., 2021).

2.1.4. Healthcare

Smart healthcare in the Gulf Region will be an AI-enabled digital transformation primer. AI jointly with other disruptive technologies will offer unique exploitation contexts for value-based healthcare: (Alsaywid et al., 2023; Lytras & Alsaywid, 2023; Lytras et al., 2023)

- Telemedicine solutions
- AI for personalized medicine
- Digital transformation
- Metaverse health
- AI models for phenotyping.

Especially in the sensitive context of healthcare critical issues related to data governance and AI ethics together with resilient regulations and policies will be required.

2.1.5. Social Inclusion

In the context of smart social inclusion, the following is an indicative list of significant interventions enabled by robust AI models and systems:

- Accessibility
- Social support platforms
- Chatbots for social inclusion

- Emotional Intelligence AI
- Social networking.

2.1.6. Civic Engagement

The utilization of the talent and the high educational level of people in the Gulf Region can be enhanced significantly by applications that engage AI's unique value proposition in the following directions:

- Community feedback platforms
- Participatory policymaking
- Co-design of strategies
- Open governance
- Opinion mining/fake news.

Fig. 2 highlights these indicative areas in only six of the pillars of the smart cities utilization framework. Similar ideas can be summarized for other areas of the framework such as:

- Smart transportation
- Smart waste management
- Smart economy
- Smart governance/E-Governance
- Smart resilience

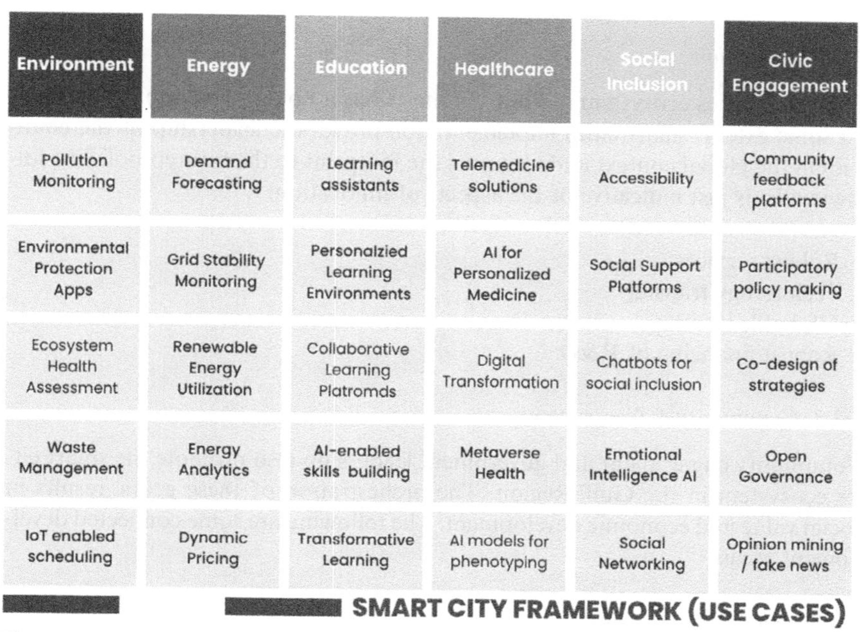

Fig. 2. Application Scenarios for the Deployment of AI in Future Smart Cities Research and Services (*Source*: The Authors).

- Smart public safety
- Smart infrastructure.

2.2. AI-Enabled Smart Cities Ecosystem in Gulf Countries

It is really quite sophisticated to try to map an emerging AI-enabled smart cities ecosystem in couple pages. In this section and in Fig. 3, we communicate a high-level abstraction of a small portion of this ecosystem. In our understanding, five bold pillars set the ground and the solid basis of the ecosystems, namely: sustainability, innovation, community and government, infrastructures and well-being. For each of these pillars, AI can be a total disruptor of the unique value proposition.

2.2.1. Sustainability

Sustainability is an overarching principle for the smart city ecosystem and has bold anchors with critical aspects including but not limited to:

- Governance
- Environment
- Social/Economic.

For the smart cities ecosystem in the Gulf countries, it is important to understand the contribution of the orchestration of strategies and projects to the environment, the social and economic development, For this purpose, a bold smart city governance is needed.

2.2.2. Innovation

The smart cities ecosystem in Gulf countries has a bold innovation component. To some extent, smart cities implementation proves the leadership of the countries in the global context and also puts the footprint to the futureproof. The following list is just indicative of the aspects of innovation:

- Robust services
- Technology transfer
- Global leadership
- Robust financing of R&D.

2.2.3. Community & Government

Community engagement and government leadership also promote the smart cities ecosystem in the Gulf Region. The orchestration of these gears, results in social value and economic development. The following are some connected development areas:

- Civic engagement
- AGI (artificial general intelligence) deployment

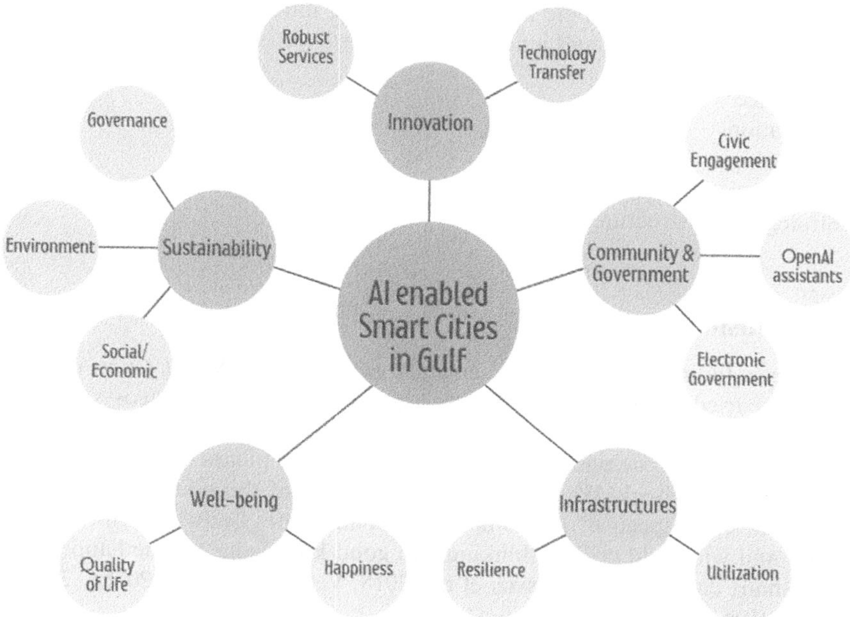

Fig. 3. An Indicative Overview of the Smart Cities Ecosystem in the Gulf Region (*Source*: The Authors).

- Electronic government
- Smart city strategy at the national level
- Smart city implementation
- Policies and regulations.

2.2.4. Well-being

The countries in the Gulf Region have developed bold strategic plans, entitled as Visions which also serve as well-being contexts for the evolution and the socially inclusive economic development of the region with a positive impact on the world economy, peace, and prosperity. Within these contexts, the following are aspects of the well-being outlook of the smart cities ecosystem in the region:

- Quality of Life
- Happiness
- Resilience.

2.2.5. Infrastructure

The infrastructure pillar of the smart cities ecosystem in the Gulf Region is robust. It is related to investments of trillions that have the capacity to transform the model of economic and social development. Megacity projects, strategic smart city projects contribute significantly to the infrastructure portfolio with significant diversification and innovation features:

- Resilience
- Utilization
- Technologies of the future
- Mega projects of smart cities
- Strategic infrastructures.

3. Conclusions

In this concluding chapter of the *Emerald Handbook on Smart Cities in Gulf Region*, we tried to communicate diverse aspects of the smart city ecosystem and to discuss indicative AI-enabled services and use cases. The contributors of this volume provide a systematic discussion of most of the pillars of the Smart City Utilization Framework. We do hope that the readers of this edition will interact with the diverse aspects of the complicated Smart Cities Ecosystem in the Gulf Region and will build critical thinking and good knowledge on the happenings in the smart cities context (Visvizi et al., 2018; Visvizi & Lytras, 2018a, 2018b, 2019a, 2019b, 2020, 2021).

References

Alhalabi, W., Lytras, M., & Aljohani, N. (2021). Crowdsourcing research for social insights into smart cities applications and services. *Sustainability (Switzerland)*, *13*(14). https://doi.org/10.3390/su13147531

Alkhammash, E. H., Jussila, J., Lytras, & Visvizi, A. (2019). Annotation of smart cities Twitter micro-contents for enhanced citizen's engagement. *IEEE Access*, *7*, 116267–116276. https://doi.org/10.1109/ACCESS.2019.2935186

Alsaywid, B. S., Qedair, J., Alkhalifah, Y., & Lytras. (2023). Research and education skills as a core part of digital transformation in healthcare in Saudi Arabia. In *Digital transformation in healthcare in Post-COVID-19 times* (pp. 205–216). https://doi.org/10.1016/B978-0-323-98353-2.00010-1

Lytras, & Alsaywid, B. S. (2023). Next generation healthcare education and research: Utilizing the talent, skills, and competencies for value-based healthcare. In *Active learning for digital transformation in healthcare education, training and research* (pp. 211–220). https://doi.org/10.1016/B978-0-443-15248-1.00005-9

Lytras, Alsaywid, B., & Housawi, A. (2023). Digital transformation and smart cities: Insights from the healthcare domain. In *Smart cities and digital transformation: Empowering communities, limitless innovation, sustainable development and the next generation* (pp. 319–325). https://doi.org/10.1108/978-1-80455-994-920231015

Lytras, Housawi, A. A., & Alsaywid, B. S. (2023). Preface. In *Smart cities and digital transformation: Empowering communities, limitless innovation, sustainable development and the next generation* (pp. xxvii–xxxvii). https://www.scopus.com/inward/record.uri?eid=2-s2.0-85163484376&partnerID=40&md5 = 19c3211d29913a9673b-7f03d3d12eb20

Lytras, & Şerban, A. C. (2020). E-government insights to smart cities research: European Union (EU) study and the role of regulations. *IEEE Access*, *8*, 65313–65326. https://doi.org/10.1109/ACCESS.2020.2982737

Lytras, Serban, A. C., Ruiz, M. J. T., Ntanos, S., & Sarirete, A. (2022). Translating knowledge into innovation capability: An exploratory study investigating the perceptions on distance learning in higher education during the COVID-19 pandemic – The case of Mexico. *Journal of Innovation and Knowledge*, *7*(4). https://doi.org/10.1016/j.jik.2022.100258

Lytras, Visvizi, A., Chopdar, P. K., Sarirete, A., & Alhalabi, W. (2021a). Information management in smart cities: Turning end users' views into multi-item scale development, validation, and policy-making recommendations. *International Journal of Information Management*, *56*. https://doi.org/10.1016/j.ijinfomgt.2020.102146

Lytras, Visvizi, A., & Jussila, J. (2020). Social media mining for smart cities and smart villages research. *Soft Computing*, *24*(15), 10983–10987. https://doi.org/10.1007/s00500-020-05084-3

Lytras, Visvizi, A., Sarirete, A., Torres-Ruiz, M., & Daim, T. U. (2021b). IEEE access special section editorial: Future generation smart cities research – Part II: Services, applications, case studies, and policymaking considerations for well-being. *IEEE Access*, *9*, 27298–27303. https://doi.org/10.1109/ACCESS.2021.3057809

Nikiforova, A., Flores, M. A. A., & Lytras. (2023). The role of open data in transforming the society to society 5.0: A resource or a tool for SDG-compliant smart living? In *Smart cities and digital transformation: Empowering communities, limitless innovation, sustainable development and the next generation* (pp. 219–252). https://doi.org/10.1108/978-1-80455-994-920231011

Sairete, A., Balfagih, Z., Brahimi, T., Amin Mousa, M. E., Lytras, M., & Visvizi, A. (2021). Editorial – Artificial Intelligence: Towards digital transformation of life, work, and education. *Procedia Computer Science*, *194*, 1–8. https://doi.org/10.1016/j.procs.2021.11.001

Visvizi, A., & Lytras. (2018a). It's not a fad: Smart cities and smart villages research in European and global contexts. *Sustainability (Switzerland)*, *10*(8). https://doi.org/10.3390/su10082727

Visvizi, A., & Lytras. (2018b). Rescaling and refocusing smart cities research: From mega cities to smart villages. *Journal of Science and Technology Policy Management*, *9*(2), 134–145. https://doi.org/10.1108/JSTPM-02-2018-0020

Visvizi, A., & Lytras. (2019a). Preface. In *Smart cities: Issues and challenges mapping political, social and economic risks and threats* (pp. xv–xvi). https://doi.org/10.1016/B978-0-12-816639-0.05001-8

Visvizi, A., & Lytras. (2019b). Reflecting on oikos and agora in smart cities context: Concluding remarks. In *Smart cities: Issues and challenges mapping political, social and economic risks and threats* (pp. 333–339). https://doi.org/10.1016/B978-0-12-816639-0.00018-1

Visvizi, A., & Lytras. (2020). Sustainable smart cities and smart villages research: Rethinking security, safety, well-being, and happiness. *Sustainability (Switzerland)*, *12*(1). https://doi.org/10.3390/su12010215

Visvizi, A., & Lytras. (2021). Guest editorial. *Transforming Government: People, Process and Policy*, *15*(2), 193–196. https://doi.org/10.1108/TG-05-2021-315

Visvizi, A., Lytras, Damiani, E., & Mathkour, H. (2018). Policy making for smart cities: Innovation and social inclusive economic growth for sustainability. *Journal of Science and Technology Policy Management*, *9*(2), 126–133. https://doi.org/10.1108/JSTPM-07-2018-079

Printed and bound by CPI Group (UK) Ltd, Croydon, CR0 4YY

17/09/2026

14964133-0001